Bombers and Reconnaissance Aircraft

The History of German Aviation

Bombers and Reconnaissance Aircraft

1935 to the Present

Roderich Cescotti

Schiffer Military History
Atglen, PA

Dedication

Dedicated to my fellow pilots, both young and old, who rose to the challenge of flying these bombers and reconnaissance planes and will continue to do so until peace is achieved throughout the world.

- R. C.

Translasted from the German by Don Johnston

Copyright © 2001 by Schiffer Publishing, Ltd.
Library of Congress Catalog Number: 00-107470

All rights reserved. No part of this work may be reproduced or used in any forms or by any means – graphic, electronic or mechanical, including photocopying or information storage and retrieval systems – without written permission from the copyright holder.

This book was originally published under the title,
Kampfflugzeuge und Aufklärer - Entwicklung, Produktion, Einsatz und zeitgeschichtliche Rahmenbedingungen von 1935 bis heute

Printed in China.
ISBN: 0-7643-1283-9

We are interested in hearing from authors with book ideas on related topics.

Published by Schiffer Publishing Ltd. 4880 Lower Valley Road Atglen, PA 19310 Phone: (610) 593-1777 FAX: (610) 593-2002 E-mail: Schifferbk@aol.com. Visit our web site at: www.schifferbooks.com Please write for a free catalog. This book may be purchased from the publisher. Please include $3.95 postage. Try your bookstore first.	In Europe, Schiffer books are distributed by: Bushwood Books 6 Marksbury Avenue Kew Gardens Surrey TW9 4JF England Phone: 44 (0) 20 8392-8585 FAX: 44 (0) 20 8392-9876 E-mail: Bushwd@aol.com. Free postage in the UK. Europe: air mail at cost. Try your bookstore first.

Contents

Foreword 9

1. In Perspective 11

Rhineland Program - *Risikoluftwaffe* - *Reichsluftwaffe* - The Bomber Arm - Douhet and the "Total Air War" - Rougeron - Developmental Trends - Dornier Do 19 - Junkers Ju 89 - End of the "Ural Bomber" Project - The "Bomber A" Project

2. Military Historical Retrospective 18

Bomber Pilots - Dive Bombing Operations - Recce Pilots - Tactical Reconnaissance - Strategic Reconnaissance - Maritime Reconnaissance

3. Proving Grounds of Spain 20

Starting Point - Germany's Initial Role - Additional Support - Initial Phase of the Civil War
Events in the Air
Bombers and Dive Bombers - Recce Planes - Fighters - Development of the Air War and Interim Balance - Air Forces of the Day - End Game
Lessons for Air Power

4. The New Generation of Aircraft 31

The Medium Bomber - The High-Speed Bomber - Thoughts Of Dive Bombing - The *Stuka*
On Cooperation

5. Germany's Aviation Industry Potential in 1937 38

Financial and Raw Material Bottlenecks - The Reality of the Developmental Period - Authority and Priorities - Production Potential Achieved in 1937

6. The 1938/1939 Rearmament Program Against the Backdrop of *"Große Politik"* 42

The 1938 Aircraft Acquisition Program - 1938 Planning Parameters - Some Examples of Cost from 1938 - Calculating the "Constant *Luftwaffe* 1942" - The Consolidated Aircraft Type Program - Pointing the Way to War

7. Taking Stock 46

Self-Deception? - Offensive Aircraft Weaponry of 1939
Ordnance - Armament - Problems of Precision Bombing - Reconnaissance Systems of 1939
1938/1939 Numbers Comparison - Assessment

8. Offensive Weapons and Reconnaissance Package Platforms Through the End of 1939 52

Medium Bombers 52
Junkers Ju 52 Auxiliary Bomber
Junkers Ju 86 Bomber
Prototypes - 0-Series - A-1 and D-1 Series - Foreign Interest - Sweden - Hungary - Portugal - Chile - Austria - South Africa - E- and G-variants - Technical Data and Details of the Ju 86G-1
Synopsis
Heinkel He 111 Bomber 59
First Prototypes - A-Series - B-Series - D-Series - E-Series - F-Series - J-Series - P-Series - H-Series - Foreign Interest - China - Turkey
Synopsis

Dornier Do 17 Bomber 66
Prototypes - E-Series - M-Series (with L- and R-variants) - Yugoslav Interest in the Do 17K - Z-Series (with S- and U-variants) - Technical Data and Details of the Do 17Z-3 - Special Foreign Interest in the Do 17Z (Do 215) - Sweden - Soviet Union
Synopsis

Junkers Ju 88 Bomber 74
Foundations - Ju 88V1 through V3 Prototypes - Changing the Concept - Ju 88V4 through V6 Prototypes - A-Series (A-0 and A-1) - Teething Troubles - Production Operations - Production Model - Dive Bombing Capability
Synopsis

Dive Bombers 85
Henschel Hs 123 Light Dive Bomber - Selection - V-types - A-Series
Synopsis

Junkers Ju 87 Dive Bomber
Foundations - First Prototypes - Fly-off and Selection - Arado Ar 81 - Ha 137 - He 118 and Ju 87 - Ju 87V4 and Initiation of A-Series

History of German Aviation: Bombers and Reconnaissance Aircraft

- B-Series - Technical Data and Details of the Ju 87B-1 - Ju 87B Dive Brakes and Automatic Recovery - C-Series
Synopsis

Strategic Reconnaissance Aircraft 95
Heinkel He 70 and He 45 Auxiliary Reconnaissance Platforms
Dornier Do 17 Strategic Reconnaissance Plane
F-Series - P- Series - Technical Data and Details of the Do 17P - S-Series and Do 17Z Multi-Role Platform
Strategic Recce Hodge-Podge
Synopsis

Tactical Reconnaissance Aircraft 101
The *Luftwaffe's* First Generation of Tactical Recce Planes - Foreign Interest - The *Luftwaffe's* Armament Situation
Henschel Hs 126 Tactical Reconnaissance Plane
From the Hs 122 to the Hs 126 - Hs 126 V-types and the A-0 Series - A-1 and B-1 Variants - Technical Data and Details of the Hs 126 - Foreign Interest - Switching to the B-1

Conclusions 104

9. An Unrefined *Luftwaffe* is Tested 105

The "Other Side's" Potential
The Quantitative Situation - The Qualitative Situation
- Penetration Ranges - Comparison Criteria - Operational Flight Performance
Assessment

Poland
The Air War - Balance
All Quiet on the Western Front
Germany's Forces - The *"Sitzkrieg"* and the Air War - Battle Plans for the West and Finland

All the More Reason To Go North! - Foundations - Occupation of Denmark/Battle for Norway

10. Initial Assessment of 1940 Following the Polish and Norwegian Campaign 112

Fw 200 Maritime Reconnaissance Plane and Auxiliary Bomber - Brief Perspective - Fw 200C in Military Service
Synopsis

Ju 86P High altitude Reconnaissance Platform and Bomber
Prototypes - P-Series - Technical Data and Details of the Ju 86P
Summary

11. The Western Campaign 119

The Netherlands
Belgium and Northern France - Dunkirk - "Operation Dynamo"
Paris and the Results
Consequences for the *Luftwaffe*

12. The Battle of Britain 122

Struggle for Air Superiority - Did Germany's Air Power Fail?
Weapons of the Air -
Bombers - Dive Bombers - Strike Aircraft - Fighters - Heavy Fighters
Interim Balance
Combat Methods -
Ju 88 Dive Bombing Methods - Bombing Naval Targets - Weapons - Outside Curve Bombing Attack - The Rechlin Method - The Liesendahl Method - Radio Beam Guidance for Bombing Through Overcast
Synopsis

13. Final Totals for 1940 Following the Western Campaign and Battle of Britain and their Effect 131

Long range Bombers 131
Focke-Wulf Fw 200
Heinkel He 177
Heinkel He 119 Test Prototype - He 177 Prototypes - He 177V1 - He 177V2 through V5

Medium Bombers 137
Junkers Ju 88 - Ju 88A-1 through A-8 (minus A-4) - Technical Data for Ju 88A-5
Summary and Other A-5 Derivatives

The Indispensable Heinkel He 111 - End of the P-Series - He 111H-2 through H-5

Dornier Do 17Z and Do 217A, C and E

Dive Bombers 145
Junkers Ju 87
The "Retailored" Fighters and *Zerstörer* - General Thoughts
Messerschmitt Me 109 and 110 As Fighter-Bombers

Strike Aircraft 148
Henschel Hs 123 and Messerschmitt Me 109
Foundations for the Henschel Hs 129

History of German Aviation: Bombers and Reconnaissance Aircraft

Long range Reconnaissance Aircraft 149
Focke-Wulf Fw 200
Junkers Ju 86 High altitude Recon Plane
Dornier Do 215B
Junkers Ju 88D

Tactical Reconnaissance Aircraft 151
Henschel Hs 126
Messerschmitt Me 110
Focke-Wulf Fw 189 -
Foundations and Proposals - Ar 198 and Fw 189 Prototypes and Selection - Ar 198 Prototypes, the Fw 189 and the Decision - Additional Fw 189 Prototypes - Fw 189 A-Series - Technical Data for the Fw 189A-1 - Assessment
Messerschmitt Me 109

1940/1941 Closing Balance 156

14. Developments in the Mediterranean and Southeast 157

First Operations in Expanding the War - Escalation in the South - Naval War From the Air - An Aircraft Carrier's Achilles Heel - Problems of Range

North Africa

The Balkans and Crete
Yugoslavia and Greece - Crete, Debate and Decision - Execution of the Plan - Results and Lessons

15. The Campaign in the East and its Effects on Air Armament 161

Course of the War on the Eastern Front to Winter 1941/1942 - Setbacks in North Africa - Japan and the United States - Status of German Air Arms in Late 1941/Early 1942 - Framework - The "Göring Program" and the Technical Leadership Apparatus - The Era of Milch

1941/1942 Aircraft Production and Subsequent Models

a) Bombers 164
Bomber "B" Prototypes - Junkers Ju 288 - Dornier Do 317 - Focke-Wulf Fw 191 - Henschel Hs 130

Old Workhorses and New Disappointments with the Heinkel He 177
Heinkel He 111H-6 - Performance Data for the He 111H-6
Junkers Ju 88A-4 - Ranges - Assessment

B) From the Messerschmitt Me 210 High-Speed Bomber and Recce Plane to the Me 410 174

Hungarian Interest and Realization - The German Me 210 - Me 210Ca-1 and Me 210C-1

c) Heavy Bombers 176
Fw 200 and the Henschel Hs 293 Glide Bomb - Fw 200F - Fw 200C-3/U4 Synopsis and Performance
Heinkel He 177 - In Production At Last! - Preproduction Series and Optimistic Appraisals - He 177A-1 and A-3 - Interim Balance and Performance for the He 177A-1
Dornier Do 217 E-Series - Models Do 217E-0 through E-5 - Basic Data for Do 217E-2 and Sub-Variants - Interim Balance and Viewpoint

d) Dive Bombers and Strike Aircraft 185
Junkers Ju 87D, F, G, H and Ju 187 - Basic Data for the Ju 87D-1 and D-7 - Ju 87G Tankbuster - Junkers Ju 187 - End of the Ju 87
Henschel Hs 129B and C - Technical Data for the Hs 129B - Final Tally and the Hs 129C

e) Reconnaissance Aircraft 193
Long range Reconnaissance Aircraft
Junkers Ju 88D - Dornier Do 215B - Focke-Wulf Fw 200C - Junkers Ju 86P and Ju 86R plus Ju 186
Tactical Reconnaissance Aircraft
Henschel Hs 126 - Messerschmitt Me 110 and Me 109 - Focke-Wulf 189 - Recce Results and the Outsider - Blohm & Voss BV 141

16. Course of the War From 1942 Onward 197

In the East - North Africa - In the West, Over the Reich and the *Luftwaffe* Leadership Crisis

17. Bomber Planning from October 1943 200

Comparison of the Bomber Situation in May 1940 with that of May 1943 - Demands of the *General der Kampfflieger*
Bomber Development to 1945
Heinkel He 177A-3 through A-7 - Japanese Interest - The *"Bomber-Zerstörer"* - Summary Including Data for the He 177A-5 - Heinkel He 274 and He 277 (a.k.a. "He 177B")
Junkers Ju 290 and the "New York/America Bomber" - Summary and the Ju 390
Messerschmitt Me 264
Focke-Wulf Ta 400
Thoughts on Six-Engines
Ju 88A-4 Follow-on Variants - Ju 88A-6 through A-17 - Ju 88B, E, and P - S-Series - Summary

Junkers Ju 188 - Prototypes - Ju 188A, E and G Bomber Variants - Ju 188A and E General Description - Ju 188D, F and H Reconnaissance Variants
Ju 188 through Ju 388 Follow-on Developments and the Ju 488 - Fate of the Ju 288 and the Ultimate Sacrifice of the Ju 188 - Ju 288C, E and G - The Last Ju 188S and T - The "Hubertus" Program - Junkers Ju 388L - Ju 388J and K - Summary and the Ju 488
Junkers Ju 88-*Mistel* - *Mistel 1* and *2* - *Mistel 3* and *4* - The "Control Planes" - Summary and the Ta 154

18. What's Left of Bomber Planning? 230

Situation Winter 1944/1945
In the East - In the South - In the West and Over the Reich
"Schnellstbomber" - Ordered but Never Materialized - Messerschmitt Me 262A-2a - Fighter-Reconnaissance Planes Me 262A-1a/U3 and A-5a - Summary
Making Due with What's On Hand
Dornier Do 217 - Do 217H, P, K, M, R Variants - DFS 228 High altitude Reconnaissance Plane - Summary
Heinkel He 111 - He 111H-7 through H-23 Variants - Summary and the He 111R - He 111Z-2 and Z-3
Messerschmitt Me 410 and Heinkel He 219 - Me 410A and B Variants - He 219A and C - He 319 - Summary

19. Last Hope - Gaining the Upper Hand through Superior Technology? 243

The True High-Speed Bomber
Arado Ar 234 - Foundations and General Project Description - Twin-Engined Prototypes - Four-Engined Prototypes - Twin-Engined Follow-On Developments and Operational Testing - Arado Ar 234B-1 - Arado Ar 234B-2 - Four-Engined C-Series - Summary and the Dornier Do 435, Do 635 and Do 335

20. Death of the German Bomber and the End of an Era 255

21. The Giant Leap to the Present 257

Road to Alliance
Allied Army, Initial Reequipment of the Air Arm and Joint Aviation Programs
Initial Reequipment of the Air Arm
Second Generation of Bombers and Reconnaissance Aircraft - Fiat G.91 Light Bomber - Bréguet Br 1150 Atlantic Maritime Reconnaissance and Anti-Submarine Aircraft - Lockheed F-104G Starfighter Medium-Range Multi-Role Strike Fighter and Tactical Reconnaissance Aircraft - F-104G Performance Characteristics

Germany's Aviation Industry Catches Up With the West

NATO's Operational Concept and Doctrine 260
The NATO Treaty as Foundation - Military Strategic Concept - General Operational Concept of NATO's Air Arm - Roles of Offensive Air Forces

Third Generation of Bomber and Reconnaissance Aircraft 262
McDonnell Douglas F-4 Phantom
Alpha Jet - Foundations, Selection and Role Establishment - Prototypes - Alpha Jet Production - Alpha Jet Technical Data and Details - Features of the Alpha Jet - Improving the Breed - Conclusions

Panavia Tornado - Foundations - V/STOL Options - From NKF to MRCA - Turbulence - General Agreement - Who Picks Up the Tab? - A Large-Scale Project Is Organized - The Definition Report and More Stumbling Blocks - Development, Maiden Flight and Production - Development and Layout - The Tornado Flies - Production Gets Underway - System Leaders - The Tornado and Operational Doctrine - Unique Features of the Tornado and Its Weaponry - The Weapons Systems - Batches 5 through 7 and Maintaining Combat Effectiveness - Tornado ECR - Determining Location and Contract

22. Bombers and Reconnaissance Aircraft from Political and Military Perspectives 288

Results of an Era - The Fighter-Bomber - Application of Reconnaissance Systems

Appendix (9 Sections) 290

1. Table of German Combat Aircraft 2. Consolidated Aircraft Type Program 3. Ordnance in 1939 4. Aircraft Armament from 1936 through Today 5. Reconnaissance Systems from 1935 Onward 6. Aircraft Utilization, Stress, and Class Groupings 7. Ju 88 Chronology 8. Performance Table for German Bombers, Dive Bombers, and Reconnaissance Aircraft 9. New Ordnance Types from 1940 Onward

Bibliography	301
Index of Aircraft Types	303
General Index	304
Index of Personnel	308
Abbreviations	309
Unit Conversion Table	310
Picture Credits	310
Author's Biography	311

Foreword

Germany's expansion of military aviation construction in the 'Thirties led to the manufacturing of a wide variety of aircraft types, peaking at some 143 variants in 1941. However, in treating the subject of bombers and reconnaissance aircraft, this large number precludes anything other than a brief mention of the numerous projects, prototypes, and testbeds which failed to reach production maturity or gain purchase orders—despite the fact that many of these stimulated considerable interest from a design and technological perspective. On the other hand, those types which proved themselves well suited to the roles of air attack and reconnaissance are examined here in detail.

This book explores the development of these aircraft types against the background of the prevailing historical situations, examining the military requirements which spawned the flying machines in a given period and which then led to the acquisition of selected aircraft.

The author's experience as a pilot, which embraces a time period from 1937 to 1980 and ranges from operating twin- and four-engined bombers and single-seat fighters in the former *Luftwaffe* to flying high-performance jet reconnaissance planes for the *Bundesluftwaffe*, actually forms only a small part of the material used in this volume. Much more valuable in creating a central framework for a verifiable treatment of the subject was the material found in the federal archives. There are gaps, however, particularly in the area of military aviation up to 1945. Despite the fact that the former Allies of WWII have since returned a large percentage of the documents confiscated after the war, the *Bundesarchiv's* military archives have only contributed about five percent of the *Luftwaffe* specific archival material. The *Bundesarchiv* and the *Militärgeschichtliches Forschungsamt* have assessed the *Luftwaffe's* written record situation between 1935 and 1945 to be poor, as more or less only fragmentary data has survived; the bulk of the files were destroyed in 1944/1945.

Which made it all the more critical to make use of credible witnesses from this period in order to provide an overall picture for such an undertaking as this book. Aside from the numerous, though not always accurate, literature on the subject, of exceptional value was the support given by the *Militärarchiv* and the *Militärgeschichtliches Forschungsamt* in Freiburg im Briesgau, as well as the *Bundesarchiv* and the *Wehrtechnisches Museum* in Koblenz. Another indispensable source of information for previously unknown or unpublished material was the special aviation collection of the *Deutsches Museum* in Munich. Finally, the works and lectures of the *Deutsches Museum's* Space and Aeronautics advisory council, as well as its authors, have all contributed to rounding out the picture.

Also worthy of note has been the support of Germany's aviation industry, particularly the Dornier and Messerschmitt-Bölkow-Blohm companies, as well as the British-German-Italian PANAVIA G.m.b.H. and its NATO contractor, NAMMA. The author expresses his gratitude to all those managers and interlocutors of the above mentioned institutions, with special thanks to the inspectors of the *Bundesluftwaffe* for making valuable background information available from their areas of responsibility.

In creating a book on bombers and reconnaissance aircraft, the picture has been fleshed out by a plethora of one-on-one conversations and correspondence over the last eight years with experienced *Luftwaffe* personnel, including *Dipl.-Ing.* Jean Roeder, as well as with pilots of the former and current *Luftwaffe* and subject matter experts like Kurt W. Streit. Despite the author's utmost efforts, no claim can be made for the absolute completeness of the work. However, all dates and figures contained herein are independently verifiable. When comparing figures it should be noted that the end balance of the initial and final data for a given period of time does not take into account day-to-day fluctuations, i.e. losses on the one hand and on-going resupply on the other. Actual aircraft loss figures are, as a rule, much higher than the purely mathematical difference between initial and final numbers, which only reflect the difference in numbers on the days in question. Nevertheless, they are an indicator for showing just how far the projected planning and production match up with anticipated and actual operational losses.

In conclusion, I wish to extend my heartfelt gratitude to all those who directly or indirectly contributed to this work. This includes scientific director Dr. Horst Boog for his critical examination of my notes, Messrs. *Dipl.-Ing. Dr.-Ing. E.h.* Ludwig Bölkow, *Dipl.-Ing.* Rüdiger Kosin, *Dipl.-Ing.* Wolfgang Degel, and Joachim Egbert for their review of the various chapters, and Messrs. Kyrill von Gersdorff and Rupprecht Sommer for their overall review—last but not least to Dr. Theodor Benecke, senior editor, the publishing house, and publisher Mr. Walter Amann for designing this volume.

Fürstenfeldbruck, Upper Bavaria, Spring 1989   R. C.

1. In Perspective

As mentioned in a previous volume, German aviation experienced unprecedented growth in the early 1930s. Its pace was determined by the political developments going on at the time. The effects it had upon German civil and military aviation, the national aviation industry, and the advances made in aeronautics are remarkable, even from a historical perspective, and drew considerable worldwide attention. This development was accelerated by the creation of a German air force, an event which occurred on 1 March 1935 and was euphemistically labeled as an "unmasking," after the *Reichswehr* had been making preparations for several years leading up to the event.

The problems and the multi-layered complexity of German bombers and reconnaissance aircraft are better understood when placed in context with the overall expansion, roles, and subsequent operations of the air force. With this in mind, a few important facts up to 1935 are provided below as a reminder; these chart the events leading up to the official establishment of the *Luftwaffe* as follows:

- On 2 February 1933—three days after Adolf Hitler had been named chancellor—*Hauptmann a. D.* Hermann Göring was appointed *Reichskommissar für die Luftfahrt* (Commissioner of Aviation) by President *Generalfeldmarschall* Paul von Hindenburg; vice commissioner was Erhard Milch, also a retired captain, who had been director of *Luft-Hansa* up to that point in time.
- On 27 April 1933 the president, acting on a recommendation of the chancellor, ordered the creation of an "air ministry." It became the highest administrative authority for aviation matters under defense minister *General der Infanterie* Werner von Blomberg; on the same day Göring covertly began working on the establishment of the *Luftwaffe*.
- On 5 May 1933 Göring's position was renamed from *Reichskommissar für die Luftfahrt* to *Reichsminister der Luftfahrt (R.d.L.)*. On Göring's recommendation, Milch became Secretary of Aviation in the new *Reichsluftfahrtministerium* (Ministry of Aviation, abbreviated to RLM)
- On 15 May 1933 the former *Luftschutzamt* (Air Protection Office) within the *Reichswehrministerium* (Ministry of Defense) was merged into the RLM

These measures served to form the framework of the organization which would guide and shape all the major impulses affecting civil and military aviation from that point until the collapse of the German *Reich* on 8 May 1945.

- On 19 October 1933 Germany withdrew from the League of Nations and left the Geneva disarmament conference at a time when it had nearly 200 military aircraft as part of its *Reichswehr*—of which approximately 100 were considered to be front line aircraft by the standards of the day. By way of comparison, the French had over seven times this number and the British nearly four times.

In effect, these actions unilaterally nullified the Paris Accords of 21 May 1926, which had returned control of the air to Germany but continued with the ban on military aviation and the construction of armed, as well as unmanned, aircraft. From this point onward, the RLM ensured that air rearmament accelerated at an exponential pace using practically unlimited means and resources.

Rhineland Program - *Risikoluftwaffe - Reichsluftwaffe*

The Rhineland Program, an emergency aircraft acquisition program running from 1 July 1934 to 31 March 1938, called for an initial acquisition of 822 bombers and 245 fighters in the first period up to 30 September 1935 alone. The final phase envisioned the construction of 2,188 bombers and 2,225 fighters. The resources of the German aviation industry, which had approximately 20,000 employed workers at the end of 1933, quintupled in less than a year.

The instrument for the prosecution of an air war arising from this program was given the internal designation of *"Risikoluftwaffe,"* literally "Risk Air Force," because it would place at risk any potential enemy considering a military intervention during Germany's rearmament period. In a dual sense, this young arm was also at risk in the sense that the antiquated aircraft types then in its inventory were neither a qualitative nor a quantitative match for a potential attack by the western powers. It was not until the Rhineland Program advanced toward realization that this internal risk gradually diminished.

As the *"RisikoLuftwaffe's"* combat strength increased, so too did its potential for intimidation, creating political pressure abroad with the goal of implementing the political will of the *Reich* Government without having to resort to war, if possible.

Following this prelude, which would have major repercussions in the history of German aviation, *Reichskanzler* Hitler announced on 26 February 1935 that:

"On 1 March 1935 the Air Force will join the Army and Navy as the third branch of service in the Reich..."

With this statement, the *Luftwaffe* emerged from the semi-darkness of its camouflaged expansion into the public spotlight of Germany and the entire world. It had a significant influence on the further course of Germany's aeronautical development, the forced expansion of Germany's aviation industry and, most importantly, on the requirements for bombers and reconnaissance aircraft in the future. After four and a half years of existence, the new *Luftwaffe* would still be an unrefined instrument of modern air warfare when it would face its ultimate challenge—a war that would rapidly snowball into total war.

At the time of this "unmasking," i.e. on 1 March 1935, the flying units of the *Reichswehr/Wehrmacht* and those organizations immediately supporting it had a personnel strength of about 900 officers and about 10,000 NCOs and crews; its inventory included just over 2,500 aircraft, of which about 800 could be considered fit for front line service.

By 1 March 1935 2,291 aircraft had been supplied to the *Luftwaffe* under the auspices of the Rhineland Program, approximately 9% less than had been planned for the first stage of the emergency program. By 30 September 1935 it had reached 94% with a total delivery of 4,021 aircraft—an enormous undertaking by the aviation industry when one considers that in early 1935 this program was overtaken by a new air armament program which placed greater emphasis on the readiness of the more expensive front line aircraft.

On 16 March 1935 Hitler announced the introduction of general conscription and stated that Germany was no longer obligated to abide by the armament restrictions of the Treaty of Versailles. In so doing, he broke the last remaining bonds of this agreement from 28 June 1919 (which, truthfully, had been violated on many occasions prior to the announcement) in full awareness of the sharp reaction it would elicit from the other signatories of the treaty. However, no pronounced response of any significance was forthcoming, as in the following year when Germany met with a barely perceptible political reaction by the former entente powers after the *Wehrmacht* moved into the Rhineland, an area which had been demilitarized by the treaty.

The Bomber Arm

The importance which Germany's air leaders placed on this subcategory of military aviation is reflected in the words of *Generalmajor* Walter Wever—at the time the Chief of the Office of Air Command (*Luftkommandoamt*), the forerunner of the Air Force General Staff in the RLM. In his remarks on 1 November 1935 at the opening of the training center of the newly established *Akademie der Luftwaffe* in Berlin-Gatow, Wever devoted part of his address before the general staff officers present to the balanced relationship between offensive and defensive strength as follows:

"...and when you are later called upon to work with these issues, never forget: the decisive weapon in an air war is the bomber! Only the State which has a large bomber force can expect its air force to play a decisive role in war."

Wever's tribute is to a large extent in harmony with the thinking of the Italian air war theoretician Giulio Douhet, as well as the armament policies practiced by the major powers of the day, which viewed the fast, well-armed bomber having the greatest possible payload as the decisive component in an air war.

Over the coming years the *Luftwaffe* would build its bomber arm into the numerically strongest and most modern in Europe. Included in the original bomber arm are a number of specialized types and roles with accordingly specific demands placed on these weapons systems. In this context, these can be broken down into:

- dive bombers, also called *Stuka* (from the German *Sturzkampfflugzeug*)
- strike aircraft
- tank busters (from 1942 onward)
- night strike aircraft (from late 1942)
- fighter-bombers, or modified fighter aircraft, also called *Jabo* (from the German *Jagdbomber*)

Douhet and the "Total Air War"

Air war theorist Giulio Douhet did not base his teachings of air power's operational versatility on theory alone. In 1918 he was at the head of the Italian Air Force. After the First World War he left, was promoted to general, and from that time until his death in 1930 devoted himself to writing books on the subject of aviation.

The core of his philosophy—published in Douhet's work "Il dominio dell'aria" in the 'twenties—is the "total air war" concept.

History of German Aviation: Bombers and Reconnaissance Aircraft

With the costly, static ground struggles typical of the 1914-1918 war still fresh in his mind, Douhet saw the wars of the future as three-dimensional, in which the massive use of air power would act as the overwhelming, dynamic third dimension. This would also dispel the former differentiation between front lines and the homeland, and between combatants and non-combatants.

The guiding principle of Douhetism is: defense and defensive operations on land and sea, massed attacks/offensive in the air. According to Douhet, a ground or naval victory was no longer the decisive factor; only by assuming air supremacy through employing the most effective means—that of an aerial offensive—could the enemy be forced to bow to one's will. Therefore, in Douhet's opinion, the air armada must make use of large-scale attacks to wring air supremacy and victory from the enemy through destruction of the enemy's air forces and their support bases, and obliteration of its material and spiritual centers. Using Douhet's philosophy, the heavy strike air force becomes the deciding weapon of war, and all means available must be sacrificed to its preservation, expansion, and constant technological advancement—even at the expense of the navy and army. In Douhet's words: "To gain air supremacy is to effectively gain the victory! To be beaten in the air, therefore, means to be hopelessly defeated."

Douhet conceded that the army and the fleet would require "support air fleets" to provide immediate coverage to these fighting branches, while the bulk of the operational air force he considered to be the actual "air fleet." He defined the air fleet as consisting of those aircraft which together would form a single, integral entity capable of achieving air supremacy; it was to include all aircraft necessary for attaining the goal of supremacy by destroying the enemy's entire aerial resources and their support facilities.

Douhet went on to say that the air fleet would naturally require suitable reconnaissance means in order to obtain data and reports on the enemy's movements. In explaining the concept of reconnaissance, he clarified that it could only be successfully accomplished if the recce mission were able to either break the enemy's resistance or avoid it. This applied equally to land, sea, and air forces. A reconnaissance under hostile conditions would have to take into account the possibility of air combat, which would require the use of the air fleet or parts thereof. Alternatively, if planning called for the reconnaissance sortie to circumvent any anticipated resistance on the part of the enemy, i.e. avoiding combat altogether, then this would require an aircraft which would need entirely different qualities than those possessed by bomber aircraft.

Douhet saw the reconnaissance mission as one to be carried out by a single airplane or, at the most, by very small formations, and felt—understandably lacking knowledge of the future development of radar technology—that these aircraft would precede air fleet formations into an area by a specified distance to avoid surprise attacks, to report on large-scale enemy action, and finally, to scout out ground targets and suitable sites for launching ground attacks.

Rougeron

Another air war theoretician, not as commonly known as Douhet, was the naval senior engineer Camille Rougeron, who developed his impressions of bomber aviation in the early thirties and published them in 1937 under the title of "L'Aviation de Bombardement."

Unlike Douhet, who saw the heavy bomber as the primary offensive weapon, Rougeron was of the opinion that the high-speed bomber—thanks to a speed which even allowed it to operate unarmed—was better suited to bringing the war home to the enemy. To be sure, the payload capacity of a high-speed bomber was limited when compared to a heavy bomber, but for most of the air forces in the thirties it was a much more practical solution than the significantly more complicated and expensive strategic bomber.

Without a doubt, Rougeron's idea of an unarmed high-speed bomber influenced certain lines of thinking and development at the time. The Dornier Do 17, with its excellent high-speed performance, corresponds perfectly to this theory and serves as but one example of his vision of a high-speed bomber. Although the Do 17 was originally conceived as a high-speed mailplane capable of carrying up to six additional passengers, subsequent prototype trials in early 1935 by the *Deutsche Lufthansa* (abbreviated DLH, with *Lufthansa* being written as one word from 1 January 1934 onward) showed that it was entirely unsuited for passenger transport. "The Flying Pencil," as the Do 17 was called because of its shape, was more in harmony with Rougeron's ideal concept of an unarmed high-speed bomber.

Developmental Trends

The schools of thought briefly mentioned above discuss the role of the bomber as the decisive air weapon, but they are by no means the sole conceivable and potential variations of this theme.

Even the words of admonishment which *Generalmajor* Wever uttered on 1 November 1935 in Gatow were not expressed in a vacuum. Wever had specific thoughts in mind when he stated that the bomber was the decisive weapon in an air war. As early as the fall of 1933 Wever had approved the development of a strategic long range heavy bomber, one which would be able to carry an appreciable payload from a jumping off point within Germany to northern Scotland or to the Urals; this formulation was not made for political reasons, rather to simply give the dimensions of the anticipated range. Wever unofficially called the bomber the "Ural Bomber," and in the summer of 1934 the RLM's *Technisches Amt* issued the specifications for what would probably be a four-engined type to the aircraft manufacturers Dornier and Junkers. These specifications were, to a great extent, in harmony with the thinking already going on within the design bureaus of these companies.

By odd coincidence, the American Army Air Corps had issued similar specifications calling for a multi-engined bomber back in May of 1934. Based on these, the Boeing company developed a four-engined aircraft, the Model 299, which took to the air for the first time on 28 July 1935 and later played a decisive role in the air war over Germany as the B-17 Flying Fortress.

Germany's heavy strategic bomber project had the full backing of the Chief of Air Command Office, Wever (by now promoted to *Generalleutnant*). He was able to silence the opposition—even from some quarters of the command staff—and ensure the successful maiden flights of Dornier's prototype, the four-engined Do 19 (on 26 October 1936), and the Junkers prototype Ju 89 (also a four-engined design, on 11 April 1937). Although neither model entered production, they deserve closer examination.

Dornier Do 19

The Do 19 was a four-engined long range bomber with a seven-man crew. Technical details are as follows:

Airframe
Fuselage: rectangular cross section with rounded corners consisting of four subassemblies; separation points ahead of the cockpit, in front of the forward wingspar, and behind the aft wing spar.
Undercarriage: retractable aft into the inner engine nacelles.
Control Surfaces: located midway up aft fuselage section. Adjustable all-metal tailplane with elevators made of metal ribbing covered in fabric. Tailplane supports two all-metal vertical stabilizers braced to fuselage, rudders equipped with trim tabs adjustable from the cockpit.
Wing Assembly: cantilever mid-wing design. Three-section twin-spar tapered wing with load bearing skin made of Alclad. Split ailerons, flaps between fuselage and ailerons.

Powerplant
Engines: 4x Bramo 322 J2 nine-cylinder air-cooled radial engines rated at 525 kW/715 hp (Do 19V1) in nacelles located along the wing leading edge. 595 kW/810 hp BMW 132F engines in same layout projected for the Do 19V2 and for the Do 19V3, and subsequent Do 19A- series the Bramo Fafnir 323A-1 (rated at 662 kW/900 hp) was planned
Propellers: three-bladed VDM variable-pitch
Fuel: 2x Cottonid tanks in the wings.

Technical Data
Performance (unarmed): maximum speed 315 km/h; service ceiling 5,600 m; range 1,600 km
Dimensions and Weights: 35.0 m wingspan; 5.8 m height; 25.5 m length; 162 m² wing area; 11,940 kg empty equipped weight; 18,500 kg takeoff weight.

Other
Armament: projected 2x 7.9 mm MG 15 + 2x 20 mm turret mounted cannons in dorsal and ventral positions.
Payload: 1,600 kg maximum.

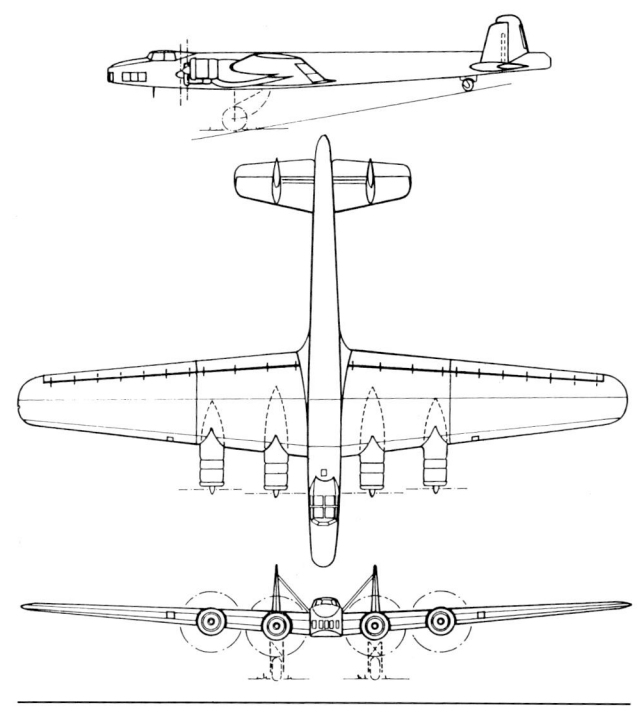

Dornier Do 19 long-range bomber (1936)

History of German Aviation: Bombers and Reconnaissance Aircraft

Dornier Do 19V1, first flown on 26 October 1936.

The Do 19V1 completed its maiden flight on 26 October 1936. The Do 19V2 and V3 were later scrapped without ever achieving flight status, while the Do 19V1 served as a troop transport from 1938 onward.

For the Do 19V3 it was assumed that suitable cannon turrets would have been developed by the time it was ready, that it would have had a takeoff weight of 19,000 kg, a maximum speed of 370 km/h, a range of nearly 2,000 km, and a ceiling of 8,000 m. Accordingly, this would in all likelihood have given the type real development potential.

Junkers Ju 89

The Ju 89 was developed as a "long range heavy bomber" (as designated by the RLM at the time) parallel to the Do 19, with a nine-man crew. Its technical features were as follows:

Airframe
Fuselage: rectangular cross section for virtually entire length of fuselage, with elliptical upper section, monocoque.
Undercarriage: retractable aft into the inner engine nacelles.
Control Surfaces: cantilever, located on the upper aft fuselage section, with dual rudders.
Wing Assembly: cantilever low-wing design. Five-section wing with an approximately 20° sweep to the leading edge. Wing center section with five tubular spars integrated into underfuselage. Two inner wing sections with five main and two auxiliary spars, with four main and three auxiliary spars supporting the two outer wing sections. Ailerons and flaps designed as the Junkers double wing in two sections, located on the trailing edge, plus split flaps situated beneath the wing center section.

Powerplant
Engines: 4x liquid-cooled Jumo 211A twelve-cylinder inline V-engines rated at 500 kW/680 hp (Ju 89V1) or 4x DB 600A in same configuration, rated at 705 kW/960 hp (Ju 89V2 and V3).
Propellers: Junkers-Hamilton PC- (Ju 89V1) or VDM variable-pitch (Ju 89V2 and V3)

Technical Data
Performance (unarmed Ju 89V2): maximum speed 350 km/h; service ceiling 7,000 m; range standard 1,600 km, maximum 2,000 km. Dimensions and Weights: 35.27 m wingspan; 7.6 m height; 26.50 m length; 184 m² wing area; 17,015 kg empty equipped weight; 22,820 kg standard takeoff weight, maximum 27,825 kg.

Other
Armament: projected for Ju 89A 2x 7.9 mm MG 15 + 2x 20 mm cannons in hydraulically-driven turrets, dorsal and ventral positions. Payload: 1,600 kg.

The Junkers Ju 89, conceived as a "long-range heavy bomber," completed its maiden flight in December 1936—nearly two months after the Do 19.
This photo shows to good effect the Junkers double wing designed to increase lift using a combination of two separate wing sections. The smaller section, basically a flap which pivoted at a point approximately 1/4 the length of its chord, was located behind and below the main wing section. During normal flight both sections were parallel, while at slower speeds the flap section rotated through about 45% so that its leading edge drew closer to the trailing edge of the main wing and with it formed a curved profile with a slotted effect.

The Ju 89V1 made its first flight in December 1936, followed by the Ju 89V2 in the spring of 1937. For the Ju 89V3 then in construction, on 2 January 1937 the RLM approved the use of this V-type for the development of a passenger aircraft, which was then given the designation Ju 90. The converted Ju 89V3 was completed as the Ju 90V-4 in September 1938 and became the pattern for ten Ju 90A-1s.

Back to the Ju 89: despite the pause in the construction program, both the V1 and V2 soldiered on as testbeds for the Ju 90 project, which saw three additional V-types built. On 4 and 8 June 1938, pilot Karlheinz Kindermann set new altitude records with a payload when he flew the Ju 89V2 to 7242 m with 10,000 kg and to 9,318 m with 5,000 kg.

Unlike the Do 19, then, the Ju 89 long range heavy bomber continued in development as the Ju 90 heavy transport, whose military use and follow-on development as a strategic maritime reconnaissance platform will be discussed in another chapter.

End of the "Ural Bomber" Project

The heavy long range bomber lost what was probably its most important advocate when *Generalleutnant* Walter Wever was tragically killed on 3 June 1936—about five months before the Do 19V-1's first flight—while attempting to take off in a Heinkel He 70. Nevertheless, Wever should not be thought of as having a one-track mind devoted solely to the strategic bomber concept; in his role as Chief of *Luftkommandoamt*, on 6 May 1936 he had confirmed his support of *Oberst* Wilhelm Wimmer's (Chief of the *Technisches Amt*) approval of construction of a high-speed bomber, and that the strategic bomber projects would take a back seat to the high-speed bomber. Wever's successor was *Generalleutnant* Albert Kesselring, who was appointed Chief of the *Luftkommandoamt* on 9 June 1936. He felt that the construction of strategic aircraft was not as critical as that of tactical aircraft at this point in the expansion of the *Luftwaffe*, and also favored the high-speed medium-bomber. Following close on its heels was Kesselring's blessing of the "Bomber A" project, a significantly more capable heavy bomber than the "Ural Bomber" and one which will be examined shortly. In spite of this, the *Technisches Amt* was able to have the prototypes of the Do 19 and Ju 89 completed—as Wever had wanted. Nevertheless, Göring personally backed up Kesselring's decision on 19 April 1937 to halt further development of the "Ural Bomber." Göring's main argument was that two or three medium range bombers could be built for the cost of a single "Ural Bomber," and that if construction of this heavy bomber were allowed to continue it would jeopardize the high-speed medium bomber project (which in the interim had settled on the Junkers Ju 88) due to the limited production capacity of Germany's aviation industry.

Thus, the first hopeful attempt to create a strategic bomber may be seen as a failure on the outside, but the operations staff by no means considered the idea dead.

The "Bomber A" Project

The *Luftkommandoamt*, the forerunner of the *Luftwaffe's* general staff, had called for a long range bomber with more than double the range of the nearly completed Do 19 and Ju 89 as early as the fall of 1936—before the "Ural Bomber's" maiden flight and its cancellation order. It was expected to travel 5,000 km with a 1,000 kg bomb payload, or 2,000 km with a 2,000 kg payload at a cruising speed of 500 km/h. These criteria may have been established based on knowledge of American advances in the field of heavy long range bombers.

Professor Ernst Heinkel, whom the RLM had asked to participate in the program along with the Junkers company, took the path of developing a long range bomber with just two propellers, each of which would be driven by two engines coupled together. Heinkel had acquired experience for such a design beginning in 1936 with a mockup (Project 1041) tailored to the "Bomber A" project, which later evolved into the He 119 test airplane. For the He 119, two Daimler-Benz DB 601 engines were arranged side by side and drove a single propeller.

Despite the revolutionary design of the He 119, the RLM did not award the Heinkel company a developmental contract. The company did, however, gain unique and fundamentally new design experience, the fruits of which would eventually lead to the Heinkel He 177. From the outset, the He 177 was conceived as the *Luftwaffe's* heavy long range bomber and entered production as such, but due to a multitude of reasons was only partially able to fulfill the high expectations placed on the design.

Regardless of how the developmental trends resulting from the various opinions are viewed today, it is quite clear that even at this early developmental stage the German *Luftwaffe* was primarily focused on offensive operations—with the goal of wiping out the enemy's air forces and, to a great extent, directly and indirectly supporting the *Heer*, or Army. Other operational needs, e.g. the fighter arm (*Jagdwaffe*), were not unreasonably neglected, but were felt to be of secondary importance.

History of German Aviation: Bombers and Reconnaissance Aircraft

The question of how the offensive force in its various components can be assessed in detail can in part be answered by looking at the armament programs which, however, were determined not only by operational considerations and the needs brought on by the *Luftwaffe's* expansion, but also to a great extent by the raw materials situation and the capacity and technological capability of the German aviation industry—armament programs which often overlapped and seldom seamlessly and logically followed each other.

2. Military Historical Retrospective

Bomber Pilots

During the First World War the main thrust of bomber operations was geared toward the direct and indirect support of the ground war, primarily on the part of the ground attack flyers. The *Luftwaffe* of 1935, however, saw the preconditions for fulfilling its mission in the availability of a strong offensive air force.

"Bearing the brunt of the air war's offensive prosecution, it is the flying arm, particularly the bomber forces, which gives the *Luftwaffe* its character,"

was printed in the *Luftwaffe* manual L.Dv. 16 "*Luftkriegführung*" (Conduct of Air Warfare) as early as its first edition in 1935. This outlined in clear and unmistakable terms the theoretical offensive role of the *Luftwaffe*. With the outbreak of war, it was expected to be an offensive instrument for its prosecution with the ability to destroy the enemy's force potential from the outset of hostilities. The opinions, differences, and the developmental trends stemming from them mentioned here cast considerable doubt upon whether or in what numbers Germany's *Luftwaffe* of the thirties, under the auspices of this generic goal, was able to penetrate far into the enemy's hinterland and thus be considered an instrument of strategic warfare.

Dive Bombing Operations

The development and realization of the dive bombing concept played a considerable role in addressing the support of ground forces from the air. When the *Luftwaffe* first began its expansion, the technological state of development of the early thirties precluded the availability of an accurate bombsight for dropping ordnance from horizontal flight at high altitudes. At the time, it was only by dive bombing that it was possible to effectively strike point targets, especially in close proximity to friendly ground forces. Although the dive bombing concept is still hotly debated, questions such as

"Could man and material withstand such strain over long periods of time? Operating at low altitudes, would dive bomber pilots be able to get through to the target against the concentrated defensive fire of every weapon available?"

highlight the lack of confidence and inexperience on the part of those making the decisions at the time, even as the *Luftwaffe's* buildup plan from August 1935 anticipated dive bombing units entering service. Nevertheless, the notion that the dive bomber was mainly planned as a weapon to engage point targets in the immediate vicinity of friendly troops must be dispelled. Also known as "quality bombing," dive bombing was just as important for striking the enemy's fixed sites having military, wartime, industrial, and other essential significance, i.e. for operations which go far above and beyond a purely tactical prosecution of air warfare.

In conjunction with the plan of incorporating dive bombing units into the *Luftwaffe*, in 1935 a developmental contract in the form of a competition was issued for a dive bomber, wherein the specifications generally paralleled the thinking of the Junkers firm. In the autumn of 1935 Junkers flew its first prototype of the Ju 87 dive bomber. Following a fly-off with prototypes from other companies (Arado with the Ar 81, Heinkel with the He 118, and Blohm & Voss with the Ha 137) one year later, the decision was made to equip the dive bombing units with the Junkers Ju 87. A year following this decision, the general staff in 1937 called for all twin-engined bombers to have the capability of dive bombing at a penetration range of 1,000 km. The twin-engined medium bomber, originally conceived as a pure high-speed aircraft capable of fulfilling the *Luftwaffe's* mission within the framework of continental Europe, thus became the backbone of the bomber arm.

The previously mentioned Dornier Do 17 was soon joined by the Junkers Ju 86 with its diesel engines, along with another large-scale production aircraft, the Heinkel He 111—designs which, when fully loaded, hardly could be considered "high-speed." The requisite dive bombing capability could no longer be realized; this was instead reserved for the Junkers Ju 88, which had excellent flight handling characteristics and whose manufacturer had already undertaken pioneering work in the field of dive bombing, particularly with the specially designed and optimally configured Ju 87. As a result the Ju 88 increasingly became the universal aerial attack weapon, capable of operating even in the dive bombing role, and whose prize for role versatility will be discussed in detail in a later chapter.

Recce Pilots

During WWI, reconnaissance pilots were the first to demonstrate the usefulness of aircraft when they underwent their baptism of fire at the Battle of Tannenberg in August of 1914, supplying senior planners with valuable reports on the movement of Russian troops. In the West, the war soon evolved into a static stalemate, and the cavalry units found themselves unable to carry out operative, tactical, or battlefield reconnaissance. The reconnaissance pilots stepped in to assume these roles. During the course of the four-year war, rapid advances in technology lightened the burden for recce pilots by providing them with better equipment and armament, while on the other hand, the enemy's increasing defensive power compelled them to search out higher or lower operating altitudes, or defend themselves in combat against enemy fighters. The reconnaissance aircraft, virtually all of which were two-seaters, carried out tactical and strategic reconnaissance, artillery spotting, and as infantry pilots mapped out the front lines and reconnoitered the battlefield. The reconnaissance forces formed the largest contingent within the *Fliegertruppe* of the First World War, a period which is covered in detail in the volume *"Bombenflugzeuge."* In building up the new German *Luftwaffe* during the mid-thirties, the number of reconnaissance units was smaller than those of the bomber and fighter units. This ratio was by default a direct byproduct of the mission assigned to the *Luftwaffe*. The reconnaissance units, including coastal patrol groups, became an integral part of the *Luftwaffe*.

Tactical Reconnaissance

Aerial reconnaissance for the benefit of the army, generally known as tactical reconnaissance, is designed to provide data on the enemy's terrain, deployment, and strength, on preparations, initiation, progress, and cessation of combat engagements, as well as on the actions and facilities in the deployment, operations, and rear areas of the enemy forces. It supports the tactical command and produces data for the employment of friendly forces, as well as provides battlefield reconnaissance for the prosecution of a battle.

At the same time, it provides target reconnaissance data for an air force's strike units operating in cooperation with the army. Air combat is a secondary role of tactical reconnaissance aircraft, as this opens the way to fighter-protected targets. Tactical recce pilots are tasked to work in close cooperation with the corps and division elements of the army and fly their missions based on the army's requirements.

Strategic Reconnaissance

Strategic reconnaissance provides data from deep within an enemy's territory (up to the limits of the aircraft's range) for the operational prosecution of the air, land, and naval war. In an air war in particular, this target reconnaissance data takes the form of information utilized to prepare for and carry out aerial attacks. As a rule, strategic recce planes avoid air combat, but if engaged by the enemy they must be capable of defending themselves effectively. They are tasked by entities within the air force, which in turn provides the army with the data—insofar as it is relevant to land forces.

Maritime Reconnaissance

Reconnaissance in support of a naval war can only differentiate between the tactical and strategic focus in specific instances. Maritime reconnaissance targets ocean areas, enemy coastal waters, and harbors. It is designed to provide data on the enemy's movements and his intentions. Large areas of the ocean are covered by several aircraft operating in so-called recce boxes. If an enemy fleet or convoy is flushed out, this is reported by radio. The recce plane must then shadow the ships until friendly air or naval forces can attack, or until another reconnaissance aircraft relieves it. Armed reconnaissance can also be employed, whereby the sighted enemy can be attacked with guns and bombs, depth charges, and aerial torpedoes. Direct and close cooperation with naval entities is therefore indispensable.

This summarizes the role definition for aerial reconnaissance at the time, and it is not difficult to see that these widely divergent reconnaissance requirements also dictated major differences in the requirements of the flying equipment. These differences will be examined in detail over the following pages.

3. Proving Grounds of Spain

Generalmajor Albert Kesselring replaces *Generalleutnant* Wever as Chief of the *Luftwaffe's* general staff (renamed from *Luftkommandoamt* in 1937) when the latter's plane crashed on 3 June 1936. He was in agreement with the armament program developed by Milch and Wever, and initially allowed it to continue according to plan.

Catching Kesselring by surprise, in late July 1936 there presented itself an opportunity to test new weapons and tactical theories regarding the prosecution of air warfare on foreign soil. The *Luftwaffe's* leadership had neither desired this opportunity, nor had the Germans provoked it politically.

Emissaries from Spain, where a civil war had broken out on 17 July 1936, brought Hitler a letter from General Franco on 26 July of that year. Franco was the self-appointed leader of a resistance movement against the Republican Spanish government. Franco had asked for approval to buy "ten transport aircraft of maximum load capacity" in order to ferry the approximately 25,000 troops available to him in Morocco, the so-called "Army of Africa," to the Spanish mainland, as the sea lanes were controlled by the Republican fleet. That same evening Hitler personally decided—over the objections of Defense Minister Werner von Blomberg and the Foreign Office—to be generous in honoring the request of the Spanish Nationalists and that double the number, i.e. twenty Ju 52 transports, was to be delivered.

Starting Point

When civil war broke out on 17 July 1936 the "Fueréas Aereas Españolas" (the air forces of the Spanish government) had approximately 100 frontline aircraft, not counting a handful of operationally insignificant airplanes. These were Spanish license-built French Bréguet XIX, a two-seat biplane used for reconnaissance and bombing missions. About a third of these fell into the hands of Franco's supporters. As far as fighters went the now fragmented Spanish Air Force had about 30 operational Nieuport-Delage Ni-52 on the Republican side—a single-seat sesquiplane of French origin also built under license in Spain—while the Nationalists had absorbed four undamaged examples of these into their ranks upon the outbreak of hostilities. Approximately one-fourth of the officers, NCOs, and crews of the "Fueréas Aereas Españolas" went over to Franco.

Both warring parties immediately undertook aggressive steps to acquire modern flying equipment and personnel from abroad: the government in Madrid solicited the Paris government, as well as the Soviet Union, and the Spanish Nationalists approached the Italians and the Germans.

Germany's Initial Role

Hitler's spontaneous approval of the delivery of 20 Ju 52 transports in lieu of Franco's requested 10—which only took place following promises of support by the French and Italians to the Republicans and Franco, respectively—led Göring to point out the Ju 52s' vulnerability to Republican fighters if they did not have their own fighter protection themselves. Hitler therefore decided to include an additional six Heinkel He 51 biplane fighters as part of the package.

On 27 July 1936 began the operation to support the Spanish Nationalists, codenamed "Feuerzauber," under the direction of the *ad hoc* established "*Dienststelle General Willberg*" (also known as "*Sonderstab W*").

Fitted with external fuel tanks, on 28 July 1936 the Ju 52s set out on a ten-hour flight to gather Franco's forces. On 29 July 1936—just slightly more than 48 hours following the decision to send them—the transports began building the air bridge from North Africa to the Spanish mainland; during the critical first 33 days of operations they ferried over 10,000 Moroccans and their weapons into Spain.

During the night of 31 July/1 August 1936 the 22,000 ton German freighter *Usaramo* slipped out of Hamburg shortly after midnight and, after passing through the Straits of Gibraltar, arrived at the harbor of Cadiz on 7 August. In its hold it was carrying the crated He 51 fighters, as well as 91 Germans who had volunteered

History of German Aviation: Bombers and Reconnaissance Aircraft

for action in Spain: ten complete Ju 52 instructor crews, six fighter pilots, technical maintenance personnel, radio operators, and medical and civilian liaison personnel.

In Seville, General Franco's representatives and the Germans decided to establish the following functional groups using the personnel and material already in place or on their way:

- a transport group in Spanish Morocco with 11 Ju 52s
- a bomber group based outside of Seville with nine Ju 52s acting as auxiliary bombers. Spanish crews were to fly the unit's missions; the German instructor crews would function as trainers.
- a fighter group with six He 51s, also stationed near Seville. The Germans would assemble the He 51s, and these would be flown exclusively by Spanish pilots following their instruction by the German fighter pilots.

Additional Support

Soviet technicians arrived in Spain on 10 September 1936 with the intent of setting up Republican airfields to accept Soviet fighters and bombers. On 13 October 1936 the first Polikarpov I-15 Chato single-seat fighters arrived via sea transport in the harbor of Cartagena. Along with these aircraft came 150 members of the Soviet Air Force, including 50 pilots.

The I-15 was a classic biplane of mixed-media construction with four fixed machine guns and cantilever undercarriage; Soviet pilots quickly showed that they were a match for the He 51 in combat when they first encountered them in the latter half of October 1936. From November 1936 onward the Chatos were supplemented by the first Polikarpov I-16 fighters.

The I-16, also known as Rata, was a low-wing fighter designed to the smallest dimensions possible, with an all-metal monocoque fuselage and a manually retractable undercarriage. It was markedly superior to the He 51 in combat, and even the Italian Fiat CR.32 had a hard time matching it.

The first Tupolev SB-2s also began operations with the Republicans in October of 1936. This twin-engined bomber, a cantilever mid-wing design, was far superior to the Ju 52 with regard to performance and roughly on par with the Italian S.M. 81. Having a top speed of nearly 400 km/h at 4,000 m, it was faster than nearly all enemy fighters—something which would not change until the later involvement of the Messerschmitt Bf 109.

Initial Phase of the Civil War

This buildup of then-modern flying equipment, which would continue throughout the conflict, set the stage for the "Spanish Proving Ground." During this civil war there was no air event independent of the land or sea war, meaning neither a Douhet-like "Total Air War," nor a graduated air war. Practically all aerial missions served to prepare for or support large and small scale land operations. All the experience to be gained by such operations were exclusively tactical in nature. This conflict was also marked by the unique conditions typical of a civil war.

In mid-October the Spanish Nationalists launched an attack on the Spanish capitol which, after initial success by Franco's troops, was beaten back in late October by a Republican counter-attack using newly acquired Soviet tanks. In early November 1936 the Nationalists renewed their drive and took the airfield of Getafe, located some 13 km from the heart of the city.

The Junkers Ju 52, which began ferrying Spanish Nationalist forces between North Africa and the Spanish mainland on 29 July 1936, was soon pressed into service as a bomber, as well.

Events in the Air

Attacks on Madrid were preceded by incessant bombing raids, generally conducted by Ju 52 formations, which rarely required fighter protection initially. On 5 November 1936 the tide turned in favor of the Republicans with the appearance of Soviet I-15 fighters and, on 15 November, of I-16 Ratas, along with the SB-2 Katyuska bombers. They shot down several Spanish Nationalist bombers approaching Madrid and forced the remaining aircraft to turn back.

The huge deliveries of Soviet aircraft seriously jeopardized a quick seizure of Madrid by Franco troops. Franco took steps to strengthen his Nationalist air forces and improve what was for him a very unpleasant situation in the air. In addition to increasing the Italian contingent, he was able to reach an agreement with Germany which obligated the latter to continuously maintain 100 front-line aircraft in Spain with German crews. Thus was born the Condor Legion, whose first commander, *Generalmajor* Hugo Sperrle, assumed his post on 6 November 1936.

The Legion was rapidly enhanced by personnel and materials, and soon had a starting aircraft inventory consisting of:

- 36 Ju 52 auxiliary bombers in three bomber squadrons
- 36 He 51 fighters in three squadrons, later supplemented by a fourth
- 12 He 70 reconnaissance aircraft in a single squadron, some of which where used for light bombing operations
- 10 Heinkel He 59 maritime reconnaissance aircraft
- several Heinkel He 45 and He 46 aircraft, used primarily for reconnaissance work

From Germany the Spanish Nationalists received reinforcements to supplement its initial delivery of 20 Ju 52s and six He 51s in the form of an additional nine He 51s and 20 He 46s, to be chiefly used in the close support role. The Spaniards soon relegated the latter to duties on secondary fronts, as they proved to be too slow and no match for the Soviet front-line aircraft.

As the air war progressed the Condor Legion was continually beefed up by more modern aircraft better suited to the combat conditions at hand. As an example, in January 1937 *Versuchsbomberstaffel* VB/88 received, or was already operating, four of the ten He 111B-0s built, four Do 17Es, and four Ju 86Ds. The latter type proved unsuitable, however, with two being lost in combat and the remaining two handed over to the Spanish Nationalists. Following the combat trials of these three bomber types VB 88 was dissolved in July of 1937.

Specifically, German-manufactured aircraft assumed the following roles during the war in Spain:

Bombers and Dive Bombers

Heinkel He 111

A Heinkel He 111 with Spanish Nationalist markings on its rudder and fuselage, seen dropping 50 kg bombs with its ventral gun position extended (in action from late 1936 onward).

The Legion's bomber evaluation squadron, VB/88, included four He 111B-0s at the end of 1936 which proved themselves well-suited to their role. When the first bomber units in Germany converted to the He 111B-1 in the winter of 1936/37, two of the Condor Legion's *Kampfgruppe* 88 (K/88 for short) quickly followed suit, with the bulk of the now redundant Ju 52s being transferred to the Spanish Nationalists.

The majority of the 30 Heinkel craft arrived in Spain in February 1937. They flew their first sorties against Republican airfields on 9 March 1937. In the fall of 1937 the improved He 111B-2s began arriving, with more powerful engines and an improved cooling system. In March 1938 the first of 45 examples of the new HE 111E-1 series began appearing in Spain. These had an even more powerful engine and a larger payload capacity, with a consequently heavier takeoff weight. The He 111's relatively weak defensive armament remained unchanged throughout its entire period of service. Due to the comparatively few losses within the He 111 squadrons in Spain, the *Luftwaffe's* leadership increasingly felt that high-speed, unescorted bombers would continue to be able to operate with relative impunity over enemy territory. Following the conflict, the He 111 remained in service with the Spanish Air Force until the late 1950s.

Dornier Do 17
Another bomber type appearing in Spain was the Do 17. During the course of 1937 an additional 20 Do 17E-1 models were assigned to the Condor Legion to supplement the Do 17s which had been

History of German Aviation: Bombers and Reconnaissance Aircraft

The Dornier Do 17E-1, which entered service with the Condor Legion in the Spanish Civil War in Nationalist markings over the course of 1937, taken from the B-Stand (dorsal gun position for the radio operator) of the lead aircraft.

serving with VB/88 since late 1936. At first, the relatively fast Do 17 bombers also remained virtually untouched by Republican fighters. Only the modern Soviet fighters posed any kind of threat to them, as the Do 17 also suffered from weak defensive armament. In August 1938 all Do 17Es, along with Do 17F and P reconnaissance aircraft, were handed over to the Spanish Air Force. A total of 13 Do 17s of all types survived to the end of the civil war in late March 1939, and these soldiered on in service for many years thereafter.

Junkers Ju 86

The Ju 86D has already been mentioned in service with VB/88. The chief reason for its negative assessment in Spain was its troublesome powerplant, poorly suited to combat conditions. The Junkers diesel aircraft engine (Jumo 205) simply did not have the flexibility and response needed for formation flying.

Henschel Hs 123

The Spanish Civil War offered an ideal opportunity for developing the dive bomber concept as the Condor Legion provided close support to the Nationalists, thereby testing the validity of the dive bombing theory in combat. To this end, the Legion's chief of staff requested the Hs 123, a light dive bomber with which the first *Stuka* groups in Germany had been equipped in the summer of 1936. The Condor Legion accordingly received five Hs 123As in December 1936 and almost immediately began operations with the type in early 1937, although primarily in the ground attack role. The Hs 123 was so successful in its *de facto* strike role that it was retained in that capacity, and the Spanish Nationalists asked to be supplied with this type. In the summer of 1938 they were ultimately provided with 16 aircraft, including those operated by the Legion when it converted over to the Junkers Ju 87.

Henschel Hs 123. The racks for 50 kg bombs can be seen underneath the lower wing of this dive bomber and strike plane (from 1937 used chiefly in the ground support role).

Henschel Hs 123s approaching their target with 50 kg bombs suspended beneath their wings (1938).

Junkers Ju 87

The Ju 87 was selected from the four aircraft participating in the standard dive bomber competition in 1936 and entered production shortly afterward. It replaced the Hs 123 in the *Luftwaffe's Stuka* groups from early 1937 onward. In late 1937 one of these units despatched a *Kette* (three) of Ju 87A-1s to Spain to serve with the Condor Legion in order to test out dive bombing technology with this new type under combat conditions. The *Kette* began operations on 17 February 1938, first with marginal results, but later showing marked improvement with bombing accuracy. They took part in the majority of Spanish Nationalist actions and were primarily used against point targets, such as road junctions and bridges, as well as ships and harbor facilities.

In October 1938 these were exchanged for five Ju 87 B-1 models. The experienced crews, almost always flying under heavy fighter protection, achieved an accuracy of +/- 5 m with 500 kg bombs and, despite generally weak enemy defenses, gained valuable combat experience—experience which the *Luftwaffe* would credit for much of its later success with dive bombing operations. All Ju 87s returned to Germany following the end of the civil war.

Junkers Ju 87 dropping a 250 kg bomb on a tank. The A-series had the easily recognizable "pantaloon" style undercarriage.

Accuracy and effect of a 250 kg bomb dropped by a Junkers Ju 87 on an intersection near Alcala del Obispo.

Recce Planes

Heinkel He 46

From September 1936 until August 1938 the Heinkel He 46c served with the Spanish Nationalist Air Force as a tactical reconnaissance aircraft and close support ground attack plane.

The Junkers Ju 87, which was combat tested in Spain as a dive bomber in small numbers (three to five) beginning on 17 February 1938.

The Nationalists' shortage of reconnaissance aircraft was initially offset by the purchase from Germany of 20 He 46Cs in September of 1936. A high-wing tactical reconnaissance plane, it was also used mainly in the ground support role.

Although the He 46 may have been a step up performance-wise with regard to the Bréguet XIX sesquiplane still found in Spanish service, by August of 1938 its performance no longer met requirements and it was pulled from front-line service. The type was then assigned to the Spanish observer's training school as a training aircraft.

Heinkel He 70
Quite a fast airplane in its day, the He 70F-2 equipped the Condor Legion in November of 1936. The Legion flew its first mission with the type as early as 6 November 1936. With its on-board guns it was also used extensively as an armed reconnaissance aircraft. The Legion's reconnaissance squadron, A/88, received a total of 18 of these aircraft designed for high operating speeds, but whose low-wing design made it poorly suited for recce operations. This was confirmed by its classification as an auxiliary reconnaissance aircraft. With the introduction of the Dornier Do 17F long range reconnaissance aircraft in Germany in early 1937, a complete Do 17

The Heinkel He 45, in Spain "demoted" from a strategic reconnaissance aircraft to tactical reconnaissance and strike duties.

The Heinkel He 70, which never fully proved its worth in Spain as a light bomber, nor a strategic reconnaissance platform. The view from the enclosed low-profile cockpit was inadequate, and the large low-wing design made spotting targets quite difficult for the observer. Furthermore, the He 70's wooden wings were quite susceptible to catching fire when hit.

squadron from the *Luftwaffe's* strategic reconnaissance group was sent to Spain to replace A/88's He 70F-2s. Compared with the He 70, Dornier's design had a significantly greater range. At Franco's urging 12 of the now redundant He 70s were transferred to his air force. Eleven of these survived with the Spanish Nationalist Air Force to the end of the civil war and continued in service until the mid-1940s.

Heinkel He 45
Although the Condor Legion's A/88 had been established in November 1936 with He 70F-2s, it also received an additional six He 45C biplanes. Originally designed as a strategic reconnaissance aircraft, this type was used almost exclusively in the tactical recce role until being replaced by six Henschel Hs 126s in the late fall of 1938. The redundant He 35s were then given to the Spanish Nationalist Air Force, which received a total of 40 He 45c models during the course of the war. The Nationalists used the type as a tactical reconnaissance platform and for supporting the ground war with its on-board guns and light bombs.

Dornier Do 17
The Do 17F-1 began replacing older *Luftwaffe* types in its long range reconnaissance group in early 1937. As mentioned in the section dealing with the He 70, the Do 17 had a much greater range compared to that of the He 70 which, when coupled with a relatively high operating speed, led to an entire squadron of Do 17F-1s being sent to Spain to replace the Condor Legion's He 70s in the spring of 1937. Its speed made the Do 17 almost immune to the best Republican fighters of the day, and thus appeared to bear out the theories of high-speed bomber and reconnaissance aircraft earlier advocated by the *Luftwaffe's* senior leaders. A/88 later received additional Do 17P-1 strategic reconnaissance models to increase its numbers and improve the stock of its inventory.

Dornier Do 17P-1 served in the Spanish Civil War from late 1937 with the Condor Legion and with the Spanish Nationalist Air Force from August 1938 as a strategic reconnaissance plane.

The air cooled BMW 132N of the P-1 was much less susceptible to enemy fire than the liquid cooled inline BMW VI of the F-1. The only real threat to the Do 17P came from Soviet Ratas which, as with the Do 17E bombers of K/88, showed a weakness in its defensive armament, particularly when attacked from below. In August 1938 the surviving Do 17Ps along with the Do 17Es of K/88, were turned over to the Spanish Nationalist Air Force.

Henschel Hs 126

The Hs 126A replaced the Condor Legion's He 45 *Kette* of A/88 in the late fall of 1938. These six machines proved effective in both

Henschel Hs 126, which replaced the He 45 in the Condor Legion as a tactical reconnaissance aircraft beginning in the late autumn of 1938.

the tactical reconnaissance role, as well as in the armed reconnaissance role using guns and light bombs. The five remaining Hs 126s were handed over to the Spanish Air Force at the end of the war.

Heinkel He 59

A formation of Heinkel He 59s. These biplane seaplanes saw extensive service in Spain from November 1936 onward in the maritime reconnaissance and night bombing roles.

The Legion's maritime reconnaissance squadron, AS/88, received a complement of ten He 59B-2s, a biplane seaplane type powered by two BMW VI engines. The unit began operations with the type in November of 1936. The squadron was mainly used for maritime reconnaissance and harbor blockade missions, as well as for nuisance bombing of roads and rail lines. Additionally, there were numerous night bombing sorties carried out against harbor facilities, whereby pilots developed the tactic of throttling back and diving from an altitude of 2,500 to 3,000 m to drop their bombs on the harbor's targets virtually silently from an altitude of about 1,000 m.

The actual strength of ten aircraft for AS/88 was maintained through immediate resupply from Germany, and for a time it included 15 aircraft in its inventory. At the end of the war seven He 59B-2s returned to Germany, while three were given to the Spanish Navy where they flew until 1945.

Heinkel He 60

The Heinkel He 60, a catapult-launched biplane seaplane, went into action beginning in November 1936 as a tactical reconnaissance plane.

The Spanish Nationalists requested this single-engined, catapult-launched tactical reconnaissance floatplane for their navy shortly before production of the He 60D ceased at the Weser Flugzeugbau company in the latter half of 1936. In November of 1936 the Spanish request was acknowledged, and six export models were delivered to the Condor Legion to be passed on to the Spanish Navy. But it was not until April of 1937 that the Spaniards finally received the He 60Ds. The Spaniards procured a total of eight of these machines; four survived the civil war and remained in service on the island of Mallorca until 1948.

Fighters

Heinkel He 51

The He 51B-1 was a single-seat biplane fighter with a 552 kW/750 hp BMW VI 7.3Z engine and was armed with two fixed MG 17 machine guns. It was the first fighter to equip J/88, the Condor Legion's fighter group, which received 36 examples of the type.

History of German Aviation: Bombers and Reconnaissance Aircraft

The Heinkel He 51 was a single-seat biplane fighter which served with the Spanish Nationalist Air Force from August 1936 and the Condor Legion from November of that year. During the course of the Spanish Civil War, it was relegated to the strike role when the Republicans' I-15 Chato and I-16 Rata Soviet-built fighters proved superior in combat.

Assembled in Spain, J/88's He 51s began operations on 6 November 1936. The Spanish Nationalists had earlier acquired several Heinkels, and received further He 51B-1s in September of 1936.

The initial superiority enjoyed by the He 51 operated by the Spanish Nationalists was short-lived. Soviet I-15 fighters engaged in air combat for the first time over Madrid on 4 November 1936 and quickly revealed themselves to be generally superior to the He 51 in terms of speed, climb rate, and maneuverability. Despite a slower rate of fire than German armament, with their four machine guns they had greater firepower than the He 51's two MG 17s. By the spring of 1937 the Republicans were able to field over 200 I-15s.

Beginning in March 1937, the He 51's increasingly obvious vulnerability led the Condor Legion to pull nearly all of them from fighter operations and utilize them instead for close combat, for which role they were fitted with racks to hold six 10 kg bombs. In November 1937 the Spanish Nationalist followed suit by changing the role of their He 51s and concentrating them into a strike unit.

Although the Condor Legion's 2. J/88 turned in their He 51s in order to pick up their first Bf 109Bs in March 1937 and 4. Staffel was temporarily disbanded, J/88's 1. and 3. *Staffel* continued operating the type—although almost exclusively in the ground support role.

Shortly afterward even 1. *Staffel* converted to the Bf 109B, and only 3. *Staffel* was left to fly their He 51s in the close air support role for the ground forces.

Messerschmitt Bf 109

The Messerschmitt Bf 109 was the outstanding fighter of the Spanish Civil War (serving in squadron strength with the Condor Legion beginning in March 1937.)

In July 1938 even 3. *Staffel* converted to the Bf 109C-1, Messerschmitt's newest model with four fixed machine guns. J/88 reactivated a fourth squadron, where it concentrated its remaining He 51s and used them exclusively in the strike fighter role. These aircraft were protected on an as-needed basis by Bf 109s. The biplane's fighter role can be considered at an end with the He 51's reclassification as an auxiliary strike plane; the biplane was forced to vacate the field in favor of the cantilever, single-engined low-wing monoplane with retracting gear. The failure of the He 51 in its original role undoubtedly contributed to the accelerated pace in re-equipping the *Luftwaffe*—to include the Condor Legion—with the Bf 109. Despite its excellent maneuverability, the He 51 was no longer a match for the superior speed and climb rate of the I-16 Rata then dominating the skies over Spain. The air superiority achieved by the Bf 109 soon changed the Spanish Nationalists' fortunes in the civil war for the better.

Development of the Air War and Interim Balance

Of dramatically increasing importance as the civil war dragged on was the aerial support of ground actions, especially direct support operations. The air forces of both sides paved the way for ground operations, and supported them both directly and indirectly through constant reconnaissance and combat missions, generally under the protective umbrella of friendly fighters. Ground attacks generally came to a standstill quite quickly where weather conditions or a lack of air power forced one of the combatants to operate without air support.

In a 1938 interim assessment ("Les Leçons de la Guerre d'Espagne," Librairie Plon, Paris), a French general by the name of Duval made the following interesting observations about the prosecution of an air war:

"Regarding the matter of aerial bombardment, one fact clearly emerges: there exists the possibility to systematically destroy a city from the air....

The air force consistently played a role in preparing for an infantry attack. The bomber generally came to be considered a cannon with the same effectiveness at all ranges, even the greatest ones...."

These views from a high-ranking French army officer indicated that air operations in Spain were no more than what the German army's leadership had always considered it to be—an extension of the artillery designed to effect the preparation of the infantry attack.

Duval also addressed the issue of the longer breaks between combat common to the civil war in Spain:

"The consumption of weapons and ammunition reaches a point where supply cannot not keep up. Industrial production capacity forces the war's rhythm. The armies must wait for the arrival of their ammunition, cannons, aircraft, and tanks—this is what causes the delay between two attacks.... Franco had to proceed so slowly because it was not possible for him to advance any faster. He was compelled to waste time, for time equals ammunition.... The greatest risk in a future war will be that a party will become involved with an enemy who has an abundant supply of material and inexhaustible resources."

In his vision of the most terrifying threat in any future conflict, Duval had concluded from the events of the Spanish Civil War up to that point that a materially superior enemy with inexhaustible resources—i.e. tanks, aircraft, ammunition, and raw materials *"en masse"* would emerge the victor.

Air Forces of the Day

When leadership of the Condor Legion transferred from *Generalmajor* Hugo Sperrle to *Generalmajor* Hellmuth Volkmann on 1 November 1937, the organization had a running inventory of approximately 100 operational aircraft whose numbers were held relatively constant by continual replenishment, as the Germans had promised General Franco. Around the end of 1937 the Legion was operating the following types:

- *Jagdgruppe* J/88: Me 109B as a pure fighter (two *Staffeln*, or squadrons), as well as He 51s (two squadrons), the latter almost exclusively in the strike role.
- *Kampfgruppe* K/88; He 111B (three squadrons), as well as a few Ju 52 aircraft for night operations.
- *Versuchsbomberstaffel* VB/88 (4./K/88): Hs 123 aircraft, also generally used as strike aircraft
- *Aufklärungsstaffel* A/88: Do 17F (four three-ship elements, or *Ketten*), often called upon to perform bombing missions, as well as He 45 (one *Kette*)
- *Seeauflkärungsstaffel* AS/88: He 59 floatplanes

In addition to the previously mentioned main types, the Spanish Nationalists also operated about a dozen other aircraft types in low numbers which had a minimal impact on the development of the air war. These included a few Dornier Do Wal flying boats and Arado Ar 68E-1 biplane fighters.

On the Republican side, the main aircraft mentioned earlier were supplemented by several designs, some of which could be considered prototypes. These were of Soviet, Czech, Polish, British, French, and American manufacture. Over the course of the war this colorful palette, consisting of the widest variety of every kind of plane, resulted in a confusing smorgasbord of aircraft types. But this well-nigh unbelievable hodgepodge neither improved the fighting capability nor the operational readiness of either side. Instead, this diversity was more of a burden to the opponents, reducing their effectiveness through a fully overburdened logistics system and barely manageable repair and maintenance program.

End Game

The final phase of the conflict continued into the 1938 Christmas season, where the Republicans found themselves facing a much stronger enemy. The ground forces of the Spanish Nationalists and their allies were numerically superior by almost a third, and an air force five times stronger. In total, on Christmas Eve 1938 300,000 men and nearly 500 front-line aircraft—including the Legion (commanded by *Generalmajor Dr.-Ing.* Wolfram *Freiherr* von Richthofen as of 30 November 1938)—opposed roughly 220,000 Republicans and 106 airplanes. The Catalonian Offensive followed in mid-January 1939. Barcelona had fallen to the Nationalists by the end of January. As late as 10 February 1939 the Condor Legion used their overwhelming aerial superiority to fly incessant bombing raids against road and rail lines in northern Catalonia in an attempt to cut off the escape routes being used by fleeing Republican troops making their way to the French border. On 28 February 1939 France, after having granted asylum to the Republican government on 5 February, was joined by Great Britain in officially recognizing Franco as the legitimate head of Spain. The battle for the capital was the last military operation of the civil war. After artillery and aerial bombardment, Madrid fell on 28 March 1939 virtually without a fight—the war had ended! On 1 April 1939 a Spanish war report succinctly noted:

> "Yesterday our soldiers achieved the last military objectives; the Red Army forces have been captured and disarmed."

Leaving Vigo on 26 May by ship, on 31 May 1939 4,770 members of the Condor Legion arrived in Hamburg's harbor—in time for a victory parade on 6 June, as well as for the beginning of the Second World War three months later!

Lessons for Air Power

What are the lessons to be learned from this civil war which has often been so emotionally interpreted in one way or another? The above overview was intentionally punctuated with figures and made as detailed as possible. What significance did it have with regard to the operations of air forces in general and of bombers and reconnaissance aircraft specifically? Was it possible to draw conclusions regarding military requirements for aircraft, air tactics, crews, and their training?

Could the events in Spain be seen as a model for a future military altercation in the air? Leaving out the political aspects and motives for the intervention of foreign powers in the Spanish Civil War, another question is whether the use of German soldiers and German products offered the *Luftwaffe* a commensurate benefit in experience gained.

The latter point is addressed by first looking at the following German losses and measurable successes:

- A total of 96 aircraft were lost, of which 40 were due to enemy action and 56 because of accidents—most of which could be attributed to winter weather conditions.
- In human life this translates to 173 killed, with a further 97 members of the Condor Legion killed in accidents and 28 deaths attributed to illness—for a total of 298 dead.
- 314 Republican airplanes confirmed destroyed by Legion members in air operations, with an additional 70 probable unconfirmed and 61 shot down by the Legion's AAA batteries—thus eliminating a total of 375 enemy aircraft (+70?).

In shooting down 314 aircraft while losing 40 of its own, the approximate loss ratio of 1:8 demonstrates a clear superiority of German products and personnel compared to those aircraft, mainly of Soviet manufacture, and Soviet (or Soviet-trained) personnel of the Republicans. However, of potentially much greater import are the intangible results of the *Luftwaffe's* involvement in Spain.

The German experience gained for a modern prosecution of air warfare was relatively unilateral, as there was no urgency involved in a strategic-operational use of air power within the framework defined and limited by the civil war. The experience restricted itself almost entirely to a tactical nature. To be sure, the bombers (both the Ju 52 auxiliary type and the newer Do 17s and He 111s) penetrated deep into tactical areas, striking enemy airfields, supply, and harbor facilities, but this quasi-strategic use in no way reflected the norm. They were directly involved in the ground war and against targets in the battle zone, exclusively providing direct support to friendly ground forces when those forces moved on the offensive or found themselves in crisis situations.

Bombing sorties were generally confined to a relatively tightly constricted area of the land surface, with the strike planes and dive bombers (He 51, Hs 123, Ju 87) being of paramount benefit to the troops tied to the ground.

One area which will be discussed in detail later are the remarkable advances achieved with regard to close interaction, or cooperation, between the ground and air forces. This related to both the operational planning and execution, as well as the precision employment of forces in total unity with the ground troops and the exact, almost immediate response time.

Of special note is the impact of *Generalmajor* von Richthofen, former chief of staff and the last commander of the Condor Legion, who was particularly influential in this development.

Where the air situation permitted, bomber sorties were flown without protective fighter cover—particularly in the early phases of the war—and losses were relatively few given the defensive guns which the bombers carried. Fighter escort for the bombers, almost always for Ju 87 dive bombing missions, was readily provided in those areas where the enemy had air superiority or where fighter opposition was expected; the bombers' relatively short penetration radius posed no problems for the endurance limits and ranges of friendly fighters, and the cooperation between bombers and fighters soon established itself into a familiar routine.

Despite their maneuverability and roll rate, the slower older-generation reconnaissance biplanes (He 45) were unable to keep pace with operational conditions, nor did the much faster, low-wing, auxiliary reconnaissance planes (He 70) match the ideal requirements for longer ranging reconnaissance missions. The single-engined high-wing design (Hs 126) was much more suited to the tactical reconnaissance role, as was the twin-engined high-wing long range design (Do 17) for strategic reconnaissance.

This summarizes the purely military consequences, which certainly had both a positive as well as—*á la longue*—a negative backlash on the continued armament of the *Luftwaffe*. The involvement in Spain can generally be considered to have held up, and been a burden to, the growth of the *Wehrmacht*. As mentioned from the outset, the *Luftwaffe* leadership had neither wanted the war, nor had the Germans provoked it.

The following notes touch on a few brief points regarding the fallout of the Spanish experience on Germany's philosophy of air power:

- The concept of the strategic bomber role as the decisive weapon in a war was shelved in Spain; there it was almost exclusively degraded into a tactical tool and thus failed to achieve its purpose.
- The predominant use of all *Luftwaffe* forces in conjunction with the actions of the *Heer* in limited areas of the front subordinated what was practically an independent branch of service to the *Heer* in a complete misjudgment of its original roles and potential.
- Little recognized or taken into account, and therefore seriously underestimated, were the potential consequences of a strong enemy air force, especially one capable of carrying out strategic action in the heart of enemy territory. Neglect of friendly air defenses was not only the direct consequence of this judgmental error, it also confirmed the offensive character of Germany's air war doctrine in L.DV. 16 *"Luftkriegführung"* from 1935, which was *de facto* to be given a higher priority at the expense of air defense.

Erroneous, unilaterally structured and utilized operational and command trends, and practices and the technological developments stemming from them—these negative products of the experience in Spain would not become obvious until a later date, when they would be felt much more painfully.

Even the statements and conclusions of the French general Duval regarding the greatest risk of a future war, made back in late 1937/early 1938, were apparently understood in their full scope by few within Germany—too few!

4. The New Generation of Aircraft

Development of new operational aircraft, some of which were mentioned in the preceding chapter, was generally tailored to the ideas of air warfare outlined in *"Luftkriegführung"* (L.Dv. 16), as well as to the experiences gained from the Spanish Civil War.

After years of tweaking, the first *Luftwaffe* service manual was published in 1935. In part, it stated that the *Wehrmacht* and Germany were constantly under threat from enemy air forces, and that this threat could not be adequately countered by defenses within Germany alone. This aerial threat to Germany compelled the use of military force against an enemy's air forces in an offensive role from the outset of hostilities. It therefore followed that offensive actions in the heart of enemy territory became essential, and thus the concept of air attack would take precedence over all others, for an attack by Germany's air forces would strike at the root of an enemy's fighting strength and the will of its people to resist. The core idea of this guideline was therefore the concept of attack, certainly not originally in the sense of an act of aggression but, due to the given geostrategic and technological situation, in the sense that the enemy's threat potential must be destroyed on his own territory before he could do the same to Germany. Consequently, relative to its peculiarity and objective within the framework of the overall prosecution of the air war, the focus of the *Luftwaffe* was clearly upon bomber aircraft. Another section of the manual defined these aircraft as "long range bomber aircraft, capable of carrying heavy payloads and operating in all weather conditions." It was obvious that such a philosophy would have an effect on air armament, since there existed a clear-cut state of interdependence between the *Luftwaffe's* long range intentions, air armament planning programs, and aircraft production.

The aircraft acquisition program running from 1 July 1934 to 31 March 1938 initially included a roughly equal number of bombers and aircraft, but the balance immediately shifted in favor of the bombers as early as the first acquisition period, which ran to the end of 1935. During this time 822 bombers were built in comparison with just 245 fighters. Delivery plan No. 1, a modification to the original acquisition program implemented in October of 1935, called for 1,849 bombers and just roughly half that number of fighters, 970 to be precise.

Although bombers and fighters are discussed here in simplistic terms, perhaps a more accurate way to refer to these would be to actually define them as offensive and defensive aircraft, or attacking and defending types. However, it is hardly possible to establish a clear demarcation between the two categories, and in some cases would even be arbitrary. For example, offensive aircraft may be used in the ground attack role for defending army units, while defensive aircraft, i.e. pure fighter and heavy fighter types, were not uncommonly found supporting air and ground offensives. Yet fluctuations in their roles do not compromise the fundamental character of these two categories.

Therefore, in addition to the traditional bombers, the category of offensive aircraft also broadly includes high-speed bomber and strike aircraft, since all of these types can be used as bomb platforms and therefore serve in an offensive capacity.

In this regard the *Zerstörer* (lit. "Destroyer"), also designated as heavy or strategic fighter, plays something of a special role.

The *Technisches Amt* first called for a heavily armed *"Kampfzerstörer,"* or bomber-destroyer, in 1934. The type was to have been a multi-role design, capable of acting as a bomber, bomber escort, and bomber interceptor, as well as a reconnaissance platform and close support aircraft.

The Focke-Wulf company developed the Fw 57 based on this concept, with Henschel offering its Hs 124 and Messerschmitt proposing the Bf 110, with the latter being laid out as a pure heavy fighter and bearing the least resemblance to the bomber-destroyer concept.

However, in 1935 doubts began cropping up as to whether such a multi-role aircraft was really a sensible idea. This line of thinking led to the multi-role concept of the *"Kampfzerstörer"* being abandoned in favor of a *"Schnellbomber"* (high-speed bomber) on the

Prototype of the Fw 57, built in 1936 but found to be too heavy and therefore eliminated from the fighter-bomber (Ka*mpfzerstörer***) competition.**

History of German Aviation: Bombers and Reconnaissance Aircraft

Henschel built two prototypes for the fighter-bomber requirement in 1936: the Hs 124V1 with a machine gun turret in the nose and 2 x Jumo 210C engines, and the Hs 124V2 with extensive nose glazing and 2 x BMW 132Dc radial engines.

Dispensing with the idea of the *Kampfzerstörer*, the three Bf 110 prototypes built for the RLM requirement (the V1 - first flown in May of 1936, V2, and V3) evolved into the classic *Zerstörer*, or heavy fighter. The Bf 110 (later Me 110) went into full-scale production in 1938 and later found use as a high-speed bomber/fighter-bomber and reconnaissance aircraft.

one hand and a specialized *Zerstörer*, or *"Flugzeugzerstörer"* (aircraft destroyer) to be more precise, on the other. Based on its conception, this newly defined *Zerstörer* would fall into the category of attacking types, since the thrust of its assigned roles—protection of friendly attacking air assets as the main one, then independent attacks against enemy aircraft over enemy territory, with defense of friendly airspace only falling into third place—would make it an offensive weapon. This volume does not include a detailed treatment of the *Zerstörer*, as the *Zerstörer* is classed as a fighter based on general agreement and its follow-on developmental models.

Numerically, reconnaissance aircraft comprised approximately 20 percent of the number of operational aircraft within the acquisition program. Just how accurate it is to classify these as offensive weapons is open for discussion; what is undisputed, however, is the fact that many of these types flew as armed reconnaissance platforms and were able to carry bombs, thus they were employed as attacking types. Nor can it be denied that long range reconnaissance aircraft were strategic-offensive in character.

The Medium Bomber

In 1934 the *Luftwaffe's* general staff called for the so-called "medium bomber" to be able to carry a 1,000 kg bomb load over a distance of 1,000 km. The priority clearly rested with payload and speed over armament and range. Relatively little emphasis was placed on range for the medium bomber, since of course the long range role was found in the requirement imposed for a strategic bomber during the same time frame, this being the previously mentioned "Ural Bomber."

The medium bomber was expected to have a radius of action of about 450 km at the given range, making Paris easily within its reach. For an operational profile of such dimensions the He 111 and Ju 86 twin-engined bombers, which had officially been contracted for, were entirely suitable. The Dornier-built Do 17 company developed civilian plane, which had drawn worldwide attention because of its outstanding performance at the time, was also classed as a medium bomber once armament and ordnance dropping equipment had been retrofitted to the type. While the debate continued over which of these three would become the standard bomber, political developments in 1935 ruled out focusing on any single type. It was felt that the conversion of manufacturing plants over to license production of a single company's product would lead to considerable delays in output.

Under this time pressure all three types initially entered series production, for even the Do 17 had been drawn into the Rhineland Program after the fact with the goal of providing the *Luftwaffe* with large numbers of bomber aircraft as quickly as possible.

Further development and production of the "Ural Bomber" (Dornier Do 19 and Junkers Ju 89) was halted once and for all in April of 1937. The long range requirement from the fall of 1936 focused more and more on a large four-engined heavy bomber, with two pairs of engines driving a single propeller each. But by the spring of 1937 the *Luftwaffe* general staff had come to the conclusion that:

- the aviation industry would have to reduce its developmental times if it were to be able to provide faster results in arming the *Luftwaffe*
- the general limited quantities of raw materials in Germany placed severe restrictions on large aircraft types, and it would therefore be more practical to build larger numbers of medium bombers
- the shortage of engines favored the twin-engined bomber over the four-engined type.

Three prototypes of the Messerschmitt Bf 162 Jaguar were built in 1937/1938 for the RLM's high-speed bomber requirement, as well as two Bf 161 strategic reconnaissance planes using the same layout in 1938; however, the *Technisches Amt* decided in favor of the Junkers Ju 88.

The High-Speed Bomber

In addition to the new long range bomber, in the spring of 1936 the general staff had also called for the development of the previously mentioned high-speed bomber. There was the clear understanding, however, that the long range bomber project would take a back seat to that of the high-speed bomber.

In actual fact, it was the high-speed medium range bomber rather than the long range bomber which more closely matched the prevailing anticipated combat picture. Conceptually, the course was thus set for prioritizing the much simpler mass production of high-speed medium bombers vice the difficult and prolonged development of heavy bombers, which in any case were felt to be out of step with the times.

The general staff's high-speed bomber requirement of the day called for an airplane having a maximum speed of 500 km/h, a payload of 500 kg capable of being carried over 2,500 km, a three man crew, and just a single flex-mounted machine gun for defensive armament.

Once Focke-Wulf pulled out of this request for tender, the *Technisches Amt* assigned the Messerschmitt, Junkers, and Henschel companies the responsibility of developing and producing three high-speed bomber prototypes each. The first prototype, the Messerschmitt Bf 162, took to the air just three months later, fol-

The sole prototype of the Henschel Hs 127, a high speed bomber development of the Hs 124 built in 1937. Unlike the Hs 124, it had a standard rudder. It first flew in late 1937 and, although already beaten out by the Ju 88, displayed excellent flight performance.

lowed by the Junkers Ju 88 on 21 December 1936 and the Henschel Hs 127 somewhat later in late 1937.

The Hs 127 was smaller and lighter than the Ju 88 and apparently had a maximum speed of 568 km/h in level flight—but by the time it had flown the *Technisches Amt* had already decided in favor of full-scale production of the Junkers Ju 88 proposal, which had become available at an earlier date. In addition, it determined that from then on Messerschmitt was to concentrate exclusively on the design and manufacture of fighter aircraft. Consequently, following Messerschmitt's construction of the three prototype Bf 162s, all further development on this type was halted, as was parallel development of the Bf 161 reconnaissance version.

Thus was the decision reached to make the Junkers Ju 88 the standard aircraft in the category of high-speed bombers.

But what had become of translating the dive bombing concept into reality in the meantime?

Thoughts of Dive Bombing

Though having been briefly addressed in a previous section, because of the significance the *Luftwaffe* had attached to it in the interim, the matter of dive bombing is worth examining in greater detail. Lingering questions include just how much the resulting dive bomber airplane could be considered to have been the ready-made solution, or whether the subsequent infatuation with the concept ultimately led to a dead end for German bomber development.

Generally, aviation literature attributes the realization of the dive bombing concept first and foremost to Ernst Udet, who "imported" the idea from the United States and effectively demonstrated it using the Curtiss Hawk II biplane. Hans Jeschonnek has received much of the credit, as well; as a result of his practical experience as commander of an operational training wing he became a supporter, and later, as chief of the *Luftwaffe's* general staff, the leading advocate of the dive bombing concept. Jeschonnek advanced the dive bomber as the predominant tactical idea within the *Luftwaffe* and saw the dive bomber as the *Luftwaffe's* only suitable bomber type.

However, the concept has a much longer history than it might first seem. During WWI the idea of dive bombing was occasionally employed, although not by specially designed dive bombing aircraft *per se*. During the interwar years the *Reichswehr* had carried out tactical trials at Lipetsk in the Soviet Union, and Germany's aviation industry had even produced dive bomber types, as well (Junkers K 47, first flight in Sweden in 1928, plus the A 48 and Heinkel He 50, first flight in 1931). To be sure, the K 47 built in Sweden and the He 50 built for Japan led to export orders for China and Japan, but in Germany the types drew no lasting attention from official quarters. It was not until the government purchased two American Curtiss-built Hawk II biplanes (the export designation of the F-11C-2 Goshawk Helldiver), which were fully dive capable, that the dive bombing idea made any kind of serious headway. In September/October of 1933 Udet had arranged for the two aircraft to be shipped from the U.S. to Germany. Even so, within Germany there was a wide range of views regarding the effectiveness of dive bombing operations, for as anti-aircraft defenses improved it was assumed that a dive bomber would be particularly vulnerable in a dive below 3,000 m, and that any such undertaking would therefore be suicidal.

At the same time, the bombing accuracy demonstrated in trials drew much attention and praise, not least because of the fact that at the time bombing accuracy from level flight at higher altitudes left much to be desired. High altitude bombing from level flight against area targets was not very accurate, and hitting a point target was purely a matter of chance. In addition, the ordnance expenditure when chain-releasing bombs from level flight was considerably higher in comparison with the single-release method employed by a dive bomber.

In view of the limited raw materials and fuel situation in Germany, the view became increasingly popular that medium and light bombers, with their increased accuracy, were viable alternatives to constructing a heavy bomber fleet of questionable bombing accuracy and ordnance-intensive chain-release bombing. To be sure, work continued at a feverish pace on eliminating the ballistic and targeting inadequacies associated with level bombing, since the bombsights of the day were simply unable to deliver any greater accuracy. But the ready-made solution to the problem seemed more and more to be achieved by sighting the entire aircraft on the target in a dive. In spite of the unsatisfactory accuracy of level bombing, the planned and growing "operative *Luftwaffe*" would still be required to fulfill its role of destroying power bases and nerve centers far behind the enemy's front lines. As a result, there increasingly grew the idea that dive bombing attacks would even be necessary deep within enemy territory.

The Stuka

But what was going on with turning the concept of dive bombing into a reality? This dream had been pursued during the *Reichswehr* era to be sure, but was now taking on a more definitive shape within the planning considerations of the *Luftwaffe's* leadership.

The emergency program for the air forces envisioned the establishment of dive bomber groups in 1933 which, because it could be done in relatively short order, were to be equipped with conventional biplanes as a stop-gap measure. To this end the Fieseler-Flugzeugbau, and the recently established Henschel Flugzeugwerke,

History of German Aviation: Bombers and Reconnaissance Aircraft

The Fieseler Fi 98 made its first flight in early 1935 under the auspices of the "*Sofortprogramm*," or Emergency Program. It was a light dive bomber, losing out in competition with the Henschel Hs 123. Construction of the second prototype was accordingly halted.

were brought into the program and asked to work out the design details for a single-seat all-metal biplane with fixed undercarriage suitable for use as a dive bomber. Henschel's design, the Hs 123, was favored over the Fi 98 proposal from Fieseler.

Henschel was awarded a contract for the construction of three prototypes, and Fieseler—with an eye toward a potential replacement program—was asked to build two prototypes. The Fi 98 completed its first flight in early 1935, carried out diving trials at the Rechlin Test Center, and drew little attention with the performance it displayed. So little, in fact, that further development was halted even before work on the second prototype had finished.

Just three days following the official announcement acknowledging its existence, Udet personally carried out the flight demonstration of the preferred Hs 123 at Johannistal, near Berlin, on 8 May 1935, something which had an enormous effect in piquing interest in the dive bomber concept, or *Stuka*. This was in complete opposition to the reaction greeting the He 50 dive bomber, a moderately performing aircraft built in 1931 by Heinkel for the Japanese, which had earlier failed to spark any such interest. Despite its clumsiness, however, the landplane version of this type (originally conceived as a seaplane) was selected to equip the first dive bomber units beginning on 1 October 1935. The He 50 was to be joined by another antiquated biplane fighter from 1931, as well, the Ar 65.

Two of the first three Hs 123s crashed during follow-on diving trials at Rechlin in the summer of 1935, with both test pilots being killed. Considerable structural improvements proved necessary before a fourth prototype, successfully diving at 80 degree angles, was finally felt suitable for series production, which began immediately. The *Luftwaffe* took its first deliveries of the Hs 123 in the summer of 1936, who was then able to replace its He 50 and Ar 65 dive bombers in the newly established *Stuka* units.

A few short months later the Hs 123 entered combat in the Spanish Civil War. In Spain the *Stuka's* originally intended role of attacking point targets in the enemy hinterland took a back seat to close interaction with ground and air forces on the front, the so-called "cooperation"; the Condor Legion—and later also the Spanish Nationalists—chiefly used the Hs 123 in the close support, or strike, role.

Appendixed to the emergency program and running parallel to the development of the Hs 123 was the second phase of equipping the units with modern dive bombers; in the interim serious progress had been made in developing the definitive *Stuka*. Between 1931 and 1934 the Junkers company had been quietly and methodically advancing the dive bombing capability of its K 47 in Sweden, where it had completed its first flight in 1928. Junkers had developed a gyroscopic controlled bombsight for dive bombers and a truly practical dive bombing system in the form of a Junkers designed dive brake. At Junkers' home plant in Dessau a mockup of a new generation dive bomber was built in 1934 incorporating the latest technology. After representatives from the RLM inspected it, this organization approved the construction of three prototypes, and a few months later published the specifications for the *Stuka* of the emergency program's second phase, the details of which closely matched the Junkers concept of a new dive bomber.

Companies taking part in the request for tender included Arado with the Ar 81 design, Heinkel with its He 118, Hamburger Flugzeugbau and its Ha 137, and Junkers with its Ju 87.

Each design was built as a prototype. The Ju 87 came out the winner of a selection competition held in the spring and summer of 1936, as it was the only machine capable of making a dive at virtually 90 degree angles almost flawlessly and most closely met the requirements outlined in the specifications. This clinched the choice of the *Luftwaffe's* standard dive bomber as the Junkers *Stuka*, and the company soon began producing the type.

However, there still remained the matter of a dive bomber capable of operating far behind enemy lines, as the Ju 87 lacked the penetration range for such tasks.

On Cooperation

The idea of a strategic air war in the sense of Douhet seldom appears in the official German writings from the thirties; instead, the school of thought tends to restrict itself to the "tactical" and "operative" use of air power. "Tactical" operations were understood to mean the direct support given to the *Heer*, while "operative" use of air power included, among other things, that segment of the operative prosecution of the air war dealing with the indirect support of the *Heer*.

Still bearing the stamp of Wever, after his death L.Dv. 16 was too often interpreted from a tactical-operative perspective at the expense of the strategic elements also contained therein. A strategic air war was not the priority, but was a subordinate function in the event that the ground front developed into a static war and no other military solution presented itself.

Despite its primary focus as a support element in a broader sense, i.e. that of cooperation, other roles directed at combating enemy air forces and power bases seemed—at least theoretically—to be on par with that of cooperation. More detailed clarifications and the order of specific functions within L.Dv. 16 reveal a prioritization in which combat against enemy air forces took precedence, followed by cooperation with army and naval forces. As far as the latter priority went, because of its less complicated execution and greater effectiveness, direct support took precedence over indirect support. Third on the list of priorities were attacks on enemy power bases, which were also chiefly viewed as cooperative and only in a limited sense as strategic-operative.

Several points should be borne in mind when attempting to understand statements regarding cooperation and its relative influence on the theories of air warfare at the time:

1. In late 1936/early 1937 *Generalleutnant* Erhard Milch, Göring's state secretary of aviation, submitted a memorandum to the *Luftwaffe*'s commander-in-chief clarifying that the *Luftwaffe* was to be primarily used for suppressing enemy air forces and to support the *Heer* and *Marine* before it would shift to "operative" air warfare. The main threat was viewed as the enemy air forces, which would need to be crippled by a surprise blow in order to ensure the mobilization and advance of Germany's armies and interfere with the enemy's ability to do the same. Following the first strike, incessant operations were to destroy the enemy's rail and road bridges, its avenues of approach, its factories, and its fuel storage facilities—in that order. Not once did Milch mention any role for the long range bomber, but he did specifically address the priority given to direct support of the army once the requisite air supremacy and/or local air superiority had been established.

2. In the matter of a strategic air war against an enemy's war economy in the spirit of Douhet's doctrine, the *Heer* did indeed take a stand; it only considered this type of air warfare justified in those instances where the *Luftwaffe* would not be needed for cooperation or wresting air superiority in the army's operating areas.

3. Up until 1937/38 the senior leadership's war planning ideas only provided a detailed plan of attack against those countries immediately bordering the German State. Having the advantage of a direct line, it was thought that these countries—individually and one after another—could be eliminated as a threat with medium-range and dive bombers.

4. Experience from the 1936-1939 Spanish Civil War indicated a favorable assessment of the concept of direct support for the army. Refuting the previously mentioned verdict of the French general Duval, the effectiveness of high altitude level bombing against population centers was considered a disappointment due to the inaccurate delivery methods, as was the effectiveness against airfields, harbors, area targets, and factories. It was recognized that low level attacks by bomber units were not economical, but instead would probably be the most costly type of operation. Operative bombing raids would only be successful in the long run if they frequently could be employed against the same target with unlimited amounts of ordnance over an unlimited period of time. Mainly developed by the Condor Legion's last commander, *Generalmajor* Wolfram von Richthofen, close combat tactics were undoubtedly one of the most enduring results of the Spanish experience, which exclusively benefitted the concept of close cooperation between land and air forces.

The *Luftwaffe* leadership's facile open-minded approach to cooperation, generally driven by WWI army doctrine or fighter concepts, was ostensibly legitimized in a memorandum by *General der Flieger* Hellmuth Felmy, dated 22 September 1938. The memorandum focused on the potential for success by the *Luftwaffe* in the event of a war with England and was based on the assumption that, of geographical necessity, such an air war would be prosecuted by the *Luftwaffe* independently and operative-strategically. Felmy concluded his thoughts by stating that such a venture would have negative consequences for Germany's *Luftwaffe* due to insufficient operative forces and a lack of strategic forces. Felmy's subsequent findings on 13 May 1939 and in the summer of 1939 on the same subject also failed to paint any more favorable a picture.

Thus, in 1939 the *Luftwaffe* was still a long way off from prosecuting an effective operative-strategic air war.

History of German Aviation: Bombers and Reconnaissance Aircraft

Increasing tensions with Poland and Germany's western neighbors that same year were also taken into consideration. In a "1939 Plan Study," which contained instructions for operations in the East, the primary role of the *Luftwaffe* was seen as eliminating the Polish Air Force. This would be followed by indirect and direct support of the army—the classic cooperation tactics—and finally, the combined attack on the capital of Warsaw.

With regard to operations in the West, *Führerweisung* (directive) No. 1 issued on 31 August 1939, i.e. just before the outbreak of WWII, spelled out that the *Luftwaffe* primarily was to focus on preventing French and British air forces from attacking the German Army and German territory. The only other plans the *Luftwaffe* was to make were for attacking British maritime shipping and its armament industry.

Under the direction of Chief of Staff Jeschonnek, the *Luftwaffe's* planning exercises in 1939 placed particular emphasis on the:

- "potential for a surprise attack against an enemy air force" and
- "potential for and prerequisites to direct cooperation with the *Heer*"

Based on the exercises, the order of priority given to the *Luftwaffe's* cooperative functions was established as follows:

- "Engaging the enemy's air forces by surprise attack, a measure which must be considered of utmost importance and, at all costs, proper and necessary for the purposes of establishing air superiority at least over the army's operating area (engaging enemy fighters is therefore considered to be the primary focus).
- Indirect support of the *Heer* through attacks on supply lines, assembly points, replacement parts storage facilities, etc. in the enemy's rear areas.
- direct support on the battlefield by using mixed tactical close combat units as well as bombers."

Although the *Luftwaffe's* aircraft resources in 1939 may have permitted such a prioritization with regard to bordering nations, it was obvious even at this date that a potential air war against England could not be won. Such a conflict would, by default, be "operative," if not even entirely operative-strategic in nature. The types of aircraft available up to this point in time had been tailored too much to the view of direct and indirect cooperation with other service branches and found expression in the selective attention given to medium/high-speed bombers and the purely tactical dive bombers.

The Heinkel He 177, a heavy/strategic bomber in development since 1937, found itself in a situation which was only conditionally suited for operative air operations. Furthermore, this situation delayed its availability far beyond the point where it could have had a measurable influence on the war's outcome, not to mention a decisive role in the conflict.

Even in this advanced stage of expanding the flying units, advocates of the doctrine of cooperation as the primary role of the German *Luftwaffe* undoubtedly had persuasive arguments as they wrestled with the armament priorities for the offensive elements within this branch. To be sure, there existed the awareness—at least among the realists in the *Luftwaffe* leadership—that an air war against an enemy such as Great Britain might not be successfully prosecuted, but this concern was never realistically taken into account in the acquisition policies of the government. Cooperation was the buzzword of the day!

5. Germany's Aviation Industry Potential in 1937

As discussed earlier in the chapter "In Perspective," Germany's aviation industry potential (expressed in the size of its work force) grew five-fold within the space of just a single year from the approximately 20,000 employees at the end of 1933. In 1939, just six years later, this number included well over 300,000 workers, which on 1 January 1939 looked as such:

- metal alloy plants	12,000
- airframe plants	138,000
- engine assembly plants	100,000
- radio and navigation plants	20,500
- avionics plants	20,000
- armament and ammunition plants	17,000
- ordnance manufacturers	10,000
total	317,500

Relatively soon after the National Socialists took power, in late January 1933 the government set up the "Rhineland Program," which called for the construction of over 4,000 aircraft by 30 September 1935. This was immediately translated into purchase contracts with the aviation industry.

With regard to the *Luftwaffe's* expansion based on this stockpiling program, on 2 May 1935 Göring declared that the formation of the German *Luftwaffe* would be so unique that one would hardly believe it possible if it were not for the documentary evidence. Uncharted territory would be mapped out. He would reject the path of a slow, gradual buildup, for this would make the *Luftwaffe* unprepared to meet the challenges of difficult situations. Rather, he would expand the technological and industrial potential to the extreme, allowing the *Luftwaffe* to be created virtually in one fell swoop.

But what were the problems lying hidden behind such boasting vis-à-vis the *Luftwaffe's* buildup and Germany's aviation industry, which had over the previous two years undergone such intensive expansion?

Financial and Raw Material Bottlenecks

In addition to its *Technisches Amt* becoming involved in all aeronautical matters, the *Reichsluftministerium* dedicated itself in large part to ensuring the necessary materials were available and, inevitably, to supporting the financing of the aviation industry. Even though in principle the industry was expected to finance itself, this was not always the case in spite of exhausting all credit opportunities. Increasingly, the *Reich* became involved in arms production, and in some cases even owned entire operations. For example, in 1934/35 the Junkers *Flugzeugwerke* and the Junkers *Motorenbau* fell into the government's hands. In this instance, not only did financial problems play a role, but there was also a deeply rooted resentment between the RLM and Professor Hugo Junkers which contributed to the latter's dismissal.

In the event that no entrepreneur could be found to back entirely new companies, government-owned facilities were made available and occasionally leased out. As the factories continued their expansion and the focus shifted to large scale production, so grew the aviation industry's economic and financial problems. Government funds also needed to be infused in this area, and these funds required close management; price checks, setting recommended prices, review of the books, etc. were the consequences which led to the RLM imposing its will on the aircraft plants to a much greater extent than had hitherto been the case. At the same time, financial requirements dictated by the expansion programs and the ongoing maintenance of the entire *Luftwaffe* had increased to such an extent that by 1936 there were serious difficulties with making the necessary budgetary resources available.

Oberst Ernst Udet, appointed chief of the *Technisches Amt* in early June of 1936, therefore called for combining airframes and engines, as well as restricting development and production to just a few aircraft and engine types, eliminating duplicate types having the same role. In so doing, he hoped to guard against overtaxing Germany's aviation industry and relieve the pressure on military supply. More will be discussed regarding the startling number and diversity of aircraft types later.

There followed a perceptible concentration on specific aircraft types selected for large scale production, although this was initially not as noticeable for the bomber category. It was not until 1937 that there were signs of movement toward a standard bomber solution.

On 11 January 1937 an increased delivery program was approved which specified that those Heinkel He 111, Junkers Ju 86, Dornier Do 17, and Junkers Ju 87 aircraft types entering or already in production would continue to be manufactured, and some of these were to be acquired in even greater quantities than originally contracted for. The plan (with a target date of 1 April 1938) envisioned the following numbers:

He 111, instead of the original 831	850
Ju 86, unchanged at	680
Do 17, instead of the original 788	1,014
Ju 87, instead of the original 264	345

With regard to bomber aircraft, this delivery requirement equated to roughly a 13% increase over previous planning. A higher production output seems to have counted for more than a consolidation of aircraft types.

In 1937, however, there arose serious problems with the procurement of raw materials, particularly for iron and steel. Just as acute was the shortage of aluminum, the availability of which was the key for fulfilling the *Luftwaffe's* delivery program. A monthly average of 4,500 metric tons of aluminum were required for carrying out the program in 1937, but the monthly average for the first half of 1937 was just 2,700 tons delivered.

Iron and steel distribution also fell behind schedule, which led to considerable delays in equipping the *Luftwaffe*. So much so, in fact, that many felt that it would be impossible to have the service fully equipped with modern, new generation aircraft before 1 April 1939.

Moreover, these bottlenecks in raw materials had a serious impact on manufacturing operations at the plants.

Ultimately, the production program for 1937 took the material situation into account and called for just over 6,000 aircraft, of which 3,376 were front line types. For this, the *Technisches Amt* required and requested a budget totaling 3.7 billion RM. A million marks, or approximately one-fourth of contract funding, of this amount was initially cut from the sum due to budgetary constraints.

In any event, layoffs within the aviation industry's work force and a drop in production numbers were averted by drawing up a more sensible forecast for the next fiscal year.

The question remained, however, of how to balance an aviation industry still tottering along in its infancy stages with the RLM's rigorous demands and tasking and the *Luftwaffe* General Staff's opinion that the industry must reduce developmental time for aircraft designs.

The Reality of the Developmental Period

By 1936 one could count 120 basic types of aircraft and their subvariants, either being operated by the *Luftwaffe* or soon to be introduced into its inventory.

A register prepared by the *Technisches Amt*, dated 8 June 1936, showed that as of 1 June 1936 there were no fewer than 61 main types. Appendix 1 is a copy of the original register, and the underlined bombers and reconnaissance aircraft, including seaplanes, alone number 34 various main types.

The number of aircraft types and their various sub-variants would peak in 1941 at a maximum of 143 different models actually constructed! Only a small percentage of those German military aircraft listed in the register will be examined in detail in the coming chapters.

Millions of developmental hours lay behind such numbers, and it is worth noting that the engineering hours devoted to refining a type for large-scale production and follow-on development are many times higher than the hours devoted to developing a new design from scratch up to maiden flight. The creation of an effective operational military aircraft was far less dependent upon achieving a prototype's first flight quickly than it was upon producing the hundredth or thousandth fully operational large-scale production machine in as short a time period as possible.

Based on the criteria of the day, what time requirements are we talking about here?

As a rule, under optimal conditions in the mid-1930s a military aircraft free of teething troubles could be made ready for full-scale production following three to five years of concentrated design work and testing. Examples—all of which were in development up until about 1936—included the Ju 52, He 111, Do 17, Ju 86, and Fw 200 (some of which had civilian backgrounds), as well as the Bf 109 and Me 110 fighters/*Zerstörers*.

The common developmental times and stages ran roughly as follows:

- first year: design and construction of a prototype/prototypes for the so-called test series, or V-series.
- second year: flight testing and start of production preparation, which would extend into year three.
- third year: construction of pre-production series, which might have overlapped the production preparation by up to 50%.
- fourth year: initiation of full-scale production

Since 1936, however, the *Reichsluftfahrtministerium* (the contractor) had been exerting a certain amount of pressure on the industry to rush development of new aircraft types as the RLM attempted to move up the "production preparation" and "pre-production" stages by a year. The result of this pressure was that both these stages were tackled at a time when the results from flight testing were not yet available or incomplete. As a result of this "time lapse" measure, acquisition planners hoped to have full-scale production of a type initiated at the two-year mark!

In practice, a lack of test results meant that new aircraft often entered large-scale production before they had reached maturity, which in turn necessitated all kinds of modifications on the assembly line. This led to massive additional time and work expenses and interrupted the production tempo. Nervousness and abrasiveness between the designers, the production centers, and the contractor (the RLM) was the result, and inevitably led to an extension of a type's production maturity and work on the pre-production series. In any event, it still took three to four years instead of the utopian two years originally expected for the first production machines to roll off the assembly lines. Examples of this rushed or interrupted development, brought on by after-action requirements or belated recognition of a specific need, include the Ju 88 and He 177 bombers, in particular. Ultimately, for reasons mentioned above, this developmental cycle was no more effective than the hitherto sequential, common cycle with its requisite time periods for each stage.

Nevertheless, the contractor's deadline planning proceeded without regard for the increasing complexity of flying machines. This can, at least in part, be attributed to aircraft manufacturers and their abbreviated or overly optimistic deadline proposals—often made under pressure of competition.

Even the test flight deadlines were shortened, ignoring the fact that saving time by cutting corners during development generally requires payment many times over with delays during the full-scale production phase.

Another aggravating point of view was that, after reaching a specified state of construction for a particular aircraft type the number of designers could be cut back, with those newly freed-up then being able to work on new developmental projects. In reality, a production aircraft requires close monitoring over a number of years by a design team familiar with the project, who can then eliminate potential problems cropping up during further flight testing or from production preparation through to large-scale production and front-line operations.

In the haste to build up Germany's *Luftwaffe*, many of these facts were criminally ignored.

Authority and Priorities

Some progress was made toward countering deficiencies in material goods and organization by employing senior executives with *carte blanche* powers directed toward overcoming bottleneck situations. One example of this was Junkers' general director, Dr. Heinrich Koppenberg, whom the *Luftwaffe's* commander-in-chief, Hermann Göring, obligated (in a written memorandum on 30 September 1938) to take all measures necessary for ensuring the quickest and highest possible output for the Ju 88. Dr. Koppenberg was specifically empowered to go even beyond the Junkers company when handling the Ju 88 program, giving orders to all those companies and subcontractors involved in the manufacturing process and in all areas developing the necessary insight for successfully running the program. Not only did he have directional and managerial authority over other companies and subcontractors, he also exercised this authority over officials and financial organizations. Ultimately, the following letter from Hitler to Göring on 21 August 1939 shows that he was also indirectly tasked by the highest authority, the *Reichskanzler* himself, to produce 300 Ju 88s per month.

With this, the Ju 88 program clearly took priority over other aircraft construction programs, and was given the same emergency status as Hitler had ordered for the buildup and expansion of the *Kriegsmarine*. Koppenberg was given free rein for issuing supplemental tasking orders within the scope of the Four Year Plan when it came to carrying out this aircraft construction program. Since Göring was simultaneously also director of the "Four-Year Plan," to a great extent he was able to intervene directly, filling the needs of the air armament industry and removing many of the obstacles facing the programs. A continuing thorn remained the imbalanced distribution of raw materials, with the consequences often being shouldered by other service branches and the civilian sector.

Production Potential Achieved in 1937

With regard to the output capacity of bombers and reconnaissance aircraft, the following picture resulted from the *Technisches Amt*'s anticipated compilation dated 24 April 1937:

Type	Manufactured by/at	Maximum Monthly Output During Peacetime		During Mobilization	
1. Bombers and Dive Bombers					
Do 17E	Blohm & Voss/Hamburg	16	=46	20	=105
	F.W.H./Halle	10		35	
	Henschel/Schönefeld	20		50	
He 11	Arado/Brandenburg	10	=87	45	=140
	A.T.G./Leipzig	20		0	
	Dornier/Wismar	25		4	
	Heinkel/Rostock	8		10	
	Heinkel/Oranienburg	16		45	
	Junkers/Dessau	8		0	
Ju 86	A.T.G./Leipzig	16	=81	25	=50
	Blohm & Voss/Hamburg	15		0	
	Henschel/Schönefeld	17		0	
	Junkers/Dessau	33		25	
Hs 123	Ago/Oschersleben	21	=36	0	=0
	Henschel/Schönefeld	15		0	
Ju 87	Junkers/Dessau	20	=23	20	=60
	Weser/Bremen	3		40	
Ju 88	Junkers/Dessau	1	=1 (still in prototype stage)		
II. Reconnaissance Aircraft					
Do 17F	Henschel/Schönefeld	25	=25	25	=25
He 46	Gothaer Waggonfabrik/Gotha	12	=12	0	=0
He 70	Heinkel/Rostock	16	=16	0	=0
Hs 126	Ago/Oschersleben	8	=12	60	=110
	Henschel/Schönefeld	4		50	

It is interesting to note that the He 46, He 70, and Hs 123 would no longer be in demand in the event of mobilization and, other than the output for the Ju 86 dropping from 81 to 50, the monthly output for all other types would rise.

Based on these figures, in April 1937 it was assumed that a production capacity would be reached which would, on a monthly basis, produce

- 214 bombers (minus the Ju 88 prototypes)
- 59 dive bombers
- 25 strategic reconnaissance aircraft
- 40 tactical reconnaissance aircraft,

meaning 338 front-line aircraft comprising the air attack components would be available for delivery to the operative *Luftwaffe*.*

A hint as to the total capacity of the German aviation industry at the time can also be found in the fact that an additional 149 front-line Ar 68, Bf 109 and 110, Fw 159/259, and Fw 187 fighters and heavy fighters were produced for the *Jagdwaffe*, plus 47 BV 138, Do 18, He 59, 60, and 114 seaplanes and flying boats, bringing the total output of front-line aircraft up to 534 per month. In addition, the industry produced a further 410 trainer, liaison, and communications aircraft, and other types not intended for front-line service, such as the Ar 66, Bf 108, Bü 131 and 133, Fi 156, Fw 44, 56 and 58, Go 145, He 42, Ju 52, Kl 25 and 35, and W 34, i.e. a total of 944 aircraft of all types per month, of which 62% were front-line types.

This is quite a remarkable feat, considering that this monthly production quota for Germany's aviation industry was achieved in just four years—even if the plethora of different types made one's head spin!

---

* With regard to the often cited term "operative *Luftwaffe*," it should be noted that L.Dv. 16 "*Luftkriegführung*" did not specify that the operative *Luftwaffe* was only to include those front-line flying units directly subordinate to the commander-in-chief of the *Luftwaffe* and that it was to include an air defense component. Certain front-line aviation elements were operationally deployed to the *Heer* and *Marine* and thus did not actually count as part of the "operative *Luftwaffe*" at the time. It was not until 1940-1942 that all aircraft subordinated to army and navy entities were returned to the *Luftwaffe's* control and made directly subordinate to that branch of service.

6. The 1938/1939 Rearmament Program Against the Backdrop of *"Große Politik"*

The 1938 Aircraft Acquisition Program

Just a year after it was discovered that the industry could support a monthly production capacity of 338 front-line offensive aircraft for the *Luftwaffe*, equating to a theoretical maximum yearly production of 4,056 front-line planes, the aircraft acquisition program within the *Technisches Amt* drew up a new list of procurement figures effective 1 April 1938. These numbers envisioned 5,254 front-line aircraft in the categories of bomber, dive bomber, and reconnaissance, broken down as follows:

Bombers
(official designation: *"mittlerer Kampf-Mehrsitzer"*)

Ju 86A/D	with Jumo 205	497	
Ju 86G	with BMW 132F	142	= 939 Ju 86
Ju 86E	with BMW 132F	300	
Do 17E	with BMW VI	299	
Do 17M	with SAM 323	200	
Do 17Z	with SAM 323	435	= 1,065 Do 17
	with BRAMO 323A	116	
Do 17U	with SAM 323	15	
Do 217	with DB 601	20	= 20 Do 217
He 111	with BMW VI	1	
He 111B/J	with DB 600	539	
He 111E/F	with Jumo 211	255	= 1,693 He 111
He 111H	with Jumo 211	564	
He 111P	with DB 601	334	
Ju 88 (after the V-6)	with Jumo 211	35	= 35 Ju 88

Total # of bombers   3,752

Dive bombers (official designation: *"Sturzzweisitzer"*)

Ju 87A	with Jumo 210	262	= 656 Ju 87
Ju 87B	with Jumo 211	394	

Total # of dive bombers   656

Reconnaissance aircraft

1. Strategic reconnaissance (official designation: *"Aufklärer F-Land"*)

Do 17F	with BMW VI	178	= 508 Do 17
Do 17P	with BMW 132N	330	

2. Tactical reconnaissance
(official designation: *"Aufklärer H-Land"*)
Hs 126 (no details provided for variants or engine data, although it was equipped with the BRAMO 323A)   338   = 338 Hs 126

Total # of reconnaissance aircraft   846

For fiscal year 1938/39 (1 April 1938 to 31 March 1939), this meant an increase totaling 5,254 aircraft for the offensive air component.

The various engine types for the same model of aircraft was not always attributable to tactical-technological reasons, but in most cases reflects the aircraft engine situation at the time, which often suffered from lack of suitable powerplants.

The difficulties arising in 1937 were brought into check after a fashion by applying the funds from the coming fiscal year, and this roughly 13% boost in production of offensive aircraft marked the beginnings of a considerable increase in the air force's offensive capabilities. But these plans would not remain in effect for long.

1938 Planning Parameters

The acquisition figures established on 1 April 1938 beg the questions of what criteria was used for planning and what was all this going to cost. It is virtually impossible to provide a definitive answer, since complete, accurate documentation covering this complex issue is practically non-existent.

In this context, it is interesting to note that, of all the file material on Germany's armed forces confiscated by the Allies after the Second World War and subsequently returned to the military archives department within the *Bundesarchiv*, only 5% actually pertains to the *Luftwaffe*. The *Bundesarchiv* and the *Militär-*

History of German Aviation: Bombers and Reconnaissance Aircraft

geschichtliches Forschungsamt have assessed the *Luftwaffe's* written record situation between 1935 and 1945 to be poor, since these archival organizations have been able to acquire only fragmentary data—about 97% of the *Luftwaffe's* records were destroyed in 1944/1945. Supplemented by eyewitness reports covering events at the time, it is only through individual surviving "snapshots" that the situation prior to 1945 can be patched together to form a complete picture.

Some Examples of Cost from 1938

For the 1 April 1938-31 March 1939 fiscal year the 13 January 1938 budget proposal listed the cost per item—as far as it affected bombers and reconnaissance aircraft—for the "1938 Aircraft Acquisition Program" supplement as follows:

Airframes
Do 17P   order of 50 from four manufacturers, RM 185,000 ea.
Do 17Z   order of 180 from two manufacturers, RM 185,000 ea.
He 111   order of 177 from five manufacturers, RM 265,000 ea.
Ju 87   order of 74 from two manufacturers, RM 100,000 ea.
The above single prices were established by the *Technisches Amt* based on the average offer price provided by the industry.
Ju 88   order of 33 from one manufacturer, RM 180,000 ea.
Estimated price by the *Technisches Amt*, which was non-binding on the *Luftwaffe* administration.

Aircraft engines
BMW 132A   order of 510, RM 15,000 ea.
Jumo 210D/G   order of 152, RM 30,105 ea.
Jumo 211   order of 291, RM 32,600 ea.
SAM 323   order of 275, RM 26,000 ea.
SAM 323   order of 20, RM 30,000 ea.

Aircraft armament
MG 15   order of 1658, RM 595 ea.
MG 17   order of 581, RM 1,250 ea.
MG-FF   order of 400, RM 8,000 ea.

Other
Autopilot   order of 200, RM 10,000 ea.
Parachutes   order of 2600, RM 700 to 730, depending on type

Calculating the "Constant *Luftwaffe* 1942"

According to "Program No. 9," the *Luftwaffe* was to have reached its final state by 1 November 1941. A "constant *Luftwaffe*," with its growth rate frozen, was apparently expected after this date.

The regenerative requirements of this air force were outlined in a meeting between the general quartermaster (Gen. Qu.) of the *Luftwaffe* general staff and the *Technisches Amt* (*C-Amt* and *LC*) on 2 November 1938.

To attain this "Constant *Luftwaffe* 1942," it was agreed that:

"The *Technisches Amt* will relay the projected numbers of specific types to be supplied to the Gen. Qu.; The Gen. Qu. will then, on the basis of the prescribed loss percentage figures, prepare graph charts for specific unit types, which will result in a projected constant for every individual unit type. For this, the Gen. Qu. will construct the standard graph (envelope) for the entire *Fliegertruppe*, which will then result in the "Constant *Luftwaffe* 1942." From this constant it is then possible to calculate proportionally the monthly resupply of individual types, which for the *C-Amt* then becomes the mobilization resupply figures for specific type aircraft."

It can only be surmised just how far the shadow of the mid-October 1938 directive calling for the *Luftwaffe* to quintuple cast itself over this agreement, for the acquisition program (the so-called "Consolidated Aircraft Type Program") stemming from the directive did not follow in written form until five days after the 2 November 1938 meeting. This directive will be discussed in detail in a later chapter.

First, however, are a few additional data relating to planning criteria: the "prescribed loss percentage figures" quoted above during wartime were assumed at the time to be:

For planning purposes, monthly losses were anticipated to be 30% within the inventory of bomber units, ground attack, tactical, and strategic reconnaissance, i.e. stock of the offensive air arm
30% for fighters, heavy fighters, transports, and catapult-launched and carrier-borne aircraft, as well
15% of seaplanes (all types) are to be written off and
3 to 4% of all trainer aircraft.

These loss percentages were binding upon planning projections for airframe resupply mobilization calculations and were based on a monthly sortie rate for specific aircraft classes and types as follows:

	Sorties by aircraft per month
Tactical reconnaissance	20
Strategic reconnaissance	15
Medium bomber	15
Heavy bomber	15
Dive bomber	20
Strike/ground attack	30

So much for the figures for the offensive air elements. The chart below shows the anticipated figures for fighters as a comparison, using the nomenclature of the day:

	Sorties by aircraft per month
Zerstörer (Bf 110)	20
Zerstörer (Bf 210)	15
leichter Zerstörer (fighter)	45

These are but a few of the planning parameters used in 1938, which little reflected the experience being gleaned from the Spanish Civil War then in full swing. In Spain, constant month-to-month stress, which for planning purposes was assumed in the tables above, was attained only sporadically and for relatively short periods of time. Nevertheless, it seems worth pointing out that, of the final German aircraft losses in Spain, barely 42% were due to enemy action; a good 58% can be attributed to accidents.

With regard to subsequent events, it cannot be denied that the planning figures from 1938 exhibited a certain realistic approach. Although no one at the time could have known it, the heavy bomber (the He 177, for which Heinkel finally got the go-ahead from the *Technisches Amt* in the summer of 1938) would not even fly the projected 13 sorties on average per month.

But let us now turn to the program for quintupling the *Luftwaffe's* flying units by 1942.

The Consolidated Aircraft Type Program

The 1938/39 planning forecast was overshadowed by the advance of German troops into Austria on 11 March 1938. This was followed on 29 September 1938 by the Munich Accords, with Prime Minister Arthur Neville Chamberlain for Great Britain, Prime Minister Edouard Daladier for France, and Prime Minister Benito Mussolini for Italy. This saw the brief cession of the Sudeten German area of Czechoslovakia, into which the *Wehrmacht* marched during October of 1938. It was undoubtedly this defining political moment which contributed to Germany's continued armament acceleration in order to be equipped for the increasingly more probable military confrontation, not only with the East, but also potentially with the West, as well.

In view of this situation and considering Britain's obvious arms buildup, in mid-October 1938 Hitler ordered his minister of aviation to immediately begin a five-fold increase in the strength of the *Luftwaffe*. This act was also a result of *Generalleutnant* Felmy's previously mentioned memorandum from 22 September 1938, in which Felmy was extremely skeptical about the *Luftwaffe's* success chances. Felmy's planning study for *Fall Grün* (Case Green), as the eventuality of an air war against England using all operative means available was known, ruled out a war of destruction with England given the state of affairs at the time.

Hitler's order led to the establishment of the so-called "Consolidated Aircraft Type Program," drawn up by the 1st Dept. (Command) of the *Luftwaffe* General Staff and signed on 7 November 1938 by *Oberst i. G.* Hans Jeschonnek, chief of command of the General Staff and simultaneously chief of 1st Dept (see Appendix 2).

Following a detailed study, the *Technisches Amt* came to the conclusion that a successful execution of this program was not even within the realm of possibility. There existed neither the production resources, nor would there be the fuel reserves by 1942, not to mention the training facilities for the flying crews needed to beef up the *Luftwaffe* to the roughly 19,000 airplanes dictated by the program, of which 10,800 were to have been active front-line aircraft with 8,200 held in reserve. Although not obvious from surviving documentation, it should also be noted that the "Consolidated Aircraft Type Program" specified that 30 of the 58 bomber *Geschwader* were to be used against England and, on the insistence of the Navy, 13 for the naval war as so-called "pirate units." The remaining 15 bomber *Geschwader* were to be utilized in other capacities.

Oberst i. G. Josef Kammhuber, chief of the organization staff, proposed an emergency program reduced by one-third, and this was approved by Chief of Staff *Generalmajor* Hans Jürgen Stumpff as an intermediate program. However, this effort was blocked by Jeschonnek, who stubbornly insisted on carrying out Hitler's instructions. After a meeting held with Göring by State Secretary Milch and Jeschonnek, Göring decided to proceed with the entire program—against all reasonable arguments. Consequently, Kammhuber requested and was given a ground position, Chief of Staff Hans Jürgen Stumpff resigned shortly afterward, and Jeschonnek succeeded him to the post on 1 February 1939 while retaining his function as chief of command of the General Staff.

Despite Göring's order and Jeschonnek's proactive involvement, this utopian program never became a reality. However, its failure was less due to the lack of material prerequisites than the fact that all these planning considerations were steamrolled by the political events of 1939.

Pointing the Way to War

Flying in the face of reason and contrary to assurances given at Munich that Germany was saturated with the Sudetenland, in March 1939 Hitler distanced himself from the Munich Accords and, by occupying the rest of Czechoslovakia, for the first time laid claim to territory in which the majority of the population was not German.

Since October of 1938 Germany had taken a much sharper tone in its demands that Poland return Danzig, and had built an extraterritorial road and rail link through a "corridor" between the *Reich* and East Prussia. These acts, and particularly the occupation of the remainder of Czechoslovakia, unsettled the Polish government which, counting on western support, on 23 March 1939 proclaimed a partial mobilization and openly rejected the German demands on 26 March 1939 for the first time. Great Britain, disillusioned by the violation of the Munich Accords, announced on 30 March 1939 that it would guarantee Poland's rights, which on 10 April 1939 was extended to include military support. These steps were designed to deter Hitler from taking the aggressive measures everyone feared. France immediately threw its support in with the guarantee for Poland, which on 13 April 1939 was extended to embrace Romania, Greece, and Turkey in an effort to stave off any potential Italian ambitions in the Balkans.

Hitler recognized that his goal of acquiring additional *Lebensraum* in the East was no longer feasible via diplomatic channels, and on 3 April 1939 gave orders to the *Wehrmacht's* leadership to prepare for "*Fall Weiß*," the attack on Poland, with 1 September 1939 as the potential date. On 28 April 1939 he declared the German/Polish non-aggression pact, in existence since 26 January 1934, to be null and void after Great Britain introduced general conscription only for the second time in its history on 27 April 1939. Hitler concluded a military alliance with Italy on 22 May 1939, the so-called "Pact of Steel," whereby Mussolini shared Hitler's view that war was now unavoidable. He did, however, cause Hitler to pause with a memorandum dated 30 May 1939 wherein he stated that the Axis Powers (Germany and Italy) would need at least three years of peace in order to make the necessary preparations.

On 21 August 1939 France began with the initial stages of mobilization, which she concluded ten days later.

Catching the West completely by surprise, a non-aggression pact was signed on 22 August 1939 in Moscow between Germany and the Soviet Union. This was an alliance of convenience which outlined the spheres of influence in the Baltic, in Poland, and in the Bessarabian areas of Romania. Japan lodged a protest, since this agreement violated the "Anticomintern Pact" of 6 November 1937 between Germany, Italy, and Japan directed against "international communism." Even Spain's head of state, General Franco, distanced himself.

Additionally, on 22 August 1939 Hitler informed the commanders-in-chief of the three branches of service that it was his irrevocable decision to risk war. On 26 August 1939 orders were given for Germany's mobilization, although these were not made public. During the afternoon of 30 August 1939 the Polish government announced general mobilization.

Junkers Ju 86D bombers flying in formation with their ventral gun positions extended (1938)—showpiece, political tool, or instrument of war?

7. Taking Stock

In the late summer of 1939, just how capable were the bombers, dive bombers, strike aircraft, and reconnaissance planes, i.e. the *Luftwaffe's* offensive power, in fulfilling their assigned roles and meeting the expectations inherent therein? Did those responsible act from realistic and realizable expectations with regard to the efficient prosecution of an air war, or were bombers considered nothing more than an extension of the artillery and reconnaissance aircraft as improved artillery spotters, as had been the case on the Spanish proving grounds?

Self-Deception?

In January 1939 the commander-in-chief of the Navy, Admiral Erich Raeder, persuaded Hitler that his *Reichsmarine* would take priority over other aspirants in all matters dealing with raw materials contingencies and industrial production. This prompted Milch to recommend to Göring in April of 1939 that the latter should arrange a demonstration of the *Luftwaffe's* operational potential for Hitler as soon as possible, so that the necessary priorities would be assigned to the "Consolidated Aircraft Type Program"—or the feasible parts of the same—for the complete buildup of the *Luftwaffe*.

In mid-1939, the parties within the *Oberkommando der Wehrmacht*, the OKW, overseeing raw materials distribution assigned the *Luftwaffe* even less of a percentage than before, prompting Udet to inform State Secretary Milch that it would no longer be possible for the air force to meet the demands placed upon it.

With this situation as a backdrop, Hitler agreed to witness a *Luftwaffe* demonstration, which took place at the Rechlin Test Center on 3 July 1939. The demonstration impressed all those present at the time; without a doubt, the weapons and aircraft on display were among the best in the world. In addition to the Heinkel He 100 and Messerschmitt Me 109 high performance fighters, the firing of a new 30 mm aircraft cannon—the MK 101—was demonstrated from an Me 110 on the ground. This machine cannon had a rate of fire ranging from 230 to 260 rounds per minute and a penetration hitherto unknown in this caliber; it had been designed to attack armored targets.

The Junkers Ju 88 was flown before Hitler, who became convinced of the type's developmental potential as the sole, fully dive capable, twin-engined universal bomber. A much overloaded Heinkel He 111 took off effortlessly with the use of RATO packs. Hitler inspected a pressurized cockpit designed for high altitude flights, and was shown the cold starting method for aircraft engines in winter conditions. He enthused over the Ju 88 and 30 mm cannon, despite the fact that these had not yet been fully tested and were still in the process of being technically refined. He was particularly impressed with the possibilities offered by the pressurized cockpit for aircraft operations at previously unheard of altitudes.

Milch warned Hitler and Göring against making any potentially erroneous political decisions based on what they had just witnessed, for everything was still in the testing and evaluation stage and would not reach front line units for years yet. Hitler reputedly answered back that there would not be any war!

Subsequent to this inspection, in the following month Göring ordered the prioritized manufacture of the MK 101 (despite not yet being ready for production), stressed the importance of the high altitude bomber, and called for accelerating the evaluation of the pressurized cockpit at all costs.

In the interim, delays had set in with starting up production of the Ju 88 on the one hand. On the other, it was intended to taper off production of the He 111 in favor of the Ju 88 and later, the He 177—which had yet to fly at this point. Coupled together, these signs pointed towards a "bomber gap" in 1940. In late July 1939, after "technical problems" had set Ju 88 production back by three months (for which Udet apologized to Göring), Udet authorized a modified *Lieferprogramm 12*, which called for 2,357 Ju 88s to be built by April 1941 and a total of 5,000 by April of 1943. For Göring, even these numbers seemed inadequate, and as a result of an inspection of the Junkers production facilities became convinced that a monthly output of 300 Ju 88s was entirely within the realm of possibility. Milch, Udet, and Jeschonnek met with Göring in early August 1939 in an effort to find realistic alternatives to the difficulties which had arisen in association with the "Consolidated Aircraft Type Program." At the expense of all other aircraft types, including transports and trainers, Göring decided to establish no less than 32 bomber *Geschwader* equipped with a total of 4,330 aircraft, of which 2,460 would be Ju 88s—all by 1 April 1941. The industry was called upon to concentrate their efforts on manufacturing offensive aircraft, i.e. bombers and recce planes. In particular, it was to focus on the Ju 88 standard medium bomber, the He 177 long range bomber (of which 800 were to be built by April 1943), and the Messerschmitt Me 210 (3,000 by the same time). The Me 210 was intended to operate as the standard multi-role aircraft, meaning as a *Zerstörer* and strike plane as well as a *Stuka*, high-speed bomber, and strategic reconnaissance platform, but in August of 1939 it had not yet even completed its maiden flight. An impressive, almost brilliantly conceived ready-made solution that, in the reality of the divergent operational requirements, had little chance of success given the multiplicity of intended roles bestowed on it.

The calamity with the Ju 88 production could not be kept hidden from Hitler, who had especially high hopes for this airplane. On 21 August 1939 he tasked Göring with increasing the monthly Ju 88 output to 300, and assigned it the same priority as he had ordered for the navy's buildup early that year. This writ, mentioned in an earlier section, may simply have been a delayed reaction to the Rechlin demonstrations of 3 July 1939. Be that as it may, Rechlin had awakened expectations at the highest levels, expectations which would be difficult to fulfill underneath the brewing storm clouds of an impending war. Unless, of course, the general mobilization were paralleled by a mobilization of the industry and a switch to wartime production. But there would be no talk of this, at least not at first!

Offensive Aircraft Weaponry of 1939

What weaponry, bombing, and reconnaissance instruments were available to the offensive aircraft, particularly the bombers? L.Dv. 16 specified that these types must carry "the war to the power bases" of the enemy, for these "would strike at the root of an enemy's fighting strength and the will of its people to resist." At the same time, they would need to defend themselves against enemy fighter attacks—a factor just as critical to survival for the reconnaissance aircraft as it was for the bomber.

Ordnance
Up to and including 1939 those 12 types of bombs, mines, and aerial torpedoes listed in appendix 3 were authorized for procurement and in production. They ranged from the 1 kg B1E electron incendiary bomb to the 500 kg PC 500 armor piercing bomb. The only ordnance heavier than 500 kg was the LMB observation mine, which weighed 1,000 kg and was used to mine bodies of water.

This was the somewhat meager arsenal of modern bombs and other ordnance which the *Luftwaffe* had available to it in 1939.

Questions on the meaning of various acronyms and abbreviations for the various types of bombs, etc. are briefly answered with the following information:

The figures following the abbreviations should cause no problems—these are simply the total weight of the individual bomb in kilograms. The acronyms indicate the effect and type of ordnance, to include aerial torpedoes and parachute-deployed mines. Based on their effect, these were categorized as either fragmentation or demolition bombs, as well as aerial mines or gas pressure bombs. Their specific designation/classification was based on the percentage of bursting charge to the total weight, as follows:

SA = bursting charge greater than 75%
SB = bursting charge 75%
SC = bursting charge 50%
SD = bursting charge 30%
SE = bursting charge 15 to 20%

The remaining abbreviations have the following meanings:

B stands for incendiary bomb
LM for mine, specifically *Luftmine*, or aerial mine (coverterm: *Wasserballon* or Water Balloon)
LT for *Lufttorpedo*, or aerial torpedo
PC for *Panzersprengbombe*, or armor piercing bomb

Provided below is a list outlining the general application of the various ordnance types. It should be kept in mind that this depended upon the type and quality of the potential target, of which the following are but a few examples:

- against fortifications of any type: SC 500/SC 250
- against heavy cruisers and destroyers: SC 500/SC 250 (filled with *Trialen*)
- against tankers: SC 500
- against merchant ships: SC 250 (*Trialen*)
- for mining harbors and coastal waters: LMA/LMB
- against airfields, airstrips, hangars, ammunition dumps, parts warehouses, repair facilities, radio stations: SC 250/SC 50
- against aircraft bordering runways and in open revetments: SD 10 mixed with SD 50 and B1E
- against aircraft in covered revetments or against supply and fuel depots: SC 50 mixed with B1E
- against vehicles of any type, underway or stopped: SD 50
- against columns of any type: SD 50 mixed with SC 50 and SD 10
- against infantry concentrations: SC 50 mixed with SD 10

History of German Aviation: Bombers and Reconnaissance Aircraft

In 1939 there were no suitable bombs or specialized weaponry to effectively attack battleships, aircraft carriers, or armored vehicles.

Armament

Although bombers and reconnaissance aircraft could only partially be considered offensive weapons, the number and quality of guns they carried were absolutely critical to the combat effectiveness of these classes. The survival of an airplane in aerial combat with enemy fighters was, to a great extent, determined by the onboard armament, which could also significantly increase the effectiveness of low-level attacks.

Introduced by 1939 were the MG 15 and MG 17 7.9 mm machine guns, with the MG 81 following into large scale production in 1939. The MG-FF was a 20 mm cannon available at the time. Brief descriptions of these weapons can be found in appendix 4, "Aircraft Armament from 1936 through Today" at the back of the book.

Problems of Precision Bombing

To be sure, this was not a problem specific to the year 1939; it is a problem independent of a particular period and, indeed, is even commonplace today. Knowledge of the basic concepts of bombing can help in better assessing the tactical-technological challenges and the bomber design layouts stemming from those challenges, and thus their operational potential.

The examples of targets listed above, with the bombs then available to be used against these targets, provide a snapshot for the character of bombing warfare. Initially, this was primarily directed against military targets and was tailored to strike and destroy the enemy's war making potential, particularly his air forces. However, the prerequisite for this was not just the bombing platform, i.e. the bomber (in a broad sense), but was much more dependent upon the precision bombing capability of the crew so that the bombs would not just land on enemy territory, but would also hit their intended target—even with the platform subject to enemy fire.

In order to hit a particular target when bombing, the bomb must be dropped from the aircraft at a specific point in time and space, known as the point of release. This point is established by calculation with the aid of bombsights for every individual bomb or salvo of bombs dropped. The sight determines the angle at which the bomb must be released in order for its trajectory to end at the target. The trajectory is affected by the initial velocity and direction of the bomb, i.e. the aircraft's movement at the point of release, and by the forces affecting the bomb as it falls (acceleration of the earth, resistance, wind, and propulsion forces, such as rocket motors). Thus, for every bomb the prerequisites for a precision hit are an accurate calculation of the initial variables (speed relative to the target, altitude, and initial direction) and an awareness of the factors influencing the bomb following release (e.g. ground lag, rocket propulsion, ballistic wind, i.e. the average value of the wind during the fall period, the earth's rotation, etc.)

The following bombing methods are classed according to the type of approach to the target:

- Horizontal release
    - high altitude
    - low altitude
- Glide release (rate of descent between 5 m/sec and 15 m/sec, glide angle 5 to 20° from horizontal)
- Dive release
    - shallow dive = 20 to 50°
    - steep dive = 50 to 90°

with both angles measured from the horizontal.

The "how" of bombing technology is, to a great extent, determined by the "Theory of Bombing," which unfortunately will not be discussed in further detail within the pages of this particular book.

Reconnaissance Systems of 1939

With regard to aerial photography, from 1935 onward a new generation of automatic aerial cameras with greatly improved performance was available, as was a new handheld camera chiefly designed for tactical reconnaissance work. The new Rb 20/30, 50/30, and 75/30 automatic cameras, primarily tailored to strategic reconnaissance, had an improved lens with F-stops between 1:4.5 and 1:6.3 and anti-reflective coating, with an angular field of view of up to 105°.

Details covering both photographic equipment and the terms and concepts used in aerial photography can be found in appendix 5 "Reconnaissance Systems from 1935 Onward."

Acquiring aerial photos at night required the use of flash bombs or flares which, despite the anti-reflective coating of the lens or the fitting of a so-called precision filter, was never really effective at providing satisfactory results. The problem with using flash bombs lay in the uniform illumination of the target area, which was only possible under favorable circumstances.

In short, it is safe to say that, other than the problems inherent in nighttime photography, the overall quality of aerial photography at the time can be considered outstanding. The limits of picture

History of German Aviation: Bombers and Reconnaissance Aircraft

resolution for enlargements and detailed evaluation depended less upon the merits of the photographic and evaluation equipment—both were of excellent quality—but rather upon the graininess of the film material and the state of film development technology.

1938/1939 Numbers Comparison

After becoming familiar with a few of the RLM's numerous acquisition programs it is interesting to discover the actual numbers of aircraft operating with front-line units based on this target inventory for bombers and reconnaissance aircraft. Drawn up based on aircraft types and, where necessary, subtypes, documents of the *Luftwaffe* General Staff show the following inventory with the effective date of:

19 September 1938

Medium Bomber	Actual Inventory		Operationally Ready	
Do 17E	328 }	430	271 }	351
Do 17M	102		80	
He 111B	272		219	
He 111E	171 }	570	141 }	468
He 111F	39		30	
He 111J	88		78	
Ju 86A/D	159 }		136 }	
Ju 86E	43 }	235	35 }	200
Ju 86G	33		29	
Total		**1235**		**1019**
Dive Bombers				
Ju 87 (total)		**247**		**227**
Strike Aircraft				
He 45		78		73
Hs 123		117		91
Total		**195**		**164**
Strategic Reconnaissance				
Do 17F	77 }	149	69 }	128
Do 17P	72		59	
He 70		73		49
Total		**222**		**177**
Tactical Reconnaissance				
He 45		58		47
He 46		189		156
Hs 126		42		35
Total		**289**		**238**

Based on this, the offensive air force at this time consisted of 2,188 front-line aircraft, of which 1,825 were operational, giving a readiness state of 83.4%.

2 September 1939

Bomber	Actual Inventory		Operationally Ready	
Do 17E	119 }		100 }	
Do 17M	40 }	371	32 }	320
Do 17Z	212		188	
He 111E	38 }		32 }	
He 111H	400 }	787	358 }	685
He 111P	349		295	
Ju 88	18		15	
Total		**1176**		**1020**
Dive Bombers				
Ju 87 (total)		**366**		**318**
Strike Aircraft				
Hs 123 (total)		**40**		**36**
Strategic Reconnaissance				
Do 17P		**257**		**213**
Tactical Reconnaissance				
He 45		14		5
He 46		67		53
Hs 126		275		251
Total		**356**		**309**

A total, therefore, of 2,195 front-line aircraft, of which 1,896 were operationally ready. This corresponded to a readiness state of 86.4%. At the same time, these numbers reflected the bombing and reconnaissance potential of the *Luftwaffe's* offensive elements at the beginning of the war, which had broken out the previous day on 1 September 1939. Compared with the numbers from the previous year, there does not seem to be any appreciable growth in the actual inventory—it remained stagnant; however, the readiness state was 3% higher at an impressive 86.4%. The direct and indirect changes to the overall combat strength can be measured as follows:

Medium bombers: A direct increase in the combat strength resulted from the marked buildup of the He 111 component, which had increased by 138% over the previous year's numbers, as well as through the introduction into front-line units of the extremely capable, although little tested, Ju 88.

Indirectly, the bomber units received a boost in their combat strength with the introduction of the more powerful Do 17Z, which replaced a large percentage of the Do 17E and M variants. Another improvement was the withdrawal from the units of the Ju 86, which no longer fit the bill as a medium bomber, and the continued use of some of these aircraft as trainers in the C-class flying schools.

Although the overall bomber strength had slipped by 59 aircraft compared to 1938, a higher readiness state was achieved by stepping up the modernization of the inventory, so that, all things considered, there were nearly the same number of bomber aircraft available for operations as in the previous year.

Dive bombers: Here there was a direct increase in combat strength as a result of a 148% increase in the number of Ju 87s over the previous year.

Strike/Ground attack aircraft: A severe reduction was caused by the withdrawal of the He 45, and at the same time reduction of the number of Hs 123 to just 34% compared to 1938.

Strategic reconnaissance aircraft: Here there was not just a quantitative growth by a good 15%, but also a markedly qualitative improvement by withdrawing the He 70, which no longer was up to the requirements of strategic reconnaissance, as well as by equipping all strategic reconnaissance units across the board with the Do 17. This latter move not only was operationally advantageous, it also made sense from training and logistical perspectives, as well.

Tactical reconnaissance aircraft: A 23% quantitative growth was noted, together with qualitative improvements by reducing the obsolescent He 45 and He 46 stock and replacing these with the more modern Hs 126, which now formed 77% of the tactical reconnaissance inventory compared to just 15% in the preceding year.

All in all, not only were the strategic and tactical reconnaissance units the recipients of a notable boost in quality, they were also assigned a total of 67 aircraft, giving them a 19% increase compared with 1938.

Assessment

In a nutshell, at the time of the war's outbreak the combat effectiveness of the offensive *Luftwaffe* can be considered as follows:

1. The numerical setback for bombers was more than offset by an improved state of readiness and the greater combat effectiveness of more modern aircraft at the same time that the inadequate Ju 86 was phased out.
2. The numerical setback for dive bombers and ground attack aircraft was virtually offset by a massive beefing up of the Ju 87 inventory and withdrawal from service of the antiquated He 45
3. A 19% increase in the number of reconnaissance aircraft and an improvement in their quality reflected the increased importance of aerial reconnaissance as the "eyes of command."

This was the state of affairs for the offensive forces of the *Luftwaffe* at the outbreak of the Second World War which, however, should not be considered entirely independent of the other operational flying elements of the *Luftwaffe*. Therefore, a brief look at these other aircraft types is in order:

	1938	1939	Result
Fighters and *Zerstörer*	810	408	-402(!)
Transports	317	552	+235
Naval planes	139	167	+28

Therefore, whereas the offensive *Luftwaffe* was highlighted by a slight qualitative vice numerical improvement in its combat effectiveness between 1938 and 1939, the fighters/*Zerstörer* suffered a marked reduction in their actual aircraft inventory. This was only in part due to the introduction of the high-performance Me 110B and C and the more modern Me 109E in conjunction with the retirement of the obsolescent Ar 68 biplane.

Airlift capability had increased considerably, although it should be noted that for the most part the J 52 transports were assigned to

Heinkel He 115, a multi-purpose floatplane used in the bomber and maritime reconnaissance role, seen here being loaded with an aerial torpedo.

C-class and instrument flying schools. In case of emergency, these would be called up and pressed into operational service at the expense of flight training operations. A 20% increase in naval aircraft was noted, along with the introduction of the He 115 torpedo seaplane and the He 111J (the latter conceived only as a stopgap torpedo bomber and still listed as a bomber just the previous year). These measures signified an increase in combat effectiveness for long range maritime operations, chiefly against shipping targets.

The total inventory of the *Luftwaffe* in front-line aircraft dropped from 3,454 aircraft in 1938 to 3,322 in 1939, whereby in 1939 the

- offensive *Luftwaffe* comprised 66% of the
- fighters and *Zerstörer* comprised 12.3% *Luftwaffe's*
- transport aircraft comprised 16.7% total stock.
- naval aircraft comprised approx. 5%

The burning question has cropped up with increasing frequency, asking just how much this *Luftwaffe*—finding itself slipping numerically and with a seemingly imbalanced, or at least stagnating buildup situation—could be considered an effective instrument for prosecuting an air war.

Was it a certain amount of blindness brought on by the easily wrested victories during the Spanish Civil War, or a result of showmanship at Rechlin which had failed to take into account what was practical and available? Perhaps it was due to the arrogance at the *Luftwaffe's* highest level (personified by Göring), or a lapdog political leadership, under the thumb of a dictatorial *Reichskanzler* who placed high hopes in his *Luftwaffe*, which had utterly failed to recognize the situation facing a half-baked air force. Whatever the reasons or combinations thereof, the die had been cast in political ignorance of the potential consequences; the war against Poland, evidently in every respect a militarily inferior enemy, was now clearly beyond the point of being stoppable!

8. Offensive Weapons and Reconnaissance Package Platforms Trough the End of 1939

Over the next several chapters this book will examine those 66% of front-line aircraft comprising the *Luftwaffe's* offensive element at the war's outbreak and their predecessors dating back to 1935. The purpose is to show the work invested by the designers and constructors, the engineers, technicians, and experts, as well as the aviation industry as a whole, plus the test and company pilots—a dramatic part of technological and German period history also containing many aircraft projects which never entered full production.

First, let us look at the offensive weapons platforms, commonly known as bombers, dive bombers, and ground attack planes, acquired by the German *Luftwaffe* from 1935 onward and listed in the order in which they entered service, covering their development up to September 1939.

Those aircraft types in these categories which were often approached with just as much creativity in design, build, and sometimes even flight testing and, for a wide variety of reasons, never entered *Luftwaffe* service, are briefly addressed to "flesh out" the picture. However, a detailed treatment and appraisal is outside the scope of this book series, which is dedicated to the great German aircraft designers or aviation companies.

Medium Bombers

Junkers Ju 52 Auxiliary Bomber

Due to a lack of better suited bomber types, the three-engined Ju 52 had entered service with *Luftwaffe* units even before 1935 as an auxiliary, stopgap bomber. As its designer, *Dipl.- Ing.* Ernst Zindel, had never envisioned the type as a bomber, its combat potential in that stopgap role can be considered something less than stellar—even after a series of modifications.

Delays in equipping the first bomber units with the twin-engined Dornier Do 11 resulted in the first flying units being stocked chiefly with the military version of the Ju 52. This continued even after the improved Dornier Do 23 became available, since by then the Ju 52 was well into production. By the end of 1935 eight of twelve, i.e. 2/3 of the *Kampfgruppen*, were equipped with the Ju 52/3m ge and the Ju 52/3m g3e.

During this time the companies of Heinkel and Junkers had started (in early 1934) parallel development of a higher performance, twin-engined bomber, the so-called "medium bomber," with the intent of replacing the Ju 52 auxiliary bomber as soon as possible. At the same time Lufthansa published a requirement for a high-speed passenger liner capable of carrying ten passengers for use on its so-called "*Blitzstrecken*," or lightning routes. Heinkel and Junkers thus developed basic airframes to meet both contractors—Heinkel the He 111, and Junkers the Ju 86. However, Junkers was handicapped by the fact that the RLM prohibited the company from installing the internally developed Jumo 205 in the Ju 86. The Jumo 205 was developed by Junkers as a heavy oil or diesel aircraft engine, type tested in 1933, and found to have a particularly economical fuel consumption rate and virtually impervious to fire. This 16.6 liter engine provided a takeoff rating of 440 kW/600 hp. For its part, the Heinkel team based its design around a higher performance engine from the outset, so that performance differences between the Ju 86 and the He 111 (to the disadvantage of the former) had been pre-programmed.

The Ju 86 completed its maiden flight as early as 4 November 1934, with the He 111 did not follow until nearly four months later. We shall therefore first take a look at the career of the Ju 86 military versions.

Junkers Ju 86 Bomber

The criteria for the "medium bomber" has already been discussed in the chapter "The New Generation of Aircraft." At this point we shall look exclusively at the military versions of the Ju 86 as a bomber and, later, as a specialized high altitude reconnaissance aircraft, as well.

In the spring of 1934 the Heinkel and Junkers companies were contracted to each build five prototypes of the He 111 and Ju 86, respectively, alternating between a military and civil layout and beginning with a bomber version.

Prototypes

The first Ju 86 prototype, the Ju 86ab1 (later designated V1) was built in the Junkers design bureau under the direction of the experienced Ernst Zindel and his assistant, *Dipl.-Ing.* Hermann Pohlmann. It was built and readied for flight in just five months, with over 270,000 man hours spent on the project. First flight was made on 4 November 1934 at Dessau. As the intended Jumo 205 diesel engines were not yet available (production having just started at Köthen), the first Ju 86 was powered by two Siemens SAM 22s, air-cooled nine-cylinder radial engines with just 405 kW/550 hp output driving two-blade fixed-pitch wooden propellers. No gun stations were fitted to this first bomber prototype. Test flights revealed problems with heaviness in the controls and a tendency to yaw about the vertical axis. During the necessary modifications the B and C gun stations (dorsal and ventral fuselage positions, respectively) were fitted, the *B-Stand* being partially exposed with a two-position sliding plexiglas cover and the *C- Stand* in a semi-retractable "dustbin" turret which, when extended, was rotatable.

The third prototype, again a bomber, was the Ju 86cb (later the V3). This was fitted with glazing in the forward nose section and a manually operated turret for a flexible mounted machine gun, the *A-Stand*.

The fifth and final prototype contracted for in the spring of 1934 was officially designated the Ju 86V5, since a standard system of numbering had been introduced by that point in time. The V5 had modified wings, a tailwheel in place of a tailskid, and improvements in the radiator design, the canopy, and the *C-Stand* turret. It flew for the first time in August 1935 and became the pattern for a pre-production series of 20 Ju 86s, for which the RLM issued a follow-on contract.

0-Series

In late 1935 Dessau initiated construction of 13 0-series, or pre-production versions, in addition to the assembly of seven civilian Ju 86s. These bombers were designated the Ju 86A-0 and were now powered by the Jumo 205C diesel engine. Beginning in February 1936 they were shipped directly from the final assembly to the *Luftwaffe* for additional evaluation.

With regard to technical details, it is worth noting that the Ju 86 initially had a tendency to stall over the wing when at high angles of attack. This was a result of too sharp a taper in the design layout of the wing, a mistake rooted in the then-prominent ideas of lift distribution for wings tapered sharply out to their tips. Careful aerodynamic studies resulted in a "bent" taper, which succeeded in preventing a premature separation of the airflow and uncontrolled wing stall.

The Ju 86 was built entirely of stressed skin design. The wing runners, however, were made of the standard Junkers tubular steel construction as on the Ju 52, with the stressed skin stiffened to prevent bowing by metal sections arranged in the direction of flight. Inside the wing itself, particularly in the area of the fuel tanks, access was provided by large, removable sections of skin. These were affixed to the skin permanently attached to the supports by means of sliding pins screwed down tight to prevent shearing. The undercarriage retracted electrically via worm gears; even the extension and retraction of the landing gear was activated by an electric switch, although the switch for halting the retraction and extension procedure initially proved temperamental. Later, the operation switched over to a hydraulic driven system with hydraulic stops which proved less problematic.

A-1 and D-1 Series

In the late spring of 1936 the first Ju 86A-1 production aircraft were delivered. Also based on the V5, the A-1 was powered by the Jumo 205C-4. Despite its improved wings, the stability along its vertical axis was never fully satisfactory. A Ju 86A-0 was pulled from the pre-production series and tested under its new designation of Ju 86V6, with the goal of improving stability. These flight trials ultimately led to a wedge-shaped extension of the fuselage end cap which extended beyond the horizontal stabilizer and was found to be an acceptable solution to the problem. With this modification and an increase in the fuel capacity, the Ju 86 then continued in production as the Ju 86D-1; it was available for duty from the fall of 1936 onward.

Junkers Ju 86D-1 with its gun positions clearly discernable, including the extended ventral position (C-*Stand*).

The Ju 86D-1 had three traversible MG 15s installed in its three gun positions and was able to carry eight SC 100 or sixteen SC 50 bombs in vertically mounted racks inside the fuselage. Its maximum speed at sea level was barely 300 km/h, and at 3,000 m was claimed to have been about 325 km/h. With a maximum permissible bomb load at an altitude of 3,800 meters, the Ju 86D-1 had a tactical range of 560 km with a cruising speed of about 230 km/h.

Soon, however, renewed concern appeared regarding the suitability of diesel engines in bombers; the finicky engines had been giving trouble even with the pre-production batch. Their slow response and rapid fluctuation in performance made them particularly unsuitable for formation flying or for making sudden tactical combat maneuvers. These concerns were borne out during the combat testing of four Ju 86D-0s in the Spanish Civil War in late 1937 and played a major role in the decision to fit another type of engine to the aircraft. This despite the fact that the Jumo 205 had proven quite acceptable for civilian long range flights with only minor fluctuations in rpm rate—despite its unfavorable performance to weight ratio in comparison with Otto engines.

Foreign Interest

Foreign parties became quite interested at a relatively early stage, not only in the civilian commercial version as a result of its spectacular long range flights, but were also particularly enticed by the Ju 86 bomber aircraft. The following countries issued purchase contracts:

Sweden

Sweden was the first government to show concrete interest. Following talks with the Junkers company about installing alternative powerplants, on 30 June 1936 Sweden ordered a Ju 86A-1 airframe, albeit with air-cooled American Pratt & Whitney Hornet S1E-G nine cylinder radial engines rated at 574 kW/780 hp. This machine, which Junkers had designated the Ju 86A-1k, was delivered on 19 December 1936 and flown from Dessau to Sweden on the same day by a Swedish Air Force pilot. Two additional Ju 86A-1k aircraft had been ordered by the Swedish government in November. From this point on Junkers used the series designation of K (for *Kampfflugzeug*, or bomber) and Z (for *Zivilflugzeug*, or civilian aircraft) for these types of foreign contracts, so that these first three Ju 86s delivered to Sweden were supplied as Ju 86K-1s*. One of these three machines was subsequently fitted with skis for testing in winter conditions.

Generally satisfied with their own flight test results, the Swedes ordered a further 37 Ju 86Ks. Those powered by the air-cooled British nine cylinder radial Bristol Pegasus III (My III) rated at 603 kW/820 hp were designated as Ju 86K-4, while those powered by the identically configured 677 kW/920 hp Pegasus XII (My XII) were known as the K-5. Both engine types were license-built in Sweden by the Nohab company. In addition, the Swedes acquired a manufacturing license for 40 Ju 86K models.

The ordered Ju 86K-4s and the total of 19 K-5s were delivered over the course of 1937, or assembled in the Trollhätten Works of the newly established Svenska Aeroplan Aktiebolaget (SAAB), where the Swedish license-building program began in late 1938. Running counter to German/Swedish planning and the governmental contract calling for 40 license-built aircraft, only 16 bombers, designated Ju 86K-13** were actually built there and delivered to the Swedish Air Force.

All told, 56 Ju 86Ks were exported to Sweden, assembled there, or built under license.

Junkers Ju 86 K-1 (Swedish designation Ju 86B3), the sole example to be fitted with skis during the winter of 1937/1938 for winter operations.

Junkers Ju 86K for Sweden (1937).

---

*   - Ju 86K-1 equated to the Swedish Ju 86B3
     - Ju 86K-4 to the Swedish Ju 86B3A and
     - Ju 86K-5 to the Ju 86B3B.
** Swedish designations for the Ju 86K-13: 1) Ju 86B3C with Pegasus XXIV (My XXIV) with 721 kW/980 hp, 2) Ju 86B3D with Pegasus XIX (My XIX) with 614 kW/835 hp.

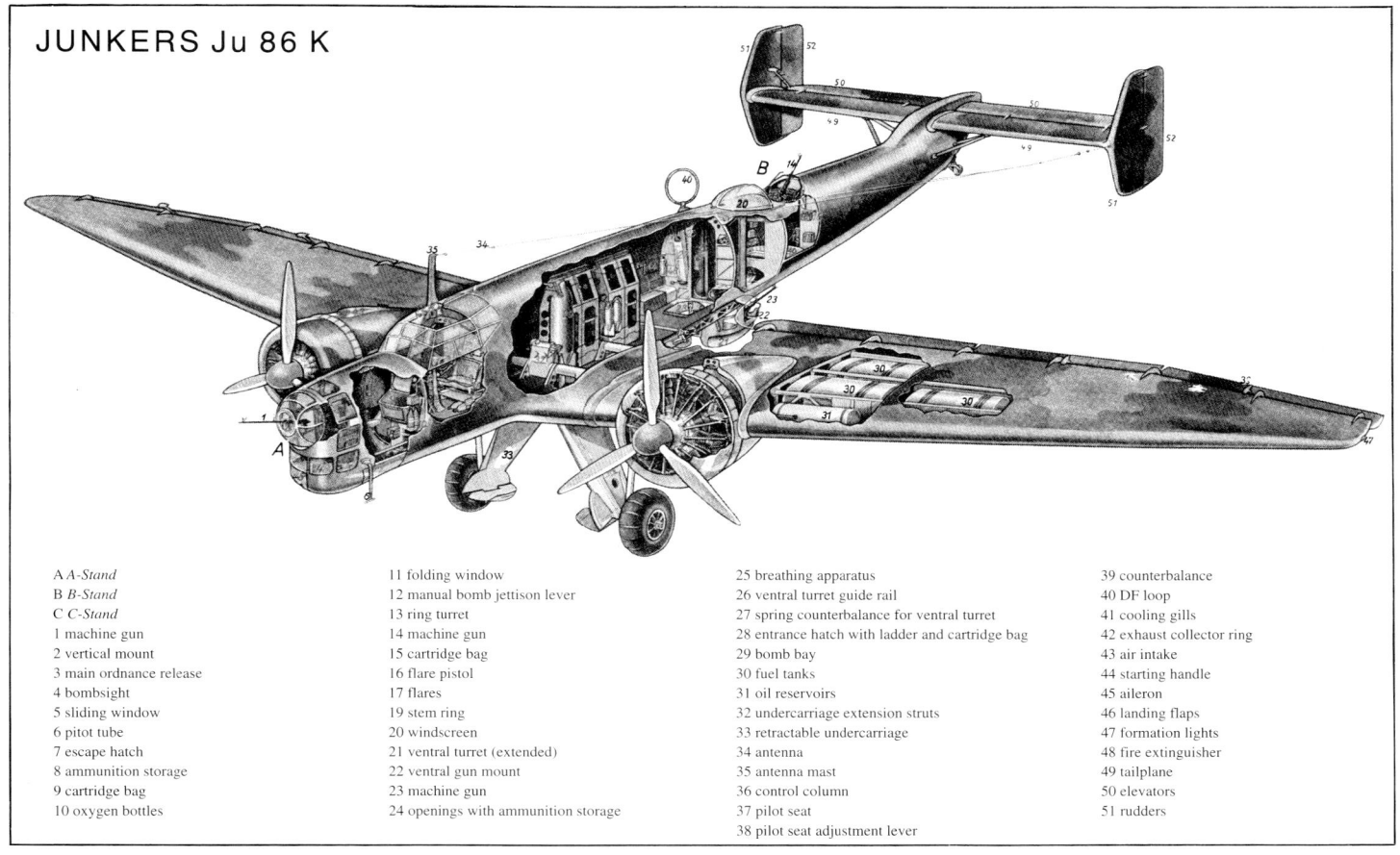

Cutaway view of the Junkers Ju 86K (1938).

The Ju 86K remained in service with the Swedish Air Force up until 1948, serving in its final role as a torpedo bomber; it subsequently found employment as a trainer and transport, with the type being phased out altogether by 1956.

Hungary
In 1936 Junkers was awarded an initial contract for 24 Ju 86K-2s—as the Ju 86 export model for Hungary was designated. This bomber version was fitted with a license-built French 14 cylinder Gnôme Rhône 14 K Mistral Major air-cooled radial powerplant, built by the company of Manfred Weiss in Budapest. The purchase order was soon raised to 66. Delays in license production of the French engines at the Weiss company resulted in the first Ju 86K-2s not reaching Hungary until early 1938, with bomber squadrons receiving the type beginning in the summer of 1938. These units attacked Slovakian positions in the Carpathian Mountains with 18 Ju 86K-2 bombers in March 1939 when Slovakian territory was occupied. The K-2 later went to war against the Soviet Union, was pulled out of operations completely over considerable objections in 1942, and served in isolated cases as a trainer.

Portugal
In 1937 Portugal also ordered ten Ju 86 bombers equipped with the American Pratt & Whitney Hornet aircraft engines. These were shipped under the export designation of Ju 86K-6.

Chile
At the same time Chile received twelve Ju 86K-6 in the same configuration as those supplied to Portugal. This type drew considerable attention in Latin America, particularly during visits in Argentina, and was considered the best bomber in the region.

Austria

Austria, too, ordered 18 Ju 86s in 1937 to modernize its bomber wing. These were E-2 variants, and in late 1937 the first of ten completed Ju 86E-2s were delivered to Austria. No further deliveries followed, however, as the "*Anschluß*" of Austria to the German *Reich* was imminent. When this became a historical reality on 13 March 1938 the contract was canceled, and political developments saw the merger of the Austrian Air Force with that of the German *Luftwaffe*.

South Africa

The South African airline company "South African Airways" (SAA) drew up an initial contract for the delivery of a Ju 86 as early as 1936. It was to be kitted out with the liquid-cooled twelve-cylinder Rolls-Royce Kestrel XVI 548 kW/745 hp engine, as there already existed overhaul facilities in connection with a British license-built aircraft engine in South Africa.

As a result, Junkers modified a civilian C-1 variant of the Ju 86 for the Kestrel powerplants. This machine first took to the air in the fall of 1936 as the Ju 86V7. Altogether South African Airways ordered 18 Ju 86s, 17 of which were civilian Ju 86Z models and one a Ju 86K-1 bomber variant, as was built for Sweden. The first Ju 86s for this contract were finished with their Kestrel engines by the end of 1936. However, while yet in Germany, South Africa expressed doubts as to the suitability of this engine and the airframes were refitted with the American Hornet engine, which had already proven itself in the Ju 86. Together with the remaining aircraft of the contract, these were delivered to South Africa as the Ju 86Z-7. The first Ju 86Z-7 arrived in South Africa in June of 1937, with the remaining following individually as they were built. After just over a year in service with the SAA, in 1939 the entire lot was turned over to the South African Air Force, which converted them to the K-1 configuration with B and C gunner stations for flexible 0.3 inch Vickers machine guns, as well as a fixed gun of the same caliber in the nose. Bomb racks, however, were not fitted inside the fuselage, but under the center section, and following the outbreak of the war these aircraft served as a combination reconnaissance bomber, initially being employed in the coastal patrol role. In December 1939 these Ju 86s flushed out the German blockade runner SS *Watussi* off the Cape coast. From may 1940 onward Ju 86s were used in Kenya, Transvaal, and Tanganyika (today's Tanzania), as well as in operations against the Italians in Ethiopia and Somalia. The remainder of the original surviving Ju 86Z-7s were combined into a squadron for coastal reconnaissance and anti-submarine duties along the South African coast in the summer of 1942 and were phased out a short time later.

E- and G-variants

The combat readiness rate of those bomber units in the *Luftwaffe* equipped with the Ju 86D-1 was, in a word, poor. The diesel engines tended to overheat, and often destroyed pistons as a result of unavoidable strain on the engine when changing rpms during tactical flight. That and wear at the welded seams on the exhaust pipes, etc. all contributed to higher than average engine outages. This prompted renewed thinking with regard to a less temperamental engine for front-line usage. A single Ju 86D-1 was thus test fitted with an American Pratt & Whitney Hornet engine, which the Bayerische Motorenwerke had been building under license for years and had already been used as an alternative engine in the K-1 variant for Sweden and the Ju 86K-6 for Portugal and Chile. The airframe fitted with this nine-cylinder radial engine, in this case the BMW 132F license-built engine, was subjected to a thorough evaluation regimen in the spring of 1937 at Rechlin under the designation Ju 86V9.

A pronounced reduction in the outage rate during military flight operations was noted compared to diesel engines, and Junkers was instructed to switch its Ju 86 production over to use of the BMW engine at the earliest possible point.

Those airplanes powered by the BMW 132F fuel injected engine, rated at 596 kW/810 hp, were assigned the series designation of Ju 86E-1 and delivered to the *Luftwaffe* beginning in the late summer of 1937. After about 30 of these had been delivered, the somewhat more powerful BMW 132N (with an output of 636 kW/865 hp) became available. The machines incorporating this engine were designated Ju 86E-2s.

The previous arrangement/layout of the canopy had proven unsatisfactory for flight operations since, among other things, the pilot had a poor forward visibility during takeoff which did not improve until the tailwheel lifted free of the ground. As a result,

Junkers Ju 86E with BMW 132 nine-cylinder radial engines (1937).

History of German Aviation: Bombers and Reconnaissance Aircraft

Junkers shifted the entire cockpit forward and gave the now fully glazed, albeit somewhat more compact, canopy a rounded shape with an MG 15 installed to the right and ahead of the bombardier.

This arrangement was tested in an appropriately modified Ju 86E-1 at Rechlin in early 1938. The aircraft was designated Ju 86V10 and was assessed to have good qualities. Those Ju 86s thus modified began rolling off the production line in late spring 1938 as the Ju 86G-1.

From this point on only about 40 machines in this series were built before Ju 86 production came to a halt in the early summer of 1938. Its demise had been brought about by the competition with the Heinkel He 111, which—due to its improved performance and greater bomb capacity—had in the meantime nudged the Ju 86 from its position as medium bomber. To be sure, this was not the final curtain call for the Ju 86, but the Ju 86G-1 was the pinnacle of this bomber's development before the war.

Technical Data and Details of the Ju 86G-1

The Ju 86G-1 was a twin-engined land based bomber powered by BMW 132F or BMW 132N engines.

Gewichte (nur für Zusammenbauzwecke)	
Flugzeugrüstgewicht	6390 kg
1 Flügel mit Motor vollst.	1600 kg
1 Motor mit Auspuff, 1 Triebwerksgerüst und Triebwerksverkleidung	750 kg
1 Ju HPC III Verstell-luftschraube	180 kg
1 Fahrgestellhälfte (Federbein, V-Strebe und Rad)	160 kg
1 Höhenleitwerk mit Abstrebung	145 kg
1 Seitenleitwerk (beide Hälften)	59 kg
1 Radsporn	26,5 kg

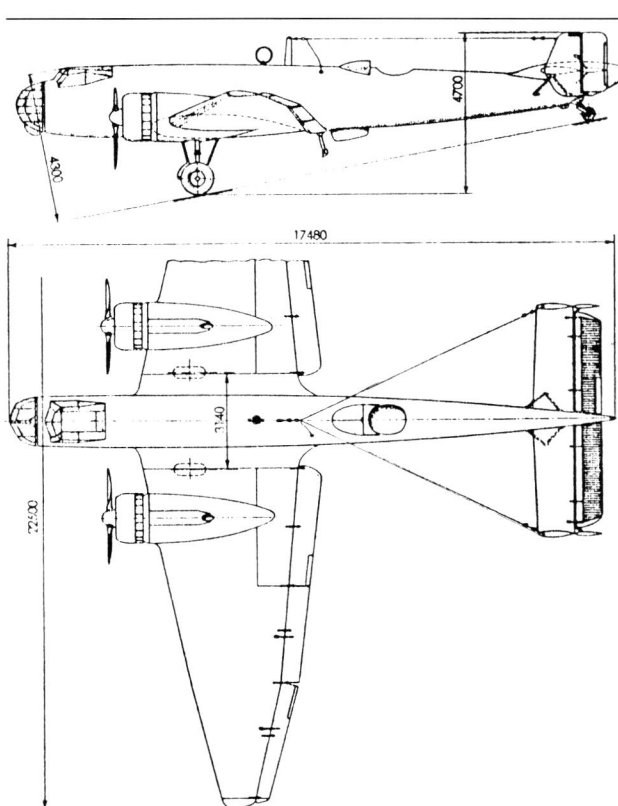

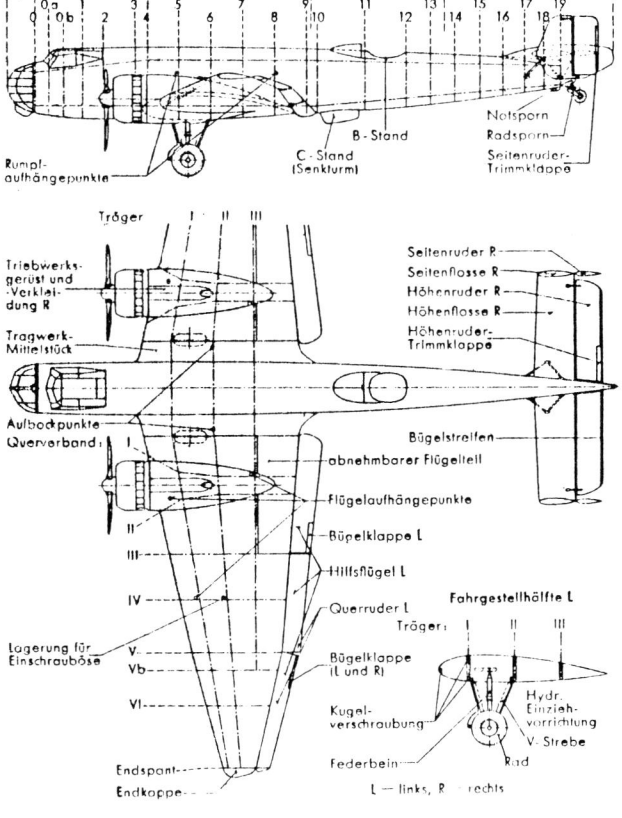

Junkers Ju 86G-1 - basic dimensions and weights, showing spars, formers, and components (1938).

Airframe

Fuselage: Oval, four-spar, stressed skin with bulkheads. Glazed nose section, removable; served as the work area of the bombardier, who operated the forward machine gun (*A-Stand*) and the bombsight. Cockpit with two adjustable seats for pilot and bombardier. Main payload area with bomb racks and entry hatch in front of bulkhead 10; behind bulkhead 10 was the lower turret (*C-Stand*), which when retracted served as the operating location for the on-board radio system, and when extended served as a firing platform for the machine gun; behind bulkhead 11 was the *B-Stand* with traverse ring for the machine gun.

Undercarriage: Non-axled, consisting of two halves hydraulically retractable into the wing. Shock absorption by means of oleo pneumatic shock absorbers. Medium-pressure tires fitted to wheels and individually braked via hydraulic brakes. A tailwheel, swiveling through 360° and capable of being locked in the direction of flight from the cockpit, was fitted to the extreme end of the fuselage.

Control surfaces: Braced double wing horizontal stabilizer with trim tabs adjustable from cockpit. Separate double wing vertical stabilizer bolted to either end of the elevators, with rudder trim tabs adjustable in flight. Separate double wing ailerons running 2/3 the span of the wing; the starboard side equipped with an aileron trim tab adjustable in flight.

Control inputs: Ailerons and elevators operated via a control column and handwheel, rudder control by means of pedals. Adjustment of elevator trim tabs, rudder trim tabs, and the aileron trim tab all accomplished using handwheels. Flaps activated hydraulically.

Wings: Cantilever design, broken down into a center section fixed to the fuselage, and outer sections attached to the center section by ball joints. The wings were the trademark Junkers double wing design; flaps were in three sections.

Powerplant

Engines: Two air-cooled nine-cylinder radial BMW 132N or BMW 132F with fuel injection.
(The second set of figures applies to the BMW 132F)
Maximum performance = 636 kW/865 hp (596 kW/810 hp)
Cruise performance = 489 kW/665 hp (485 kW/660 hp)
Maximum permissible rpms in flight = 2,400 min^{-1} (2,300 min^{-1})
RPM in flight at cruise = 2,150 min^{-1} (2,100 min^{-1})
Fuel consumption rate at cruise = 320 g per kW/hr /235 g per hp/hr (313 g per kW/hr /230 g per hp/hr)
oil consumption at cruise = 1-6 kg/hr.
Fuel/Lubricant type:
Fuel: A2 87 Octane Aviation fuel
Oil: Intava Rotring-100M or Aeroshell Medium

Propellers: Three-bladed two-position Ju HPC III variable pitch metal airscrews. Pitch adjusted hydraulically. Planetary transmission to blades with ratio of 1.61:1.

Tanks: One protected tank with 920 liter capacity carried in fuselage. One protected auxiliary tank of 250 liter capacity in fuselage. One oil tank of 60 liter capacity carried in each wing.

Performance
Maximum speed = 380 km/h at 3,400 m altitude (4,200 m)
Cruise speed = 350 km/h at 4,500 m altitude (4,000 m)
Speed at best sustained climb rate = 160 km/h (at low altitude)
Landing speed = 100 km/h
Glide ratios:
High-speed configuration, landing gear retracted 12.9
Flaps in landing configuration, landing gear extended 8.5

Dimensions

Wingspan	22.5 m	Wing area	82 m^2
Height	4.7 m	Takeoff weight	8,500 kg
Length	17.48 m	Wing loading	100 kg/m^2
Wheel track	3.14	Power loading	4.74 kg/hp (5.05 kg/hp)

Of interest is the design and layout of the ventral "dustbin" turret (*C-stand*) situated between bulkheads 10 and 11. In its retracted position it functioned as the radio room for the on-board radio operator, but when extended it served as a gun turret. Such a design for a modern bomber, even at the time, can only have been considered of limited value in the hands of crews during combat operations.

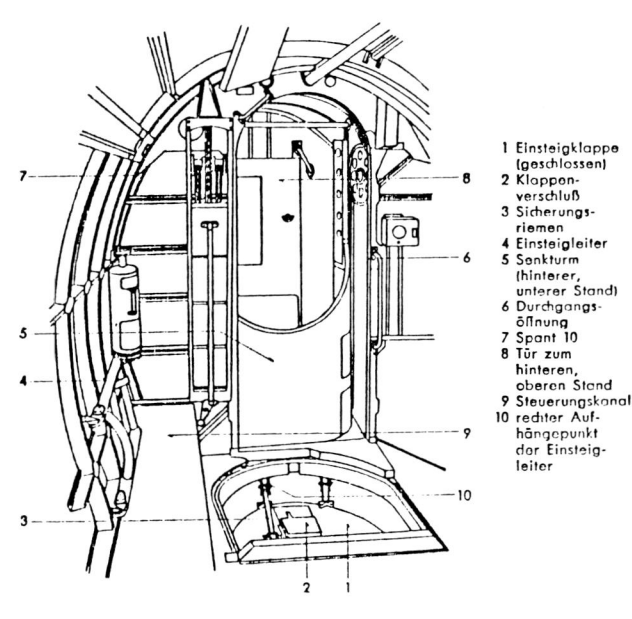

1. Einsteigklappe (geschlossen)
2. Klappenverschluß
3. Sicherungsriemen
4. Einsteigleiter
5. Senkturm (hinterer, unterer Stand)
6. Durchgangsöffnung
7. Spant 10
8. Tür zum hinteren, oberen Stand
9. Steuerungskanal
10. rechter Aufhängepunkt der Einsteigleiter

Junkers Ju 86G - boarding hatch and ventral gun position.

History of German Aviation: Bombers and Reconnaissance Aircraft

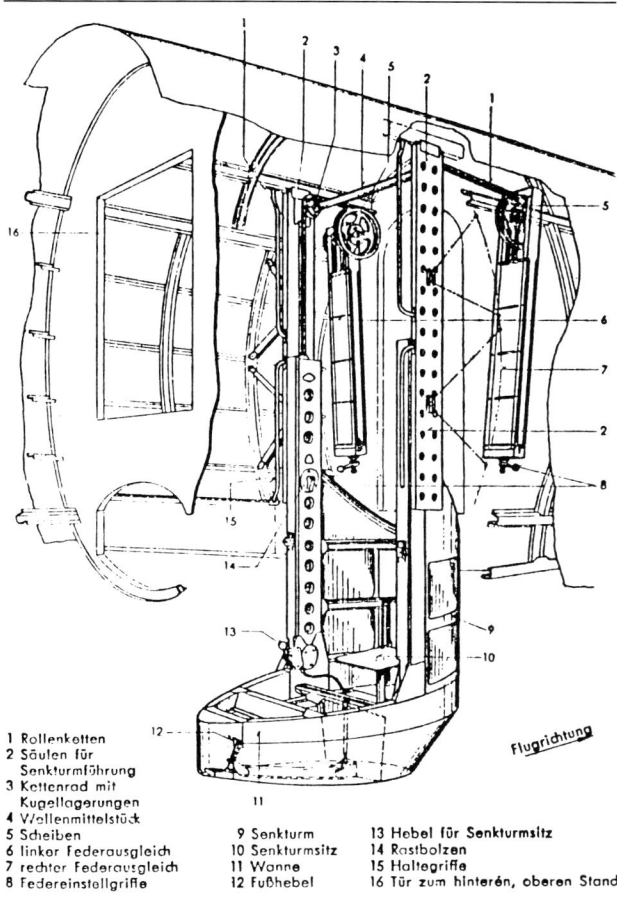

Junkers Ju 86G - ventral gun position extended, showing identifying features.

1 Rollenketten
2 Säulen für Senkturmführung
3 Kettenrad mit Kugellagerungen
4 Wellenmittelstück
5 Scheiben
6 linker Federausgleich
7 rechter Federausgleich
8 Federeinstellgriffe
9 Senkturm
10 Senkturmsitz
11 Wanne
12 Fußhebel
13 Hebel für Senkturmsitz
14 Rastbolzen
15 Haltegriffe
16 Tür zum hinteren, oberen Stand

By the time Ju 86 series production ceased in the early summer of 1938, Junkers had built a total of about 390 Ju 86s, mostly bombers. This number includes those assembled in Sweden, but overlooks the aircraft which were built under license in that country.

Heinkel He 111 Bomber

In the spring of 1934 the RLM, with the blessings of *Deutsche Lufthansa* (DLH), awarded contracts to the Heinkel and Junkers companies for building five prototypes each of a twin-engined aircraft. The design was to be capable of functioning both as a high-speed passenger liner, as well as a medium bomber. Although the DLH airliner contract may possibly have been easier to fulfill from a time standpoint, it became quite clear at an early stage that the focus would be on the military version. It was decided to start with the bomber, then alternate between civil and military versions, exactly as had been the case with the Ju 86's three military and two civilian prototypes.

Of the exceptional designers at the Heinkel company, Siegfried Günter and his twin brother, Walter (who passed away in September of 1937), it was Siegfried who was given overall responsibility for the Heinkel He 111 project (although both worked extensively on the design). Siegfried was noted for his earlier creation of the Heinkel He 70. The director of the Heinkel design department, Karl Schwärzler, created the final design from the draft, which had leaned heavily on the He 70. Prior to acceptance, models had been subjected to thorough wind tunnel testing and aerodynamically refined accordingly.

Synopsis

All in all, outfitting the *Luftwaffe* with the Ju 86 bomber version peaked in the fall of 1938 with a maximum of 235 aircraft of all variants, of which 159 were Ju 86A and Ds, 43 were Ju 86Es, and 33 were Ju 86Gs. From that point on the numbers of Ju 86 bombers dwindled at a steady pace, and from 1939 this pace increased markedly to make room for the Heinkel He 111 and Dornier Do 17 bombers. By the time war broke out, only a single IV *Gruppe* was still in operation, serving as a conversion unit for retraining young crews in their bomber duties. Therefore, it could not be considered a front-line unit. It was equipped with about 30 Ju 86G-1s, with other Ju 86s serving in the C-class flight training schools to familiarize apprentice bomber pilots with the flight handling of the heavier front-line bomber types.

First Prototypes

The airplane, conceived as an all-metal cantilever low-wing design, radiated elegance and beauty even from the drawing board with its harmonious lines and elliptical wings.

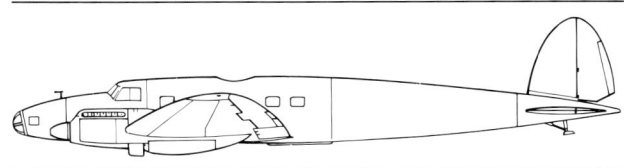

Heinkel He 111a (later He 111V1) minus weapons and ventral gun position; maiden flight completed on 24 February 1935.

The first prototype, the He 111a (later designated V1 with the newly introduced numbering system) completed its maiden flight on 24 February 1935 at Rostock-Marienehe with Heinkel test pilot Gerhard Nitschke at the controls. This was almost four months after its competitor, the Junkers Ju 86, had taken to the air for the first time.

From its first flight on, the He 111's flight handling characteristics were considered completely satisfactory, particularly its landing qualities.

Nevertheless, experimentation was carried out on the wing design which, like the Ju 86, had a tendency to stall out at higher angles of attack. Its length was reduced from 25 m on the V1 to 23 m on the civilian V2 and V4, which had an area of 88.5 m^2, then to 22.60 m and a wing area of 87.6 m^2 for the military V3 and V5. The purely elliptical wing shape of the V1 saw the trailing edge on subsequent prototypes generally straightened out, and the wing tips were rounded to a varying extent. Initially, the He 111 prototypes were fitted with the BMW VI 6.0 Z series of engines, which proved too weak to power a design that had proven robust from the outset. The BMW VI had been around in one form or another since 1925, and was a reliable water-cooled twelve-cylinder V inline engine with an output of 485 kW/660 hp driving a three-bladed variable-pitch propeller. Only the He 111c (later V2) was initially fitted with a two-blade fixed pitch propeller. Flight performance was considered relatively marginal to poor, although it was somewhat better than the Ju 86 prototypes with diesel engines. Counting as they were upon more powerful engines anticipated by the contractor party, the He 111 prototypes were felt to offer plenty of opportunity for further development.

The gun stations had not yet been installed in the He 111V1, whose installation areas were covered over in sheet metal. Only the nose, with its bomber-like glazing, gave any hint as to the type's future combat role. During its first evaluation period the V1, despite its relatively weak engines, reached a maximum speed of nearly 350 km/h, a ceiling of 5,400 m, and covered a maximum radius of almost 1,500 km. These figures were in no small part thanks to its aerodynamic shape and limited equipment carried on board.

A short four weeks after the maiden flight of this first He 111, the civilian V2 (originally the He 111c) took to the skies. The V2, and the V4 which subsequently flew in late 1935, were eventually handed over to the DLH and flown to Berlin-Staaken where *Lufthansa* carried out its own familiarization and long range testing program. On 10 January 1936 at Berlin-Tempelhof the public was given a most convincing demonstration of the V4, flown yet again by Gerhard Nitschke. On 1 July 1935 Nitschke had been promoted to chief test pilot of the Heinkel works. The V4 was the prototype for the commercial airliner version of the He 111 built later, although in relatively few numbers compared with its military cousin.

The He 111V3 (originally the He 111b) was finished as a bomber prototype shortly after the civilian He 111c. It became the prototype for a pre-production batch of ten He 111A-0 bombers contracted for by the RLM in the latter half of 1935.

Independent of the development of these four V-types, Heinkel had taken the He 111V5 with the same airframe as the military V3 and anticipated the installation of more powerful engines. The company was successful in procuring two samples of the Daimler-Benz DB 600 aircraft engine, with an output of 735 kW/1,000 hp, and began flight testing with these in the early part of 1936.

The V5, weighing in at 8,600 kg with its more powerful engines, reached a maximum speed of 360 km/h and a best cruising speed of just over 340 km/h with a full combat load. Thanks to the DB 600 engines, these performance figures led to the He 111V5 becoming the prototype for the large-scale production of the He 111B later contracted for by the RLM.

A-Series

The He 111A-0 was the first pre-production series, and from it was diverted the He 111A-02 and -03 in the early spring of 1936 to the *Luftwaffe's* flight test center at Rechlin, where they underwent official flight and weapons trials. For these, they were fitted with three gun stations for flex-mounted MG 15s, with the *C-Stand* in the ventral position being an extendable turret in the same vein as the design found on the Ju 52 and Ju 86. The A-series had a payload capacity of 1,000 kg of bombs, the same amount as the Ju 86. Performance, however, proved somewhat disappointing at full military weight, which had now increased to 8,220 kg over the 7,700 kg takeoff weight of the V3. With a full load the He 111A-0 became cumbersome and tail heavy when compared with the prototypes. At its optimum flying altitude and with maximum engine output and retracted ventral "dustbin" turret it reached a maximum speed of barely 310 km/h and a best cruising speed of 270 km/h.

Based on these performance figures the test center returned a verdict against the He 111A-0 as unsuitable for *Luftwaffe* service. The A-series did not see further development.

E-Stelle Rechlin's rejection of the A-series pre-production models did not affect the Heinkel company particularly much, as the better performing He 111V5 had been given a favorable assessment in the interim. This was soon followed by a contract for 300 He 111B models. Experience gained in manufacturing the pre-production batch had been put to good effect for the large-scale production contract.

B-Series

For the B-series, too, there was first a pre-production batch of ten He 111B-0s, the first examples of which went to the test center at Rechlin in the fall of 1936. During flight testing, it was discovered that the control of the aircraft around its longitudinal axis left something to be desired with regard to the aileron effect in certain speed envelopes. This defect was quickly corrected, along with a few modifications to the combat gear. In a remarkably short period of time the type was certified for use by the *Luftwaffe*, and the first of 300 production machines left the assembly line as the He 111B-1 variant. They reached the bomber units beginning in January 1937, where they replaced the lethargic and cumbersome Dornier Do 23G and the Ju 52 auxiliary bomber.

By now, the He 111B-1 had a maximum weight of 9,315 kg and was fitted with the DB 600Aa, a 735 kW/1,000 hp low-level blower engine. Subsequent sub-variants in the series, however, received the less powerful DB 600C across the board. Although it only had an output of 625 kW/850 hp, it was fitted with a boost control which gave it virtually the same performance at 4,000 m as at low levels.

Although still not entirely satisfactory from a performance standpoint, four machines from the B series pre-production batch made their initial debut with the bomber evaluation squadron in Spain, VB/88, as early as late 1936. These were followed in February 1937 by two full bomber squadrons from *Kampfgruppe* K/88 with a total of 30 He 111B-1s. These have been described previously, and favorably, in the chapter "Proving Grounds of Spain," which related how the combat experience of these machines was given an overly optimistic assessment with an eye toward the future. Among the lessons learned was to extend the "dustbin," the *C-Stand* turret, in combat only when enemy fighters were in the immediate area, as the drag caused a noticeable loss in speed and made formation flying difficult.

In May 1937 the Marienehe plant's aircraft output was supplemented by the newly built Heinkel Works at Oranienburg. Both factories were now producing the B-2 sub-variant, which was fitted with the previously mentioned DB 600C supercharged engine with 625 kW/850 hp output and the DB 600G with 700 kW/950 hp output, which also had nearly the same performance at its maximum pressure altitude. Due to the cooling problems which had

Heinkel He 111B-2 (built from May of 1937).

cropped up with the B-1, the B-2 was given supplementary surface coolers mounted on the wing leading edge on both sides of the engines.

Each new production model of the DB 600 aviation engine was at once followed by He 111 design modifications to accommodate it. As soon as the DB 600Ga (770 kW/1050 hp takeoff output and 700 kW/950 hp at altitude) became available in 1937, an He 111B-0 was immediately fitted with one and went into flight testing as the He 111V9 in the summer of 1937, at first retaining the additional surface radiators of the B-2. With the flight test results fully acceptable, the V9 became the pattern for a new He 111D series. This new series was to be equipped with the same DB engines, although in the interim the entire cooling system was given a complete makeover with the goal of eliminating the drag inducing surface radiators.

D-Series

In the fall of 1937 the first pre-production D-series model, an He 111D-0 with DB 600Ga engines, entered flight testing alongside the He 111V9. The surface radiators had by now disappeared, and in their place the ventral radiators beneath the engines, now semi-retractable, were enlarged. The individual exhaust ports were combined into one on both sides, and the engine housing design was refined. The machine's total weight was now at 8,810 kg, and the official performance increase brought about by this more powerful engine was remarkable. Maximum speed at an altitude of 4,000 m was 410 km/h, with extended ventral turret 370 km/h—the same speed the B-2 had with its turret in the retracted position. Near the end of 1937 a planned joint production of the He 111D-1 at the Marienehe, Oranienburg, and Dornier-owned Wismar facilities failed, however, shortly after the first D-1 machine had been completed, because a bottleneck in powerplants began to call into question the entire He 111 program; the Bf 109 and Bf 110 fighter programs had priority with regard to the DB 600 engines.

Recognizing this bottleneck situation, Heinkel had already attempted to find an alternate engine, and flight testing began in the first half of 1937 with the fuel injected version of the Junkers Jumo 210Ga (volume of 19.71:l) in a He 111B-0 airframe. This He 111, designated V6, was from a powerplant standpoint clearly handicapped, as the Jumo 21G had an output of just 535 kW/730 hp; it was not capable of even approaching the performance of those He 111s fitted with the DB 600.

In the meantime, the larger version of the Jumo 210 had become available. The Jumo 211 (with a volume of 35 liters) in its A-1 variant had a takeoff rating of 735 kW/1,000 hp and still retained 706 kW/960 hp at an altitude of 1,500 m.

The V6 was switched over to this more powerful engine without delay. Satisfied with the results of flight testing the Jumo 211, the RLM ordered that the He 111D-1 production be stopped at once so that the program could shift to a new Jumo 211 equipped E-series of the Heinkel He 111. Those few He 111D-1s built prior to this decision never made it to the field units, instead being retained as flying testbeds. The actual testbed for the future He 111E was a modified He 111D-0, which flew successfully as the He 111V10. The test aircraft for the Jumo 2111, the He 111V6, was given to the Junkers company for testing its new line of variable-pitch propellers.

E-Series

In January 1938 the first He 111E-0s of the pre-production batch were available and kitted out with the Jumo 211A-1. By February 1938 the first He 111E-1s had followed, having a total weight of 10,600 kg and capable of carrying a bomb payload now increased to 2,000 kg internally. In March 1938 the first of what was to be 45 He 111E-1s began arriving in Spain, so that the two B-2 equipped squadrons and the remainder of K/88, i.e. a total of four squadrons, might convert over to the new model.

Identified by several, mostly internal, equipment modifications, an E-2 sub-variant was soon replaced by production of the E-3, under which designation the series continued on. An E-4 followed as a variant capable of carrying a 1,000 kg bomb internally and 1,000 kg on an external rack under the fuselage.

At the same time an E-5 sub-variant series was built having the same layout, although with a supplemental 835 liter fuel tank in the fuselage in place of the port side bomb racks. With this, the range of the E-5 increased to 1,820 km. Of the E-series, the E-3

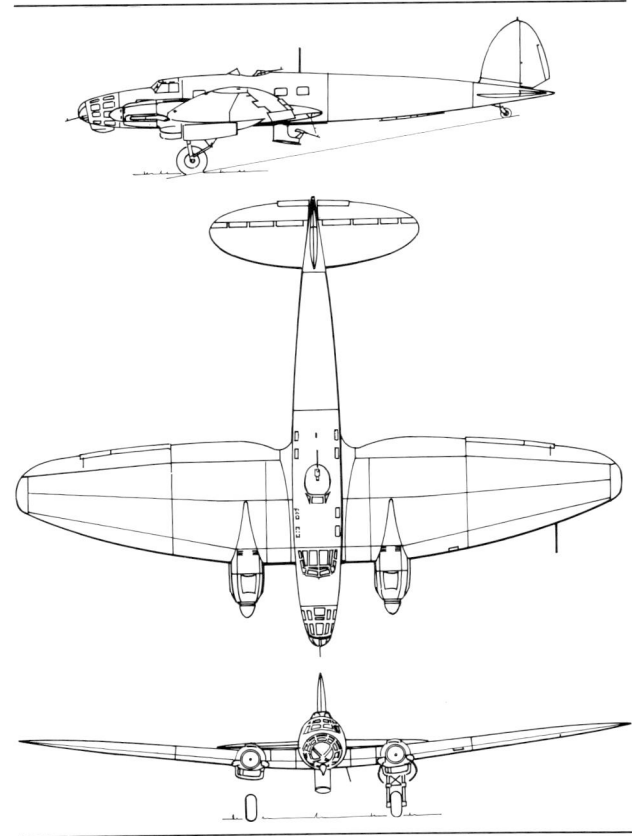

Heinkel He 111E-3 three view, showing the curved leading edge of the wings and ventral gun position (1938).

Heinkel He 111E-4 - an atypical example modified for courier duties.

was built in the greatest numbers, while the E-4 and E-5 only saw limited numbers produced. The E-3 reached its maximum speed of 420 km/h at 4,000 m, flying a maximum 325 km/h at sea level. Its greatest range with a 1,000 kg bomb load was 1,500 km, and it had a service ceiling of 7,200 m.

F-Series

The F-series was characterized by a new wing planform for the He 111. Although quite elegant in appearance, the elliptical shape of the wing was not altogether that easy to create, and as early as 1936 there were studies undertaken regarding a simpler technological wing design while retaining its area and span. In the summer of 1936 a new wing with a straight leading edge was tested on an appropriately modified He 111B-0, with the test aircraft being designated the He 111V7.

The new wings, which proved themselves most effective relative to both the He 111's flight handling characteristics, as well as from an aerodynamic perspective, were initially only introduced on the civilian He 111 types (the G-series). The bomber version, however, was postponed indefinitely at the behest of the *Technisches Amt*, which felt that the delays incurred on the production line by such modifications were unacceptable in the autumn of 1936. It was not until the summer of 1937 that the matter of wing standardization for the bomber was seriously broached. The forerunner of the F-series became an He 111B-1 with its DB 600C engines, fitted with the new wing and flown for the first time as the He 111V11 in July of 1937. This machine formed the basis for an He 111F-1 series, appearing in late 1937. This sub-variant was powered by the Jumo 211A-3, which had a takeoff rating of 809 kW/1,100 hp and offered roughly the same range as that of the He 111E-5, although with a maximum permissible takeoff weight of barely 11,000 kg. However, the *Luftwaffe* never took delivery of F-1 models, these being supplied exclusively to Turkey.

A batch of forty almost identical He 111F-4s were, in fact, supplied to the *Luftwaffe* in the spring/summer of 1938. These had the bomb configuration of the He 111E-4, meaning 1,000 kg carried internally and 1,000 kg externally, but unlike the F-1s delivered to Turkey they did not have a supplemental fuel tank in the fuselage.

Heinkel He 111F-4 (built from the spring/summer of 1938) - also with the ventral gun position, but now having a new wing design with a straight leading edge.

J-Series

In the summer of 1938 Daimler-Benz found itself in a better position to supply its engines, and the DB 600C and G versions once again became available for bomber aircraft. As a result, in conjunction with the He 111F-4 Heinkel was contracted to build an He 111J torpedo bomber designed around this powerplant, with two external hardpoints for aerial torpedoes. Thus, the pre-production He 111J-0s first built had no type of bomb bay.

Prior to starting with the J-1 series, however, plans for introducing the He 111J as a torpedo bomber were shelved, and it was decided to install a bomb bay after all. With the exception of its DB 600C/G engines, the 90 examples of the He 111J-1 delivered to the *Luftwaffe* over the course of the summer of 1938 were therefore virtually identical to the He 111F-4 being supplied at the same time.

P-Series

The He 111 was given a complete makeover beginning with this variant, which followed up on the standardization of the wing planform as a tapered design with straight leading and trailing edges (this latter introduced as early as the first V-types) initiated with the F-series. With regard to crew visibility, especially that of the pilot, the He 111 canopy up to this point was considered anything but ideal. As a result, a Heinkel team had been studying the most effective way to improve matters since 1937. The result of this work

gun position for the C-*Stand* was replaced by a ventral "tub," and the semi-enclosed *B-Stand* (dorsal gun position) for the radio man was now fully enclosed (beginning in 1938).

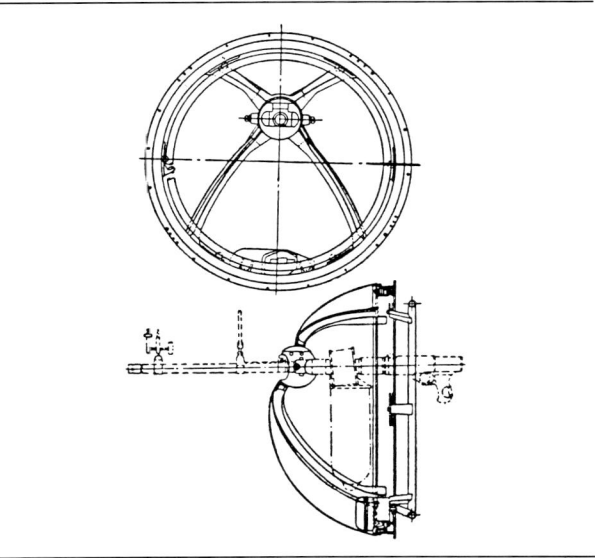

"Ikaria" gun position, from which the observer had a 120% forward field of fire.

was a fully glazed canopy with hemispherical, unbroken lines. The design underwent flight testing in a suitably modified He 111B-0 airframe as the He 111V8 starting in January 1938, with positive results.

The somewhat asymmetrically designed canopy had an Ikaria nose cupola offset to starboard to provide the bombardier with better operation of the cupola-mounted MG 15. The flight instrumentation was suspended from the canopy roof, so that no instrument panel would interrupt the pilot's forward and downward visibility, giving an altogether ideal view.

The canopy flooring was made of flat glazed panels to provide the bombardier with an undistorted view, while the remaining panels of the canopy were curved. This disadvantage was more than offset by the advantage of having a good all-around view. This redesigned and aerodynamically sound canopy layout resulted in the aircraft being shortened by 1.10 meters. In the spring of 1938 the new canopy was also fitted to the He 111V7, which two years earlier had also served as the test platform for the new tapered wings.

Another aerodynamic improvement to the V7—and a long overdue correction to the ventral gunner's position—was a blended ventral gondola in place of the awkward "dustbin." The gondola contained a ball socket mount for the rearward firing MG 15, which was capable of traversing through 90° and was fired by the gunner from a prone position. The semi-exposed dorsal gunner/radio operator's position also was finally fitted with full glazing. With all these improvements, some of which were based on experience gleaned in Spain, the V7 became the prototype for the new P-series.

The P-variant was equipped with the more powerful fuel injected DB 601Aa (860 kW/1,175 hp takeoff rating) from the beginning of its He 111P-0 pre-production batch. Production began at Marienehe in the fall of 1938 when the last of the He 111Js left the assembly line. Due to the uncertainty of the DB 601 engine's future availability, Heinkel developed the Jumo 211 powered H-series parallel to the P-variant. In any event, deliveries of the DB 601 continued to remain steady, and the RLM prioritized the P-series. Not only was it being manufactured at Marienehe and at Dornier's Wismar facilities, but was also produced at the Arado works in Warnemünde.

The He 111P-1 was supplied to the *Luftwaffe's* bomber units beginning in the early spring of 1939, when they started replacing the He 111B machines now considered to be past their prime. Performance figures for the P-1 were quite impressive, with a maximum speed of 475 km/h and a cruising speed of 7\370 km/h at an altitude of 5,000 meters. However, with a full bomb load these values dropped to a maximum of 325 km/h and 305 km/h, respectively! 2,000 kg of bombs could be carried inside the fuselage. Initially, efforts were made on the P-series to carry the bombs horizontally due to the better ballistic drop characteristics, but these were abandoned in favor of the standard vertical arrangement of the ESAC magazines due to the considerable structural modifications involved. Surprisingly, for what was quite an advanced version of the He 111 no effort was made to increase the armament, with the configuration remaining at the three flex-mounted MG 15s. In May of that year there appeared another sub-variant, the He 111P-2, which differed from the P-1 only in its improved radio system.

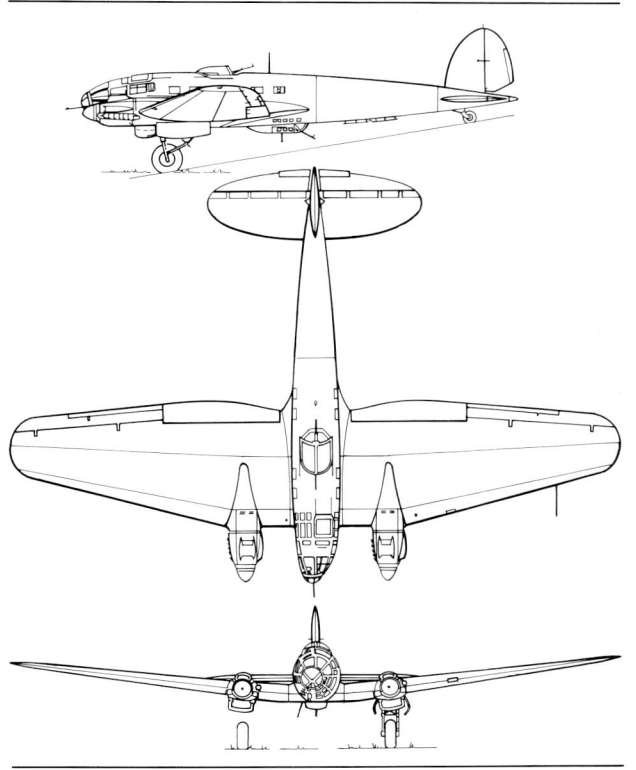

Heinkel He 111P-2, from May of 1939.

In the same month Oranienburg began rolling out H-series machines, followed a few weeks later by the Junkers plant in Dessau and the new ATG production facilities in Leipzig. By the time war broke out the P-series production had been overtaken by an energetically driven H-series output, so that by 2 September 1939 there were 400 He 111H models serving with the *Luftwaffe's* front-line units, compared to 349 P models.

H-Series

The parallel development of the H-series and P-series under a different set of priorities has already been mentioned. The prototype for the H-series, the V19 (modified V7), completed its maiden flight in January 1939, driven by Jumo 211 engines. Both the pre-production He 111H-0s, as well as the first variant tailored for field service (the He 111H-1), were completely identical to the He 111P-2—with the exception of the choice of powerplant, of course. At the beginning of the war the manufacturer was just switching over from production of the H-1 to the H-2, the latter being identical to the H-1 except for its more powerful Jumo 211A-3 series engine. The Jumo 211A-1 installed in the He 111H-0 and H-1 had a takeoff rating of 743 kW/1,010 hp, whereas the A-3 engine in the He 111H-2 increased this to 809 kW/1,100 hp.

Foreign Interest

China

After the failure of the He 111A-0 in early 1936 during *E-Stelle* Rechlin's evaluation and the contractor's rejection of the machines, the company found it quite easy to obtain the RLM's approval for exporting the pre-production series to China. Already a Heinkel customer—about a dozen He 50 dive bombers (Chinese export designation was He 66) had been sent to China in late 1935—a team of Chinese experts had also shown interest in the He 111 during a visit to Heinkel's Marienehe plant. Heinkel succeeded in selling all ten pre-production models to the Chinese government of President Chiang Kai-Shek. Once radio systems, bombsights, armament, and self-destruct systems had been removed, the He 111A-0s were disassembled and shipped by sea to the port of Canton. A few months later two were lost in combat with the Japanese during the second Sino-Japanese War, which had broken out in 1937. By and by, the remaining eight became victims of accidents at the hands of the inexperienced Chinese and foreign volunteer pilots.

Turkey

In 1938 the Turkish government expressed interest in the He 111. Following a demonstration by an He 111F-0 in Ankara the Turks, after a series of long, drawn-out negotiations, contracted for 24 He 111F-1 series aircraft equipped with two Jumo 211A-3s (809 kW/1,100 hp takeoff rating) and the additional 835 liter fuselage tank. They were designated as bombers and courier aircraft. and delivered to Turkey between summer and the end of 1938. The machines equipped three bomber squadrons and flew in air force service until 1946.

Synopsis

It is worth noting that Heinkel, together with the license-production companies, managed to build about 400 He 111s within the space of four months.

This, then, was the state of development and production of the Heinkel He 111 medium bomber in the fall of 1939, whose unexpectedly long continued service and further development will be charted later.

As of 2 September 1939 the *Luftwaffe's* front-line units operated 787 He 111s, i.e. 67% of the bomber inventory had been provided by the Heinkel company and its licensees, with 38 He 111E models, 400 He 111H, and 349 He 111P sub-types. With these numbers, the type formed the backbone of the operative *Luftwaffe*.

Dornier Do 17 Bomber

The third leg of the new generation modern bomber triad in the mid-1930s was the Dornier Do 17, when it was incorporated after the fact into the medium bomber category of the "*Rheinlandprogramm*" alongside the Junkers Ju 86 and Heinkel He 111 in 1936. The first production variant fully equipped as a bomber was the Do 17E-1, which flew in May of 1936. The Do 17 had gone through several interesting evaluation stages up to that point, and these will be examined here in the context of the type's various bomber versions. A later section will be dedicated to the great role of the Do 17 in aerial reconnaissance, with the type's multi-role capabilities also being addressed in the section on Do 17 reconnaissance aircraft.

Prototypes

The first three Do 17 aircraft in development in 1933 were conceived exclusively for *Deutsche Lufthansa* as high-speed passenger airliners and mailplanes. Later designated V1 through V3, these were unusually advanced all-new designs of a pleasant aerodynamic shape.

The Do 17V1 took to the air for the first time on 23 November 1934 with Dornier chief test pilot Egon Fath at the controls.

Those aircraft handed over to DLH for further evaluation were rejected. On the one hand, the two cabins planned for six passengers were too narrow and low—in addition to being most uncomfortable—and on the other hand, the Do 17 would have been too expensive to operate purely as a mailplane. The three airplanes were therefore returned to Dornier.

Around the middle of 1935 one of these prototypes was flown at Dornier's company airfield at Löwenthal by DLH pilot Robert Untucht, the test pilot and record-setter in the He 70, who was a former Dornier company man. He was quite impressed with the maneuverability and speed of the Do 17 and its water-cooled BMW VI6.0 inline engines (485 kW/660 hp). In his role as liaison between the DLH and the RLM he succeeded in stimulating the *Technisches Amt*'s interest in the promising Do 17. After test pilots

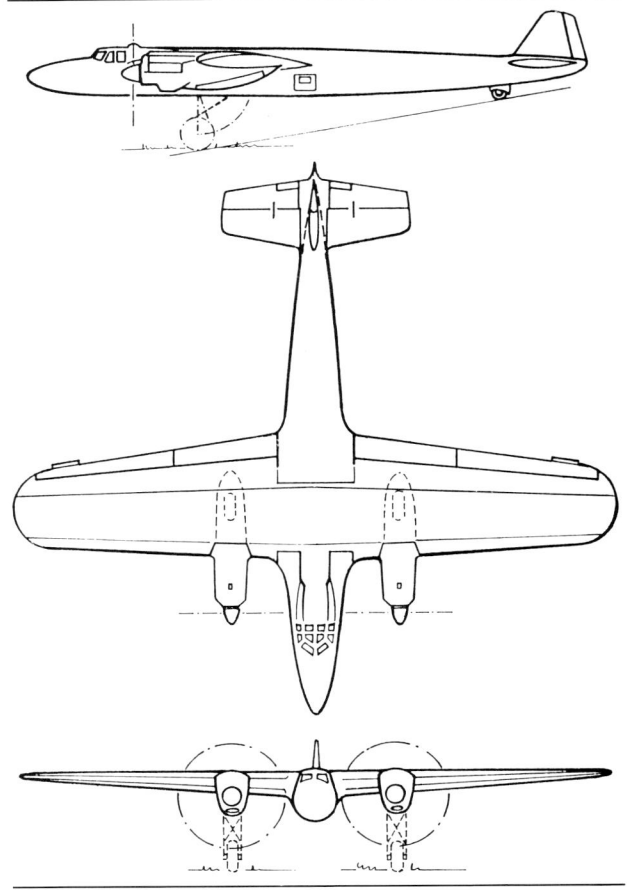

Dornier Do 17V1 (1934).

Dornier Do 17V1 seen shortly before its maiden flight (23 November 1934). This was the first "civil" prototype and, like the three subsequent V-models, was fitted with a standard rudder. It was not until the "military" Do 17V4 that the type was fitted with the characteristic dual rudders.

from the RLM had flown the machine and found the type to have considerable developmental potential, Dornier was awarded a contract for building a fourth prototype with a series of modifications for military evaluation.

The resulting Do 17V4 flew in the late summer of 1935. Externally, it differed from the V1 civilian prototype by having twin rudders. The single rudder of the Do 17V1 led to an instability around the yaw axis, which Untucht had also been critical of, whereas the V4 had much better lateral stability thanks to its new twin rudder arrangement. An improved field of fire may also have contributed to this solution, although there was little thought of armament initially given for this relatively fast machine. By dispensing with the constricted two and four man cockpit, the radio operator was given his own work area, and a bomb bay for a horizontally carried bomb load could be fitted.

Parallel to the V4 were built the V5 and V6 with the intent of expanding the test base. These two joined the flight test program beginning in the fall of 1935. While the V6 was practically identical to the V4—both were powered by the BMW VI—the V5 received two water-cooled Hispano-Suiza 12Ybrs twelve-cylinder inline engines, each rated at 570 kW/775 hp at sea level and 633 kW/860 hp at 4,000 meters altitude. This gave the V5 a top speed of 390 km/h.

It is worth noting that at that time the Royal Air Force was just beginning to field the Gloster Gauntlet biplane fighter, which could barely muster 370 km/h!

Despite its speed advantage, the Do 17 was indeed fitted with a flexible gun, the MG 15, in the following V7 prototype. The aft-firing machine gun was installed in a lens mount inside a bubble-shaped cupola on the upper fuselage and was operated by the radio operator. The Do 17V7 ultimately became the pattern for the Do 17E initial operational series mentioned earlier, in particular the Do 17E-2.

In all, 21 evaluation prototypes of the Do 17 were built up to the end of 1937, including civil, reconnaissance, and command versions. These will be explored over the next few pages, as they pertain to production versions of the bombers and reconnaissance aircraft for which they were evaluated.

Before continuing, we should perhaps mention the role of what at the time was designated the Do 17V8. According to official information, it was powered by the DB 600A, rated at 736 kW/2,000 hp, and was an improved prototype of the bomber version. With its more powerful engines it had been planned from the outset for the International Flying Meet at Zurich-Dübendorf, where it logged a top speed of 425 km/h—the highest speed recorded for all participating military aircraft.

Dornier Do 17V6 (1935) showing the dual rudder configuration.

The Dornier Do 17M-V1 (alias Do 17V8), here lacking its registration markings on the fuselage, was built as a competition aircraft. At the Zurich International Flying Meet in 1937, its "tailored" DB 600A engines (which were actually pre-production DB 601A-0 with an output of 805 kW/1,100 hp) powered it to a remarkable speed of 425 km/h, enabling pilot Polte to win the competition and attracting worldwide attention to the design. The M-V1 is unlike the subsequent M-series in that, aside from the different DB engines, it had a long, conservatively glazed nose. The Do 17M-V2 and M-V3, on the other hand, were powered by the Bramo 323A-0 radial engine (660 kW/900 hp) beneath NACA cowlings and had a somewhat shorter, rounder, fully glazed nose, making these the prototypes for the Do 17M-1 series of medium bombers.

In actual fact, this record-setting airplane had been powered by fuel-injected DB 601Ao pre-production engines rated at 805 kW/1,100 hp. For competition certification, however, it was given as a common carbureted engine (DB 600). The performance of the Do 17 at the time remains an impressive feat nonetheless.

The Do 17V8 was subsequently designated internally as the Do 17MV1, while the V8 designation was applied to the prototype of the Do 17F-1 strategic reconnaissance version.

E-Series

The canopy shape of the twin-engined mid-wing design of the prototype underwent numerous modifications to accommodate the three man crew. The first production version, the Do 17E-1, had a stumpy, semi-glazed nose, and instead of the BMW VI made use of the somewhat more powerful BMW VId-series with a 7.3 compression ratio (552 kW/750 hp takeoff rating). In early 1936 Dornier began making preparations for full-scale production of this model at its Allmansweiler, Löwenthal, and Manzell bei Friedrichshafen facilities. To this end, the airframe was broken down into its main components, consisting of the actual fuselage, the cockpit area, the control surfaces, and the wings, thus making it easier for sub-contractors to be included in production operations.

At first, this bomber version also had only one MG 15, although this was soon joined by another. This gun was mounted in the cockpit's windscreen, and could be locked for fixed sighting by the pilot or unlocked and made into a flex-mounted forward firing gun. On 20 May 1936 the Do 17E-1 completed its maiden flight, and production quickly followed once minor problems with some of the sub-contractors were alleviated. Field unit testing for this airframe began in early 1937.

Combat evaluation of the Do 17 in Spain was risked as early as 1936, when four machines were detached to VB/88. There followed 20 Do 17E-1s for the Condor Legion's *Kampfgruppe* K/88 in 1937. To be sure, the Do 17E carried just 500 kg of bombs in combinations of 10 SC 50s, 4 SC 100s, or two SC 250s, but over short distances this load could be increased to 750 kg. Unloaded, with a maximum speed of 356 km/h it was capable of running away from most of the enemy fighters it encountered, yet even here its lack of suitable firepower was occasionally felt. At maximum military load the E-1 had a tactical penetration range of 500 km, and in an unloaded configuration reached a service ceiling of 5,100 m.

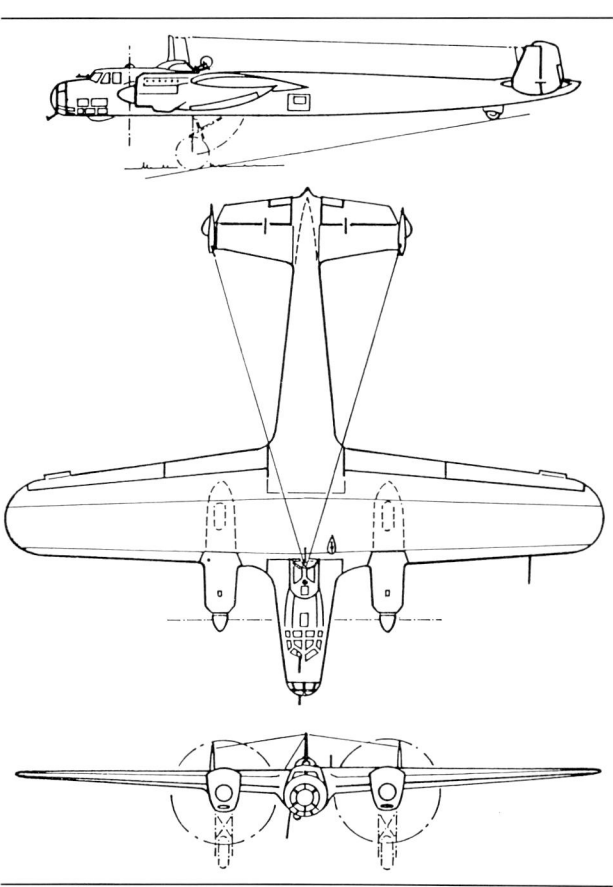

Dornier Do 17E (1936).

A Dornier Do 17E on a compass swinging base, designed to determine and compensate for magnetic interference caused by airframe components and other sources (the difference between magnetic and compass headings).

M-Series (with L- and R-variants)

The Do 17MV1(V8) has already been discussed; two additional V-types were the Do 17MV2(V13) and the MV3(V14), both powered by nine-cylinder radial engines. The MV2 was equipped with the Bramo-Fafnir 323A1 engine, which provided 660 kW/900 hp on takeoff and 736 kW/1,000 hp at 3,000 meters, and the Do 17MV3 had the Bramo 323D fuel-injected engine built by the Brandenburgische Motorenwerke (formerly Siemens). The MV2 served as a testbed for evaluating the K4Ü, whereas the Do 17MV3 became the actual pattern for the M-series, which was equipped standard with the Bramo 323A with the same performance as for the MV2. The first of 200 production machines flew its maiden flight on 7 April 1937.

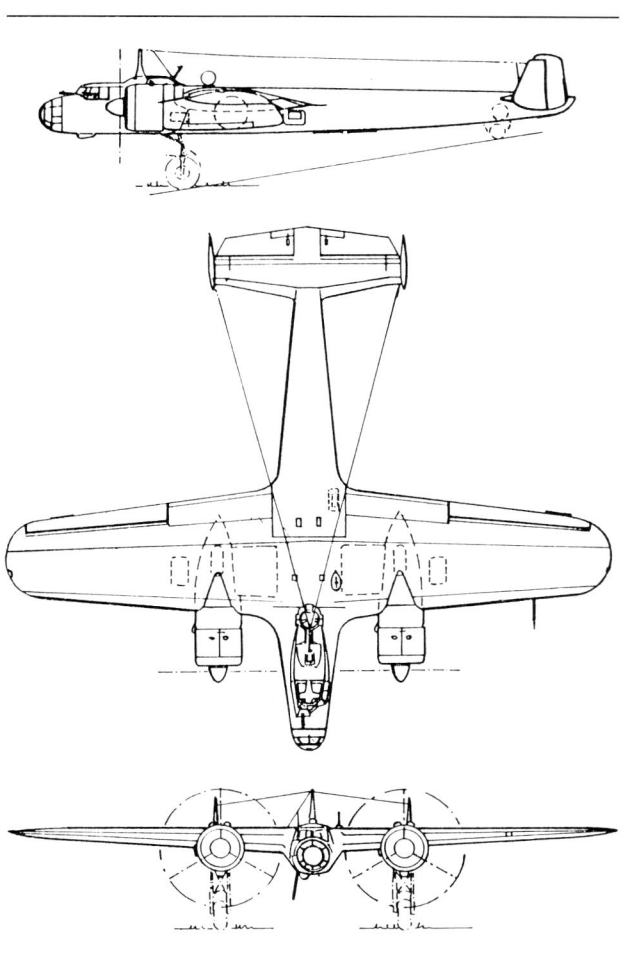

Dornier Do 17M (1937).

The Do 17M-1 had a top speed of 365 km/h, a tactical penetration range with maximum combat load (1,000 kg bomb) the same as that of the E-1, i.e. 500 km, and a service ceiling of 6,700 m. While the "E" had a takeoff weight of 7,040 kg, for the M-series this had climbed to 8,135 kg. The wings of the Do 17M were covered entirely with Dural (earlier models of the Do 17, including the E-series, were covered in fabric on the wing undersides), the undercarriage retracted electro-mechanically (or manually in emergencies), the wheels braked hydraulically, the tailwheel was retractable, and two armored fuel tanks in the wings and one in the fuselage provided a total capacity of 1,910 liters.

Over the course of 1938 the "M" gradually began replacing the E models in the bomber units, and as of 19 September of that year these units had 328 Do 17Es and already 102 Do 17Ms in their inventory. This meant that the Do 17 comprised 35% of the *Luftwaffe's* bombers at the time. Two other parallel developments to the M-series—the reconnaissance versions will be discussed later—are worth mentioning here:

1. An L-series, conceived as a pathfinder/command and control plane with a fourth crew member, of which only two were built (the Do 17LV1, or V11, and the Do 17LV2, or V12) and
2. The R-series, designed as a high-speed bomber, which also included only two airplanes:
- the Do 17RV1, initially fitted with BMW VI engines, then converted to DB 600G (772 kW/1,050 hp), served primarily as an evaluation platform for bombsights and powerplants, and
- the Do 17RV2 with the DB 601A (809 kW/1,100 hp) also served chiefly as an engine testbed. However, with the shift of Daimler Benz engines to fighter production the series production of the R model was canceled.

Yugoslav Interest in the Do 17K

Following a visit by a team of experts at the 1937 International Flying Meet in Zurich, Yugoslavia was so interested in the Do 17 bomber that the Yugoslav government soon began negotiations with their German counterpart with the intent of acquiring this aircraft for its Royal Yugoslav Air Force. Just two months later came a purchase order for 20 Do 17Ks (the designator for the Yugoslav export model), followed by the simultaneous acquisition of a copy license by the Yugoslavs.

For driving the Do 17K, they supplied the air-cooled French Gnôme-Rhône 14 N1/2 radial engine, built under license at Rakovica, near Belgrade. The first of these Do 17Ks made its first flight on 6 October 1937 at Manzell, and on 25 October 1937 was personally flown to Belgrade by Dornier's chief test pilot, Egon Fath. The Do 17K was supplied in three versions, all of which had the longer, somewhat boxier canopy of the MV1, and built under license in Yugoslavia from 1939 onward:

- Do 17Kb-1 for the pure bomber role with a 1,000 kg payload
- Do 17Ka-2 and Ka-3, mainly in the reconnaissance role fitted with various aerial photography systems, although could operate in a dual role as a bomber with reduced payload.

The Gnôme-Rhône engines provided 721 kW/980 hp at 4,500 meters and gave the Do 17K a top speed of 357 km/h at sea level and 417 km/h at an altitude of 3,500 m; maximum range for the reconnaissance version was 2,400 km. Its armament was much heavier than that of its contemporary German Do 17.

License production of the Do 17K began at the state-owned aircraft factory in Kraljevo during the course of 1939, with the Yugoslav Air Force taking initial deliveries of the type in early 1940; 40 to 50 Do 17Ks were built there under licence.

When German forces advanced into Yugoslavia on 6 April 1941 the Yugoslav Air Force had 60 Do 17Ks, whose airfields became the main targets of German air raids. Almost half of these machines were destroyed in the first attack.

The remaining few Yugoslav Air Force Do 17Ks captured by German troops were repaired as needed and provided to the newly formed Croatian Air Force in early 1942. They were later supplemented by a small number of Do 17Es.

Z-Series (with S- and U-variants)

The Z-series, along with the S- and U-, was characterized by a completely new, larger, and more angular glazed canopy. In early 1938 a Dornier team began working on a new cockpit tailored exclusively to operational requirements. The various preceding versions had proven too narrow for the three man crew to effectively carry out their functions, and the problem of inadequate armament was difficult to resolve due to the space constraints of previous designs, primarily because there was simply no room for an additional gunner.

Thus, with what may have been a general disregard for an aerodynamically aesthetic planform, was created a cockpit—or more appropriately, a forward fuselage section. Its features included a raised, glazed canopy having a *B-Stand* gun mount, a deeper gondola extending back even with the wing root, with a ventral rearward firing flex-mounted machine gun operated by a fourth crew member—the gunner/flight engineer. The fully glazed nose canopy had angular lines and consisted of a number of small flat glass plates.

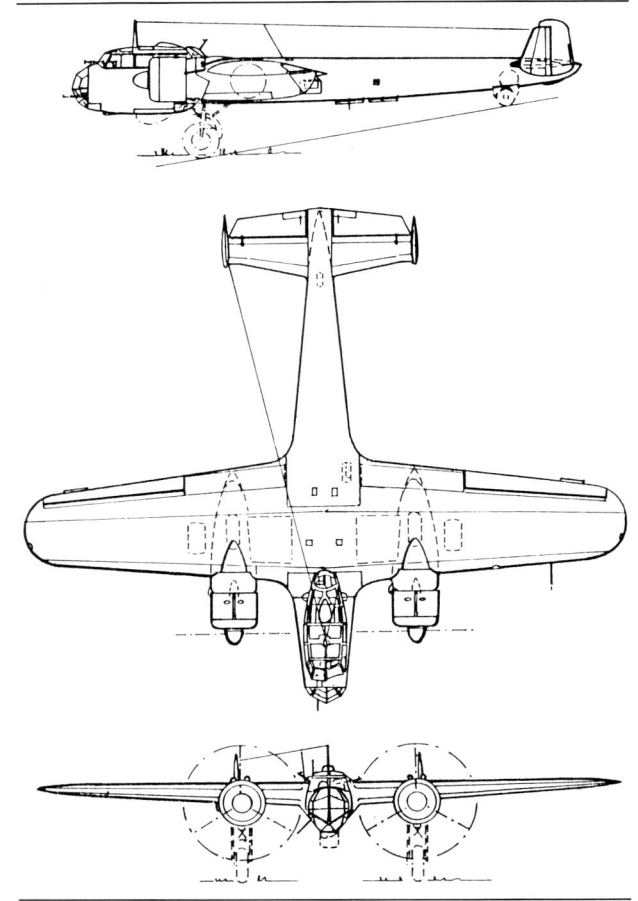

Dornier Do 17Z (1938).

However, the new cockpit was not first immediately fitted to a Z-series machine, but to a high-speed reconnaissance version known as the Do 17S-0, of which only three examples were built. The DB 600G engine used in the type was no longer available—fighter production took priority!

A pathfinder/command aircraft version begun at the same time was the Do 17U, whose three Do 17U-0 pre-production machines and 12 Do 17U-1 production aircraft were also fitted with the new cockpit to accommodate the five to six-man crew. The U-series had an additional second radio operator on board especially for navigational functions, a work area for the formation commander, and was assigned to the command *Ketten* within the Do 17 bomber groups to act as a formation command ship.

From an engine standpoint, the Do 17U—like the M-series—was powered by the Bramo Fafnir 323A, weighed a maximum of

8,505 kg, had a maximum speed of 417 km/h, and a service ceiling of 6,000 meters. Its fuel capacity gave it a range of 2,945 kilometers.

But let us now turn to the Dornier Do 17Z, the most advanced bomber version of the Do 17 prior to the outbreak of the war.

With its new cockpit, the version was produced in five bomber sub-variants beginning in 1938, all powered by the Bramo Fafnir 323—the Do 17Z-1 with the 323A (668 kW/900 hp), and the others with the more powerful Bramo 323P rated at 736 kW/1,000 hp.

Generally, the Do 17Z was similar to the Do 17M, but like the S- and U-series had the new forward fuselage section with the all-around view canopy and a four-man crew.

The Do 17Z-1 completed its maiden flight on 1 March 1938. Production took place at the Dornier plant in Munich (Oberpfaffenhofen), as well as at three license production facilities. Whereas the pre-production batch, the Do 17Z-0, still was armed with just three MG 15s, the Z-1 series was initially fitted with a fourth MG 15 in the nose. This was later brought up to the standard of the Do 17Z-3's armament discussed shortly. Nevertheless, its payload capacity was restricted to 500 kg.

The Do 17Z-2 version was designed with the advent of the more powerful Bramo 323P. With this engine, the plane had a take-

A Dornier Do 17Z dropping a stick of 50 kg bombs. The advantage of carrying the bombs in the bay in a horizontal position, a feature of all Do 17 versions (as well as the later Do 217), lies in the fact that the bombs' trajectory has only a minimal trail angle in comparison with bombs carried in the vertical position (such as on the Heinkel He 111), which first must nose over into the trajectory path (taking some five seconds to do so) and therefore incur a greater trail angle.

Dornier Do 17Z, the most advanced bomber version of the Do 17 to see service prior to the Second World War, showing its fully glazed, multi-panel nose (from 1938 onward).

off weight of a maximum of 8,840 kg, a top speed of 421 km/h, and a service ceiling of 6,900. It was armed with six MG 15s.

The bomb capacity of the Z-2 was—as with the M-series—yet again a maximum of 1,000 kg, with 20 SC 50s or 4 SC 250s being a typical load. However, its tactical penetration range with maximum combat payload was just 330 km at normal fuel capacity; with an additional 896 liter fuselage fuel tank and reduction to 500 kg payload, this range increased to 1,150 km. This example highlights the interdependence of fuel quantity and bomb load at a given maximum permissible weight. The bulk of Do 17Zs built were of the Z-2 variety.

Technical Data and Details of the Do 17Z-3

The Do 17Z-3 model described below is possibly the most interesting and versatile version of the Z-series, and should therefore be examined in greater detail. It, too, enjoyed what was at the time considered to be adequate armament in the form of six MG 15s, whose arrangement can be seen in the diagram on the following page. Some Z-3s were subsequently refitted with the heavier 15 mm MG 151 in the nose. It could carry 1,000 kg of bombs, wherein it had racks for two SC 500s (*Beladefall I* in the following drawing). With a fuselage tank (*Beladefall II*) its payload potential fell to 500 kg with a corresponding increase in range, as mentioned in

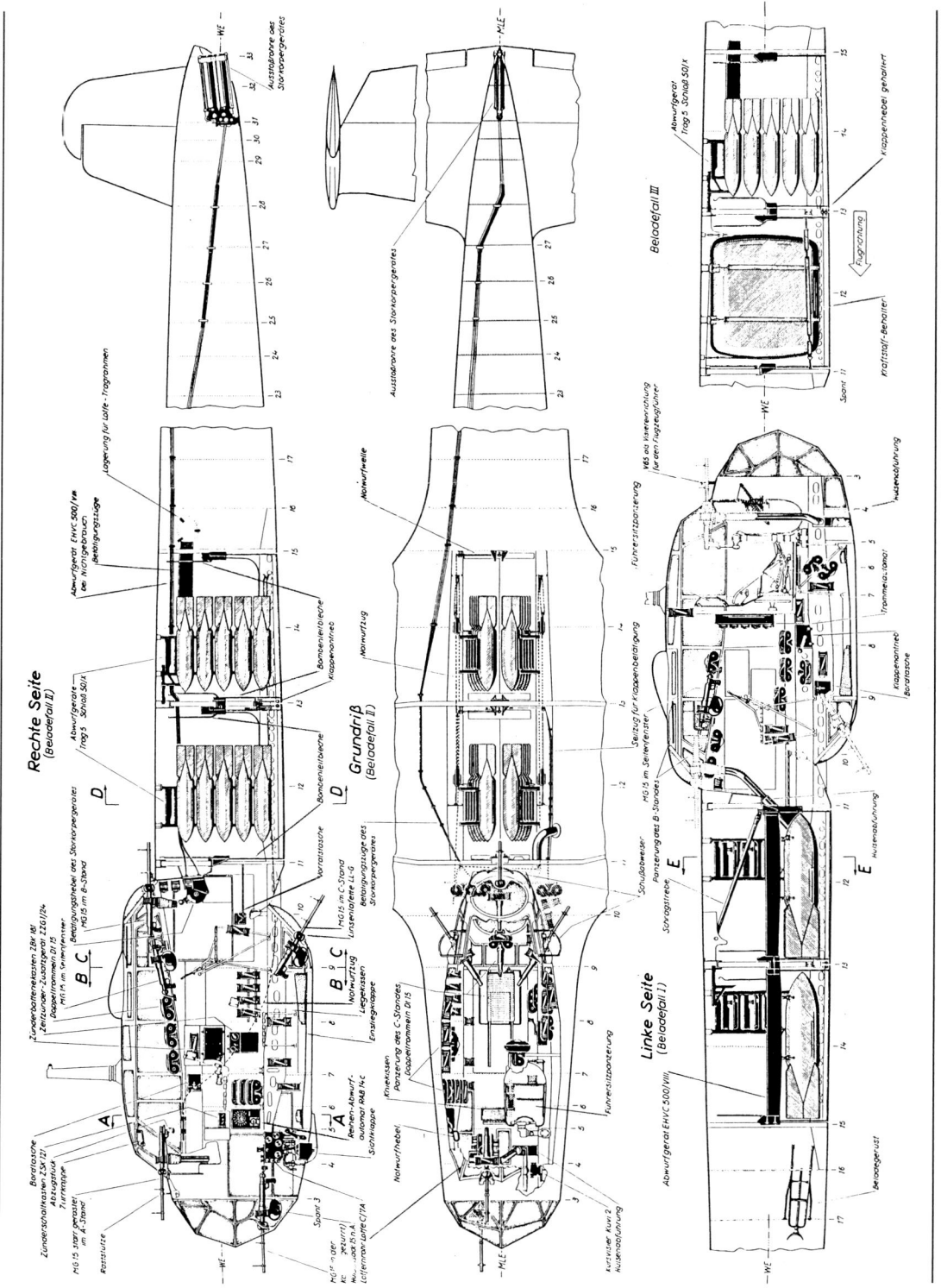

Dornier Do 17Z-3, cutaway showing the armament and ordnance carried.

Do 17 Z
Übersicht der Bewaffnung
(Baureihe 3)

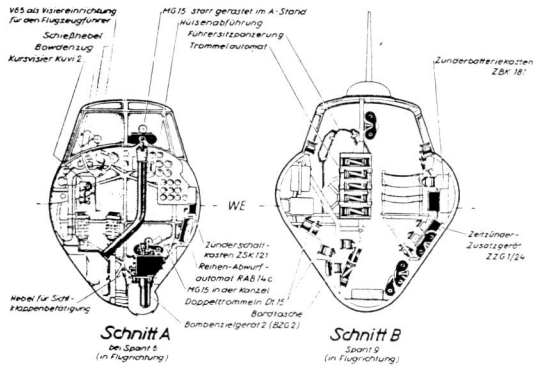

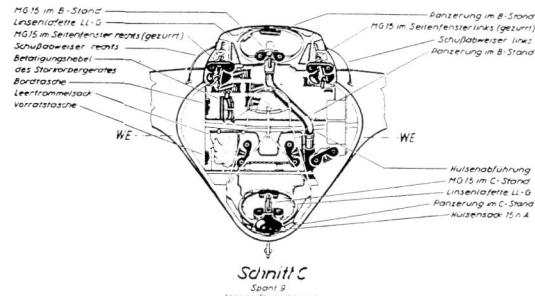

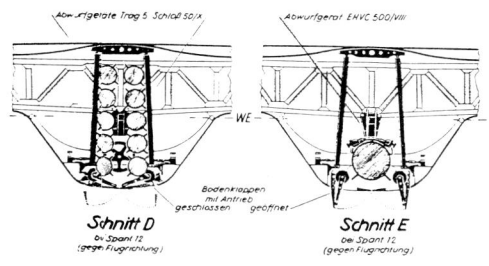

Detailed features associated with the drawing on the previous page.

the paragraph on the Z-2. Its versatility was enhanced by installation fittings for additional photography equipment, in which capacity it was produced in limited numbers for operations as a HQ squadron scout plane for those bomber wings equipped with the Do 17.

The fourth bomber version of the Do 17Z series was the Do 17Z-5, which was virtually the same as the Z-3, but was designed for overwater operations. The Z-5 had drift and ground speed gauges, and carried inflatable floats in the forward fuselage and in both engine nacelles to increase buoyancy in the event of ditching. It also had additional water survival gear. Only forty examples were built.

In addition, a little known fifth model was designed as the Do 17Z-9. Produced in limited numbers, it was tailored especially to low level operations. In place of the two Träg 5 Schloß 50/X and EHVC 500/VIII racks fitted into the two bomb bays, the Z-9 had 16 Elvemag 5C10 racks installed, each capable of carrying five 10 kg fragmentation bombs (SD 10) vertically. To this end, the bomb bay doors were removed and replaced by cover plates with openings. The fragmentation bombs were released using a RAB 14c bomb release system, with blind bombing and jettisoning controlled by the radio operator from bulkhead 11.

To round out the picture, two additional versions should be mentioned. Both of these were night fighters (the Do 17Z-7 Kauz 1, a two-man version with the MG 15 in the window and *C-Stand* removed, and the Do 17Z-10 Kauz 2, with somewhat better armament and a three-man crew) with the unglazed nose replaced by a battery of four fixed weapons. These machines were also built in limited numbers (Kauz 2: 9 examples)

So much for the Do 17Z series, although it should be mentioned that there was no Do 17Z-8 sub-variant built.

Special Foreign Interest in the Do 17Z (Do 215)

Sweden

The export version of the Do 17Z was designated the Do 215, which differed from the German *Luftwaffe's* Do 17s mainly in having a more powerful Daimler Benz DB 601A (809 kW/1,100 hp) engine and a few equipment modifications. Sweden ordered 18 of these bombers, known as the Do 215A-1 series. However, due to the outbreak of war on 1 September 1939, these were never delivered, instead being used by the *Luftwaffe* primarily in the reconnaissance role after modification. They were supplied to the reconnaissance units as strategic reconnaissance platforms under the designation Do 215B-0 and (from January 1940) the Do 215B-1.

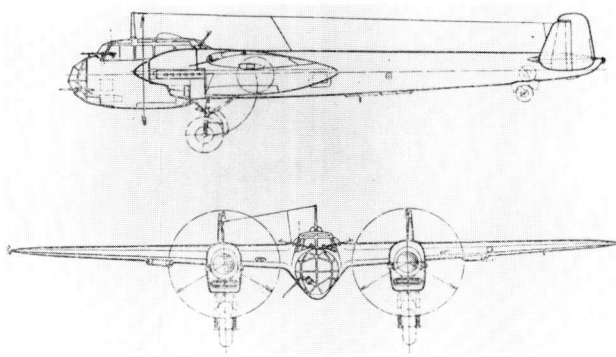

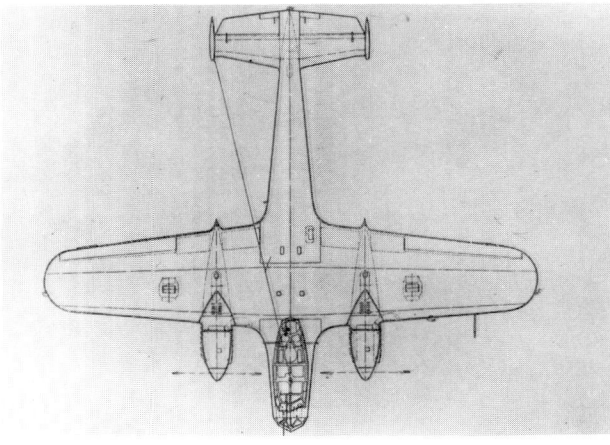

Dornier Do 215 (from late 1939).

Dornier Do 215B (1940) originally was conceived as an export version of the Do 17Z for Sweden (as the Do 215A-1). A so-called "reconnaissance bomber," it mainly saw duty with the *Luftwaffe's* strategic reconnaissance units. Particularly noteworthy are the clean lines of the nacelles for the Daimler Benz DB 601A engines.

Soviet Union

According to agreements made in the wake of the German-Soviet non-aggression pact signed in 1939, the Soviet Union received two stripped-down machines from the Do 17Z production line. Delivered to the Russians as the Do 215B-3, no further information is known as to their subsequent usage or fate.

Synopsis

In all, the Dornier Do 17 was considered quite a reliable airplane with good handling characteristics among the flying crew, and was praised by ground crew for its ease of maintenance. It proved to be a versatile weapons platform, but was clearly inferior to the He 111 with regard to bomb capacity and to the Junkers Ju 88 with regard to speed—despite its initial superiority and successes in this area. The Junkers design was the last medium bomber developed before the war and began service with the field units in 1939. In light of this development, Do 17Z production began tapering off in the fall of 1939, and by the spring of 1940 had stopped altogether. By then a total of 913 Do 17Zs had been built, broken down as follows:

- 420 at the Dornier-Werk München (Oberpfaffenhofen)
- 99 at Seibel in Halle/Saale
- 320 at Henschel in Berlin-Schönefeld
- 74 at Hamburger Flugzeugbau

As of 2 September 1939 the *Luftwaffe's* front-line units had 371 Dornier bombers on hand, i.e. 31.5% of the bomber inventory was comprised of Dornier supplied aircraft—119 Do 17Es, 40 Do 17Ms, and 212 Do 17Zs.

Junkers Ju 88 Bomber

Foundations

The chain of events leading to a superior high-speed medium range, medium load bomber has already been discussed earlier. The reader will remember that the principle of a universally applicable "bomber-destroyer" from 1934 had, just one year later, been abandoned in favor of a pure "destroyer" (heavy fighter) on the one hand, and a lightly armed high-speed bomber on the other. What had since become of this high-speed bomber project which, from the outset, had been conceived as a purely military aircraft lacking any type of civilian application?

History of German Aviation: Bombers and Reconnaissance Aircraft

According to a first-hand account—the renowned aviatrix Marga von Etzdorf returned with the information after an extended stay in the United States—the American company of Glenn L. Martin had developed a high-speed twin-engined bomber which reputedly had an impressive maximum speed of 450 km/h*. This would have left the medium bomber being developed by Junkers (Ju 86) well and truly in the dust! And in mid 1935 the Ju 86 was still in prototype stage, with the Ju 86V5 prototype for the pre-production series scheduled to enter production in late 1935, only flying in August of that year.

Marga von Etzdorf's eye-opening report prompted the Junkers company sometime in mid 1935 to begin drawing up its own proposals and designs for a bomber having a top speed of at least 500 km/h and a payload of about 1,000 kg. This work was carried out by the design department under the direction of *Dipl.-Ing.* August Quick.

Shortly thereafter, the RLM began making serious enquiries of not only Junkers, but also of several other companies (Focke-Wulf, Henschel, Messerschmitt/Bayerische Flugzeugwerke) as to their ability to build a medium load bomber only marginally slower than the new fighter aircraft then in development. The thinking was based on the tactical idea that the attacking bomber would, after a successful attack, already be leaving the area before the pursuing enemy fighters could climb to its operating altitude. It would be a long chase before the fighters could get within firing range, and their relatively limited range would force them to break off and return to base either before or after making their first attack.

The RLM's *Technisches Amt* issued requests for tender which had some pretty ambitious specifications for the time. They called for a high-speed bomber with a 700 to 800 kg payload, a maximum speed of 500 km/h sustained for up to 30 minutes, a 450 km/h cruise speed, a capability of climbing to an altitude of 7,000 m with normal payload in less than 25 minutes, optical bombsight for level bombing, advanced radio and navigational systems, and a de-icing system. Takeoff and landing distances were specified, and this covers only the more important elements of the high-speed bomber request for tender.

To this end the command staff had published its tactical requirements as early as spring of 1935, upon which basis the *Technisches Amt* drew up its detailed technical-tactical specifications and—as just explained—passed on to the aviation industry as a request for tender.

With speed being of paramount importance to the high-speed bomber concept, the design requirements were structured so that the bomber, if attacked, would exercise the better part of valor by fleeing. This meant that the focus would be upon weight, which played a key role in an aircraft's performance. Defensive armament, with the exception of a single MG 15 and 500 rounds of ammunition in the upper rear firing *B-Stand*, was dispensed with, as were heavier fuel tanks, armor, etc., and only a three-man crew was projected.

Junkers took its data from preliminary research begun in August 1935, and work on the project began in earnest on 15 January 1936. In late December 1935 the company had hired two design engineers with experience in alloy stressed skin construction, Alfred Gassner and Heinrich Evers from the Fairchild Company, who respectively became type director and administrator for the timely completion of the detailed Junkers high-speed bomber design.

At the time work was initiated there was a conceptual difference of opinion between the *Technisches Amt* and *Dipl.- Ing.* Ernst Zindel, the chief designer and director of the design department who was responsible for the basic conception and principal shape and layout of the Ju 88. This disagreement took place over the wing size and the type of bomb load configuration. For the roughly 8,000 kg initial design, Junkers had proposed a wing area of 52 m^2, giving it a wing loading of about 160 kg/m^2. Despite being half again as much as the roughly 100 kg/m^2 common to other bombers of the period, the *Technisches Amt* eventually accepted the proposal. As far as ordnance went, the *Technisches Amt* decided—over the objections of Zindel, who also felt that 250 kg bombs should be carried within the fuselage—that the capability of carrying 50 kg bombs would be adequate for ensuring the smallest fuselage dimensions possible. Detail planning then centered around ordnance of either ten 50 kg bombs and an auxiliary fuel tank of about 950 liter capacity, or in place of the fuel tank an additional eight 50 kg bombs at the expense of range.

Negotiations with the RLM over the final version of the design—two designs were debated, the Ju 85 with twin rudders and the Ju 88 with a single rudder—dragged on into May 1936. Then, when the cockpit and fuselage mockups with bomb bays had been built, there followed one inspection after another by those in the *Technisches Amt* responsible for the project, as well as by military officials all the way down to the future chief of the general staff, Jeschonnek, who at the time was still a *Staffelkapitän* and *Major* and, from 16 March 1936, commander of the *Fliegergruppe*

---

* This refers to the predecessor of the later Martin B-26 Marauder, which in fact reached a maximum speed of 500 km/h. Over 5,000 of these were built for the U.S. Army Air Force between 1941 and 1945.

Greifswald. Not until the design for the Ju 88 was finally accepted did the design work and preparation of factory blueprints for the first Ju 88 prototype actually begin.

The completely new design team came together under the supervision of Gassner and Evers, supported by the director of prototype construction, Ritter. They succeeded in making the first Ju 88 prototype available for its maiden flight on 21 December 1936, shortly after the competing high-speed bomber design by the Bayerische Flugzeugwerke, the Bf 110, virtual lookalike designated the Bf 162V1, had completed its first flight.

Henschel's competing design, the Hs 127V1 (originally designated the HS 124 and conceived for the "bomber-destroyer" role), lagged considerably in its development. The Focke-Wulf company had, in the meantime, withdrawn from the competition.

Ju 88V1 through V3 Prototypes

The Ju 88 had its origins within the Junkers design department in studies for a high-speed bomber in the 500 km/h class dating back to mid 1935; preliminary studies were made under the project designation EF 59. Preparatory design work on the Ju 88V1 began in January 1936 with Alfred Gassner as the project director, along with Heinrich Evers, under the supervision of Junkers design department director, Ernst Zindel. The first two designs, already mentioned, were the twin rudder Ju 85 (similar to that of the Dornier Do 17) and the Ju 88 with standard single rudder. The as-yet unselected version was to be produced within nine months at a labor expense of 30,000 man-hours. The *Technisches Amt* ultimately decided in favor of the Ju 88 design, so that actual work finally could begin in May 1936.

The technical details of this low-wing cantilever twin-engined aircraft of all-metal stressed skin construction will be examined in greater detail at a later, more representative period of its development.

On 21 December 1936 the Ju 88V1 was certified for flight and successfully flew for the first time with pilot Karlheinz Kindermann, Junkers' chief pilot, at the controls. It was powered by two Daimler Benz DB 600Aa (735 kW/1,000 hp takeoff rating) and took 105,000 man-hours to build.

Assembly was accelerated for the Ju 88V2, which began its test flight program on 10 April 1937. Other than minor changes in the radiator location, it was practically identical to the V1; it enjoyed a slight advantage in speed due to the fact that the oil coolers formerly mounted under the engines had now been integrated into the annular radiators.

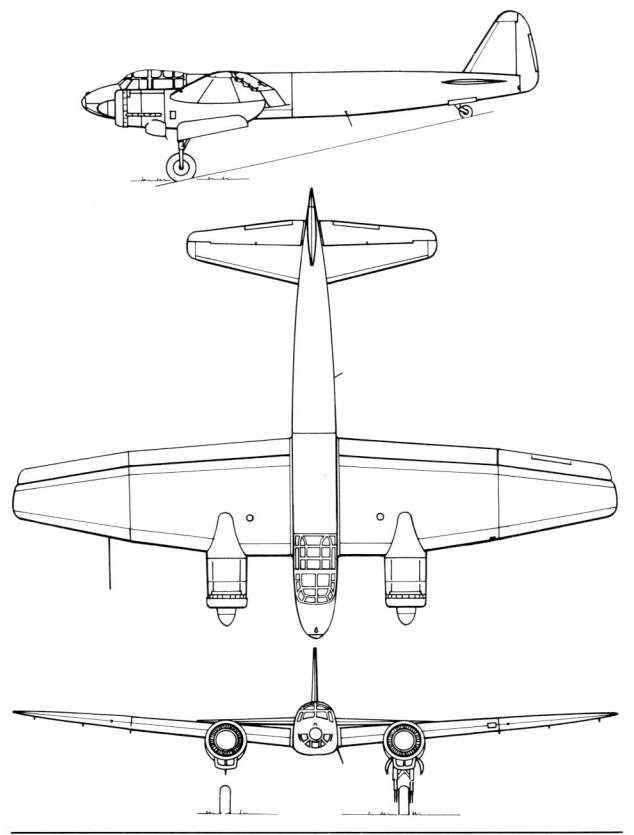

Junkers Ju 88V1, first flight 21 December 1936.

Work on the third prototype, the Ju 88V3, began in January 1937; it was powered by the roughly equivalent Jumo 211A (736 kW/1,000 hp), had a rearward firing MG 15 and was later fitted with a canopy bubble for a bombsight. On 13 September 1937 this prototype also took to the air. It had profited aerodynamically from the test flights of the V1, and especially the V2, and on subsequent test flights not only achieved the requisite performance requirements, but exceeded them with a maximum speed of 520 km/h. Unarmed and with a takeoff weight of 7,000 kg, it was able to maintain this speed for 30 minutes, and had a maximum sustained speed of 504 km/h. Fully equipped, this was still 450 km/h—not unimpressive. It was 55 km/h faster than the V2, which had attained a maximum speed of 465 km/h, and had a takeoff weight of 8,482 kg. Many records were to have been attempted with this third prototype, but on one such attempt an engine malfunction forced an emergency landing on 24 February 1938 at Nuremberg-Fürth and the plane was a total write-off. Pilot Limberger and a test engineer were killed in the crash.

In spite of this, the Ju 88V3 would have been the ideal example of the high-speed bomber, but the fly-off against the competing Bf 162 and Hs 127 models was still pending. Based on the V3's excellent performance and inherent development potential, Junkers was given a contract to build three additional prototypes for further evaluation in order to accelerate the test program of the V1 and V2. This follow-on contract was quite encouraging for those working on the project, but as summer turned into autumn, there was still no decision made on when production of the bomber should begin, despite strong pressure from Dr. Heinrich Koppenberg, the CEO of the Junkers Werke mentioned earlier. Indeed, in August 1937 the general staff directed that the aircraft was to be able to make attack runs at a 30-degree angle, a decision that would have a major effect on the future development of the Ju 88.

Two factors practically cleared the field for the Junkers team. The first was the delay of the Henschel Hs 127*, and the second was a decision by the RLM that Willy Messerschmitt's Bayerische Flugzeugwerke should concentrate on fighter development and abandon the not altogether satisfactory Bf 162 high-speed bomber project. In the meantime, the *Luftwaffe* senior command had been rethinking its strategy, as well. The RLM asked the team to submit a paper showing planning for full-scale production of the Ju 88 with the involvement of license-building companies. Junkers submitted its planning proposal to the RLM in the fourth quarter of 1937 as a "Production Memorandum" and found general approval and support of its proposals. Yet the previously mentioned rethinking strategy was to bring along with it more drastic conceptual changes for the Ju 88 high-speed bomber.

Changing the Concept

During the preceding months doubts were raised within the RLM and the senior *Luftwaffe* ranks as to the military value and suitability of the high-speed bomber. The *Luftwaffe*, distancing itself even further from Douhet and Rougeron, was shifting its line of thinking to the importance of effective and accurate strikes on militarily significant point targets, particularly with large caliber heavy bombs. This would naturally require the development of heavy dive bombers capable of carrying greater payloads over longer distances than those possible with the single-engined Junkers Ju 87. Chief advocate of converting the Ju 88 to a heavy twin-engined dive bomber, vice simply a high-speed bomber with the capability of diving at 30-degree angles, was the chief of the *Technisches Amt*, *Generalmajor* Ernst Udet. Accordingly, Junkers was asked whether the company would be able to redesign the Ju 88 as a dive bomber and put it into production in short order, i.e. in a few months.

After a few weeks of intensive blueprint work, the decision was made at a meeting in the RLM with Udet on 23 December 1937. Junkers entered into a binding contract to tackle the redesign of the Ju 88 into a dive bomber. Consequently, in January 1938 all available resources were brought to bear in order to bring to pass conversion of the high-speed bomber into a heavy dive bomber, to be equipped with dive brakes and automatic recovery system, while at the same time making preparations for full-scale production.

Ju 88V4 through V6 Prototypes

In the meantime, prototype production continued with unchanged urgency, and Ju 88V4 flew for the first time on 2 February 1938. In place of the aerodynamically pleasing, rounded cockpit shape of its predecessors, it had an angular, glazed canopy and a ventral gondola beneath the cockpit, now designed to accommodate a four-man crew and three machine guns. In view of the type's anticipated dive capability, the V4 had a few changes made to the airframe, as well, e.g. dive brakes and some reinforced structural areas, and was given a thorough flight workout at Rechlin. Despite having the same powerplants, its performance lagged behind that of the V3.

The machine designated the Ju 88V5 was based on the V4, but reverted to a more aerodynamic albeit unglazed nose section, and was fitted with a canopy having a flatter profile. Equipped with the

---

* The smaller and lighter Hs 127 did not complete its first flight until late 1937, although it soon revealed an unconfirmed maximum speed of 568 km/h(!).

History of German Aviation: Bombers and Reconnaissance Aircraft

Junkers Ju 88V5, first flight 13 April 1938.

more powerful Jumo 211B-1 (883 kW/1,200 hp), it made its first flight on 13 April 1938 and was unarmed.

Approximately eleven months later, after a regimented testing program, the V5 vindicated the planned record attempts of the V3. On 19 March its crew, company pilots Ernst Siebert and Kurt Heintz, succeeded in breaking the existing world's record when they flew a distance of 2,000 km with a 1,000 kg payload before official witnesses at an average speed of 517 km/h over a Dessau-Zugspitze-Dessau route. On 30 June the pair set another record when they flew the same route at an average speed of 500.8 km/h with a 2,000 kg payload. The V5 had a takeoff weight of 8,990 kg.

These prestige flights took place at a time when the fate of the high-speed bomber had long since been sealed in favor of the heavy dive bomber. Yet these records were passed off to the public—and ultimately the world in general—as the performance of a standard production bomber—a similar bluff as had occurred with the Dornier Do 17 at the International Flying Meet in Zurich back in 1937!

On 18 June 1938, long before the V5's record-setting flights, the Ju 88V6 entered its flight testing program. By then work had started on building the jigs at the Junkers Werk Schönebeck for full-scale production, which would be sent to Aschersleben for airframe construction, Halberstadt for wing assembly, and Leopoldshall for the rudder and control surface construction. Assembly of larger components, the final assembly, and acceptance flight operations were being readied at Bernburg.

The V6 was designed from the outset to be dive capable; its dive brakes, similar in design to those on the Ju 87, were situated beneath the wings outboard of the engines, and the automatic recovery system was built into the elevators. Both components were hydraulically driven. It also had a newly designed hydraulic single strut landing gear that retracted through 90°, and thereby disappeared completely within the engine nacelle. Between fuselage and nacelle there were two ETC bomb racks under each wing for carrying heavy external bomb loads. The V6 was also driven by the more powerful Jumo 211B-1 engine as had been used for the V5's record setting flights, and reached a service ceiling of 8,500 m unloaded. It was armed with three MG 15s and had a four-man crew. The nose was in the final form for full-scale production, with angular lines formed by flat glazing panels, and a ventral gondola was set to starboard with an entry hatch and mount for the *C-Stand* machine gun. Specifically tailored to dive bombing operations and categorized in utilization and stress group H 3*, the V6 saw its wing load factor increase from 5.5 to 10 and had a takeoff weight of 10,250 kg. Compared to the barely 8,500 kg weight of the speedy V3, these were major handicaps, and were magnified by other drag inducing features which made the V6 some 40 km/h slower than the V3 in spite of the former's more powerful engines. Despite performance figures which were not nearly as impressive as before, the Ju 88V6 became the pattern for the first Ju 88A-0 and A-1 series established by the full-scale production contract signed on 3 September 1938, and which designated the type as a level and dive bomber. The tables on page 79 chart this almost negative development of the Ju 88 occurring between the V3 and V6, brought on as a result of the type's role expansion.

In looking at this comparison, it should be noted that it traces the jump in development from a single-role V3 high-speed bomber prototype geared for speed to a multi-role heavy dive bomber, the V6. Whether this was worth the cost involved remained to be seen.

Those numerous test versions either derived from the Ju 88V6 or following it—the V25 flew for the first time in September of 1940—will be treated over the coming pages insofar as they relate to bomber and/or reconnaissance series, for which evaluation or use they were built.

---

* See Appendix 6 "Aircraft Utilization, Stress, and Class Groupings"

A-Series (A-0 and A-1)

Teething Troubles

The Junkers Company, under the energetic leadership of general director Dr. Koppenberg, immediately set about starting full-scale production once it had been decided that the Ju 88V6—flying since mid 1938—was to be the mold for the standard bomber of the *Luftwaffe*. As a level and dive bomber, sometimes called a *"sturzflugfähiger Gleitbomber"* (literally, a dive-capable glide bomber), it was expected to fulfill all anticipated tasking in the operative arena and replace the Ju 86, Do 17, and He 111.

The *Luftwaffe's* leadership had hoped to have a total of 8,300 Ju 88s built by 1 April 1942; the Junkers Company aimed its sights at a lower, more realistic number in what was still quite an optimistic counterproposal when it stated in October 1938 that 6,800 aircraft could be built by the date specified.

Awareness of the anticipated bomber production of England and France led to the assumption that these two countries would be able to manufacture about 6,300 bombers by the end of 1942.

Against this backdrop, the Junkers Werke drew up detailed plans for full-scale production from November 1938. Several license-building companies had been contracted with in the meantime (joined into four production groups), and together these would have to produce 300 Ju 88s and 730 Jumo 211B aircraft engines per month following an initial spin-up period. The combined potential of these companies, with a total of about 100,000 employees the program would require, was considered sufficient for meeting this production goal—assuming that the 35,000 personnel (including 16,000 skilled workers) still needed could be found. On the other hand, the capability of raw material and parts suppliers for such a large-scale program was felt to be wholly inadequate if additional supply sources could not be located.

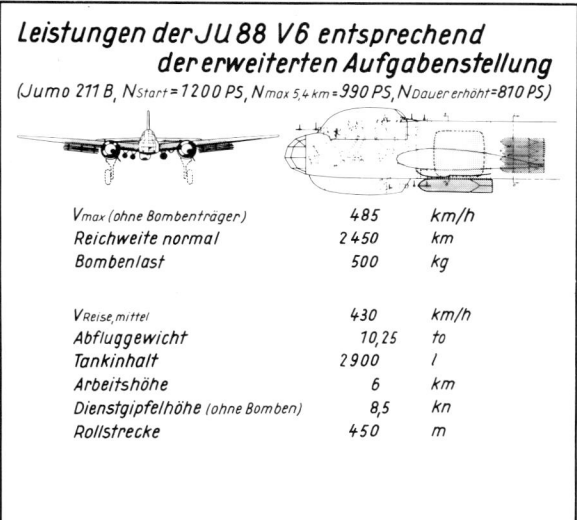

Evolution of Ju 88 performance resulting from expansion of its roles.

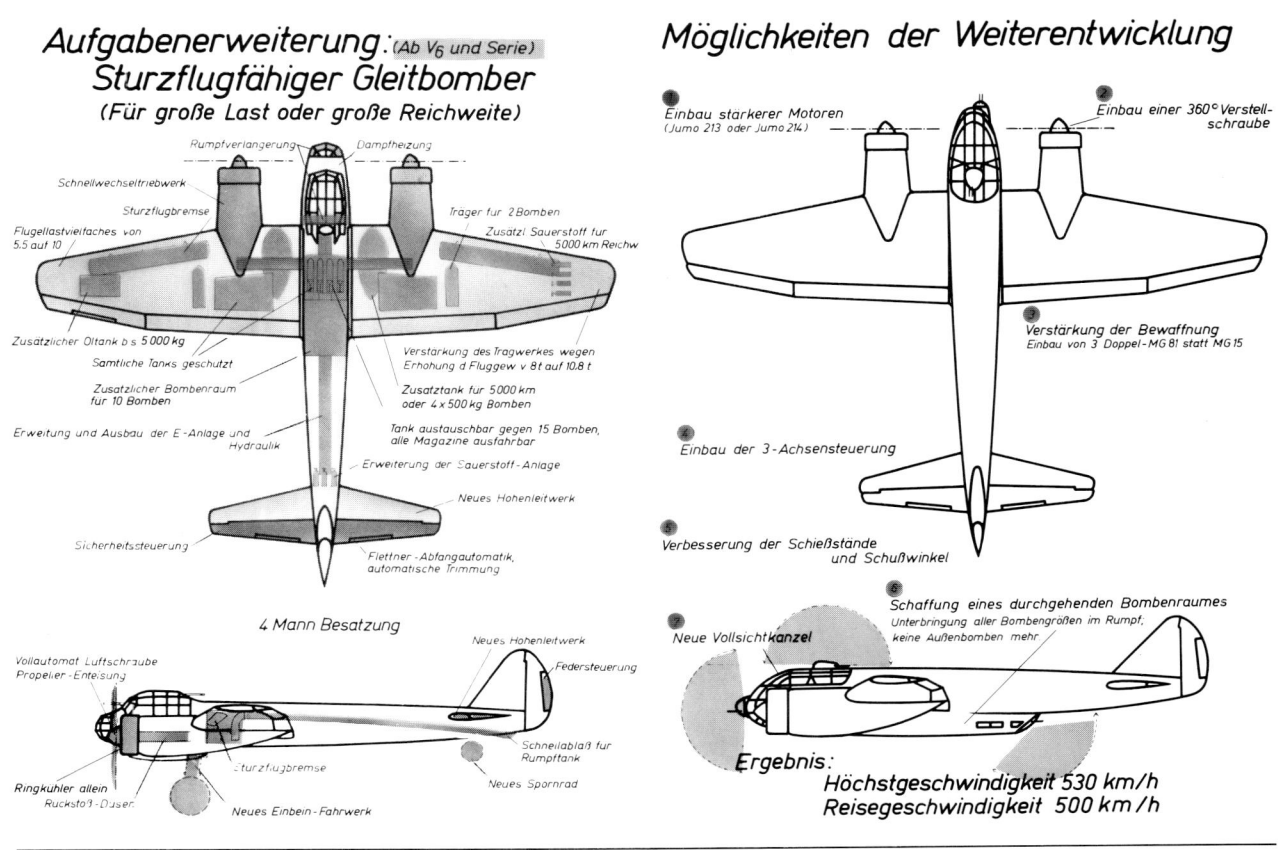

Role expansion (beginning with the Ju 88V6) and potential follow-on developments under consideration.

Production Operations

Within the four license production groups and the parent company in Dessau (planned production: 65 Ju 88s per month), the specific breakdown of tasking was as follows:

Group I: Arado - wings; Henschel - airframe; AEG - empennage Monthly output: 80 Ju 88s

Group II: Heinkel/Oranienburg - wings; Dornier/Wismar - airframe and empennage Monthly output: 70 Ju 88s

Group III: Dornier/Friedrichshafen - airframe, wings, empennage and engine mounts Monthly output: 35 Ju 88s

Group IV: ATG - airframe; Siebel - wings and empennage Monthly output: 50 Ju 88s

The problems with expanding the aviation industry prior to the war's outbreak have already been touched upon in an earlier chapter, as has the special authority given to both the Junkers management and Dr. Koppenberg personally, as well as Hitler's orders to Göring to ensure that at least 300 Ju 88s would be produced each month.

All these extra measures came relatively late in the day. And they were not effective until the foundations for a continual program development *in toto* had been laid through normal personnel and materiel appropriation channels that best concentrated on specific production areas.

Nevertheless, an almost unimaginable effort was put into hiring programs and technical training, procurement of construction materials to expand the infrastructure within the factories and, last but not least, the acquisition of raw materials necessary for the actual assembly of the aircraft and engines; all these measures slowly but surely began to whittle away at the delays. By the beginning of the second quarter of 1939, ten months after the first flight of the Ju 88V6 prototype, production of the so-called parent company series could begin. Completing just one aircraft in the first month of production, this figure rose to an output of 27 in the last month of 1939, still a far cry from the anticipated monthly output of 65.

Actual full-scale production did not start until August 1939, some 14 months after the V6's maiden flight, and came at a time when the Ju 88's general flight testing program was winding down. Production at the first license company (Henschel) began shortly before the war, i.e. about 15 months after the first flight of the V6; in October 1939 its output was three aircraft, but within a year this number had climbed to 40 planes by October 1940—although still just 50% of the planned target for Group I. All other license companies did not begin production of their components until late 1939/early 1940.

Adjusting itself to these events, the RLM's delivery plans were drastically reduced in comparison with the *Luftwaffe* leadership's originally hoped for 8,300 Ju 88s by 1 April 1942.

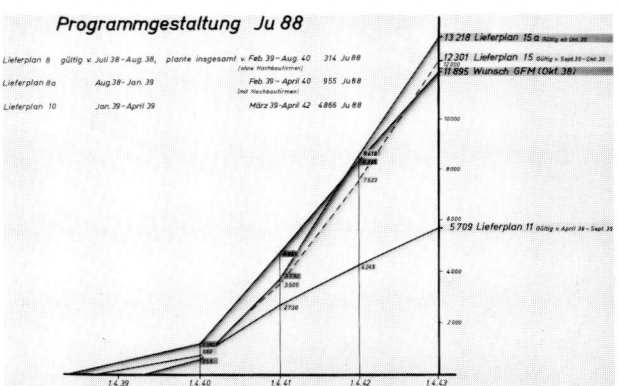

Ju 88 program graph charting delivery plans 8, 8a, 10, 11, 15, and 15a. The inconsistencies between these are remarkable.

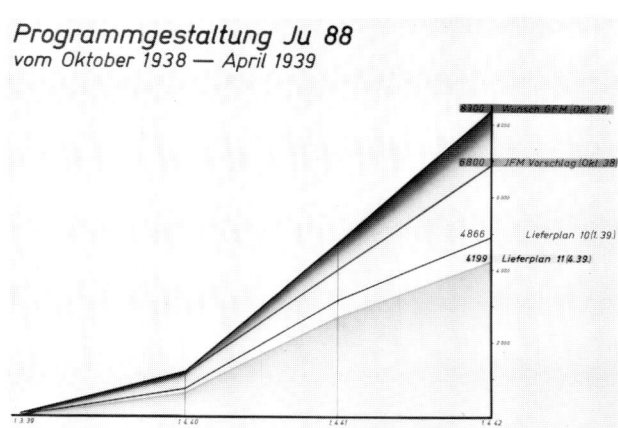

Ju 88 program planning from October 1938 to April 1939.

- Delivery plan 10 from January 1939 called for 4,866 planes, and just three months later
- Delivery plan 11 from April 1939 reduced this to 4,199 planes to be built by the same original date.

For the acquisition goals, these would not be the final adjustments made to accommodate the economic realities under which Germany's aviation industry was forced to work.

Production Model

The Ju 88A-0 began rolling off the assembly line in March 1939. It came standard with the underwing slotted dive brakes tested intensively on the Ju 88V8 and V9 (first flights 3 and 31 October 1938, respectively), and had the ETC bomb racks for carrying external loads, which had undergone thorough evaluation mainly with the Ju 88V10.

The A-0 aircraft were supplied to *Erprobungskommando Ju 88*, a combat evaluation detachment specially drawn from an active bomber unit. The detachment put the Ju 88 through its paces and used the experience to work out guidelines for training future bomber crews who would be going to war in the new type.

Erprobungskommando Ju 88 formed the core of the first field unit to be equipped with the Ju 88, which was established in August 1939 with the newly available Ju 88A-1. This was the *Lehrgruppe Ju 88*, activated on 22 August 1939 at Greifswald. The primary function of this tactical development unit was to train crews, as well as refine dive bombing techniques and tactics.

Deliveries of the Ju 88A-1 occurred at a much slower pace than official planning, constantly under revision, was calling for. Among other things, this was caused by the plethora of technical modifications necessitated by evaluation results. The chart "Ju 88 Chronology" (see appendix 7) provides an overview of the aircraft actually accepted by the contractor from the parent company and full-scale production series, and shows that of the 34 Ju 88s delivered by 2 September 1939, only 18 A-0 and A-1 variants and their trained crews were available to front-line units on that date.

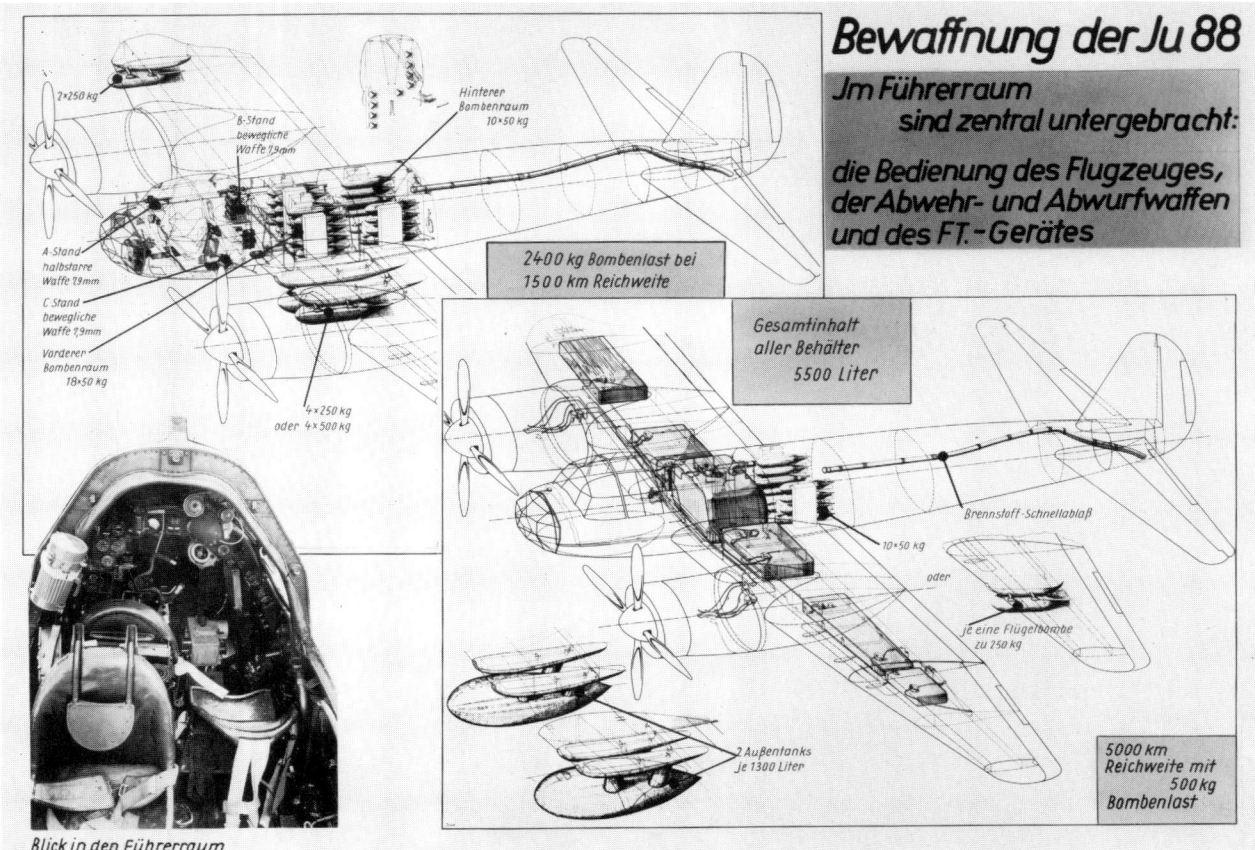

Ju 88 armament.

Thus, at that time the Ju 88 made up just 1.5% of the total operational bombers within the *Luftwaffe*. However, month by month, the supply situation started to improve, and over the course of time front-line units began receiving relatively continuous deliveries.

Quite apparent in the drawing shown here is the Ju 88's versatility, pointing the way forward for the role it would play as a bomber and (soon) as a reconnaissance platform (as well as a heavy fighter and night fighter, which this volume does not cover).

Dive Bombing Capability

And how had the requisite dive capability materialized for the Ju 88?

The horizontally slotted dive brakes fitted beneath each wing and the automatic dive recovery system have already been briefly mentioned as the main components for this type of attack method. That the forces built up in a dive place severe restrictions on the ability of the human hand to mechanically operate these components goes without saying. The Junkers designers, with the involvement of the experienced *Dipl.-Ing.* Hermann Polhmann (the creator of the Ju 87 dive bomber) would therefore have to redesign the Ju 88's hydraulic system to handle the dive bombing capability of this aircraft and ensure a general automatization of the functions needed for a controlled dive bombing attack. This was done in order to compensate for the weaknesses of the human element, or to be specific, the pilot, and to keep his workload within acceptable bounds.

The drawing on page 83 shows the operation of the Ju 88's hydraulic system and, in particular, its function during a dive with automatic recovery. Also shown is the dive bombing tactic developed especially for this aircraft type. Both of these were determined to a large extent by the type and operation of the automatic dive recovery system itself.

History of German Aviation: Bombers and Reconnaissance Aircraft

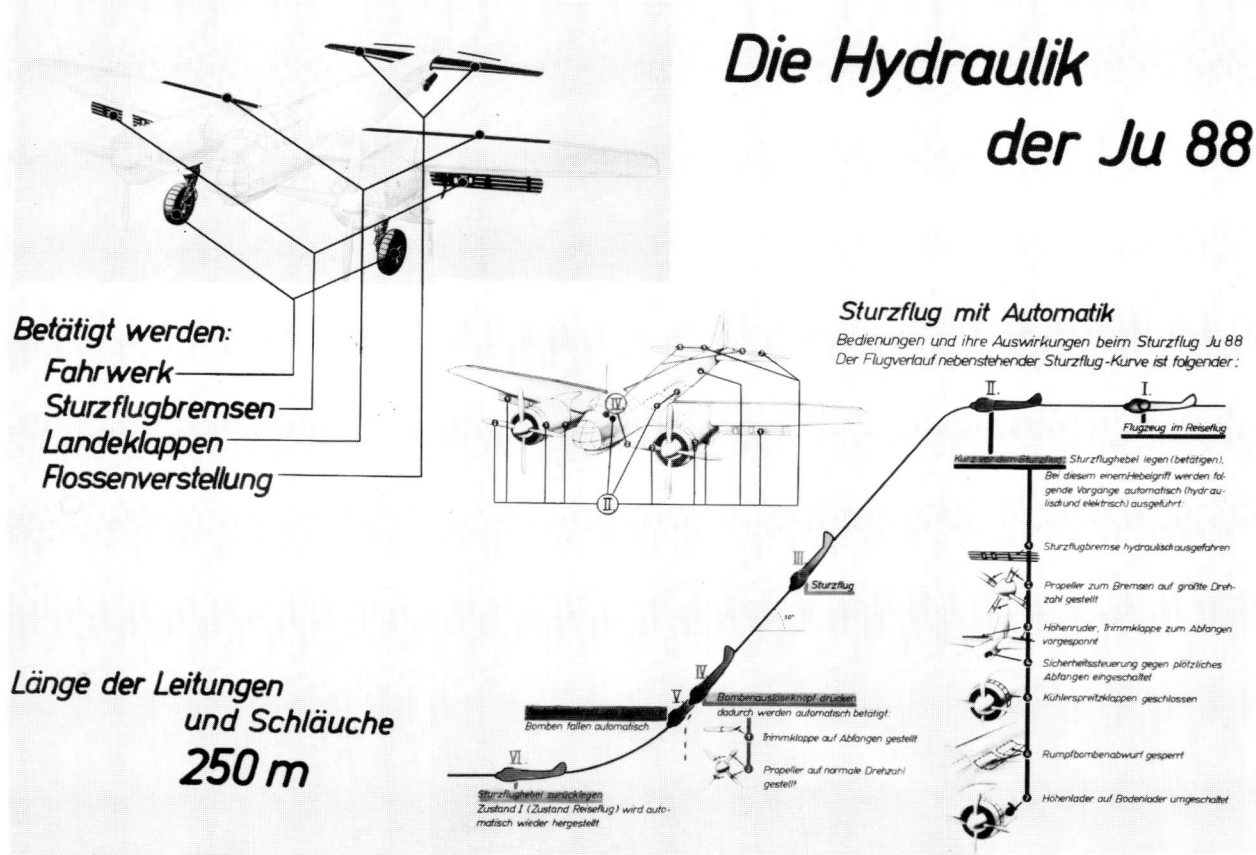

Ju 88 hydraulic system.

The unique attack profile developed especially for the Ju 88, which theoretically allowed crews of average experience to dive at angles up to 80°, differed in several important points from the earlier methods used, primarily those by the Ju 87. These points will be addressed in more detail with regard to the Ju 87 and in view of the development of the operational requirements established later.

Synopsis

Taking into account the conceptual and production hurdles which first had to be overcome, the Ju 88 bomber entered front-line service relatively soon after the initial flight of the Ju 88V6 prototype, a date which serves as a key reference point for the series production. It is therefore no wonder that, due to this short interval, the Ju 88 was not yet a fully matured front-line design. Nevertheless, its dive bombing capability opened new perspectives for the operative conduct of the air war and provided a highly interesting supplement to the available potential offered by the Do 17 and He 111 level bombers.

Comprising just 1.5% of the total number of bombers with front-line units at the start of the war, the Ju 88 can be considered more of a "*quantité négliable*" than a real increase in the fighting potential. And its crews—with few exceptions—were still a long way from reaching a point in their training, not to mention experience level, where they could play any type of effective role in the event of war breaking out.

Despite this, the Ju 88's star was rising and, once it had reached its development and combat potential, in the future the type would become a powerful element within the operative *Luftwaffe*

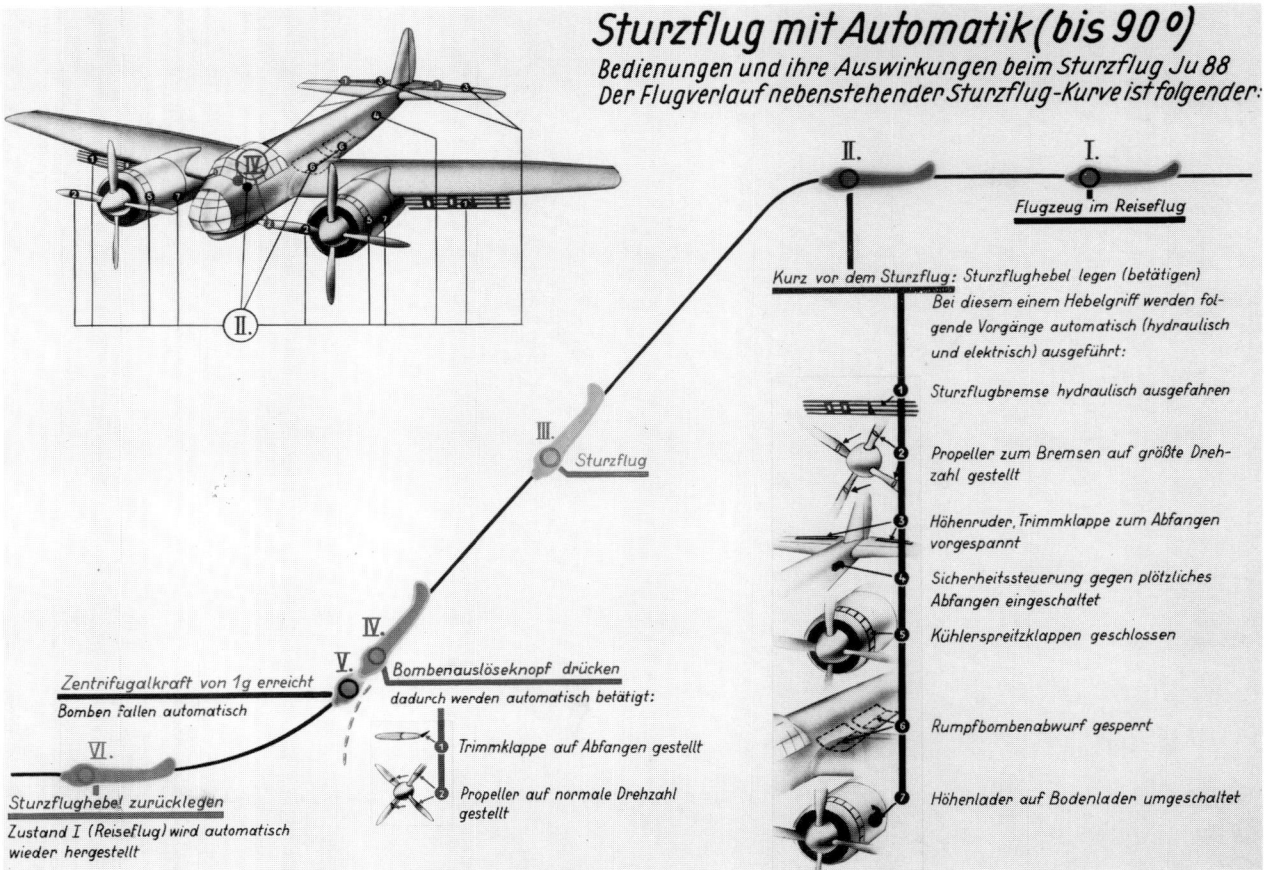

Chart showing a typical dive profile of a Ju 88 with automatic pullout (up to 90%).

Junkers Ju 88A-1 production model as used by *Lehrgruppe Ju 88* from August 1939. Werk-Nr. 299 shown here is destined for the *Edelweiß-Geschwader*, which converted to this type following the Western Campaign in 1940.

History of German Aviation: Bombers and Reconnaissance Aircraft

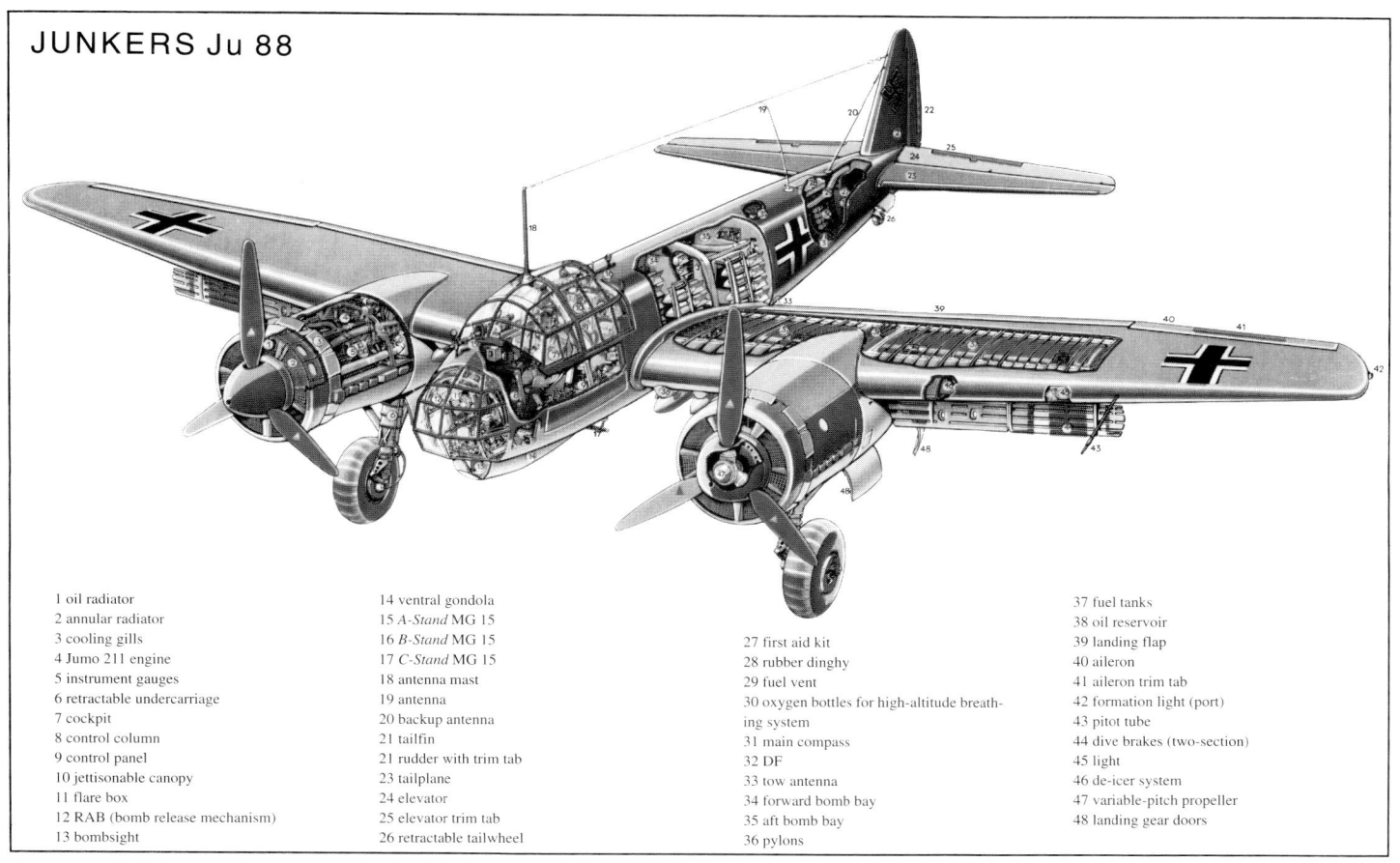

JUNKERS Ju 88

1 oil radiator
2 annular radiator
3 cooling gills
4 Jumo 211 engine
5 instrument gauges
6 retractable undercarriage
7 cockpit
8 control column
9 control panel
10 jettisonable canopy
11 flare box
12 RAB (bomb release mechanism)
13 bombsight
14 ventral gondola
15 *A-Stand* MG 15
16 *B-Stand* MG 15
17 *C-Stand* MG 15
18 antenna mast
19 antenna
20 backup antenna
21 tailfin
21 rudder with trim tab
23 tailplane
24 elevator
25 elevator trim tab
26 retractable tailwheel
27 first aid kit
28 rubber dinghy
29 fuel vent
30 oxygen bottles for high-altitude breathing system
31 main compass
32 DF
33 tow antenna
34 forward bomb bay
35 aft bomb bay
36 pylons
37 fuel tanks
38 oil reservoir
39 landing flap
40 aileron
41 aileron trim tab
42 formation light (port)
43 pitot tube
44 dive brakes (two-section)
45 light
46 de-icer system
47 variable-pitch propeller
48 landing gear doors

In late 1939 the Ju 88 had an unmistakable shape and design. The main components are shown to good effect here.

Dive Bombers

Henschel Hs 123 Light Dive Bomber

Selection

The Fieseler Fi 98 and Henschel Hs 123 dive bombers were participants in a 1935 competition under the auspices of the emergency program and represent the first stage of the planned interim stocking of the newly established dive bombing units. These planes would then be phased out once the ultimate *Stuka* became available during the course of the second phase expansion.

Both craft, the Fi 98 and the Hs 123, completed their maiden flights in early 1935; it is worth noting that the person at the controls during the Hs 123's first flight on 8 May 1935 was none other than Ernst Udet. We have already looked at his role in connection with the dive bombing concept in general, and the *Stuka* choice in particular (see pp. 35-37), but his influence was not the only thing determining the outcome. Despite both competing types being powered by the same air-cooled BMW 132A-3 nine-cylinder radial engine (478 kW/650 hp), the Hs 123 turned in a better performance than the Fi 98 almost from the outset, and Fieseler's design soon bowed out of the competition when further testing at Rechlin only magnified its weaknesses. Follow-on development of this braced biplane design ceased even before the second Fi 98b prototype had been built following a failed attempt to sell the design to the Japanese Navy.

V-types

Following the maiden flight of the Hs 123V1 on 8 May 1935, the initial contract issued to Henschel by the RLM's LC department (*Technisches Amt*) for the construction of three prototypes was fulfilled with the completion of the Hs 123 V2 and V3 in the early summer of that same year. All three prototypes were subjected to a rigorous flight test program at Rechlin, which particularly focused on the type's dive bombing capability. A suction effect initially caused problems with exhaust vapors in the cockpit, but this problem was fixed by sealing the control line feed openings. In spite of this attention to detail, two of the test aircraft broke apart in midair during terminal velocity dive speed trials, with both pilots being killed. Investigation revealed a weakness in the cabane, the center section of the upper wing, whose N-struts came loose or separated as a result of torsional stress when pulling out of a dive—leading to the entire wing to break off. Careful studies of the surviving prototype confirmed this finding.

A new contract was issued for an Hs 123V4, which had a reinforced wing center section, braces replacing cables for the empennage, and a few detail modifications to ease production. Over the late summer of 1935 the V4 was put through its paces at Rechlin. It now had the structural soundness required to perform 80-degree dives and pullouts, and Henschel was given a contract to immediately start production of the Hs 123A-0 and A-1 based on the V4 at its Johannisthal plant, and the newly completed facility at Schönefeld, both situated near Berlin.

A-Series

Hs 123A-0 production began in 1936, with the first deliveries being made to the newly established *Stukagruppen* or to existing *Stuka* units retiring their He 50s and Ar 65s in the summer of 1936. The Hs 123A-0 differed from the V4 in having a more powerful BMW 132Dc fuel-injected engine rated at 647 kW/880 hp at takeoff and 640 kW/870 hp at 2,500 meters altitude. It was armed with two MG 17s in the upper fuselage and was fitted with four ETC 50 racks for carrying four 50 kg bombs under the wings. Between the two fixed undercarriage struts was another rack for carrying a 250 kg bomb or a drop tank.

In general, the Hs 123 was a robust and pleasant looking sesquiplane of smooth skinned monocoque construction. Upper and lower wings were quite sound; cable bracing as on the Fi 98 was superfluous—both wings were joined solely by a wide I-beam strut on each side. Ailerons were on the upper wing, and the flaps were located on the stubbier lower wing. Beneath the open cockpit was a small bomb bay. The spatted undercarriage was attached to the lower wing. A NACA cowling enclosed the engine. This was different than the smooth, somewhat larger cowling of the V1 in that it fit tightly around the engine and had bumps covering the cylinder head covers, giving the Hs 123 its characteristic "face."

Applicable data included: 10.50 m wingspan; 8.58 m length; approx. 1,400 kg empty weight; 2,100 kg takeoff weight; and 350 km/h maximum speed.

The Hs 123's operational testing in Spain beginning in December 1938 has already been discussed, with reference to the type's reclassification from a light dive bomber as a ground attack/strike

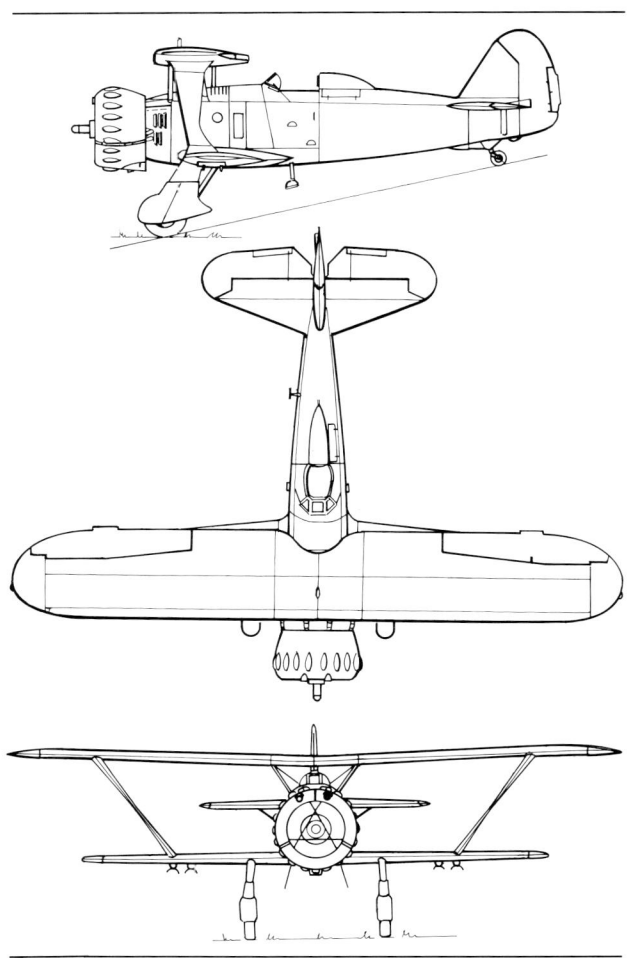

Henschel Hs 123A-1 (1936). First flight of the Henschel Hs 123V1 took place on 8 May 1935.

History of German Aviation: Bombers and Reconnaissance Aircraft

Henschel Hs 123A with a drop tank which could be carried in place of a 250 kg bomb (from 1936 onward).

plane. While the heavier *Stuka* of the second phase, the Ju 87, was introduced into the *Stukagruppen* in the spring of 1937 and began replacing those He 50s and Ar 65s not yet sidelined by the Hs 123, all Hs 123s were combined into the *Schlachtfliegergruppen* (ground attack groups) established on 1 August 1938 and assigned directly to army support.

Hs 123 production ceased in the summer of 1938, although a B-series had been planned—the Hs 123V5 with a 706 kW/ 960 hp BMW 132K had already flown as the prototype for the series. There was even talk of a C-series with enclosed cockpit and heavier armament, with the V6 being the prototype. Both of these were abandoned in favor of the Ju 87 production just getting underway, however.

Synopsis

In September 1938 there were still 117 Hs 123s serving in ground attack units alongside 78 Heinkel He 45s, which meant that 60% of the strike aircraft inventory was comprised of the former type. By 2 September 1939 this number had dropped dramatically, to just 40 aircraft. By then the He 45 had been completely withdrawn from front-line units and only served in the training capacity. With regard to the He 45, it should be noted that this biplane was originally conceived as a strategic reconnaissance platform, but as better reconnaissance aircraft became available in Spain this type was increasingly relegated to the ground support role, in effect becoming reclassified as a strike plane. Even then, by the end of 1938 it had outlived its usefulness with the German *Luftwaffe*—unlike with the Spanish Nationalist Air Force—even in this role so closely associated with battlefield reconnaissance.

To be sure, at the war's outbreak the Hs 123 was being phased out, but this robust ground attack plane would later serve in an entirely unexpected capacity.

Junkers Ju 87 Dive Bomber

Foundations

The *Luftwaffe's* buildup plan from August 1935 envisioned a whole group of dive bomber units. Bit by bit, from the summer of 1936 onward the previously mentioned Henschel Hs 123 began replacing those aircraft which had improvised as dive bombers when the *Stuka* units were first activated on 1 October 1935. These aircraft included the clumsy Heinkel He 50 biplane dive bomber and the antiquated Arado Ar 65 fighter, as well as a few He 51 biplane fighters. Yet as a single-seater even the Hs 123 was only planned as an interim solution for the *Stukagruppen* under the auspices of the emergency program's first stage, simply filling a gap until the arrival of the "real" heavier, two-seat aircraft of the second phase.

For this second phase the RLM also called upon several companies to develop prototypes; among these were Arado, the Hamburger Flugzeugbau, Heinkel, and Junkers, which in January 1935 received the official request for tender/specifications for a two seat dive bomber.

Arado developed an all-metal biplane, the Ar 81, with fixed undercarriage. The Hamburger Flugzeugbau built the Ha 137, already mentioned in context with the Hs 123, although this was a single-seater and therefore could only be considered as having an outside chance in the selection of the heavier dive bomber. For its part, Heinkel developed a quite speedy, two-seat dive bomber, the He 118, which also had retractable landing gear.

The RLM's specifications to a large extent corresponded to the ideas of the Junkers company, which had not only been intimately involved in the theory of dive bombers for years, but had also acquired much valuable practical experience from the K 47. This aircraft, designed by *Dipl.-Ing.* Karl Plauth, first flew in 1928 in Sweden, and the experience came not only from that country, but from additional testing at Lipetsk in the Soviet Union.

Dipl.-Ing. Hermann Pohlmann had been employed by Junkers since 1923, was a former pilot of the "*Bomberversuchsabteilung*," and had worked on the design of the K 47. Indeed, following the death of Plauth after an accident involving a sportplane on 1 November 1927, Pohlmann assumed responsibility for follow-on development and flight testing of the plane, which was subsequently built in Germany as a civil sportplane and testbed under the designation A 48. Thus, had Junkers ensured continuity of work on the dive bomber concept.

In 1933 Pohlmann drew up the basic design concept for the Ju 87 as a two-seat, low-wing monoplane fitted with dive brakes and automatic recovery system. Even at this early stage, the type had its characteristic "bent-wing" feature, with the lowest point serving as the attachment point for the cantilever, spatted undercarriage. Pohlmann had a mockup built, as well.

In 1934 this Ju 87 mockup was inspected by representatives from the RLM, who shortly afterwards—several months before the official request for tender had been issued—approved the construction of three prototypes with this layout. Thus, Junkers was given a significant advantage over its competitors, even when the belated issuance of the specifications was most likely attributable to the speculation that the RLM had wanted to ensure as much safety as possible was incorporated into the requirements.

First Prototypes

With this advantage, Junkers was able to complete the first Ju 87V1 by the late summer of 1935. On 17 September 1935 it made its maiden flight at Dessau with Willi Neuenhofen at the controls. Lacking a suitably powerful engine, the V1 was powered by the British Rolls-Royce Kestrel V motor, a liquid-cooled 12-cylinder V-type engine with a takeoff rating of 368 kW/525 hp and 471 kW/640 hp at 4,270 meters/14,000 feet, driving a two-bladed wooden propeller. Like the K 47, the type had dual rudders. However, with just a few flights the prototype crashed during dive testing—it had not yet been fitted with dive brakes—after the aircraft began fluttering and the empennage broke up on 24 January 1936 at Dessau. Both pilot Willi Neuenhofen and engineer Kreft were killed in the crash.

Following a detailed investigation into the cause of the crash, the Ju 87V2 (which had been frozen in its construction phase for this very reason) was fitted with a simple enlarged and reinforced vertical stabilizer, the Junkers dive brakes (based on the principle of the Junkers patent #665316) underneath the wings, and the Jumo 210Aa engine just becoming available. This powerplant, with an output of 449 kW/610 hp at 2,800 meters, now drove a three-bladed propeller and was nearly as powerful as the Kestrel engine in the V1.

On 25 February 1936 the V2 made its first flight and went to Rechlin for further testing in March 1936.

Built to nearly the same layout pattern, the Ju 87V3 followed with its maiden flight on 27 March 1936 and also went off for further testing to Rechlin in early May of that same year.

Fly-off and Selection

In early June 1936 the decisive fly-off between the rival dive bomber types took place at Rechlin.

Arado Ar 81

Compared to the Ha 137, He 118, and Ju 87, this was the most conventional design. Although having a good upward field of fire, as the only biplane layout it was not able to match the performance of the other competitors.

The Arado Ar 81, eliminated in fly-offs from the 1936 two-seat dive bomber competition for which it was designed. Only three prototypes were built, the V1 and V2 with dual rudders (which tended to flutter), and the V3 with its resulting single standard rudder.

In addition, the dual rudders of the V1 and V2 prototypes tended to vibrate and flutter. The V3 attempted to alleviate the problem with its single rudder design, but the Ar 81's flight handling still continued to lag behind that of the others nonetheless. It was only considered a backup at the fly-off, which was attended by Arado's chief test pilot Carl-August von Schönebeck. Following another flight (flown by Ernst Udet personally), it was considered a non-starter and therefore only these three prototypes were built.

Ha 137

Conceived as a single-seater by Hamburger Flugzeugbau chief designer Dr. Richard Vogt to no official requirement, the Ha 137 was thus outside of the new specification standards. Nevertheless, it was an advanced design similar to the Ju 87 in layout, but in a lighter class with its 2,400 kg takeoff weight, and therefore also of a lighter design than the robust 3,400 kg Ju 87.

The single seat Ha 137 dive bomber offered by the Hamburger Flugzeugbau, of which six V-types were built (from 1936 on); the relatively light Ha 137 proved to be no match for the Ju 87 in the fly-offs at Rechlin in June 1936.

The Ha 137 participated in the competition virtually without a rival, since in the meantime the *Technisches Amt* had decided to consider it a potential candidate for the ground attack role. Follow-on development, however, officially ended at the six prototypes built. One of these prototypes was configured as the *"Projekt 11"* for use from the planned aircraft carrier, but its limited range and single-seat layout excluded it from this program, too. Another V-type of the Ha 137 was later used for testing rocket weapons.

He 118 and Ju 87

Actually, only the He 118 designed by Heinkel's Siegfried Günter and Hermann Pohlmann's Ju 87 designed at Junkers could be considered serious competitors at the fly-off. This had become obvious at an early stage, and the RLM had awarded both companies contracts to build ten pre-production machines each.

Heinkel took part in the competition with its V2 prototype of the He 118, as did Junkers. The He 118, flown by Heinkel test pilot Hinrichs, compared unfavorably to the Junkers Ju 87 despite the former's higher speed, since without dive brakes Hinrichs was only able to demonstrate dive angles of a maximum of 50°. On the other hand, the Ju 87—the only competitor to have been fitted with the brakes indispensable for dive bombing—was flown by *Dipl.- Ing.* Hesselbach from the Junkers flight test group with precision accuracy at truly steep angles up to and including a vertical profile. Following his disappointing flight in the Ar 81, Udet also test flew the Ju 87 and was satisfied with its flight handling characteristics.

Udet nevertheless continued to show an interest in the He 118, which was fitted with the more powerful 30 liter aircraft engine (DB 600C with takeoff rating of 625 kW/850 hp and 669 kW/910 hp at 4,000 meters); at the time the Ju 87V2 was still powered by the previously mentioned, less powerful 20 liter Jumo 211Aa motor.

The Heinkel He 118 which, despite its higher speeds (although lacking dive brakes), lost out to the Ju 87 in the 1936 fly-offs.

Because of this interest Udet flew the He 118 after the Rechlin fly-offs during the same month at Heinkel's Marienehe facility. Due to operator error, Udet oversped the propeller, which tore from the shaft and destroyed the elevator assembly. Udet, however, succeeded in baling out of the doomed plane.*

This accident practically ensured the decision in favor of the Ju 87, particularly as Udet, in the rank of *Oberst*, became chief of the *Technisches Amt* on 9 June 1936 and was now the decision-making authority. A few days after the crash of the He 118, Junkers was given official approval to build the standard dive bomber of the *Luftwaffe*.

Ju 87V4 and Initiation of A-Series

The Ju 87V4 was completed in June 1936, as well, completing its first flight on the 20th of that month; the V5's maiden flight took place on the 14th of August, 1936, and by the end of the year all ten Ju-87A-0 pre-production aircraft had been built. These were assigned to field testing, which concluded in the spring of 1937.

The Ju 87 underwent rigorous bomb release trials at Rechlin starting in November 1936 and became the pattern for the A-1 production series. It had a 7.9 mm MG 17 fixed in the starboard wing, a somewhat lower engine profile for improved pilot visibility, and the cowling tapering downward more dramatically. The single rudder was enlarged compared to that of the V2 and V3, the rear cock-

---

* Additional details, plus the fate of the 13 He 118s built, can be found in volume 5 of this series in German.

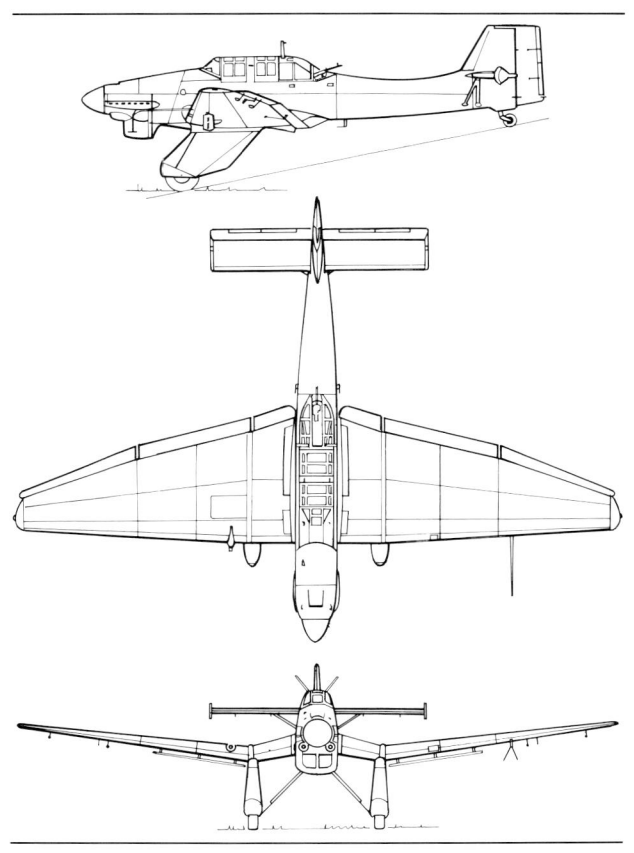

Junkers Ju 87A-1 (1937).

Also in late 1937 the improved Jumo 210Da became available which, fitted with a two-stage turbocharger, provided a takeoff rating of 500 kW/680 hp and 493 kW/670 hp at an altitude of 4,200 m. Those aircraft fitted with the Jumo 210Da were designated the Ju 87A-2 series. Production took place at full steam in Dessau up until the late spring of 1938, finally drawing to a close in the early summer of that year. Even with this series' minor improvement in performance the Ju 87's flight characteristics—despite the type's accuracy in Spain—were not considered optimal. Loaded with a 500 kg bomb, it generally had to fly without a second crewman and was barely able to reach speeds of 300 km/h.

Therefore, two aircraft were pulled from the A-0 series, the V6 and the V7. Beginning in late 1937 the pair were converted over to accept the much more powerful 30 liter Jumo 211 aircraft engine, requiring many changes before the much heavier engine could be fitted. Until this Ju 87B series became available, a total of 262 Ju 87As were built. 192 were produced at the Dessau parent factory and 70 under license at Weserflugzeugbau in Bremen, which the RLM had designated as a license company for the Ju 87 due to the fact that Dessau was preparing to initiate the Ju 88 program.

B-Series

The Ju 87V6 became the forerunner of the B-series and flew for the first time in early 1938. It was followed by the V7, which had undergone further modifications, so that only the wings and the empennage were fully identical with the A-series.

The pre-production series for the B model, the Ju 87B-0, was fitted with the same type of Jumo 211A engine that had been tested in the V7. It provided 736 kW/1,000 hp for five minutes at takeoff and 680 kW/925 hp at 4,500 meters with an rpm of 2,300 min^{-1}. Over the course of 1936 the fuel injected Jumo 211Da, with 883 kW/1,200 hp takeoff and boost rating at 2,400 min^{-1} became available in place of the carbureted version. It became the standard engine for the Ju 87B-1 production. This variant was a major advance over the A-series; it was also tested in Spain without delay, with five of the first production Ju 87B-1s being sent there in October 1938 to replace the three Ju 87A-1s already serving.

The Ju 87B-1 not only had the more powerful Jumo 211Da, but also an improved canopy with fore and aft sliding sections. The latter incorporated a lens mounting for the rearward-firing flexible MG 15. The B-1 also had a second fixed MG 17 in the port wing and a hydraulic dive recovery system like the one which had been tested in the Ju 88 in late 1938. An improved variable pitch propel-

pit was designed better, and the aft fuselage narrowed to increase the field of fire. The wing leading edge was kept straight to facilitate ease of production, and the dimensions of the "pantaloon" undercarriage were reduced somewhat. The V4 was fitted with a bomb release trapeze under the fuselage, which was used to guide the centrally carried 250 kg or 500 kg bomb away from the fuselage and clear the propeller arc.

The Ju 87A-1 production machines were therefore generally based on the V4 and began rolling from the assembly line in early 1937. Six dive bomber groups, each with 39 aircraft (12 per squadron plus three for the group staff flight) were to have been equipped by these aircraft.

In late 1937 a flight of three Ju 87A-1s from one of the groups was detached to Spain for combat testing with the Condor Legion (see p. 23).

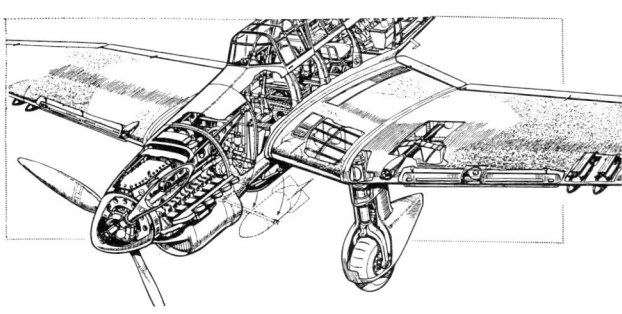

Junkers Ju 87B-1 showing the bomb trapeze fitted to Ju 9\87s beginning in 1936. The trapeze was designed to swing the centrally mounted bomb away from the fuselage and clear the propeller arc. The cutaway sections clearly reveal the two 50 kg bombs under each wing, the dive brakes beneath the wing leading edge, and the two fixed 7.9 mm MG 17s just outside of the undercarriage legs. The port machine gun was not fitted until the Ju 87B-1 and its more powerful Jumo 211Da arrived on the scene.

ler in combination with a constant speed governor for the engine reduced the chances of overspeeding the engine in a dive. The dive brakes kept the speed to between 500 and 600 km/h, depending on dive angle, not only ensuring a constant dive rate, but also an accurate alignment with the target by the entire aircraft.

The higher weight of the more powerful motor meant that the airplane's average center of gravity shifted, necessitating a change in the wing-fuselage join. The already emaciated trouser legs were

Junkers Ju 87B-1 in factory painted markings.

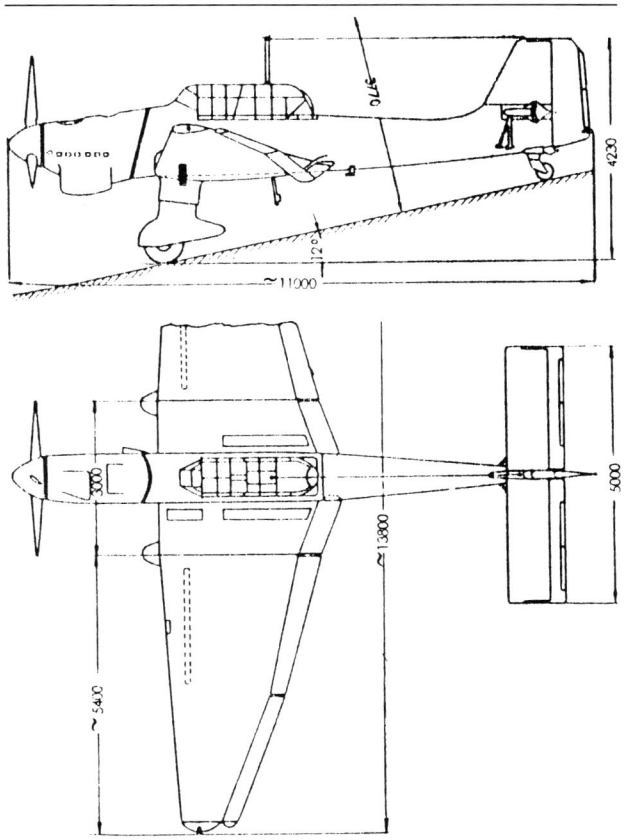

Main dimensions of the Junkers Ju 87B-1.

replaced by smaller, more aerodynamic coverings for the wheels and struts. Thus modified, this Ju 87 was able to regularly carry two crewmen and a central underfuselage 500 kg bomb, or a 250 kg bomb plus four 50 kg bombs under the wings.

Now fully front-line capable, the Ju 87B-1 was the standard dive bomber of the *Luftwaffe*.

Technical Data and Details of the Ju 87B-1

The Ju 87B-1 was a single-engined land-based dive bomber powered by a Jumo 211A engine.

> **Airframe**
> *Fuselage*: Monocoque, oval, stressed-skin design with four spars and vertically arranged bulkheads. Fuselage and wing center section joined.
> The operating area (pilot and gunner cockpits) of the fuselage was blocked off at the engine by a firewall and at the aft section by a solid bulkhead. Pilot seat was adjustable manually; observer's seat could

swivel. Both seats were designed to accommodate seat parachute packs. Both pilot and gunner sections were each covered by a removable canopy section.
Pilot and gunner sections were separated by a strong anti-roll bulkhead.

Undercarriage: Landing gear was designed as a single strut type. Shock absorbers cushioned the wheels and tailwheel. Medium-pressure tires fitted to wheels and individually braked via hydraulic foot brakes.
The undercarriage halves were interchangeable down to their individual components.
A 360-degree swiveling, shock absorbed tailwheel was located at the aft end of the fuselage which could be locked in place from the cockpit.
Control surfaces: Braced double wing horizontal stabilizer with trim tabs adjustable from cockpit. Tailplane hydraulically adjustable. Trimmed automatically when hydraulically driven landing flaps activated. Central, tapered vertical stabilizer with trim tab adjustable in flight.

Control inputs: Ailerons and elevators operated via a control column, rudder control by means of pedals. Flaps and tailplane adjusted hydraulically by means of levers on the left fuselage sidewall. The dive brakes mounted beneath the wings were also hydraulically operated using a lever on the left fuselage sidewall. Elevator deflection was reduced by a stop (safety control), preventing overly sharp, unauthorized pullouts.
The dive recovery system, "armed" automatically with the activation of the dive brakes, initiated the pullout sequence via the movement of the elevator trim tabs upon bomb release.

Wings: Cantilever design, broken down into a center section fixed to the fuselage, and outer sections attached to the center section by ball joints (four each side). The wings utilized the Junkers double wing design; flaps were in three sections.

Powerplant

Engine: Jumo 211A liquid-cooled twelve-cylinder four-stroke engine, 60° inverted V-type. Fuel injected with turbocharger and automatic pressure regulator. The turbocharger could be switched from low-level to high altitude via a two-gear transmission.
Gearing to propeller was 1.55:1
Maximum performance at ground level = 588 kW/800 hp
at 5,500 meters' altitude = 662 kW/900 hp

Propeller: Three-bladed Junkers variable pitch metal airscrews (Hamilton type) with automatic pitch control. Diameter = 3.4 meters, pitch variation = 20°.

Tanks: One self-sealing fuel tank in either side of wing center section with a total capacity of approx. 480 liters. One non-protected oil tank between bulkheads one and two with a capacity of approx. 37/47 liters.

Performance

Maximum speed
    at sea level = 340 km/h
    at 4000 m = 380 km/h
Cruise speed
    at 4,600 m = 282 km/h
    Service ceiling = 8,000 m
    Fuel consumption rate @ cruise (5,500 m) = 225 g/hp/hr
    Range @ cruise (5,500 m) = 550 km

Dimensions

Wingspan	13.8 m	Wing area	31.9 m²
Height	4.27 m	Takeoff weight	4,235 kg*
Length	11.10 m	(normal)	
Wheel track	2.96	Wing loading	132.5 kg/m²
		Power loading	4.7 kg/hp
*empty weight	2,745 kg	oil	36 kg
add'l equip.	325 kg	payload:	
crew (2 man)	200 kg	bombs	500 kg
fuel	380 kg	ammunition	49 kg

Ju 87B Dive Brakes and Automatic Recovery

The most pronounced feature of this rather unattractive and fearsome looking aircraft was the Ju 87's complete and uncompromising layout geared exclusively toward dive bombing as designed by Hermann Pohlmann and his team. One prerequisite for this capability was the design's extremely rugged construction, classed in the stress group HK5 (H was for *Hochleistungsflugzeug*, or high-performance aircraft, K stood for *kunstflugtauglich*, or aerobatic class, and the 5 represented the highest authorized stress classification). Another prerequisite was the virtually automatic rpm control for the engine and propeller, thus avoiding overspeeding the propeller. But the third—and really '*Stuka*-unique—prerequisite was found in the functional dive brakes and a reliable dive and recovery system to ease the load placed on the crew by the acceleration forces.

The purpose of the dive brakes was to keep the speed during a dive within a very narrow window, and as constant as possible. It made it easier for the pilot to approach close to the target, thereby increasing bombing accuracy.

The dive brakes were fitted as slats to the wing undersides of the Ju 87, as can easily be seen in the drawings and photos. Just

History of German Aviation: Bombers and Reconnaissance Aircraft

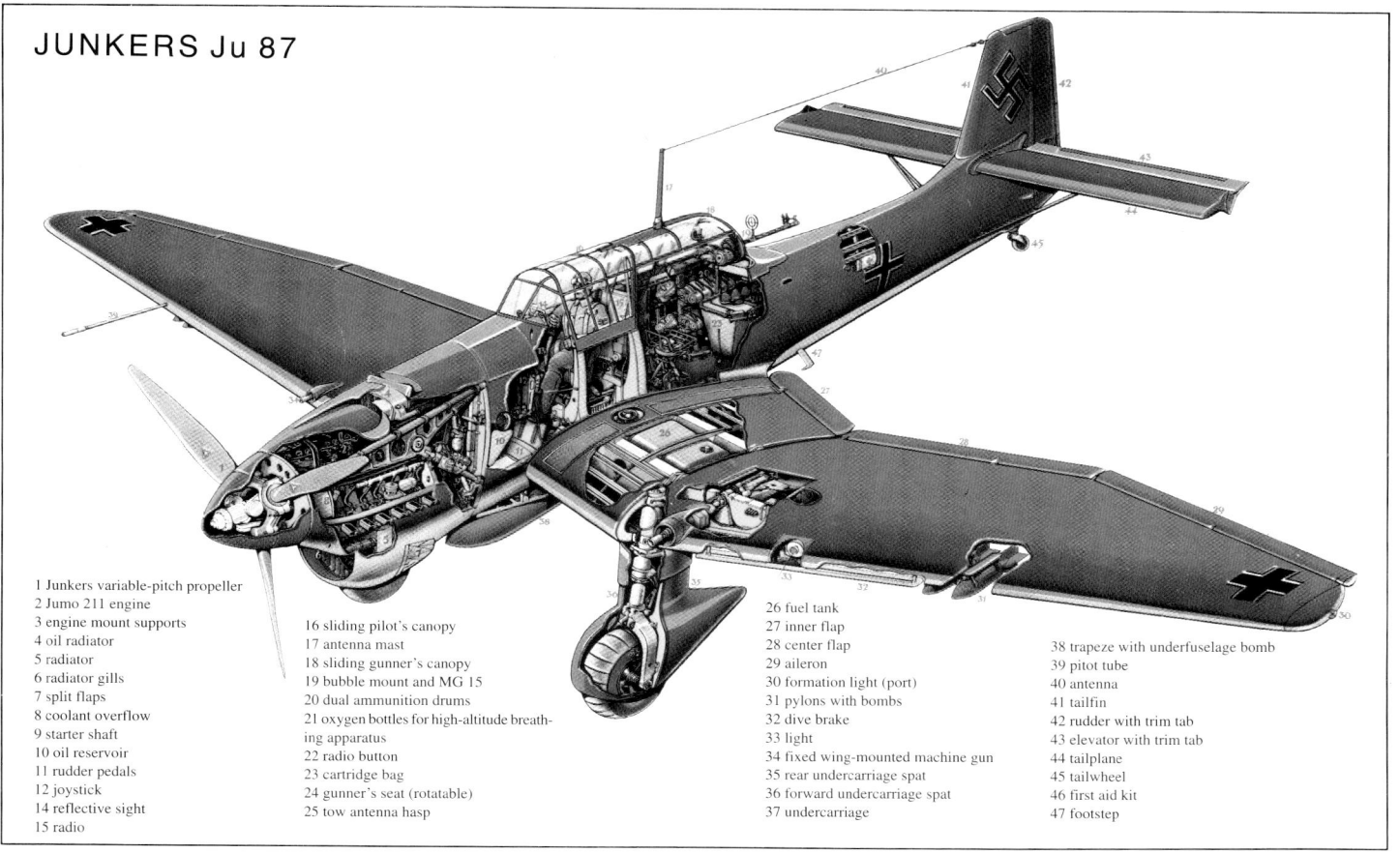

JUNKERS Ju 87

1 Junkers variable-pitch propeller
2 Jumo 211 engine
3 engine mount supports
4 oil radiator
5 radiator
6 radiator gills
7 split flaps
8 coolant overflow
9 starter shaft
10 oil reservoir
11 rudder pedals
12 joystick
14 reflective sight
15 radio
16 sliding pilot's canopy
17 antenna mast
18 sliding gunner's canopy
19 bubble mount and MG 15
20 dual ammunition drums
21 oxygen bottles for high-altitude breathing apparatus
22 radio button
23 cartridge bag
24 gunner's seat (rotatable)
25 tow antenna hasp
26 fuel tank
27 inner flap
28 center flap
29 aileron
30 formation light (port)
31 pylons with bombs
32 dive brake
33 light
34 fixed wing-mounted machine gun
35 rear undercarriage spat
36 forward undercarriage spat
37 undercarriage
38 trapeze with underfuselage bomb
39 pitot tube
40 antenna
41 tailfin
42 rudder with trim tab
43 elevator with trim tab
44 tailplane
45 tailwheel
46 first aid kit
47 footstep

Cutaway drawing of the Junkers Ju 87 showing breakdown of components.

before bunting over into a dive, the pilot hydraulically activated the brakes by moving a lever, which in turn moved a slide valve controlling the oil pressure for the brakes' retraction struts. Prior to this, the pilot throttled back, closed the radiator gills and, if necessary, switched the engine to low-level turbocharger and the propeller to maximum pitch. When the brakes extended, the automatic dive recovery system moved a trim tab on the right elevator to the up position, making the airplane nose-heavy, i.e. by activating the dive brakes the airplane had a tendency to naturally go into a dive as a result of the nose heaviness induced by the recovery system.

Once the predetermined release altitude had been reached, the pilot dropped the bomb(s), and the dive recovery system automatically moved the trim tab back into its normal position. This made the aircraft tail-heavy and initiated the pull-out. After recovering from the dive and retraction of the brakes, the radiator gills had to be opened up again so that the motor, which had been running full bore on approach to the target, would not become overheated. It goes without saying that the propeller pitch would need to be adjusted accordingly.

Another feature worth mentioning is the safety mechanism built in to prevent too sharp of a pull-out. This so-called safety control was activated when the dive brakes extended and was a simple means to prevent excessively hard pull-outs from dives. Once the safety control had been set, the control stick could only be pulled back some 5° from its normal position, which deflected the elevators only marginally upward. In emergency situations which required a sharper pull-out, the pilot could override the safety control by manually applying a force of about 30 kg.

History of German Aviation: Bombers and Reconnaissance Aircraft

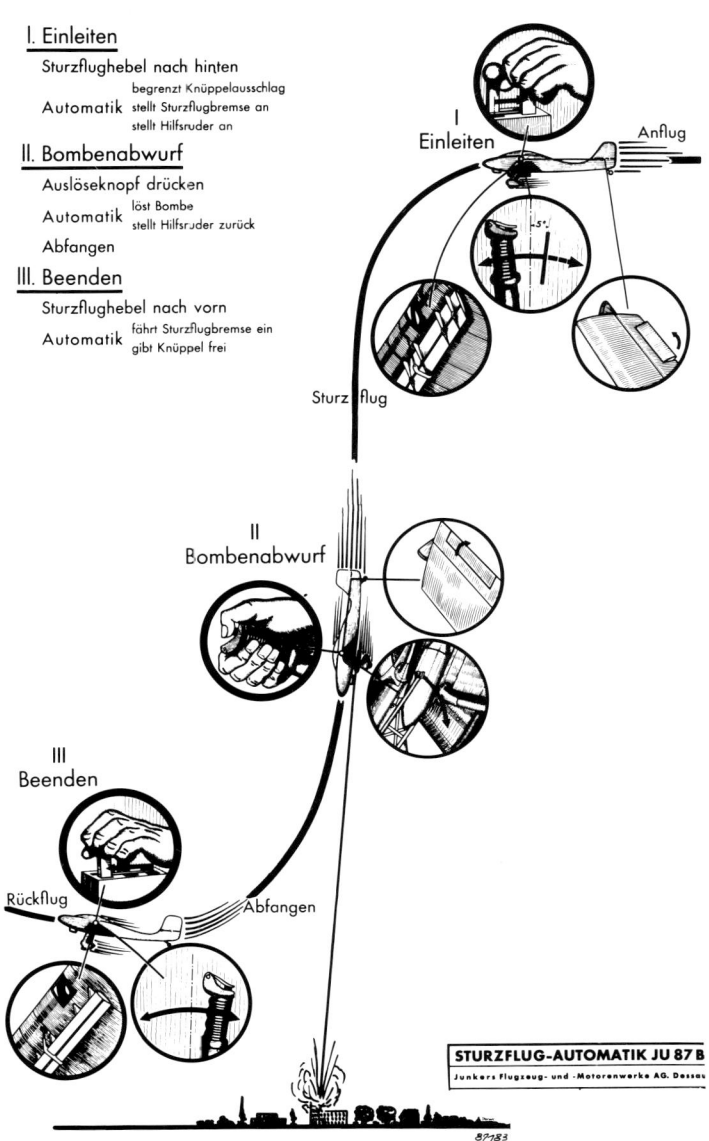

These company drawings show the automatic dive system of the Ju 87B in action and provide a better understanding of its operation.

I. Initiation
Dive lever pulled back
limits stick movement
Automatic System activates dive brakes
activates auxiliary controls
II. Bomb Release
Push release button
Automatic System releases bomb
moves auxiliary controls back
Pullout
III. Completion
Dive lever moved forward
Automatic System retracts dive brakes
releases control stick

Junkers Ju 87C with folding outer wings in anticipation of this variant's use on aircraft carriers.

C-Series

Finally, we take a look at an interesting development from 1938: for the "*Flugzeugträger A*" (Aircraft Carrier A, which rolled from the slipway in Kiel on 8 December 1938 as the Graf Zeppelin) Junkers developed a special, carrier-capable version of the Ju 87 B. Designated as the Ju 87C, this version was kitted out with catapult gear and an arrester hook for carrier landings, jettisonable undercarriage for ditching, fitted flotation devices, and manually folding outer wing panels. It went into production in the summer of 1939 at Weser Flugzeugbau, where the decision was made to equip the Ju 87C-1 version destined for carrier operations with electrically folding wings, the ability to carry a torpedo in place of bombs, and additional fuel tanks in the wings. Those few C-0 variants rolling off the assembly line went to Kiel-Holtenau, where a *Stuka* squadron had been formed in December 1938 with Ju 87As as part of the carrier group for the Graf Zeppelin. There they began practicing carrier landing techniques on simulated carrier decks. By early September 1939 this squadron had expanded and been refitted to include 12 Ju 87B-1s and C-0s.

With regard to the B- and C-series, by this time production of the Ju 87B-1 at Weser Flugzeugbau in Bremen was in full swing, reaching a monthly output of over 60 aircraft in mid-1939. By September 1939 all *Stuka* groups had converted over to the Ju 87B-1 after the Hs 123 was reclassed from a dive bomber to a ground attack aircraft on 1 August 1938.

Synopsis

The Ju 87 had been selected as the standard dive bomber in the summer of 1936 as a result of the direct and unrelenting involvement of Ernst Udet. Its robust design, fully controllable dive capability, and its proven record in Spain (albeit under limited conditions) aroused a well-founded hope within the *Luftwaffe* leadership with regard to the type's combat potential.

Despite the refined perfection of dive bombing techniques, the Achilles' Heel of accurate dive bombing was (and remains) the necessity of plunging directly into the enemy's AAA firepower around the target—with virtually no hope of avoiding it. Also, a fully laden *Stuka* formation—despite the good maneuverability of a single plane—was relatively slow, equipped with weak defensive armament, and thus as a rule dependent upon close fighter escort.

A tragic event on 15 August 1939 was an evil omen for things to come: on that day a *Stuka* group was to carry out a dive bombing demonstration using smoke bombs at the Neuhammer training grounds in Silesia. The weather forecast for the training area had predicted good weather below 7/10ths cloud cover, which extended from 900 to 2,800 meters above ground level. The formation leader planned on approaching the target area with his 30 Ju 87s flying in close formation at 5,000 m, diving through the cloud layers and then apparently releasing the smoke bombs at an altitude of 300 m. This traditional version of the attack seems less probable in light of the fact that a realistic, average release altitude at the time was somewhere around 1,000 m above ground. As this was a routine training demonstration without any defensive fire in the target area, the intended release altitude could safely have fallen below 1,000 m, but just 300 m? A lone airplane might just have been able to pull out, but this was a close flying formation of 30 planes. Be that as it may, the report of a ground fog which had suddenly appeared in the target area, making the weather situation much worse, failed to reach the approaching formation in time. During the final phase of the formation dive, the formation leader recognized the radically changed situation; for many of the crew, however, his warning came too late, and for 13 planes the dive ended on impact with the ground and explosion—26 men died a flyer's death!

This loss did nothing to detract from the type's effectiveness. It had been due neither to poor maneuverability nor to inadequate training, for the planned close formation dive by a group of aircraft through cloud layers required remarkable abilities in tight formation flying, or split-second timing in separation while maintaining precise headings.

Compared with France, which had over two dozen aircraft classed as dive bombers, and Great Britain, which had a good dozen of the type, Germany's dive bomber arm can be considered a one-of-a-kind instrument of war designed primarily for the tactical role and embodying many strong points, as well as inherent weaknesses.

With a total inventory of 366 Ju 87Bs on 2 September 1939, the type comprised 23% of the available operative and tactical bomb platforms (bombers, dive bombers, and ground attack aircraft).

Thanks to the Ju 87's good takeoff and landing qualities, even from unprepared fields, and its ease of maintenance and technical reliability, the *Stuka* provided Germany's air arm with the mobility it needed to support its doctrine.

Strategic Reconnaissance Aircraft

Heinkel He 70 and He 45 Auxiliary Reconnaissance Platforms

The Heinkel He 70 was not an original military or reconnaissance airplane. Its military role, as well as that of Heinkel's He 45 reconnaissance platform, can be found in greater detail elsewhere. Since these two planes, both employed in the strategic reconnaissance role, had their origins prior to 1935 they will only be covered briefly here. The use of both types in the Spanish Civil War has also been discussed, with references to their unsuitability in the strategic reconnaissance mission.

In 1937 a good 80% of the aircraft in the strategic reconnaissance groups (*Aufklärungsgruppen [F]*) were either He 45s or He 70s (a total of 135 of the former and 45 of the latter). Early that same year an export version of the He 70F-3 went to Hungary, where it was given the designation of He 170. All told, 18 He 170As were supplied to Hungary equipped with license-built French engines between September 1937 and February 1938.

On 19 September 1938 there were 73 He 70s in the inventory of Germany's strategic reconnaissance groups. However, there remained no He 45s, as these had already been withdrawn from service and over time replaced by the reconnaissance version of the Dornier Do 17. By this time there were already 149 Dorniers flying in front-line service.

Just one year later the He 70 had suffered the same fate, for by 2 September 1939 it could no longer be found in the inventory of the flying units. With the arrival of additional Do 17s, starting in late 1938 these had gradually been pulled from the inventory and replaced. Surviving He 70s generally soldiered on in the A and B class pilot training schools, the aerial reconnaissance school, and in courier squadrons.

Thus ended, for the time being, the era of the He 70 auxiliary reconnaissance aircraft in Germany. Heinkel was not even able to generate interest from either the military or the civilian sector with an improved performance He 70, designated the He 270V1, which had flown its maiden flight in the spring of 1938.

Dornier Do 17 Strategic Reconnaissance Plane

By 2 September 1939 the strategic reconnaissance groups were virtually all equipped with the Do 17P-1, with the Do 17Fs having been withdrawn from the units. When the war broke out there were 257 aircraft of this variant in service with reconnaissance units, which—aside from a few specialized aircraft with the

"*Aufklärungsgruppe des Oberbefehlshabers der Luftwaffe*" (Reconnaissance group of the Air Force Commander-in-Chief)—were stocked exclusively with Dornier-supplied products. How did this development occur?

F-Series

The history of the Do 17 bomber has already been covered in a previous section. Parallel to the bomber was the Do 17V8, developed as the pattern for a strategic reconnaissance version, the Do 17F-1. The Do 17V11 served as the prototype for the Do 17F-2 reconnaissance version, and both were built starting in 1936 (albeit the F-2 in limited numbers). The F-series was produced at the Dornier plant in Munich-Oberpfaffenhofen, under license at Siebel in Halle a.d. Salle, and at Hamburger Flugzeugbau.

Reconnaissance versions, like the E-series bombers, were fitted with three-bladed wooden airscrews driven by BMW VI engines (these being the 9-series without gearbox). In place of the bomb racks in the bomb bay, these variants carried three cameras of varying focal length and formats. The Do 17F also flew with a three-man crew and, like the E-series bomber, lacked any type of heating system—making for an uncomfortable flight and limiting its poor weather usefulness. With a top speed of 356 km/h it was just as fast as the E version. However, it had just two flex-mounted MG 15s installed, one with a field of fire to the rear and up, and the other with a limited rear field of fire down. Externally, the F reconnaissance version differed from the E bomber variant only in the absence of a single machine gun in the nose (*A-Stand*), and by having round windows in the fuselage underside for the cameras, the lenses of which were protected by covers. Further equipped with a hand-held camera, the Do 17F was suited to the tactical reconnaissance role and was also assigned to army reconnaissance groups (*Aufklärer H*) in limited numbers. With the experience gained in Spain, it was also developed as a strategic reconnaissance platform for nighttime operations. For night missions, the large camera (Rb 50/18 or Rb 50/30) was removed, and a magazine called an *Elvemag 1000/IX* for carrying six flares was fitted in its place.

Beginning in the first months of 1937, by April of that year the first strategic reconnaissance group had been fully equipped with

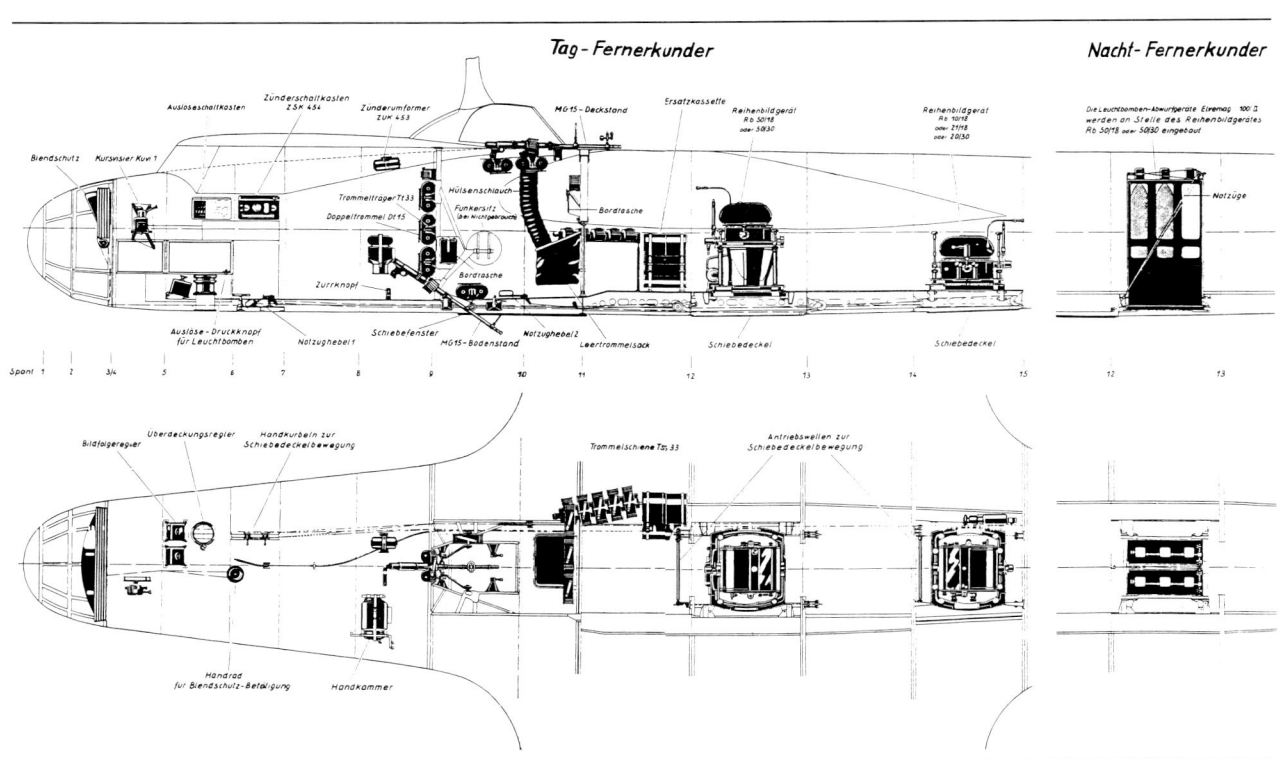

Dornier Do 17F, a long-range reconnaissance version which carried the armament shown here in March 1938, the "armament" of a reconnaissance aircraft being chiefly its aerial cameras.

36 Do 17F-1s. That same spring a complete squadron from this group, 15 Do 17F-1s in all, went to Spain to test the type under operational conditions. This became the *1.Staffel* of A/88 under the Condor Legion and replaced the Heinkel He 70 as a strategic reconnaissance platform. The Do 17F's clearly positive experience in Spain has already been mentioned, as has that of its follow-on replacement, the Do 17P-1 series. Ten of the latter type's initial production batch had been sent to Spain in 1938, where they replaced the F-1.

P- Series

This was the reconnaissance version developed in tandem with the M-series bomber. The DO 17P made its first flight on 18 June 1938. It was built at Dornier, but also under license at Siebel, Henschel, and the Hamburger Flugzeugbau. With its more powerful motors (the nine-cylinder BMW 132N radial engine rated at 636 kW/865 hp on takeoff at 2,450 min^{-1} rpm and 1.35 *ata* pressure), the variant steadily replaced the less powerful Do 17F types in the units. Altogether 330 examples of the Do 17P were built, eight of these at Dornier, 73 at Siebel, 100 at Henschel, and 149 at Hamburger Flugzeugbau.

Armament and photo equipment for the Do 17P as it was available in August 1939 is shown in the drawing on page 98. This was at a time when the P had virtually replaced all the Do 17F-1s in the field.

Notable is not only the third machine gun installed in the *A-Stand*, but also the fact that the Do 17F-1's interior bomb bay for carrying six flares was exchanged for external racks (ETC 50/VIIIc), mounted underneath the center fuselage of the Do 17P for carrying four LC 50F parachute retarded flares, each weighing 20 kg. However, this less practical aerodynamic "feature" was created at the expense of using the large *Reihenbildner* 50/30x30 camera, since the bomb racks were directly underneath this camera.

Technical Data and Details of the Do 17P

The Do 17P generally resembled the Do 17M bomber version, which—as previously mentioned—had been relatively quickly replaced by the Do 17Z after only 200 had been built. The following details apply to the Do 17P reconnaissance version, of which 330 were built. The Do 17P was a reconnaissance airplane with a crew of three, conceived as a twin-engined cantilever mid-wing design with retractable landing gear of all-metal stressed-skin construction. Its technical details are as follows:

Airframe

Fuselage: Oval, nearly circular cross-section of monocoque construction. Longitudinal members followed the contours of the bulkheads up to the four main spars, which passed through the frame section. Fuselage (crew compartment and camera bay) heated.

Undercarriage: Electro-mechanical landing gear, retracting into the engine nacelles. The tailwheel, swiveling through 360°, retracted into the fuselage. Undercarriage and tailwheel could be retracted manually via a hand crank.

Control surfaces, consisting of:
a) ailerons, fabric-covered slotted type with anti-torque metal leading edge. Counterbalance in the leading edge. When landing flaps activated these could be deflected to a 15 degree angle, thus supplementing the flap effect.
b) landing flaps: between aileron and fuselage, split flap type. 60° maximum deflection.
c) horizontal stabilizer: cantilever, twin-spar tailplane. Pivoted at the rear spar attachment point for trim effect. Counterbalanced elevator controlled from fin.
d) vertical stabilizer: double keeled capping the tailplane. Same construction as the horizontal stabilizer

Control inputs: Control rods running to the rudder/elevator control levers. Flaps coupled with elevator trim. Controlled either manually or electro-mechanically. Rudder trim controlled mechanically by means of a rotating knob.

Wings: Twin-spar, stressed skin covering with heated anti-icing leading edges.

Powerplant

Engines: Two air-cooled nine-cylinder radial BMW 132N, fuel injected with single stage planetary gearing.
Reduction ratio to propeller was 0.62:1

Performance boost endurance	kW	hp	min^{-1}	*ata* boost pressure	
at low level:	636	865	2,450	1.35	1 min.
	563	765	2,350	1.25	5 min.
	427	580	2,150	1.05	no limit
at 4,500 m:	489	665	2,150	1.05	no limit

Propellers: Three-bladed VDM airscrew with basic position at the 11.20 o'clock position, corresponding to 25 degree angle. 3.7 m diameter.

Tanks:
a) Protected fuel tanks broken down as follows:
- two wing tanks, 760 liters each = 1,520 l
- one fuselage tank = 350 l
- one fuselage tank = 225 l

total 2,095 l

including approx. 50 liters unusable fuel
b) Two protected oil reservoirs, each with a capacity of 95 l = 190 l

Performance

a) Speed (at standard atmosphere)

altitude	maximum speed
sea level	350
1,000 m	367
2,000 m	383

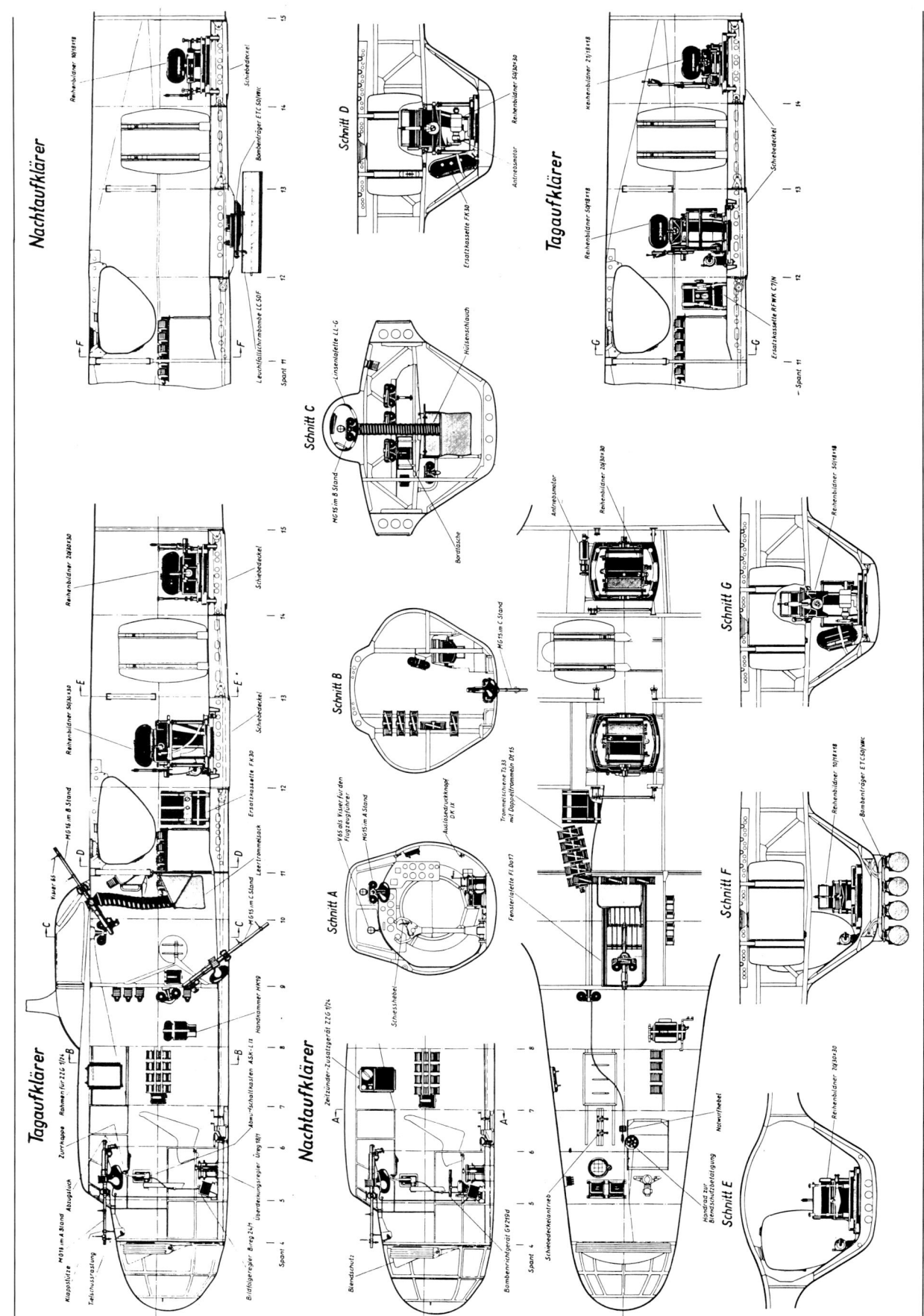

Dornier Do 17P, an improved long-range reconnaissance platform. Overview of the variant's armament and camera layout.

History of German Aviation: Bombers and Reconnaissance Aircraft

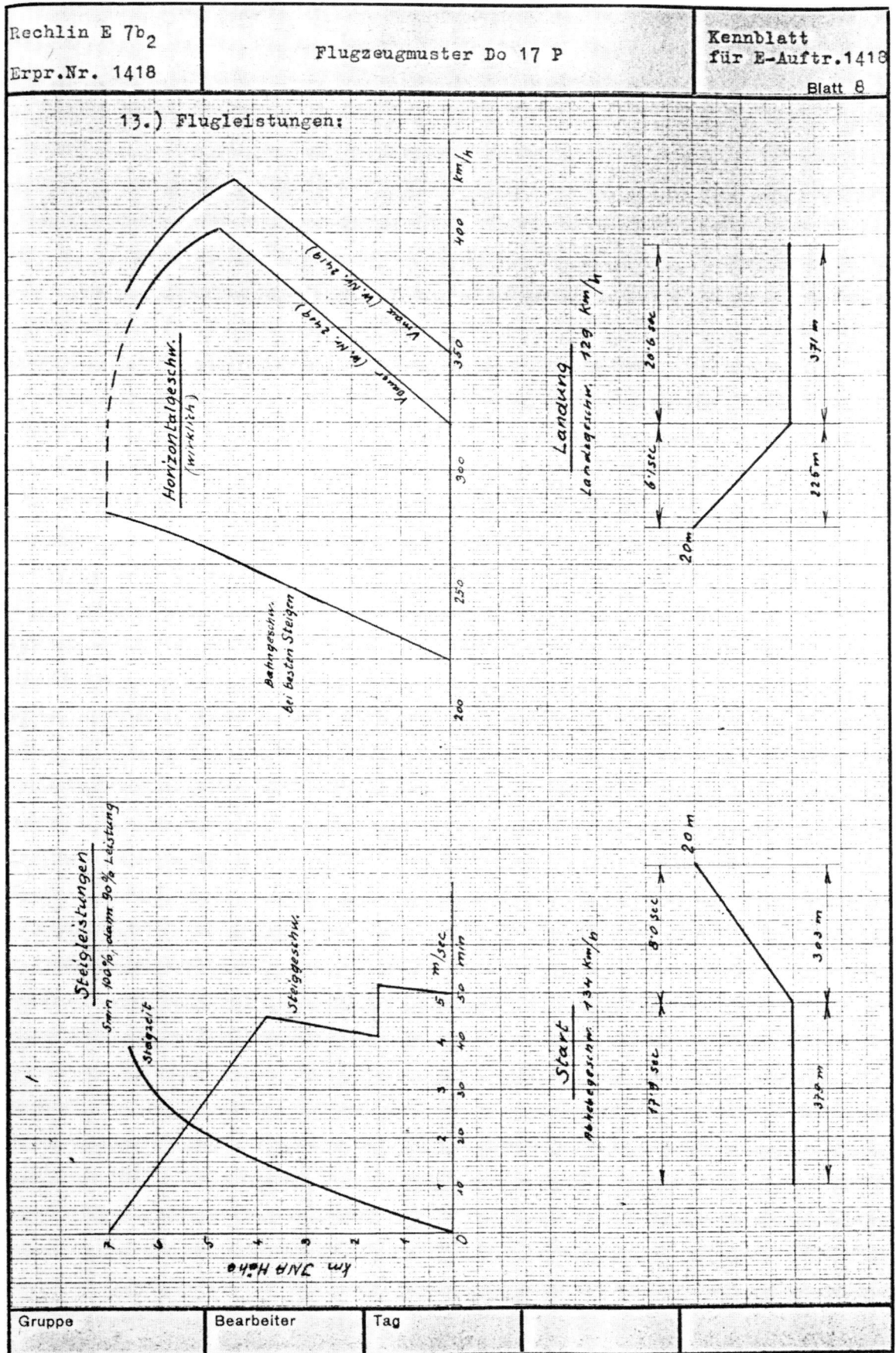

The performance data for the Do 17P test flown at Rechlin are shown on this sheet, part of evaluation contract no. 1418. These values apply to the airframe tested there, which, for example, reached a maximum speed of 416 km/h at 4,000 m (422 km/h at 4,300 m), while Dornier had claimed only 410 km/h maximum speed at 4,000 m in its company data. This was not always the case, as shown in appendix 8, which shows the realistic performance values obtained by the *Luftwaffe* General Staff. For evaluation purposes, the *Luftwaffe* based their figures on standard aircraft flown by the conversion unit and at the Rechlin Test Center.

3,000 m   400
4,000 m   416
5,000 m   415
6,000 m   394 (all speed values are with engine rpm at 2,350 min^{-1} and boost pressure at 1.25 *ata* [up to 4,000 m altitude, then based on max throttle setting] and optimum propeller pitch setting)

b) Climb performance (for 5 min at 100% engine power, then at 90% power)

altitude	climb rate in min.	climb speed in m/s	actual speed in km/h
sea level	-	5	220
1,000 m	3.5	5.1	229
2,000 m	7	4.2	238
3,000 m	11	4.4	248
4,000 m	15	4.2	257
5,000 m	21	2.8	267
6,000 m	29	1.5	270

c) Takeoff and landing characteristics

Best takeoff: from static to 20 m altitude = 682 m, separation speed was 134 km/h at 25 degree flap setting.

Best landing: from 20 m altitude to stop = 596 m, landing speed 129 km/h at 55 degree flap setting.

d) Ranges (based on 2,095 liter fuel capacity (minus unusable 50 liters) for climbing to given altitude, level flight at same altitude, and landing descent with optimal propeller setting)

altitude	engine load	ata	rpm min^{-1}	speed	endurance (hr:min)	range (km)
sea lvl	constant	1.05	2,150	320 km/h	4:25	1,400
	max range	0.95	2,000	290 km/h	5:20	1,550
2000 m	constant	1.05	2,150	355 km/h	3:55	1,350
	max range	0.85	1,900	300 km/h	5:55	1,750
4000 m (max pressure alt)						
	constant	1.05	2,150	390 km/h	3:25	1,250
	max range	0.80	1,800	310 km/h	6:00	1,875
6000 m	constant (at full throttle)	0.87	2,150	380 km/h	4:45	1,700
	max range	0.80	2,050	355 km/h	5:25	1,825
6700 m	constant/					
	max range	0.80	2,150	360 km/h	5:10	1,725

From this intentionally detailed performance overview it becomes apparent what factors have a bearing on the maximum performance value in question, here a range of 1,875 km, in the majority of tables.

Dimensions

Wingspan	18.00 m	Wing area	55 m²
Height	4.55 m	Wing loading	140 kg/m²
Length	16.25 m	Power loading	4.81 kg/hp

* The 7,730 kg takeoff weight was broken down for the night reconnaissance version as follows:

empty weight	4,925 kg	fuel in fuselage tank I	262 kg
add'l equipment	625 kg	fuel in fuselage tank II	172 kg
crew	300 kg	oil	160 kg
fuel in wing tanks	1,150 kg	flares and cartridges	13 kg
payload: a) bombs - 80 kg, b) ammunition - 38 kg			

S-Series and Do 17Z Multi-Role Platform

With its newer, more spacious forward fuselage section, four-man crew, and improved armament, the Do 17S-0 high-speed reconnaissance version has already been addressed in context with the Do 17Z bomber. As only three pre-production aircraft of this series were ever built, despite its effectiveness it had absolutely no impact whatsoever on the strategic reconnaissance potential of the *Luftwaffe*.

Also mentioned in detail was the Do 17Z-3, a variant best classified within the Do 17Z program as a combined bomber and strategic reconnaissance aircraft. Only 17 of the type were built from late 1939 onward as HQ squadron reconnaissance platforms.

Strategic Recce Hodge-Podge

In the summer of 1934 a strategic reconnaissance unit was formed from within the flight readiness department of the RLM. This unit was christened with the name "*Fliegerstaffel z.b.V.*" (lit. "Special Operations Flying Squadron") and was based at Staaken near Berlin. Leader of this unit was *Hauptmann a. D.* Theodor Rowehl, who had previously flown as an observer in the Imperial Navy. He gained his initial experience carrying out clandestine strategic reconnaissance missions for the *Reich* Navy, flying in Junkers F 13 catapult floatplanes. This type had been operating from ocean-going steamers since 1929, and some of them were later fitted with automatic aerial cameras (as was, for example, the Junkers W 33).

The "*Fliegerstaffel z. b. V.*" flew the widest variety of aircraft types, initially operating exclusively with civil registration markings, and was renamed the "Hansa-Luftbild GmbH - Erprobung" (Hansa Aerial Photography, Inc. - Evaluation) in 1936, although in fact it had nothing to do with the actual "Hansa-Luftbild GmbH"

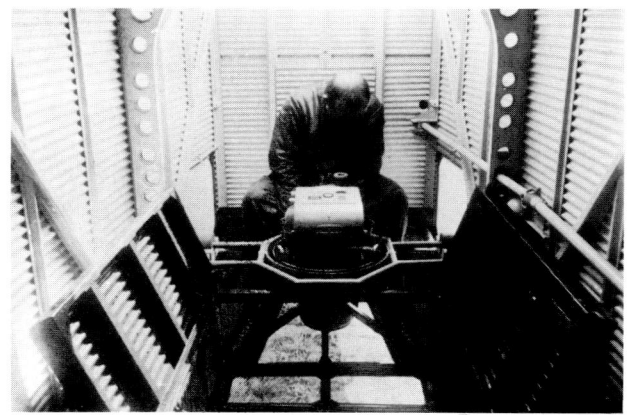

Camera installation in the Junkers W 33 (built from 1926 onward), a follow-on development of the Junkers F13.

History of German Aviation: Bombers and Reconnaissance Aircraft

company. Included in its inventory were the single engine He 70F-2 and He 46F-0, as well as the twin-engined Ju 86V9, He 111V22, and the Do 17V14. The last two types went to the Rowehl unit in 1938, where by the war's outbreak they were reinforced by the He 111V2, three examples of the He 111J-1 reconnaissance aircraft (modified from its original torpedo bomber design), plus a former civilian He 111C-0, as well as the Do 17R-2, a Do 17F-1, and an additional Ju 86C/D. The bomb bays and torpedo racks fitted to the He 111s were removed by Rowehl's unit and replaced by automatic aerial cameras on the one hand and additional fuel tanks for increasing range on the other.

It was with this motley collection of disguised reconnaissance aircraft, whose crews still continued in civil service following the "unmasking" of the *Luftwaffe* in 1935, that strategic reconnaissance missions were repeatedly flown over eastern and western Europe. Using special high-resolution cameras, the missions in the east—especially over Poland—provided planners with extremely accurate data.

By 1939 this reconnaissance organization, operating in the clandestine reconnaissance role even in peacetime, had blossomed into the *"Aufklärungsgruppe des Oberbefehlshabers der Luftwaffe"* (AufklGr Ob. d. L., or CinC AF Reconnaissance Group) with several squadrons. These units continued to function under the command of Rowehl, flying high altitude reconnaissance missions over France and England and enabling remarkably accurate target data and mapping to be compiled for the aerial assault units.

Synopsis

The Do 17P was virtually standard equipment for all strategic reconnaissance groups and possessed excellent photographic equipment on board for day and night operations. On the eve of WWII, the *Luftwaffe* was undoubtedly well equipped in this particular area for conducting an operative war. With the AufklGr Ob. d. L., created for the purpose of secretly performing strategic reconnaissance missions, it had a capable and experienced special unit at its disposal for reconnoitering the sensitive target areas of potential enemies, even in peacetime.

Tactical Reconnaissance Aircraft

The *Luftwaffe's* First Generation of Tactical Recce Planes

The He 46 was the embodiment of the motto of the *Luftwaffe* leadership of the day, which stated that a device of limited usefulness readily available was better than no device at all. The type survived for a remarkably long time in the inventory of the reconnaissance units, although in many respects it was severely lacking. The Heinkel He 46 was a type which metamorphosed from a biplane into a high-wing monoplane design. There follows here just a brief sketch from the history of this robust and long-lived aircraft design, not unjustifiably called by its pilots by the nickname of *"Rüttelfalke"* (Kestrel, a bird of prey known for its ability to hover), despite its shortcomings.

In early 1935 the He 46 was the standard plane of the tactical reconnaissance squadrons, which had 84 of the type at the time.

Foreign Interest

In 1936 three foreign air forces simultaneously expressed an interest in this airplane.

The Spanish Nationalists were the first to receive the type, purchasing a lot of 20 He 46C models in 1936. However, these were pulled from front-line service in 1938 when the machines proved to have inadequate flight performance.

Also in 1936, the Bulgars took delivery of 18 examples of specially modified He 46C-2s, which were shipped early that year with a more powerful British engine under the designation He 46eBu (for Bulgaria).

The Hungarians, too, were shipped a total of 36 planes, designated He 46eUn (for Hungary). These were fitted with a French engine built under license in Hungary.

The *Luftwaffe's* Armament Situation

In 1937 75% of the tactical reconnaissance units within Germany's *Luftwaffe* were equipped with the He 46C, operating a total of 162 aircraft of this type—twice the number of the previous year. Beginning in the spring of 1938, these were complemented by the first of the new Henschel Hs 126 reconnaissance platforms. On 19 September 1938 the number of He 46s in front-line service had risen to 189, supplemented by another biplane first conceived in 1931, the He 45. This latter type had originally been intended as a strategic reconnaissance airplane, and was represented by 58 examples serving in tactical reconnaissance squadrons and a further 78 flying with ground attack squadrons.

At this time, the Henschel Hs 126 was represented in front-line units by 42 examples of the A-0 and A-1 series, and were greatly loved by those who flew and benefitted from them. These front-line units were part of the *Aufklärer H* organization, as those tactical reconnaissance units acting primarily on behalf of the army were designated at the time.

In the fall of 1939 the armament scale had tipped further in favor of the better performing Hs 126. Of the 356 tactical reconnaissance aircraft on hand on 2 September 1939, 77% (275) were Hs 126s and just 19% (67) were He 46s, with the remaining 4% (14) made up by the He 45.

Henschel Hs 126 Tactical Reconnaissance Plane

From the Hs 122 to the Hs 126

The first prototype of the Hs 126, the Hs 126V1, took to the air on its maiden flight in the autumn of 1936, but the type had its roots in the two-seat Hs 122 tactical reconnaissance airplane.

There were calls for an improved successor to the previously mentioned He 46 as early as 1933. The requirement of the *C-Amt* (later *Technisches Amt*) specified a design which was more advanced both structurally and aerodynamically, with better slow-flying characteristics and improved STOL features than evidenced by the He 46. Henschel accordingly built such a design under the supervision of its chief design engineer, *Dipl.-Ing.* Nicolaus, the first prototype of which took to the air in early 1935 as the Hs 122a with a British inline engine. The second and third prototypes followed shortly thereafter, powered by German engines, and by the summer and fall of 1935 the He 122's testing and evaluation was well underway in Faßberg.

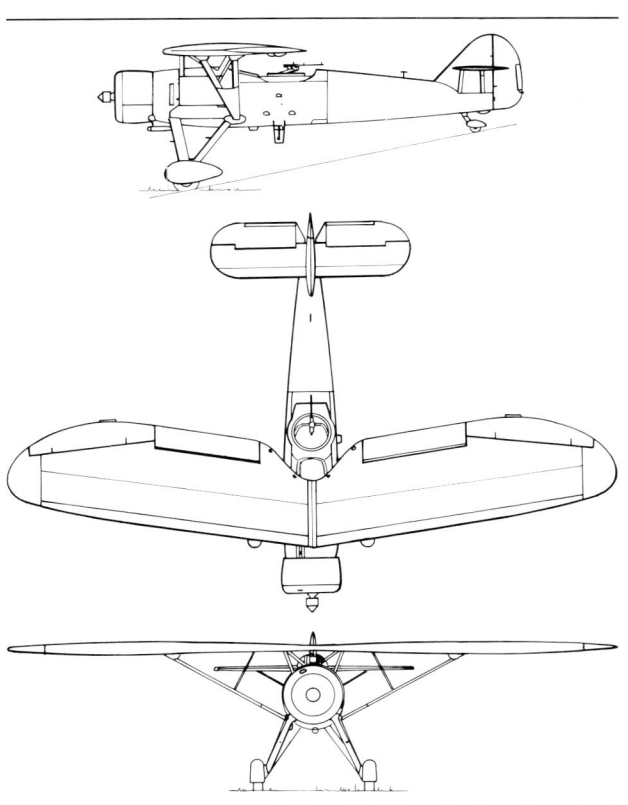

Henschel Hs 122B-0 (1936).

With regard to its STOL and slow flying characteristics, the results were better than the requirements specified. At 250 km/h its maximum speed, however, barely exceeded that of the He 46C at 210 km/h.

The *Technisches Amt* therefore proposed that the Hs 122 be developed further to accommodate the imminent and more powerful Bramo 323 Fafnir engine. The *Amt* issued a contract for a pre-preproduction run of seven Hs 122B-0s to be used for conducting additional evaluative flight studies. These were assembled in the late spring and summer of 1936, with the fourth airframe being modified to become the pattern for the prototype of the improved-performance follow-on development, designated as the Hs 126. This machine was built in the fall of 1936 and resumed flight testing as the Hs 126V1.

Hs 126 V-types and the A-0 Series

As the Bramo Fafnir was not yet available at the time the V1 was first fitted with the Junkers Jumo 210C engine, the reputable, liquid-cooled twelve-cylinder inverted-V inline engine.

The V2 and V3 followed in the spring of 1937 and were powered by the Bramo 323A-1, an air-cooled nine-cylinder radial type providing 610 kW/830 hp at an altitude of 4,000 m and driving a three-bladed VDM variable-pitch propeller. Over the summer of 1937 the V2 and V3 were subjected to a rigorous flight test program. They were pleasant to fly, had somewhat sluggish reaction, but had good stall handling characteristics. The type's short takeoff and landing features were quite impressive. It was of a robust design, particularly with regard to the single-strut landing gear which proved to have excellent cushioning and absorbed even the hardest landing impacts. The struts were attached to the fuselage sidewalls and were designed to spread apart when subjected to load stress. The actual shock absorbers within the fuselage (so-called Uerdinger annular springs) responded based on the principle of lever action. Beginning with the V3 the undercarriage legs were positioned perpendicular to the fuselage centerline, whereas before they had been angled somewhat aft.

While the three prototypes continued with their testing, Henschel began work on then Hs 126A-0 pre-production aircraft powered by the Bramo 323 Fafnir engine. All these were finished by the end of 1937, and some of them went to tactical reconnaissance units for evaluation under field conditions.

A-1 and B-1 Variants

Immediately upon completion of the A-0 series the Henschel Works at Schönefeld and Johannisthal began laying out the jigs for the A-1 series, with the first machines being delivered to the *Luftwaffe* in the spring of 1938. The following three-view drawing shows this variant, which now corresponded to the idea of a tactical reconnaissance platform:

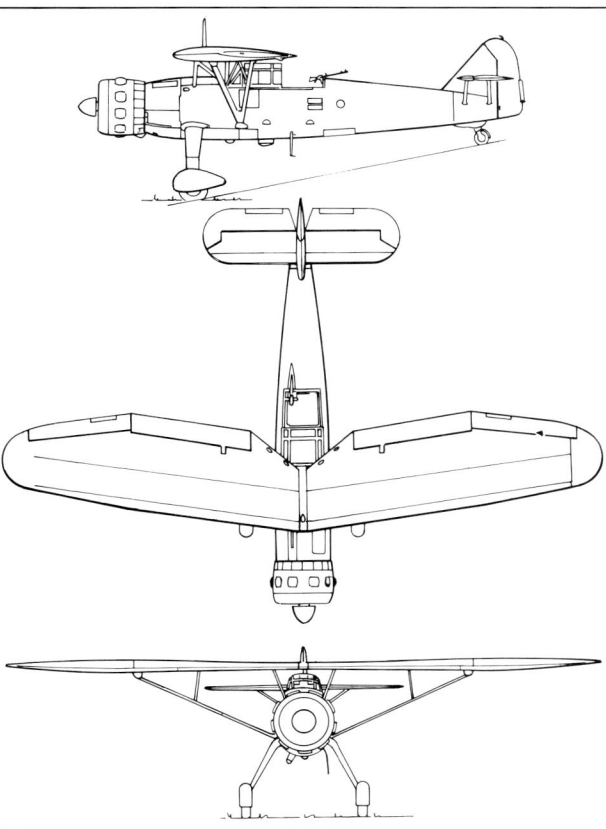

Henschel Hs 126A (1938).

Technical Data and Details of the Hs 126

The Hs 126A was a tactical reconnaissance aircraft with a two-man crew, conceived as a single-engined, braced high-wing design with fixed, single-strut undercarriage and of an all-metal, stressed-skin construction.

Airframe
Fuselage: Monocoque, oval cross-section and heavily glazed sliding canopy for both pilot and observer.
Undercarriage: As previously described for the Hs 126V3. Wheels initially aerodynamically spatted.
Control surfaces: Single rudder, at extreme aft section of fuselage. Metal covered tailplane, fabric-covered rudder. All control surfaces balanced with trim tabs adjustable in flight.
Wings: Two-piece, twin-spar wings of all-metal construction.

Powerplant
Engine: Due to the insufficient availability of the Bramo Fafnir, the Hs 126A-1 series was also initially fitted with the air-cooled BMW 132Dc nine-cylinder radial engine, rated at 647 kW/880 hp takeoff output, which provided 640 kW/870 hp at 2,500 m.
Propeller: Three-bladed VDM variable pitch airscrew.
Tanks: Fuel tank in forward fuselage ahead of cockpit with a capacity of 543 liters.

Armament: One fixed MG 17 on the starboard upper fuselage decking with 500 rounds of ammunition, plus a flexible rearward firing MG 15 in an Arado gun mount and 975 rounds of ammunition carried in 13 dual drums.
In the aft fuselage area were two bomb magazines, each with a capacity of five 10 kg bombs. Bomb capacity increased by fitting an external bomb rack on the port side, braced to the wing and fuselage spars, which could carry a single 50 kg bomb.

Camera: A fully automatic camera, with a lens protected by a cover operated by the observer. A hand camera was carried on the port side of the observer's cockpit.

Other: Takeoff weight was approx. 3,200 kg. Maximum speed was 320 km/h. Landing flaps, cooling gills, oil cooler slats, and landing light all functioned hydraulically.

Henschel Hs 126A-1 (1938). For the observer, holding the hand camera in such a manner was quite a drafty business.

The type's successful service has already been mentioned elsewhere in connection with the Spanish Civil War, where six Hs 126A-1s were sent in the late autumn of 1938 and served until the end of the war.

Foreign Interest

Impressed with the performance of the Hs 126, the Greek government ordered this aircraft for the Royal Greek Air Force. 16 Hs 126A-1 models were delivered to the air force in early 1939 and equipped an army support squadron.

Switching to the B-1

The B-1 was now fitted with the BMW-built Bramo Fafnir 323A-1 and A-2, as well as the later Q-1 and Q-2 versions, which had a two-stage turbocharger and fuel injection.

The 323A-1 and Q-1 provided an output on takeoff of 625 kW/ 850 hp at 2,450 min^{-1} rpms and the A-2 and Q-2 662 kW/900 hp at 2,500 min^{-1}; maximum pressure altitude for these engines was 4,200 meters. Compared to the engine of the Hs 126A-1, this was a marked improvement in higher altitude performance and quick start capability.

The Hs 126B-1 had a maximum takeoff weight of 3,385 kg and reached a maximum speed of 355 km/h.

Conclusions

This sums up the overview of the offensive weapons and reconnaissance platforms up to and including 1939. Nothing can disguise the fact that there was a noticeable gap in the area of deep penetration bombers and reconnaissance aircraft. Everything that had been designed and built up to this point in time directly or indirectly served in cooperation with the two other branches of service, especially the army.

Perhaps this state of affairs should not be considered a direct result of a false or unilateral air war doctrine, rather more as symptomatic of the limits Germany had placed on what it felt to be economically feasible—limits that in utopian overconfidence were all too often ignored in general, and in particular by the highest ranking *Luftwaffe* authorities.

9. An Unrefined *Luftwaffe* is Tested

To be sure, against the backdrop of political pressures brought to bear upon foreign powers, the *Luftwaffe* was a power factor to be reckoned with—although this weight had yet to be seriously put to the test. During the Spanish Civil War it had displayed its operational potential to good effect, but to use this as a litmus test for a complete and fully functional instrument of air power required overlooking certain factors. These included the tempo of its expansion and the many shortcomings associated with this growth, not least of which was the overly hasty training programs involved.

What was the state of comparable air forces of Germany's potential enemies in 1939, particularly with regard to area of offensive aircraft?

The "Other Side's" Potential

In the spring of 1939, the *Luftwaffe* General Staff's Chief of Intelligence for "foreign air forces" provided what was generally a negative assessment of the air power of Germany's potential enemies. Their seeming strengths were apparently somewhat unilaterally offset by what was accepted as certain technological advantages inherent in German front-line aircraft, the superiority of its on-board weaponry and in the not insignificant area of sealed fuel tanks.

The Quantitative Situation

Overlooking antiquated types, those first-class land-based front-line aircraft roughly on par with Germany's inventory in 1939 were numbered as follows:

Poland:
- 130 bombers
- 175 multi-role reconnaissance aircraft
- 90 tactical reconnaissance aircraft
- 30 first-class fighters

(of a total of 315 fighters), which was viewed in conjunction with an anti-aircraft organization considered to be weak.

France:
- 390 bombers and strategic reconnaissance aircraft
- 150 strike/heavy fighter aircraft
- 400 fighters, which combined with what was felt to be an obsolescent AA organization formed the air defense component of the French Air Force.

Great Britain:
- 500 bombers
- 30 dive bombers
- 45 strategic reconnaissance aircraft
(all Army aircraft were considered second class)
- 200 fighters with an entirely underarmed AA organization.

Belgium and the Netherlands: Their air arms were considered inadequate.

Denmark, Sweden, Baltic States, and the Balkan Countries: With regard to their air potential these were all felt to be insignificant; only Sweden and Finland were of interest due to their ability to supply raw materials.

USA: The air forces of the United States were considered numerically weak, but they had modern aircraft types.

European Soviet Union:
- 800 bombers and strategic reconnaissance aircraft of the first class (all 250 tactical reconnaissance aircraft were categorized as second class)
- 350 strike aircraft
- 1,200 fighters

At the time, Germany's leadership anticipated that it would need to deal with Poland, France, and possibly Great Britain. Based on the above simplified, quantitative analysis these three main contenders would be able to field the following modern land-based aircraft:
1,240 bombers, strategic, and multi-role reconnaissance aircraft
180 dive bombers/strike-/heavy fighters
90 tactical reconnaissance aircraft, i.e. a total of
1,510 operationally available first-class land-based aircraft.

In September 1939 the German counter to the operative air power of these three nations consisted of (including older types):
771 fighters
408 heavy fighters, for a total of 1,179 front-line aircraft functioning in the air defense role, with another 2,600 ground-based heavy AA guns and roughly an additional 6,700 light and medium guns, plus about 3,000 searchlights added to that number.

On the German side, the 1,510 operative front-line aircraft of these three nations were balanced by:
1,180 bombers
336 dive bombers
40 strike aircraft
379 strategic reconnaissance aircraft
342 tactical reconnaissance aircraft, i.e. a total of 2,277 operative land-based planes, which would have to face an enemy fighter defense of 630 first-class fighters of generally adequate performance capabilities.

The numerical superiority enjoyed by Germany's *Luftwaffe* in comparison to the modern components of the air forces of these three nations was therefore set at 2,277:1,510 on the offensive side and 1,179:630 on the defensive. These figures, coupled with the advantage of disrupting the inner line led Germany's highest levels of leadership, not entirely without reason, to consider themselves at an advantage in the air. Particularly since Hitler still felt he would be able to keep Great Britain out of any open conflict.

The Qualitative Situation

With regard to quality, the general consensus has been that Germany's front-line forces were in possession of the most modern airplanes at the time. However, reference to the 1937 performance data shown in an earlier chapter paints a somewhat more sobering assessment picture, one which necessitated additional studies in the interim. Viewed in a neutral light, this is particularly applicable to the range data for offensive aircraft, which had spawned false impressions as to their operational potential. This qualitative factor, which strongly influenced the operational potential, is shown in somewhat more detail here before we attempt a qualitative comparison with the aircraft of Germany's presumed enemies (see Appendix 8).

Endurance and range are fundamental aspects of a type's operational quality. The data for these factors, normally determined theoretically by engineers, was constantly being revised in light of front-line experience. Tactically, the sole planning factor which could really be compared was the penetration range of a given aircraft type.

Penetration Ranges

The penetration range was calculated from the given theoretical range at an altitude of 4,000 m at cruising speed with a bomb load, from which 10% of the range and 30 minutes' flying time were subtracted as a so-called "safety buffer" to take into account potential weather and enemy action which may extend the mission distance.

This resulted in the following realistic operational planning figures for those offensive aircraft available to Germany in September 1939:

bombers	payload in kg	penetration range in km
Do 17M	1,000	325
DO 17Z	1,000	440
He 111E	1,000	325 (no fuselage tank)
He 111P	500	935
dive bombers		
Ju 87B	500	195
strike aircraft		
Hs 123	200	130
strategic reconnaissance aircraft		
Do 17F	-	820
Do 17P	-	565

These figures speak for themselves, primarily when one views them in relation to the various range data included in the specific aircraft type descriptions.

Comparison Criteria

A sensible approach involves examining the observation methodology in assessing the performance figures of enemy types, particularly those figures compiled by the General Staff and considered as "probable operational data for combat applications."

In arriving at these figures, the penetration range was assumed to be half of the overall range at cruising speed with payload and maximum takeoff weight, minus 20%. The value for the cruising speed was accordingly based on engine power set at 65 to 75% of the maximum output. This in turn was derived from the maximum rating at the most favorable engine operating altitude, generally considered to be between 4,000 and 5,000 meters. Because some of these operational data were based on estimated performance, the provided figures had an error of plus or minus 10%.

Operational Flight Performance

Based on this, on 1 September 1939 the General Staff released relevant data in its L.Dv. 900 "Flight Performance" (the "other side") report. This report provides the following snapshot of several typical primary aircraft types either recently introduced or still under construction:

Nation	Type	Engine HP	Cruise Spd (km/h)	Payload (kg)	Penetration Range (km)
Bombers					
Fr	Bloch 131/133	2x900	340	1,500	400
Fr	LeO 45	2x1,100	410	1,200	1,000
Fr	Bloch 174	2x1,050	440	500	600 (also recce)
GB	Fairey Battle	1x1,050	320	500	600
GB	Armstrong-Whitworth Whitley IV & V	2x1,050	345	800	800
GB	Bristol Blenheim I & IV	2x840	350	450	1,200
GB	Handley-Page Hampden	2x925	350	500	1,100
GB	Handley-Page Hereford	2x1,000	425	700	800
GB	Lockheed Hudson	2x900	345	400	600 (U.S. supplied)
GB	Vickers Wellington I	2x940	340	1,000	800
	Wellington III	2x1,150	380	500	1,500
SU	SB-2	2x860	360	600	800
SU	ZKB-26	2x850	360	800	800
SU	TB-6	4x1000	335	3,000	1,200
USA	Boeing B-17B	4x1,000	335	3,000	800*
USA	Douglas B-18	2x900	300	1,000	3,000
	as reconnaissance platform				3,900
Fr	Potez 63	2x650	?	-	520
Fr	Bloch 174	2x1,050	?	-	720
GB	Westland Lysander	1x890	240	180	400
GB	Lockheed Hudson	2x900	345	-	800

*4,800 km without payload

Since as a rule penetration range was relative to payload, even discounting the +/- 10% deviation these figures were naturally variable up to a limit, something which logically applied to the data for German aircraft, as well.

To be sure, the French and British both had several twin-engined bombers whose "operational data for combat applications" not only matched, but in some respects exceeded that of their German counterparts. Yet they, too, lacked a heavy bomber capable of carrying a large payload over great distances—a serious qualitative gap for all three countries! Only the Americans and the Soviets had types of this kind serving in their front-line units, with the most striking—and amazing—miscalculation in this official report being the estimated penetration range of the B-17 at slightly over 800 km with a 3,000 kg payload.

Assessment

Despite the "opposing forces" having some qualitatively better twin-engined bombers—Germany's Ju 88 had not yet entered the picture—in view of the numbers of these modern types available to front-line forces the offensive strength of Germany's *Luftwaffe* was assessed to be on par with, and in some cases better than, that of the combined combat forces of France and Britain in every category. The realm where the *Luftwaffe* was clearly superior was in the areas of dive bombers and strike aircraft. The opposing reconnaissance potential could be considered equal, but it went without saying that Poland's offensive strength was considered markedly inferior to that of Germany.

Thus were the cards on the table. The German self-assessment was that its *Luftwaffe* enjoyed a clear advantage with regard to its operational potential. It was this consideration alone which aroused within the highest tiers of Germany's leadership the irrational feeling that it was no longer necessary to avoid armed conflict.

Poland

The Air War

On 1 September 1939 Germany struck out against Poland. In virtually unceasing operations, a combined large scale *Luftwaffe* attack against the Polish Air Force destroyed the bulk of the enemy's aircraft within the first 48 hours, catching the greater number of Poland's planes on the ground. By 5 September 1939 Poland had only about 40% of its original operational strength left—its air and air defense components had been all but crushed—and those fighters and reconnaissance aircraft spared in the first raids were unable to prevent the *Luftwaffe* from maintaining its quickly achieved absolute supremacy of the air over the Polish battleground. Following this first strike, which in light of its planning and execution paralleled Douhet's doctrine of an operative—if not altogether strategic—venture, the *Luftwaffe* was almost exclusively used in a tactical role in support of ground operations. On 9 September 1939 the Poles were able to temporarily regain the initiative with cavalry and armor at Kutno, but on 11 September massive waves of German dive bombers and strike aircraft struck hard and brought the Polish counterattack to a standstill. Even the attacks on the 24th and 25th of September against an encircled Warsaw, which saw approximately 400 German bombers drop 486 metric tons of high explosive bombs and 72 tons of incendiary ordnance onto the city, could be considered generally tactical in nature by this stage of the war. The bulk of what remained of the Polish Army, with about 12,000 men, were still putting up a fight in the capital, but were forced to capitulate on 28 September 1939. On 6 October 1939 the remaining Polish troops in the field followed suit, thus ending the campaign against Poland.

It is apparent from documents of the "*Studiengruppe der Geschichte des Luftkriegs*" (Study Group of the History of the Air War) at MGFA that on 1 September 1939 a total of 897 bomb-carrying offensive aircraft (with 879 belonging to the operative *Luftwaffe*) were used by the Germans in the air war against Poland. This was countered by a total of no more than 745 Polish aircraft, including its hopelessly outclassed and antiquated inventory.

From 1 to 28 September 1939 the Germans lost a total of 385 front-line aircraft, which included:

> 78 bombers   ⎫ = 109 aircraft = 12.2% of the bomber
> 31 dive bombers, plus ⎭   force
> 63 reconnaissance aircraft = 24.1% of the 261 strategic, tactical, and command-subordinated reconnaissance force.

Relatively speaking, the reconnaissance aircraft suffered nearly double the losses as the bombers and the dive bombers combined. The offensive elements (bombers, dive bombers, and reconnaissance platforms) lost a total of 172 aircraft, or 60.4% of the 285 number of aircraft lost altogether, which in turn was 14.8% of the 1929 combat aircraft used against Poland in total.

413 flying personnel were either killed or numbered as missing in action, with a further 126 wounded. Total loss: 539 men.

The French Air Force attaché in Warsaw, General André Armengaud, reported to Paris on the air war:

"I must underscore that the German *Luftwaffe* conducted itself according to the rules of armed conflict. It only attacked military targets."*

Balance

Developed and tested in the Spanish Civil War, the cooperation between Germany's *Luftwaffe* and *Heer* was an outstanding contribution and proved itself fully in the Polish campaign. Quantity and quality of the Polish Air Force was so inferior to the *Luftwaffe* from the outset that they would never have weathered the struggle for air superiority even if a better command structure had earnestly sought to prosecute the struggle. A decisive factor in the failure of the Polish Air Force as a whole may have been the fact that Poland had no independently functioning air force. From the outset, this organization neglected to carry out an operative air war against rear area nerve centers and bases of the German offensive, instead concentrating its aircraft exclusively in tactical support of the army, to which it was subordinated—distributed among the various armies—from the time of mobilization onward. Thus, the Polish Air Force lacked a unified command, which may have enabled it more to focus on the struggle for air superiority, instead of squandering itself away in a purely defensive role. Other than a few fighter successes on the part of Polish pilots, there was no truly serious battle for the skies in this campaign: victory was literally handed to the Germans!

The material loss of Germany's *Luftwaffe*, at just 15% of the aircraft used, could be replaced by ongoing industry production without difficulty—somewhat more problematic was the replacement of the 539 flying personnel.

All Quiet on the Western Front

France and Germany officially mobilized their forces on 1 September 1939, and due to their treaty obligations to Poland declared war against the German *Reich* on 3 September 1939. Thus began the expansion of the war Germany had unleashed against Poland, although little or virtually nothing was initially undertaken in the West to relieve Poland militarily.

This period, lasting approximately eight months, was called the "Sitzkrieg" in the West, but such an appellation was applicable to the air forces of both sides in only a limited sense. To be sure, the German air units in service on the western front on 1 September 1939 were relatively weak compared to those *Luftwaffe* units advancing against Poland, but it was certainly no "Sitzkrieg" for them during this period.

The previously mentioned "Study Group of the History of the Air War" reveals that on 1 September 1939 only 285 of a total 1,182 "bomb platforms" were available on the German side in the West, thus severely limiting any type of offensive action in this sector.

Germany's Forces

We should first perhaps take a look at Germany's situation with regard to the air: the data for East and West alone does not entirely show the overall strength of Germany's air power, since it does not

---

*   Molloy Mason, Die Luftwaffe, Paul Neff Verlag, Vienna, Berlin, 1973, p. 212 (Heyne book ref. no. 5575).

History of German Aviation: Bombers and Reconnaissance Aircraft

include naval aviation. This sub-branch totaled 180 operational aircraft on 1 September 1939, i.e. approx. 6% of the total inventory of 2,955 front-line aircraft. The figure of 180 included weather reconnaissance aircraft, transports, and those courier planes provided to the *Heer*. In view of the approx. 6% of naval aviation, the remaining approx. 94% were broken down as follows on 1 September 1939:

1,182 "bomb platforms"	= 40%
897 reconnaissance aircraft	= 30.5%
516 fighters and heavy fighters	= 17.5%
168 transports and courier types	= 6%

From these figures we can deduce that the offensive section of the German Air Force including naval aviation comprised 76.5% of the overall inventory of front-line aircraft, while the single-engined fighters specially reserved for protection of the homeland made up just 12.4% of the total strength, with 366 aircraft. The latter's role cannot just be considered homeland protection, however, as they were also expected to fly over the given operational area. The question is, whether such a distribution of forces was in harmony with the "balanced ratio between offensive and defensive forces" *Generalmajor* Walter Wever had promised on 1 November 1935 and had linked to what was possibly a misinterpreted admonition that only the State which had a strong bomber force could expect its air force to play a decisive role in war. With a pronounced accent on offensive air components primarily tailored to cooperation with the *Heer*, the *Luftwaffe* deviated little from Wever's intentions and thus pursued a path of fateful developments leading to a fighter component too weak to meet the needs of air defense.

The *"Sitzkrieg"* and the Air War

In the main, the *Luftwaffe's* activities during this time involved ongoing aerial reconnaissance over the areas adjacent to the Maginot Line and the West Wall. There were engagements between reconnaissance aircraft and French fighters, as well as fighter-against-fighter combat, from which some practical benefit was gained. However, there was no real significant experience gleaned for the later course of the air war. The French behaved almost discretely reserved. These front-line reconnaissance flights, read tactical reconnaissance, were supplemented by strategic reconnaissance missions deep behind enemy lines for scouting out targets and monitoring potential advance movements, plus monitoring the North Sea and English Channel. Attacks on shipping targets were also sometimes made.

The British showed themselves to take their declaration of war seriously and used their twin-engined Blenheim bombers to attack German warships off the coast of Wilhelmshaven and Cuxhaven just a day later, on 4 September 1939. It was not a successful mission, and they lost seven aircraft in the attempt. Another attempt, on 3 December 1939, this time against German warships near Heligoland, was also a failure. An attack with 24 Wellington bombers on 18 December 1939 in the area of Wilhelmshaven cost the British a total of 12 aircraft at the hands of German fighters. From that point on the Royal Air Force ceased its daytime operations without fighter cover and shifted over to nighttime bombing raids virtually for the remainder of the war. The British, however, had gained the experience needed for their nighttime missions over Germany within the first few weeks and months of the war by frequently taking their heavy bombers deep into German airspace and dropping propaganda leaflets in an attempt to influence the civilian populace. In doing so, they suffered only minor losses, but the experience gained with regard to navigating over enemy territory at night, the location and effect of German anti-aircraft defenses, and the effectiveness of a German night fighter force still in its infant stages was of enormous value for its subsequent large-scale bomber offensive.

Compared to Great Britain, German activity restricted itself in the air to preparing for an aerial assault against the island nation using an intensive reconnaissance program designed to reconnoiter potential target areas. It was not until the period of 7 to 9 October that Germany carried out its first, albeit unsuccessful attacks on the British fleet, followed by raids on British warships in the Firth of Forth in Scotland and the Scapa Flow in the Orkney Islands on 16/17 October. In this latter raid, three Ju 88s were lost, along with a handful of He 111 bombers. The attack profile of the new dive-capable medium bomber was thus introduced to the British from early on, before even larger formations of Ju 88s were used. In January 1940 German bombers, mostly He 111s flying at both high and low altitudes, sank a total of twelve ships off the eastern English coast using bombs.

In late February 1940 an advance unit of the German fleet, under air cover, struck out against Allied convoys between Great Britain and Scandinavia. This operation, taking place under the coverterm "Nordmark" from 18 to 20 February, met with virtually no success. It is noteworthy that during the German-British prosecution of the air war during this "Sitzkrieg," both sides took pains to avoid population centers when making their attacks, limiting themselves to coastal military targets, especially shipping.

Nevertheless, the losses incurred during this "Sitzkrieg" and its unspectacular battles were perceptible, as evidenced by the following internal figures recorded by the Ob. d. L. These figures were for the time period from October 1939 to March 1940, i.e. after the conclusion of the Polish campaign and prior to the start of the Norwegian campaign in April of 1940, and included the following losses:

- 246 bombers (including dive bombers)
- 34 strike aircraft
- 67 tactical reconnaissance aircraft
- 62 strategic reconnaissance aircraft

409 front-line aircraft for the offensive role total. Here, too, was the high loss rate of reconnaissance aircraft at 129 aircraft, an indicator of the criticality of aerial reconnaissance.

It is interesting to note that in the same period of time 433 fighters and heavy fighters were lost, approximately the same number of aircraft.

On the basis of the equivalent aircraft/personnel loss quota in the Polish campaign, we can estimate that personnel losses during this period amounted to about 1,500 to 1,600 flying crewmen. Even in this relatively unproductive period of the air war known as the "Sitzkrieg," the combat potential of the *Luftwaffe* continued to be whittled away, while the *Heer* found itself able to use the time to regenerate and regroup.

Battle Plans for the West and Finland

Behind the relative peace during this time, in early October Hitler had made the decision to launch the invasion of France before the outbreak of winter in the West. Originally planned for the 12th of November 1939, it was postponed for a multitude of reasons no less than 29 times—until May 1940!

In the meantime, Finland was attacked by the Soviets on 30 November 1939 without any prior declaration of war being made—the Soviet-Finnish Winter War began with the bombing of Helsinki. On 5 February 1940 the Anglo-French Allied War Council decided to support Finland and, to disrupt the ore supplies from northern Sweden to Germany, intended to send troops to Narvik in northern Norway. From these two countries and from many others, Finland received material assistance of every kind, yet this was not sufficient to prevent it—after stiff resistance—from signing the Peace of Moscow with the overpowering Soviets. This agreement was a strategic victory for the USSR.

The Soviet troops did not acquit themselves well against the numerically far inferior Finnish defenders in this war, which lasted some three times longer than the Polish campaign; they seemed poorly trained and led, and based on the impressions from the 1939/1940 Winter War, Germany's leaders felt that the Red Army was an easily beatable and poorly organized force.

All the More Reason To Go North!

Foundations

The decision of the Allied War Council on 5 February 1940, which called for intervening militarily in the north in favor of Finland and to the detriment of Germany, was actually nothing new. Back on 19 September 1939 Churchill indicated in a memorandum that Germany should be cut off from the supply of Swedish ore, and planning began accordingly. In October 1939 those in Germany responsible for the prosecution of the naval war pointed to the danger of having the enemy gain a foothold in Scandinavia and thereby get the upper hand in the Baltic, with all the military and economic consequences such an action would spawn.

On 16 January 1940 the Allies began preliminary work for military action in Scandinavia. On 27 January 1940 Hitler ordered the OKW to plan for the occupation of Danish and Norwegian airfields, since seizing these points was considered vital to ensuring the link to the North. There developed a neck-to-neck race between Operation Weserübung, the study derived from this order and signed by Hitler on 1 March 1940, and the Allied preparations for a similar operation in Scandinavia.

Although repeatedly delayed, the campaign against France was still on. But it was now in a conflict of priorities with Operation Weserübung. On 26 March 1940 the German High Command bumped Weserübung slightly higher on the timetable prioritization than the western offensive, delayed yet again and now reworked into a "sickle cut plan" under the codename of "*Fall Gelb*" (Case Yellow).

Occupation of Denmark/Battle for Norway

So it was that on 2 April 1940 the German attack against Denmark and Norway was set for the 9th of April. For technical reasons, on 5 April 1940 the Allies delayed the departure of the first squadron of their expeditionary corps, originally scheduled for that date, to 8 April 1940. On 7 April the first German convoys with army troops on board set out in a northerly direction. Following an announcement by the British and French governments on 8 April, British and French warships mined Norwegian waters, whereupon the Norwegian government protested "most strenuously and solemnly against this open breach of the rights of the people" in London and Paris that very day.* Nonetheless, the Allied expeditionary corps plowed through the water in the direction of Norway.

---

*   Jahrbuch für Auswärtige Politik, 7th annual edition (1941), p. 219 (Norska Telegrambyrå from 8 April 1940).

But the Germans beat them by a nose and, without a declaration of war, on 9 April 1940 at the early hour of 5 am, began their execution of Operation Weserübung for the occupation of the neutral countries of Denmark and Norway. Denmark, without a struggle, resigned itself to the fate of being occupied by an overly powerful enemy.

That same day, German bombers attacked units of the British Home Fleet which had been spotted by reconnaissance aircraft in the waters off Bergen, in western Norway. The British ships had been attempting to engage German warships and transports in the harbor of Bergen and seize the port for itself. A British destroyer was sunk, three cruisers, and the battleship *Rodney* were damaged. German attacks continued for nearly three hours, during which 41 He 111s and 47 Ju 88s took part.

On this and the following day British air power and submarines sank two German light cruisers and several transport ships off south and west Norway, while on the first day of the German invasion the Norwegians succeeded in sinking the heavy cruiser *Blücher* in the Oslo Fjord with torpedoes.

From the 10th to the 13th of April 1940 the British attacked German destroyers near Narvik with superior naval forces, sinking them or forcing their crews to scuttle them after the German ships had run out of ammunition and an air umbrella from nearby airbases never materialized. On 14 April the Allies (British, French, Polish) landed near Narvik, on 15 April near Namsos, and on 18 April near Andalsnes. Following 14 days of bitter fighting against the Allies and the Norwegians, with the advantage shifting from one side to the other and the Allies under constant attack by German bombers, the landing was finally deemed a failure. On 2 and 3 May 1940 Namsos and Andalsnes were vacated by the Allies. In textbook co-operation with the *Heer* and *Marine*, the *Luftwaffe* had played a major role in this success, continuing on with their attacks on the allied maritime supply lines (sometimes operating from the frozen seas in the early phases). The Allies suffered major losses to their shipping, as German bombers incessantly harassed the expeditionary corps' supplies.

In spite of this, the Allies succeeded in taking the harbor and city of Narvik in a renewed attack on 27 May 1940 with the support of their superior naval forces; they held on until 8 June. Supply to German troops at the time was entirely inadequate, with materials being flown in by a small number of Fw 200s and Ju 90s. Moreover, there were now far fewer bombers able to support the conflict on the ground—the bulk of these had been transferred to the West where the large-scale offensive against France had finally gotten underway on 10 May 1940. The progress of this offensive ultimately led the Allies to evacuate Norway altogether, which they did from 3 to 8 June. As a result, on 10 June 1940 the commander of Norwegian forces in northern Norway capitulated.

The air and naval bases in Denmark and Norway were firmly in German hands following a two-month battle. The danger of losing the supply of ore from Sweden and the unrestricted use of the Baltic had passed: the way to the Atlantic was clear!

10. Initial Assessment of 1940 Following the Polish and Norwegian Campaign

The chronological overlap—by a month—of the campaign in Norway with that of the offensive against France, which finally got underway on 10 May 1940, initiated a development in the war which Germany most certainly had not figured into her war-planning calculations. Although the British and French were compelled to abandon their intentions in the North, the tying up and wearing down of forces there did not go unfelt on the part of the Germans. Chiefly, the *Luftwaffe* units operating under difficult conditions were only able to provide assistance in stages to the main combat areas of the land and naval forces. This was due to the limited range of bomber and transport aircraft, and continued even after paratroopers and airmobile units had beaten down Norwegian resistance and secured for them the airbases in southern and central Norway. They were generally able to achieve their operational goals in spite of these difficulties, since they nearly always operated with total air superiority (depending in part on the area and time). As in the Polish campaign, in close cooperation with ground forces the *Luftwaffe* had played a decisive role in destroying or driving back the Allied expeditionary corps in Norway, as well. It also demonstrated that bomber formations were entirely capable of hindering and effectively disrupting naval operations.

The aircraft inventory used had remained virtually unchanged from that at the beginning of the war—only the Ju 88 appeared in quantity for the first time. However, with England's entry into the war in the early days of September 1939 an air war over water became unavoidable. With the occupation of Norway the northern Atlantic routes were open, emphasizing even further the need for overwater operations and leading to increased attacks on shipping on the high seas. To this end, the Ju 87 and Ju 88—even with their limited range—would have been good candidates thanks to their dive bombing capability, and even level attacks by He 111s operating at low altitudes would have sufficed, but none of these three types were adequate for long range warfare in the Atlantic. Such thinking led to the sobering conclusion that a suitable aircraft for long range maritime reconnaissance and bombing would not be available for the foreseeable future. The only potential solution on the horizon was the Heinkel He 177 project, a risky concept fraught with inherent complications, but this was still a long way from frontline service suitability. A further handicap for strategic reconnaissance over France and Great Britain was discovered when it was learned that the service ceiling of German reconnaissance aircraft, meaning mainly the Dornier Do 17, was no longer adequate for avoiding modern fighter defenses—particularly the threat posed by the British Spitfire.

Also, the bombs used in Poland and Norway against both land and naval targets proved to be inadequate with regard to their penetration and explosive force capabilities against heavily armored naval and ground targets.

The consequences of these weaknesses were forthcoming quite quickly, in some respects even being initiated immediately following the British declaration of war, and led to the developments described over the following pages.

Fw 200 Maritime Reconnaissance Plane and Auxiliary Bomber

Brief Perspective

The Focke-Wulf Fw 200 Condor was designed by *Dr.-Ing. e. h.* Kurt Tank, going from initial contract signing to Lufthansa service in the remarkably short time of one year. It took to the air for the first time in Bremen, on 27 July 1937, with Kurt Tank at the controls.

In 1938 the Japanese ordered five Fw 200Bs for a commercial airline company, while at the same time the Imperial Japanese Navy contracted for a sixth machine to be built as a maritime reconnaissance platform. This can be considered the first step in the militarization of the Fw 200 Condor. Focke-Wulf modified the Fw 200V10 test prototype to the specifications of the Japanese military, incorporating two MG 15s as defensive weaponry, two aerial cameras for vertical shots, and additional fuel tanks in the fuselage, but the work had not been completed by the time the war broke out.

With the He 177 plagued with major problems, the RLM now took a keen interest in the maritime reconnaissance version of the Fw 200 ordered by Japan. The *Technisches Amt* awarded Focke-Wulf with an urgent contract for researching the expansion of the Fw 200's role from that of a reconnaissance platform to a maritime bomber. In this context, there was also talk of a so-called "*Kaperflugzeug,*" or "naval raider," for which role the Fw 200B was to have been modified from the second production aircraft onward. After exploring the possibilities, Focke-Wulf confirmed that it would be possible to accordingly modify the Fw 200B, the first example of which was just being completed as the war began.

This military version, designated as the Fw 200C, was eagerly accepted by the RLM, and in September 1939 a pre-production batch of ten Fw 200C-0s was contracted for. These would be built from Fw 200B airframes then in production, while in conjunction with this the jigs would be set up for production of the C-1 series. However, the first four B-series airframes planned for the Fw 200C-0 batch were so far advanced in their construction that they could not be fully converted over to strategic reconnaissance/bomb platforms; they were therefore delivered as unarmed transports beginning in January 1940. These four machines were used for crew conversion and training, as well as on transport missions during the Norwegian campaign, especially for supplying the German troops trapped in Narvik for several weeks.

The remaining six aircraft of the pre-production series were delivered to the *Luftwaffe* starting in the early spring of 1940, complete with their defensive armament and bombing systems. At first, these too served as crew training platforms, but soon successfully began operating from Denmark over the north Atlantic as armed reconnaissance aircraft and for strikes against British convoys and individual ships.

At the beginning of the French campaign only two of the Fw 200C-0s were in operational front-line service, however, and these were pulled out in July 1940 when the unit re-equipped with the Fw 200C-1 then being delivered.

Fw 200C in Military Service

What did the German military version of the Fw 200 look like at the time? The data corresponded to that of the Fw 200V11 as follows:

Armament: initially equipped with only one 7.9 mm MG 15 in a bubble mount located in a hydraulically driven upper turret just behind the cockpit, another MG 15 in the B-station cupola on the aft upper fuselage, plus a third MG 15 in a ventral bay with a field of fire downward and to the rear. An additional 20 mm MG-FF was planned for a forward-firing station, but was not initially fitted.

Ordnance: for its operational role of armed reconnaissance, a 250 kg bomb was carried in the aft section of each of the outer engine nacelles, as well as two additional 250 kg bombs on racks located beneath the wings just to the outside of the two outer engine nacelles—for a total of 1,000 kg maximum bomb payload. In the ventral bay there was space allocated for a 250 kg cement bomb, which served as a marker bomb for calibrating the bombsight prior to dropping the four live bombs—a rather primitive approach to targeting and dropping ordnance. It was soon abandoned in favor of an easily removable fitting of the Lotfe C7C in the latter half of April 1940. The C7C could be swapped out for the less accurate high altitude GV 219 bombsight.

Fuel tanks: for increased range—the aircraft was to have a range of 5,000 km at 4,000 m with a speed of about 290 km/h—a total of eight self-sealing tanks were fitted into the existing space in the wings. These consisted of two large and two small tanks (i.e. two complete sets) from the Messerschmitt Me 110 heavy fighter type. In the fuselage five self-sealing tanks from the Do 18 flying boat were installed, providing a total capacity of about 5,300 kg of aviation fuel.

Engines: the engines of the Fw 200B, the BMW 132H-1, were kept without any modifications, but by using 100 octane fuel for takeoff these provided 735 kW/1,000 hp each—necessary to compensate for the higher takeoff weight of the C-series. This was assumed to be 20,500 kg, some 3,000 kg over the maximum authorized takeoff weight for the Fw 200B. The previously used Junkers-Hamilton 20-degree variable-pitch propellers were replaced by VDM feather-pitch propellers offering a minimum of drag when properly adjusted in the event of an engine failure.

Structure: structural soundness tests were initiated when it appeared that the previous authorized stress limits, based on the P3 criteria, would no longer be applicable given the potential 17% higher takeoff weight for the C-series. On 22 September 1939 the RLM's stress analysis center in Berlin-Adlershof came to the unmistakable conclusion that, among other things:

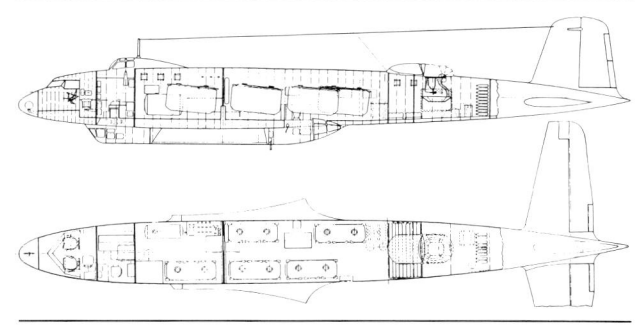

Focke-Wulf Fw 200C - standard fuel tank arrangement.

- the weakest section of the outer wing area came in at approx. 5% under the 1.35x safety margin for the stress limit
- the fuselage was able to withstand stresses during flight, but did not meet safety margins on landings
- the landing gear was considered adequate for normal safety considerations, but was not fully acceptable during overload testing

Given these restrictions, the Fw 200B was certified as being suited for the role of "auxiliary bomber" (= Fw 200C).

Contract and Costs: On 25 September 1939 Focke-Wulf confirmed for the RLM that it would construct and deliver 36 Fw 200s, indicating the following costs in its payment plan:

1st through 8th airframe	each RM 706,000
9th through 12th airframe	each RM 572,000
13th through 17th airframe	each RM 528,000
18th through 21st airframe	each RM 501,000

Ultimately, the developmental stage of the Fw 200C was finalized in June 1940 as follows (D[LUFT]T. 2661/1 from June 1940):

"With regard to its role as an auxiliary bomber the Fw 200C is a follow-on developmental modification of the well-known Fw 200 commercial airliner, designed for the H3 utilization and stress group. Its four-engined layout enables it to continue with the mission even in the event of a loss of one or two given engines.

- maximum flight weight	22.7 metric tons
- stress load factor on pull-out	3.3
- maximum horizontal speed at sea level	355 km/h
- maximum horizontal speed at 1,100 m	370 km/h
- maximum descent speed from 0 to 3,000 m	450 km/h
- maximum permissible weight for safe landing	17.5 metric tons
- maximum permissible weight for landing with limited safety	20.5 metric tons

The aircraft is cleared for unrestricted IFR flying.
Aerobatics are prohibited!
Fuel tanks are self-sealing to protect against gunfire. Armament is located at six different points, two on the dorsal fuselage, two in the fuselage gondola, and one on each side of the fuselage."

Synopsis

With the conversion of the long range civilian Fw 200 Condor airliner into a strategic maritime reconnaissance platform and auxiliary bomber in a relatively short time, a stopgap solution was found to close the sensitive gap in the weaponry needed for long range overwater operations. Certainly the Fw 200C was not bad as an

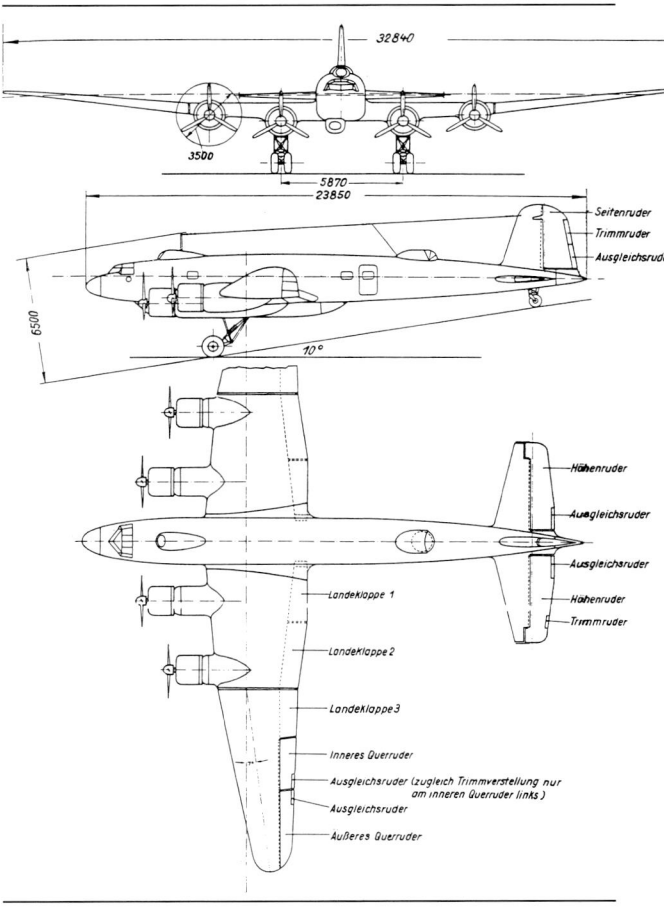

Main dimensions of the Fw 200C. The paired wheels shown in this set of drawings had been introduced earlier on the Fw 200B-1 and B-2 civilian versions built for Lufthansa, although the Fw 200C military version had larger tires because of its increased weight.

auxiliary bomber, but was ultimately not the perfect solution. With regard to its structural soundness, which in some areas was at its limits, the cautious advice given in late September 1939 indicates that the aircraft would almost assuredly hold some rather unpleasant surprises under constant combat usage. It was with good reason—and as a direct result of specific instances—that the excerpt from the operation manual printed above included the warning "Aerobatics are prohibited!" Any follow-on development of the military Fw 200 would undoubtedly be influenced by its operational performance and its reliability, as well as the point in time at which an improved long range bomber would become available.

History of German Aviation: Bombers and Reconnaissance Aircraft

Ju 86P High altitude Reconnaissance Platform and Bomber

In early September 1939 the strategic reconnaissance branch of the *Luftwaffe* High Command, whose task it was to fly the high altitude reconnaissance missions which would spawn target area mapping for bomber units, issued a requirement for a long range reconnaissance aircraft which would be able to fly so high that neither AAA nor fighters would be able to hinder the reconnaissance airplane in its assigned tasking.

The pressurized cockpit necessary for such extreme operating altitudes had already been developed by the Junkers company for the Ju 49 high altitude testbed in 1931 and the EF 61 in 1937. The cockpit was demonstrated to senior leaders in Rechlin on 3 July 1939.

Parallel to the pressurized cockpit, creation of an aircraft engine for high altitude operations had begun in 1932 with the testing of a Jumo 205 diesel engine fitted with pre-activated exhaust turbine. From these efforts came the Jumo 207 in 1939, with two centrifugal turbochargers in tandem. It seemed that a suitable high altitude engine would be available.

The Junkers EF 61V2, the most noticeable feature of which was the bulky looking pressurized cockpit/altitude chamber.

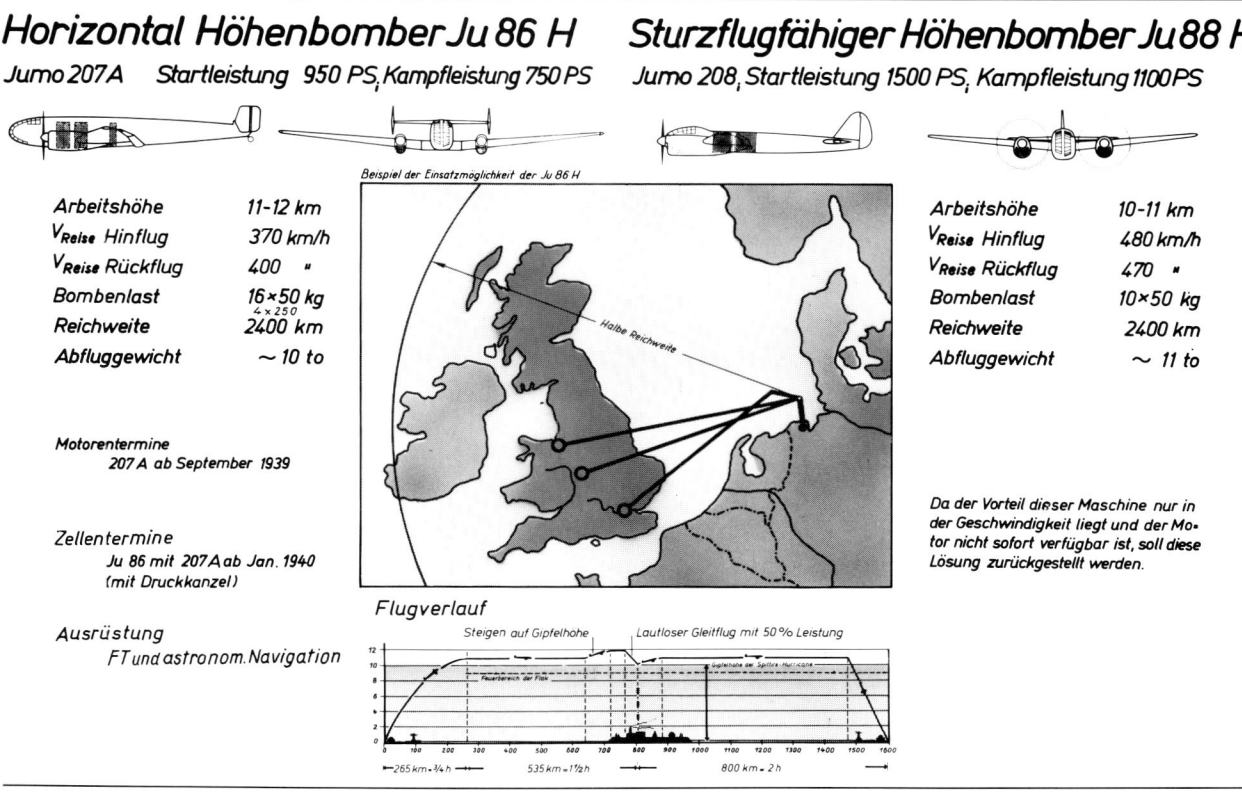

Comparison of the two projects showing the claimed operational profiles of each. The data for the engine performance of the Ju 86H's Jumo 207A, at 950 hp (about 700 kW) is somewhat exaggerated.

Prototypes

Also in September 1939 the RLM contracted with the Junkers company to build three high altitude prototype aircraft for strategic reconnaissance and nuisance raids, powered by the Jumo 207. The company designation for the project, to be based on the Ju 86D airframe, was the Ju 86H. It is interesting to note that it was within this context that Junkers began toying with the idea of developing the high altitude Ju 88H dive bomber. However, this idea never came to fruition because of the matter of a powerplant, since the more powerful Jumo 208 intended for the design was not yet available. The Ju 88H's bomb capacity fell short, as well, with its 10x50 kg payload less than that of the Ju 86H. All in all, with the availability of the Jumo 207A high altitude engine and the pressurized cockpit, the Ju 86H was a project which could be realized much more quickly.

The RLM decided on the development of the Ju 86H, but the modified Ju 86 was now designated officially as the Ju 86P. Initially, three prototypes would come from the D-1/D-2 series, the Ju 86PV1, PV2, and PV3, with two-seat pressurized cockpit and two Jumo 207A-1 engines each delivering 647 kW/880 hp of power. Flight testing began in February 1940. The wingspan of the Ju 86P-1 bomber and JU 86P-2 reconnaissance versions, production of which began that same year, was increased to 25.60 m after the Ju 86PV3 had reached a maximum altitude of 12,150 m with the same wingspan. In September 1940 the Jumo 207 powered prototype for these aircraft finally reached the maximum pressure altitude of 10,000 m specified in the contract.

P-Series

A total of 40 aircraft of both versions, the P-1 and the P-2, were contracted for and built from suitably modified G-series Ju 86 airframes. Following delivery to the *Aufklärungsgruppe Rowehl*, they flew in their intended roles over France and (mainly) Great Britain, then over Egypt and the Soviet Union, flying unarmed at altitudes initially out of reach of fighter defenses. During combat testing the Ju 86P-2 flew over Great Britain during the summer of 1940 at an altitude of 12,500 m and remained completely untouched.

Technical Data and Details of the Ju 86P

The Ju 86P-1 series was a twin-engined land-based high altitude bomber, while the Ju 86P-2 was a twin-engined land-based strategic reconnaissance platform for high altitude operation and had a two-man crew; both versions were powered by the Jumo 207A-1 diesel aircraft engine. Technical details differing from the G-series include:

Airframe
Fuselage: pressurized forward section, removable, divided into nose/glazing and cockpit. Cockpit with two adjustable seats for pilot and gunner/observer, the latter also acting as radio operator.
Main payload compartment:
Ju 86P-1: compartment fitted as bomb bay, entry hatch ahead of bulkhead 10.
Ju 86P-2 (strategic reconnaissance): compartment with camera equipment vice bomb bay; entry hatch ahead of bulkhead 10 and ground access hatch between bulkheads 2 and 3 (second access point to fuselage).
Undercarriage: As for G-1, but with swiveling tailwheel.
Control surfaces: As for G-1, but with port aileron trim tab adjustable in flight.
Control inputs: As for G-1, but landing flaps activated via hydraulic pressure.
Wings: As for G-1, but with three-piece landing flaps.

Powerplant:
Two reduction gear liquid-cooled six-cylinder dual piston inline diesel engines with exhaust turbine and high altitude blower, model type: Jumo 207A-1.

Max. output (takeoff rating) at sea level	647 kW/880 hp
Climb and combat rating at 10,000 m	552 kW/750 hp
Max. sustained output at 10,000 m	478 kW/650 hp
Recommended cruise output at 10,000 m	370-440 kW/500-600 hp
Max. Rpm in flight	2,800 min^{-1}
Fuel consumption rate at recommended cruise	238-252 g/kWh (175-185 g/hp/hr)
Oil consumption	5-9 kg/hr.

A) Fuel: standard petrol/oil (octane rating 0.84 to 0.88); b) Oil: Aero-Shell for summer and winter
Propellers: Two three-bladed VDM variable pitch airscrews, blades electro-mechanically adjustable via integral controller. Control gearing electrically heated. Heating activated simultaneously with activation of turbines at altitudes greater than 2,000 m. Shaft reduction: airscrew = 1.58:1
Fuel/Oil cells:
- unprotected, insulated tank with 1,290 liter capacity (forward tank on Ju 86P-1, aft on P-2)
- additional fuselage tank with 1,130 liter capacity (aft on P-1, forward on P-2); both tanks refillable only to 1,100 liters
- one oil reservoir in each of the wings with approx 152 liter capacity.

Performance

Classified into category H (bombers and reconnaissance aircraft) and stress group 3, the aircraft's performance was as follows:

Max. permissible speed in a descent
- over 6,000 m altitude — 300 km/h
- under 6,000 m altitude — 360 km/h

Cruising speed at low altitude — 270 km/h

Best climb speed
- 0-8,000 m altitude — 165-175 km/h
- over 8,000 m altitude — 155-165 km/h

Service ceiling 12,000 m, range 1,200 km, endurance approx. 6.5 hours

Weights and Dimensions

wingspan	25.50 m
length	16.46 m
wing area	84.8 m²
height	4.70 m
wheel track	3.14 m

Max. permissible weight (with variable-pitch propellers)
Ju 86P-1	10,400 kg
Ju 86P-2	9,500 kg

Wing loading
Ju 86P-1	124 kg/m²
Ju 86P-2	112 kg/m²

Performance load
Ju 86P-1	6.9 kg/hp
Ju 86 P-2	6.3 kg/hp

The P-series was approx. 1 m shorter than the G-series, had a 3 m larger wingspan, and at 9,895 kg assembled weight was (corresponding to today's standard empty weight IAW DIN-Norm 9020) some 585 kg heavier than the G-series. The maximum weight of the P-1 was 10,500 kg, and 9,500 kg for the P-2 (the latter being equipped with 2-3 wide aperture RB 75x30 aerial photo cameras in place of the bomb bay).

Design Features

Of interest is the fuselage design incorporating the high altitude pressurized cabin.

Fuselage: The fuselage consisted of the pressurized cockpit, the center section, and the tail section. The fuselage was integral with the wing center section, which was standard Junkers low-wing construction design. The main components of the fuselage frame consisted of four longitudinal formers and bulkheads arranged perpendicular to the longitudinal axis. The outer skin covering consisted of thin sheet metal and was reinforced by braces. Up to bulkhead 17 it was attached by means of flush rivets, while from bulkhead 17 back round head rivets were used. The high altitude pressurized cabin comprised the nose glazing, the cockpit, and the cockpit canopy. Its cross section conformed to that of the fuselage. It was hermetically sealed from the rest of the fuselage by bulkhead 1a; it was also sealed using

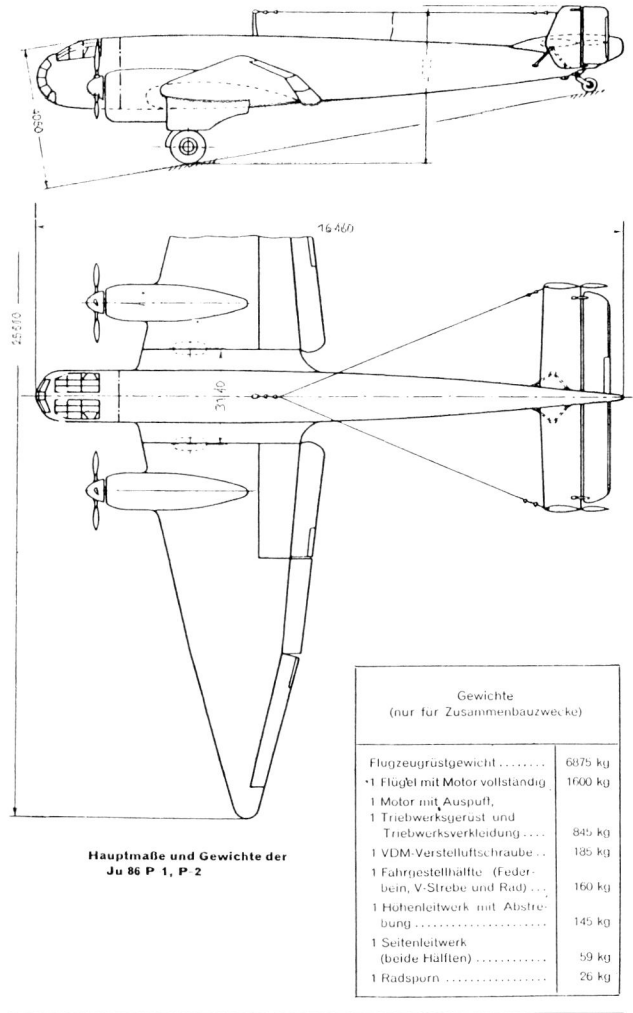

Dimensions and component weights for the Ju 86P-1 and P-2.

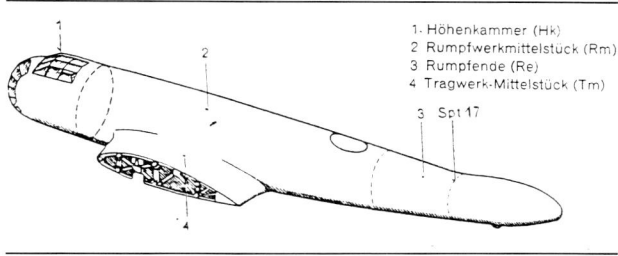

Fuselage assembly of the Junkers Ju 86P.

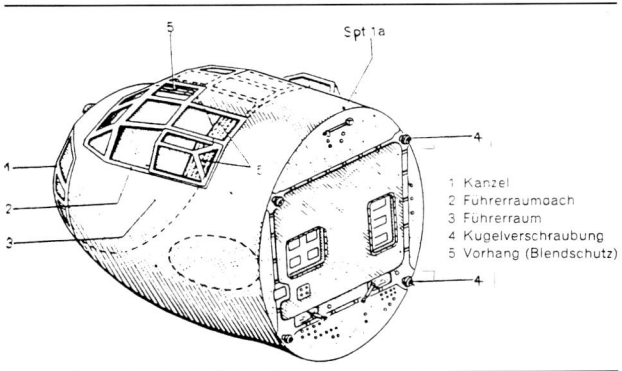

Pressurized cockpit of the Junkers Ju 86P.

an insulating coat of plywood and *Alfol* (aluminum foil). The cabin attached to the fuselage center section by means of four ball joints. Within the cabin the pressure remained constant regardless of the altitude of the airplane. It was ventilated using charged air bled from the engines, with the increased pressure reduced and the ingressing air cooled and filtered before entering the cabin. Ventilation was by means of fans. Cockpit pressure was regulated automatically at approximately 0.72 ata, corresponding to an internal pressure equivalent of between 3,000 and 35,000 m. Operation of all systems and equipment was carried out from within the pressurized cabin. The control rods for the various systems ran through airtight openings in the aft wall of the cockpit.

Armament: The Ju 86P was initially built and flown without any defensive armament—the previous gun stations were removed and the openings covered. Due to its limited payload capacity (maximum of 1,000 kg = 4 x 250 kg or 16 x 50 kg bombs) the Ju 86P-1 was used exclusively on one-plane "nuisance raids"; the majority of the limited number of P-series aircraft flew as high altitude strategic reconnaissance platforms.

Summary

With the Ju 86P-1 and P-2 a high altitude bomber and reconnaissance aircraft had been created in short order which met the demands of the war situation in 1940, providing a type able to operate at previously unattainable altitudes. Junkers had broken new ground with the type, despite the RLM only contracting for a limited series of roughly 40 planes. The amazing fact that the progress of the air war led to even further operational improvements in what was already an obsolescent basic design will be discussed later.

It should be mentioned that, despite the Ju 86P-series at least in part resurrecting the halted Ju 86G assembly program, the decision to cancel Ju 86 production in the late summer of 1938 in favor of the Heinkel He 111 and Ju 88 had left pre-cut and finished Ju 86 components as nothing more than scrap metal. The idea of continuing Ju 86 production for increased use as a trainer in the C-class and instrument flying schools would, at the time, most certainly have provided perceptible relief to Ju 52 elements—in spite of the fact that the Ju 86 was by no means the ideal trainer platform. The importance of the Ju 52 for combat operations was recognized by the Norwegian campaign. Shifted from campaign to campaign, and later from crisis to crisis, it became common to pull the Ju 52s, in the main with their instructor crews, from pilot and instrument training centers and press them into combat—a practice which would increasingly have a negative impact on the supply of well-trained crews for bomber and reconnaissance units in the future.

11. The Western Campaign

On 10 January 1940 two *Luftwaffe* officers, flying a Messerschmitt Me 108 *Taifun*, were forced to make an emergency landing when their engine failed and they found themselves in poor weather conditions. The plane, which came down near Mechelen, Belgium, was only partially destroyed in the landing, and what fell into Belgian hands was nothing less than the plans for *Luftflotte*'s operations in the Western campaign. Despite this setback, by 24 February 1940 the so-called "Sickle Cut Plan" had been finalized, with the main focus being an armored spearhead in the middle of the Western Front driving toward northern France; the Netherlands and Belgium were part of the operational planning. The paratroopers and airborne assault forces employed so effectively in Norway would be used to seize key positions over the Maas and mouth of the Rhine. Following a total of 29 postponements in the start of the attack, by early May it was finally time to give the order to attack—in spite of the ongoing conflict in Norway.

The Netherlands

On 10 May 1940 at 0535 hrs the assault on the West began. Air transport units, comprised of hundreds of Ju 52s, dropped paratroopers onto their assigned targets which had been deemed critical to the operation. They ran into fierce opposition from the Dutch, who were well prepared against such tactics following the example in Norway, and suffered considerable losses. By the end of the first day the nearly 120 front-line aircraft of the Dutch Air Force had been eliminated, and throughout Belgium, Holland, and northern France there was hardly an airfield—over 70 were attacked—which had not been hit by German bombs. As in Poland, the *Luftwaffe*'s aim was to establish its air superiority in the skies over Holland and Belgium. Within the space of three days the British had sacrificed half of their Blenheim and Fairey Battle bombers, among other types based on the Continent, mostly to low level attacks by Do 17s on their airfields or to the guns of German fighters.

In Holland, the battle was virtually decided with the dramatic fall of Rotterdam. On 14 May, with surrender negotiations underway, Rotterdam was attacked by 100 He 111s beginning at 1500 hrs. Some of these (43 to be precise) turned back without dropping their ordnance after spotting the standard signal for breaking off an attack (red flares which had been fired off by the formation leader). The remaining 57 planes dropped their HE bombs on the northern part of the city, the triangular section of the old city bordering on the Maas bridges, which burned almost completely to the ground barely two hours before the surrender. Dutch resistance collapsed after just five days of fighting; on 15 May the Dutch capitulated. This quick victory was bought at a heavy price, however—nearly two-thirds of the 430 Ju 52s taking part in the operation never returned from Holland.

Belgium and Northern France

Here, too, began a simultaneous attack on 10 May 1940 with concentrated air strikes on Belgian airfields, generally crippling the enemy's air defense and counterattack potential. Paratroopers seized important bridges, and Fort Eben Emael, overlooking the Albert Canal, was captured in the first use of DFS 230 assault gliders towed by Ju 52s close to the target area before being released. Advancing in waves, German ground forces were supported from the air in their drive over the Maas on 13 May 1940 as they fought French and Belgian troops. On 14 May the Allies threw their remaining air assets into the Battle of Sedan. In what turned out to be the first large-scale air battle in the West, lasting the entire day, 89 enemy aircraft fell to the guns of German fighters, and a further 112 were brought down by AAA fire. For all intents and purposes, these irreplaceable losses broke the back of the French bomber forces; even the British lost some 60% of their bombers participating in the actions of 14 May, another painful bloodletting for the Royal Air Force. On 20 May 1940 advancing Allied armored columns, caught in the long, unprotected flanks of the "sickle cut," were neutralized by Hs 123 strike planes and Ju 87 dive bombers flying under fighter cover. These aircraft continued to provide unceasing support during the subsequent breakthrough of German armor units in their drive to the Channel coast. On 24 May 1940 German armored troops advanced west and south of Dunkirk. In doing so, they exposed themselves to enemy fighters whose range now permitted them to operate from bases in England with fresh crews. Here the German armored attack was ordered to halt on 25 May 1940, giving those Allied forces cut off from the rest of France the time to pull out of the massive encircling movement in northern Belgium and evacuate across the Channel via Dunkirk. On 28 May 1940 Belgium surrendered.

However, in the interlude the German dive bombers and fighters had difficulties with forward deploying to airfields in the conquered areas, the majority of which were heavily damaged. Their limited range made such airfields necessary for continued operations by the *Luftwaffe*. In addition, supply problems only aggravated the combat readiness state for the exhausted units. Into this state of affairs wormed a serious change which was to prove most unfavorable to an air force which had been flying non-stop for the last two weeks. It had entered the Western campaign with

501	strategic and tactical reconnaissance aircraft (Do 17s, He 46s, Hs 126s)
1,120	bombers (Do 17s, He 111s, Ju 88s)
342	Ju 87 dive bombers and
42	Hs 123 strike aircraft,

but these numbers had seen a steady reduction over the course of the conflict.

Dunkirk

In such a weakened state Göring, with Hitler's approval, tasked his flying units to destroy the British Expeditionary Forces at Dunkirk and prevent them from fleeing across the Canal. During this period the *Heer* would regroup for the second phase of the Western campaign, the offensive drive toward Paris as part of the plan known as *Fall Rot* (Case Red). The British were now compelled to use every means at hand to rescue at least their living troops from the impending catastrophe, if necessary abandoning their materials and equipment in the process.

"Operation Dynamo"

Poor weather favored the British evacuation efforts, which began with a menagerie of ships and boats on 26 May as Operation Dynamo. The numbers of rescued troops increased as the days went on, for the *Luftwaffe's* available forces could only operate sporadically against Dunkirk in the time period between 25 May and 2 June 1940; their bomber and dive bomber units had been diverted to stop a massed French armor attack on 25 May along the long southern flank of the German units near Amiens. The bulk of these flying units "prepped" Calais for storming on 26 May, forcing the city to capitulate that same afternoon. During the decisive days of the evacuation, from 29 May to 1 June (when some 50,000 Allied forces were pulled off the beach each day) German air attacks were hampered by foggy weather over northern France to the point where their influence was negligible. Only on 27 May, the late afternoon of 29 May, and on 1 June was the *Luftwaffe* able to conduct truly effective strikes against Dunkirk and the Allied shipping. Despite their success, the branch suffered high losses among its bomber forces, primarily the slow Ju 87s. Only the high-speed, maneuverable Ju 88 making lone dives on shipping targets remained relatively untouched by British fighters, yet even these bombers (operating from bases in Holland) did not escape unscathed. The British fighter squadrons of the home defense, hitherto unbloodied, rested, and at full strength, were equipped with fighter aircraft on par with the Me 109. Repeatedly they were able to wrest air superiority from an enemy weakened by three weeks of incessant fighting and suffering from a lack of personnel and materiel. British fighters formed a protective umbrella over Dunkirk and the embarkation points, and thereby fended off attacking German bombers and dive bombers, inflicting severe losses in the process.

On 4 June 1940 the British declared Operation Dynamo to have ended. Approximately 338,000 men—225,000 British and 113,000 French and Belgian troops—had been evacuated and were thus able to avoid internment, but had lost 85% of their equipment in the process (although this could—and was—later replaced). Among the numerous ships sunk by the *Luftwaffe* were nine destroyers. Dunkirk fell the same day, and the battle ended while the severely taxed bomber units of the *Luftwaffe* had flown major attacks against airfields and aircraft factories in the Paris area. With further raids on 5 June 1940 these forces opened the final act of the Western campaign.

The limits of the *Luftwaffe* had become plainly visible at Dunkirk for the first time. Since the beginning of the war it had logged one victory after another, enhanced by the fact that up until now it had been fighting against enemies inferior in both their equipment and their training. When the *Luftwaffe* was ordered to single-handedly carry out an operation of the magnitude of Dunkirk, however, it found itself clearly overtaxed in the face of the counter-operations of all three opposing branches of service—including the largest naval power on earth at the time. Despite considerable losses in personnel and an almost complete loss of equipment, the enemy was not destroyed on land or on sea as had been directed. The Dunkirk operation had prevented the flying units from enjoying a much needed breathing period—something the army was able to experience to a great extent between the two operations of *Fall*

Gelb and *Fall Rot*. The Dunkirk operation effectively meant that, for *Luftwaffe* units, the two operations merged into each other without a break and thus formed an uninterrupted, merciless operation for the flying units, with Dunkirk providing the least amount of victory bought at the highest price.

Paris and the Results

During the continuing struggle of *Fall Rot* the *Luftwaffe*, in addition to maintaining air superiority, was assigned to support the army in its advances, thwart any attempt by the enemy to form pockets of resistance, prevent movement of enemy forces, and secure the west flank of the front. Aerial support of the army was flown against an enemy whose air defense forces were already paralyzed in the face of overwhelming air superiority, after the highest levels of the French government had vainly sought for British fighter support. The concept of *Luftwaffe/Heer* cooperation played out well, as it had in Spain, Poland, Norway, and in the initial phase of the Western campaign. After heavy air raids against military facilities on the outskirts of the French capital, Paris fell on 14 June 1940 without a fight—ten days after the Allies had successfully concluded their Operation Dynamo. By 17 June the entire northern and western coasts of France were in German hands, and German forces had reached the border with Switzerland.

Shortly before France's military collapse, Italy (up to that point not particularly combat-minded) declared that effective 11 June 1940 a state of war existed between Italy on the one hand and France and Great Britain on the other. Too late to effectively participate in the Blitzkrieg, which was already well into its fourth week. On 25 June a German-French-Italian ceasefire went into effect with the well-known consequence of having France divided into an occupied and unoccupied zone.

Consequences for the *Luftwaffe*

It can be safely stated that the six week campaign was successful thanks to the full support of the *Luftwaffe*. The French had learned nothing from the Polish campaign, which for Germany's leaders had served as a model of success for future lightning campaigns with regard to the combined use of *Luftwaffe* and *Heer*. The French bomber arm was weak; a large percentage of its air force had been destroyed on the ground or never employed in combat, the latter a by-product of organizing squadrons under army commanders and thus eliminating any type of effective central command and control for the air arm as a whole. Nevertheless, the losses suffered by the attacking German air units were considerable. The total of 2,694 front-line aircraft losses* occurring in the period from April 1940 (thus including the Norwegian campaign) to June 1940 were broken down as a percentage of total aircraft lost as follows:

329 tactical reconnaissance aircraft	= 8.6%
105 strategic reconnaissance aircraft	= 3.9%
974 bombers	= 36.2%
186 dive bombers and strike aircraft	= 6.9%

Compare the above figures with the percentage lost by other aircraft classes:

794 fighters and heavy fighters	= 29.4%
165 maritime and liaison aircraft	= 6.1%
241 transport aircraft	= 8.9%

The 55.6% portion of the attacking elements and the 29.4% portion of fighters and heavy fighters lost compare to the 15% figure for transport, maritime, and liaison aircraft. The figures speak for themselves; it clearly highlights the 85% loss rate suffered by those units involved directly in combat, to which should certainly be added at least a portion of the 90 maritime aircraft lost mainly in Norway and their 3.3% part of the overall number of losses.

An additional indicator of the force distribution for attacking elements is evidenced by a comparison of the approximately 500 aircraft used in this capacity in Norway and the nearly 2,000 used in the Western campaign (with some of these being carried over from the Norway numbers), with the overall loss of nearly 1,500 offensive aircraft—approximately 60% of the total number of aircraft used. These losses would need to be compensated for if the war was to continue.

The loss of crew members associated with these material losses were somewhat cushioned by the fact that approximately 400 pilots, most of which were shot down by British fighters, had been made French POWs. Following the ceasefire with Germany they were repatriated, and for the most part were available to fight in the upcoming conflict against Great Britain or elsewhere. One of these was *Oberst* Josef Kammhuber, who had been shot down over France while serving as a *Kommodore* of a bomber wing and had petitioned for front-line service in early 1937.

---

*   Michaelis, Schraepler, Scheel, Ursachen und Folgen vom deutschen Zusammenbruch, vol. 15: Kriegführung gegen die Westmächte 1940, p. 366.

12. The Battle of Britain

Despite being driven from the Continent, Great Britain gave every indication that it would continue the fight alone. Winston Churchill, the country's prime minister from 10 May 1940, ignored a German offer of peace on 19 July and instructed his foreign minister, Lord Halifax, to reject it on 22 July. At this point the war against Great Britain entered into its decisive phase. In mid-July 1940 Hitler had ordered preparations to be made for a landing in England, Operation Sea Lion (Seelöwe), which was canceled on 12 October 1940. What were the reasons behind this important decision, one which effectively determined the future course of the war?

Struggle for Air Superiority

In keeping with an Ob. d. L. directive dated 30 June 1940, during the first three weeks in July only nuisance raids were carried out against England and the shipping lanes around the island nation. *Führerweisung* No. 17 from 1 August 1940 changed that, and instructed that "the Royal Air Force was to be crushed," the aviation industry and AAA production sites attacked, British foodstuffs sources destroyed, and their merchant marine and warship fleets crippled; London was to remain hands off! Much has been written about the so-called "*Adlertag*," the beginning of large-scale attacks against the Royal Air Force in the vicinity of London, as well as against airfields in central England from bases in Norway. It only needs mentioning that the attacks beginning on 13 August 1940 had been preceded on 12 August by Me 110 fighter-bomber strikes against British coastal radar stations, with minimal success, and that the air offensive had to be temporarily called off on 18 August because of the weather. But just in these few days 225 German aircraft had been shot down compared to only 103 British fighter losses, whose pilots were "recyclable" for the most part. Consequently, the Ju 87 and Me 110 were both generally pulled from the conflict due to their vulnerability to Spitfire and Hurricane fighter defenses. On 23 August 1940 the attacks resumed, but by then the British had abandoned their fighter bases in southern England and pulled back to airstrips on the periphery of London. They knew that the *Luftwaffe* did not have long range fighter escort capability for its bomber formations. On par with British fighters, the Me 109 was certainly able to reach London, but over the capital it had virtually no appreciable loiter time to provide effective cover for bomber formations in the target area.

On 25 August 1940 a handful of RAF planes bombed Berlin for the fifth time in ten days in retaliation for German bombers accidentally dropping their ordnance on London a few days previously (the parties guilty of this error in judgment were disciplined). On 27 August 1940 there followed the first of four large-scale German raids against Liverpool-Birkenhead, with a total of 629 bombers taking part. On 1 September 1940 German bomber formations reported weaknesses in the British fighter defense; on 4 September the senior *Luftwaffe* commanders reckoned that the RAF had only about 420 fighters left. In reality, however, the number was closer to 650, for despite the concentrated bombing attacks British fighter production output remained significantly higher than that of the Me 109. Beginning on 5 September 1940 London's docklands and oil storage facilities along the Thames were bombed.

Influenced by the Berlin raids and the apparent hesitation in British fighter defenses, on that same day Hitler ordered the attack on London to begin. This was initiated by 68 bombers during the night of 5/6 September 1940. On 7 September a large-scale daylight raid comprising 625 bombers hit the docklands of London in an attempt to force the British fighters out to fight.

With the prerequisite for an invasion of England—the establishment of air superiority—having failed to materialize thus far, it remained to be seen if the *Luftwaffe* would draw closer to achieving this goal through their daylight raids on the British capital. It requires no great judgmental skills to realize that the capabilities of the escorting Me 109s and their pilots were barely able to keep up with such a tactic.

Subjectively viewed intermittently as a weaker or weakened British fighter defense, which was discovered to include experienced Polish and Czech pilots, was less due to a lack of materials than a problem of organization and control.

The RAF's fighter control—despite radar guidance—was generally unable to bring the fighter squadrons based north of London (some of which were intentionally held in reserve) to bear in concentrated force against the German bomber formations in a timely manner.

A few more daylight raids followed and met with varying success, with a lack of fighter escort forcing German bombing operations to be broken off over the south coast of England and highlighting the true air picture. On 15 September 1940 a new large-scale daytime attack on London was ordered, during which the Brit-

History of German Aviation: Bombers and Reconnaissance Aircraft

ish finally succeeded in bringing their entire fighter force to bear against the raiders. Spitfire squadrons primarily concentrated against the German escorts with their combat capability limited purely by their limited range, while the Hurricanes focused on the bomber formations. The loss of 56 German aircraft on this day was high, but the existence of an intact British fighter defense, which Germany had considered all but finished, was even more unsettling. On 16 September 1940 the British fighter arm was given an unexpected recuperative period, for weather prevented the *Luftwaffe* from mustering any large-scale attack. The season was too far gone to count on good weather and forced a switch to night operations.

Heinkel He 111H-3 during a daylight raid on London. On 7 September the Docklands were hit, including the Royal Victoria Dock and the King George V Dock. This photo, taken over the Thames, clearly shows from left to right the Outer Dock, South Dock, and the two basins of the West India Dock.

Dornier Do 17 bombers over the Thames (September 1940).

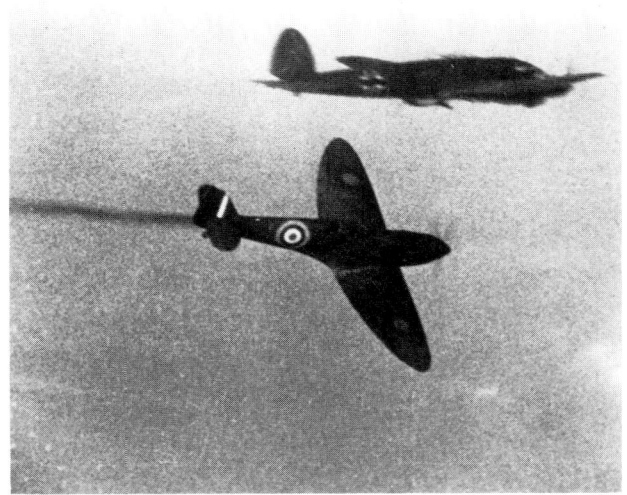

Hunter and hunted during the Battle of Britain. Trailing smoke from a hit in its engine, an attacking Spitfire fighter cuts through a formation of He 111s (September 1940).

History of German Aviation: Bombers and Reconnaissance Aircraft

From that point on, in 86 almost uninterrupted nights, raids were flown against British industrial cities with the goal of crippling the country's armament industry and to break the will of the British people to resist. In addition, hand-picked crews flew individual strikes against key production sites, such as aluminum processing plants, ball-bearing factories, and aircraft engine production centers.

One noteworthy example from the initial phase of this period is the unexpected success of a fighter-bomber test group, which had converted from the Hs 123 to the Me 109E after the Western campaign. On 20 September 1940, using the element of surprise, it flew its first successful bombing raid against London, providing first-hand evidence for the soundness of the order which directed that part of the fighter arm be converted to fighter-bomber operations. Nevertheless, even these victories, which soon turned into costly sorties, were short-lived and were only pinpricks—a single 250 kg bomb or 4 x 50 kg bombs per airplane were too little to have any lasting destructive effect.

The goal of establishing air superiority over the Channel and southern England—an indispensable prerequisite for the army and navy's Operation Sea Lion—was not achieved, despite grueling operations on the part of Germany's offensive air units and their often inadequate fighter escort. On Hitler's orders, 12 October 1940 saw the OKW postpone the landings in England until the spring of 1941, limiting its preparations for the invasion to political and military pressure on Great Britain.

In practice, however, this meant that Germany had effectively abandoned the operation, particularly since even before the Battle of Britain Hitler had decided to attack the Soviet Union. He continued to stand by his decision even after his hopes of British negotiators arriving following his peace offer of 19 July 1940 failed to materialize. The British would not be eliminated from the war by such political moves. They entered the year 1941 unbeaten in the air, with deliveries of American aircraft beginning in October 1940, unvanquished and with no German invasion attempt of their island. Thus was pre-programmed the massive two-front war!

Did Germany's Air Power Fail?

We must look closer in order to answer this difficult question. First let us take a look at the German methods and equipment available for the Battle of Britain.

Weapons of the Air

The Do 17, He 111, and Ju 88 bombers had an actual stock on hand of 1,458 aircraft and 1,514 crews at the beginning of the conflict, of which 981 aircraft (67%) and 1,015 crews (also 67%) were operationally fit. It is interesting to note that on 14 September 1940, i.e. one day prior to the air battle which precipitated cessation of large-scale daylight bombing raids against London, the bomber numbers were broken down as follows:

	On Hand	Operationally Ready
Do 17Z	256	195 = 76%
He 111H	369	199 = 54%
He 111P	231	167 = 72%
Ju 88A	534	351 = 66%

Conversion to the planned Ju 88 standardized bomber was already well underway by this point; it had almost reached the He 111's total numbers of 600 in front-line service, whereas the numbers of Do 17s continued to drop.

During the period from 3 August to 28 September 1940 719 bombers (49% of the original total on hand) and 400 crews were lost.* On 28 September 1940 818 out of a total 1,420 aircraft were operationally ready, equating to 58%, and of 1,663 crews 1,074 (65%) reported in as operational. From these figures, which were average values drawn from documents of the general quartermaster of the Ob. d. L. on display in the so-called Karlsruhe Collection, it is clear that the heavy material and personnel losses suffered during the Battle of Britain were constantly being replaced and indeed could practically be fully offset.

The Ju 87B, R, and M dive bomber models stood at 446 aircraft total at the beginning of this phase of the war, of which 351 (79%) were operational; of 388 crews there were 336 (87%) operationally ready. The loss of 97 machines, i.e. about 22% of the original number on hand, suffered between the 3rd of August and the 28th of September were replaced without difficulty; crew assignments not only compensated for losses, but even heralded an increase by 8%. It should be noted, however, that as early as mid-August the Ju 87 was only used in extremely limited numbers and under full fighter protection.

The actual number of strike aircraft on 3 August 1940 consisted of 35 Messerschmitt Me 109Es serving with a fighter-bomber trials unit which had reequipped on the Me 109 from the Hs 123 after the Western campaign. Of these there were 34, practically

---

* Aircraft losses were defined as either total or as sustaining greater than 10% damage, which were sometimes reassigned to a unit following repairs.

History of German Aviation: Bombers and Reconnaissance Aircraft

100%, deemed operationally ready. The 61 pilots of this group were reported as non-operational, since they were still in conversion training for fighter-bomber missions on what was for them the new Me 109 type. On 28 September 1940 there were still 31 Me 109s on hand, of which 26 (84%) were serving operationally. With regard to pilots, 45 were still listed, of which 35 (78%) were operational. Despite constant resupply of crews and aircraft during this period of fighting, in which this trials unit was not fully committed until the latter half of September, the number of aircraft had dropped by 11% and the number of pilots by 25%. This negative development was symptomatic for the high losses among the fighters and heavy fighters, also listed below, and is necessary for a realistic assessment of the supply situation at the material and personnel levels.

Fighters: These fell from 760 Me 109 aircraft on 3 August 1940 to 414 on 28 September, 54% of their original stock, of which only 276 were declared as true fighters and 138 (exactly half) as fighter-bombers.

Heavy fighters: Of the 230 Me 110s listed on 3 August, by 28 September 1940 there were still 100 on hand, 43% of their original number.

Interim Balance

These, then, are the tools which played a decisive role in the struggle for air superiority on the German side. The strategic and tactical reconnaissance aircraft were not included in these figures, since they were of minimal value as a real combat factor in the Battle of Britain.

During the two key months of August and September 1940 forced resupply kept bomber and dive bomber strength on a relatively even keel, despite the loss of almost half the bombers and nearly a quarter of the dive bombers—with the latter being pulled out of the battle in short order after suffering heavy initial losses. This was not the case with the Me 109 and Me 110 units, where it was not possible to compensate for material losses, even though the Me 110—like the Ju 87—was used only sparingly because of its weaknesses in combat; the bloodletting that was the Battle of Britain could not be offset by fighter production.

Combat Methods

The large percentage of Ju 88As at this stage has already been mentioned. At 534 aircraft, on 14 September 1940 the type formed 38% of the total inventory of 1,390 bombers available on this day for the Battle of Britain. At first glance it differed little from the other bombers, and not always in its favor when used to bomb from higher altitudes. It could carry a maximum of 1,400 kg of bombs internally, whereas the He 111 carried a 2,000 kg payload and the Do 17 no more than 1,000 kg. Its irrefutable advantages lay in its 50 km/h speed advantage at 5,000 m and its maneuverability when compared with the He 111, but mainly in its fully integral dive bombing capability. To this end, it is worth examining just how far it had gone to achieving its operational potential with regard to its dive bombing capability.

Ju 88 Dive Bombing Methods

As previously mentioned in the section discussing the Junkers Ju 88, the aircraft had an inherent dive bombing capability thanks to its automatic recovery system. The diagram showing the procedures when using the automatic recovery system of the Ju 88 had, in the interim, evolved into a refined dive bombing method that, when executed correctly, had proved itself most effectively on attacks against shipping targets and airfields (see p. 126).

A drill-type method for this had been developed in order to enable an average qualified crew to fully exploit the technological capabilities which the Ju 88 now embodied. The method was tailored to the Ju 88 and no longer bore resemblance to the steep diving angle of the Ju 87, whereby the ordnance was released just prior to the aircraft pulling out of its dive. An example of the Ju 88 methodology can be seen in the drawing titled *"Vorteile des Ju 88-Bombenabwurfverfahrens aus dem Sturzflug"* (Advantages of the Ju 88 Dive Bombing Method, see p. 127).

This method took into account the horizontal bomb arrangement in the Ju 88's bomb bay, the possibility of stick bombing and, finally, the minimum loss of altitude when pulling out of a dive. At an assumed diving speed of 600 km/h the altitude loss during a steep dive angle (60 to 90°) was about 800 m, calculated at a constant pullout rate of 3.5 g. On the other hand, at a medium angle of 50° only 290 m was used. This dive bombing technique became the standard attack method for the bulk of Ju 88 crews, with some individual and target defense induced minor variations.

Naturally, this raises the question of whether the structural expenditure and associated weight increase necessary for executing almost vertical dives and pulling out could actually be justified in light of the fact that in actuality dive attacks were almost always made from angles of 50°. The odd crew routinely making steeper dives on shipping targets was the exception rather than the rule when it came to dive bombing sorties using the Ju 88 in practice. Without a doubt, by limiting the tactical-technological requirements to a 50-60 degree dive angle, the machine would have undoubtedly

History of German Aviation: Bombers and Reconnaissance Aircraft

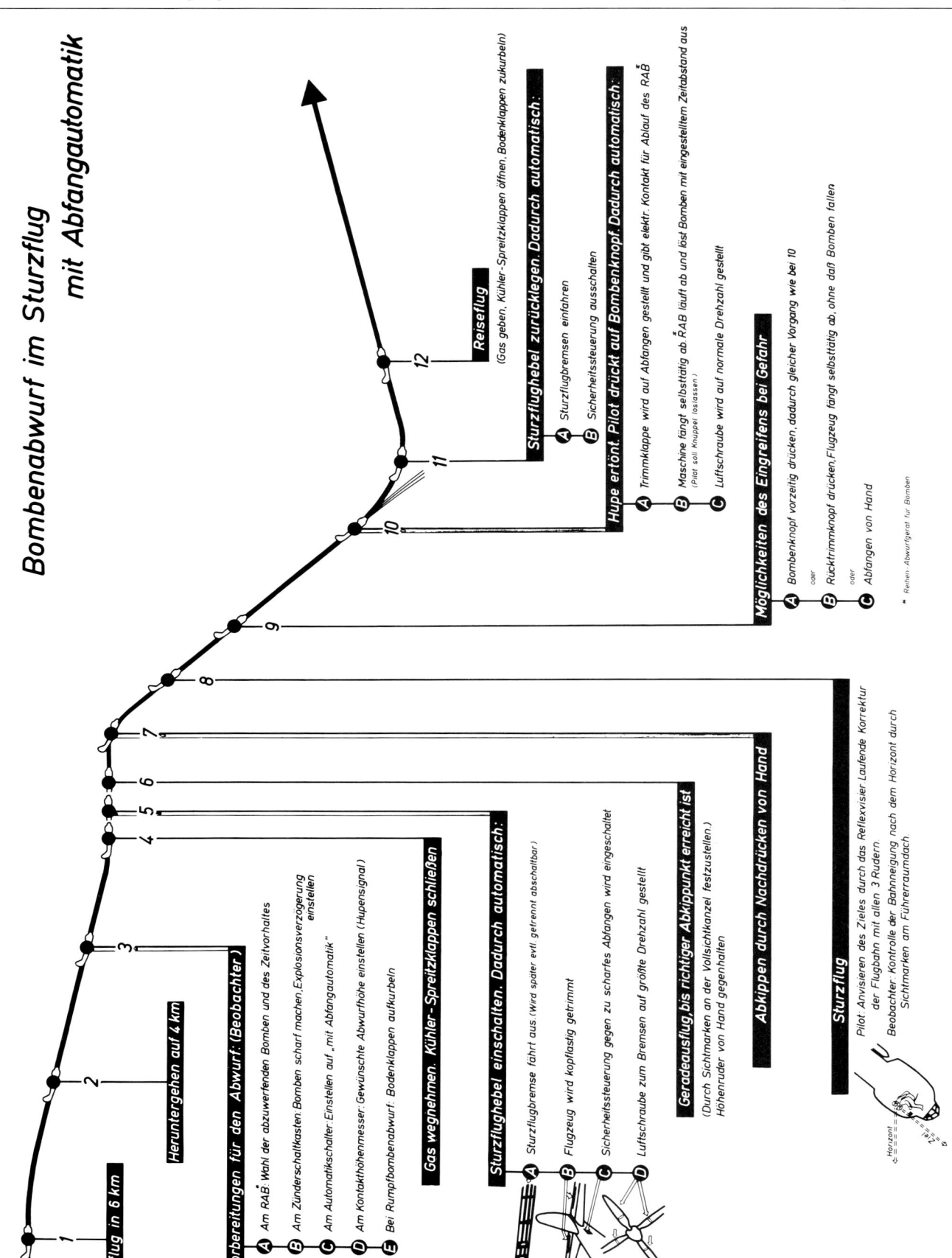

The twelve points of a Ju 88 dive bombing profile using the automatic pullout and the reflective gunsight as a bomb aiming device for the pilot.

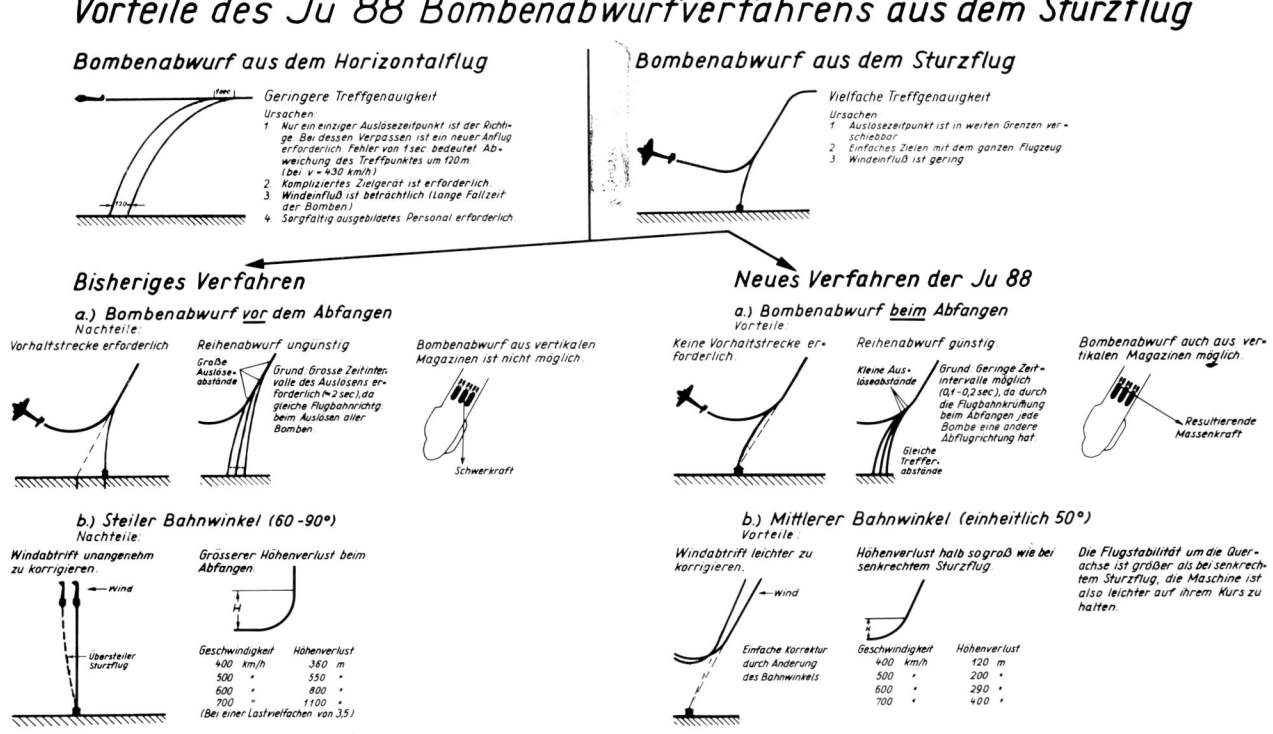

more closely approached the concept of a true high-speed bomber. However, this potential profit with regard to the Ju 88V5's performance was eclipsed by the insistent demands of the General Staff and Ernst Udet, Chief of the *Technisches Amt* at the time, that the type be capable of 90 degree dive angles. Nevertheless, the Ju 88's dive capability in conjunction with the above-mentioned attack profile was an enormous plus with regard to operational flexibility, one which the enemy was unable to counter with a similar aircraft of its own.

However, even the best dive bombing method was of little effect if the ordnance was unable to pack the punch needed to knock out the intended target, e.g. a ship.

Bombing Naval Targets

Weapons

Against merchant ships and surface warships without armor protection, as well as submarines, the commonly available SC 250, SC 500, and SC 1000 high explosive bombs were used. Larger ships, those over 8000 BRT, were more effectively engaged with SD bombs of the same caliber. As a rule the attacking formation flew a horizontal attack profile at altitudes over 2,000 m up to their service ceiling, or low altitude attacks at altitudes of between 20 and 50 m, flown by fighter-bombers using the Rechlin or the Liesendahl methods (see page 129).

The PC 500 armor piercing bomb introduced in 1939 and in good supply by 1940 was quite effective against destroyers and torpedo boats, but did not have adequate penetration capability against larger, more heavily armored warships. As a result it was fitted with a rocket set in order to provide better penetration force, and was employed as the PC 500 RS (for *Raketensatz*, or rocket kit). It was effective against larger warships, such as battleships, up to 35,000 t. Other than the PC 500 RS, in 1940 there was no other better-penetrating, more effective bomb for use against the large warships, i.e. cruisers, battleships, and aircraft carriers. The explosive charge of the PC 500 RS was relatively small, limiting its effectiveness and only enabling it to sink a warship when dropped with precision accuracy.

Outside Curve Bombing Attack
The so-called offset dive bombing attack profile was developed especially for the Ju 87 using the reflective gunsight as the bombsight. Ju 88 pilots also made use of this attack method.

Not every bomb dropped from a dive is a direct hit, as the ship under attack naturally makes full use of its maneuverability and defensive firepower, and the pilot has to be able to do much more than just fly his plane if he wants to hit his target effectively.

Particularly fast ships, both warships as well as merchant ships, attempt to prevent the dive bomber from lining up—generally along the ship's longitudinal (x) axis—for an effective dive attack run by making full or semi-circles (snaking movements). The outside curve method enables the bombs to be dropped along the ship's axis even when the ship is turning. If the attacker does not enjoy complete and sure surprise, then the enemy must be tricked or forced into turning, because once initiated the ship (due to its enormous mass) must continue movement in a set direction for a long period of time. The enemy often encounters the start of the outside curve attack when he is peacefully moving in a straight line and, upon catching sight of the diving aircraft, begins to make a turn at full power. The goal is to force the attacker to dive toward the ship's inside curve and minimizes the chances for a hit considerably. With formation attacks and a straight moving target, a *Kette* or *Schwarm* was detached (ignoring fighter defenses) with the intent of enticing the ship into a turn, at which point the target could easily be handled by the massed attack of the main formation using the outside curve method. The ship finds itself in a turning movement which it must continue with for a time. If at this point the aircraft, which by now has taken defensive action and is within about 2,500 m over the target, begins its attack dive it must by default wind up in the inside curve, i.e. inside the circle of the ship's movement (assuming a dive of at least 10 seconds from initiation to bomb release). In this

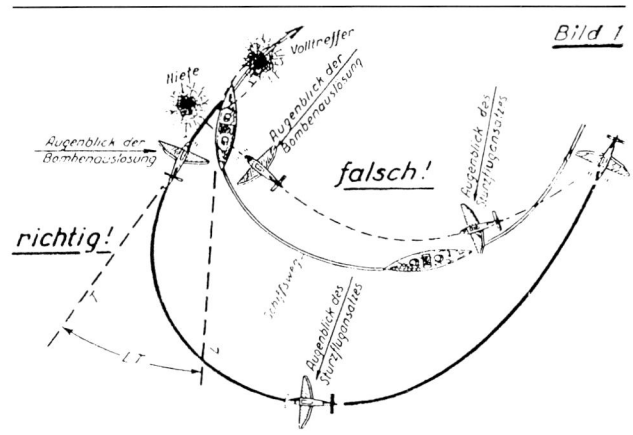

The correct (*richtig*) and incorrect (*falsch*) manner of executing an outside curve attack.

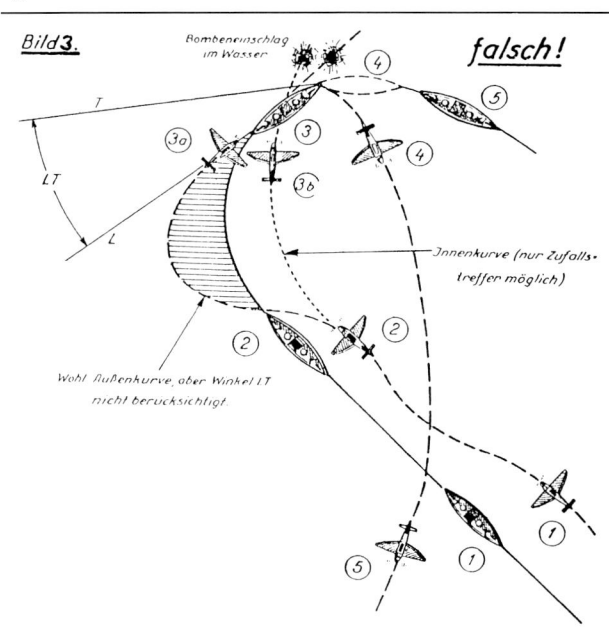

The incorrect method of dive bombing a moving naval target.

instance the attack would only be a successful one if the bomb were to hit the ship purely by chance, as shown in the drawings opposite (pictures 1 and 3). In all probability the bomb impact, assuming everything else is properly done, would fall outside the ship's turning arc. The same thing would happen if the outside curve attack were executed without heeding the LT angle (compare pictures 1 and 2). The correct profile for the outside curve bombing attack

History of German Aviation: Bombers and Reconnaissance Aircraft

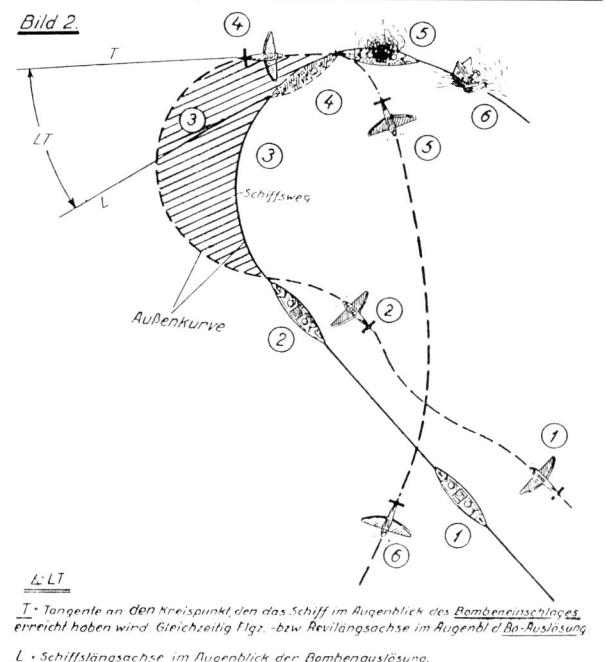

The proper way to execute an outside curve attack.

Loading a Heinkel He 115 with an aerial torpedo. Originally the *Luftwaffe's* main torpedo bomber, by the summer of 1940 this multi-purpose floatplane was proving unable to meet the demands of the air and naval war, suffering major losses as a consequence. Only when it had the element of surprise or when attacking convoy stragglers could the He 115 enjoy an element of success.

method is shown in picture 2 (with angle clarification: T = tangent at the point of the arc which the ship will have reached at the moment of bomb impact. Corresponds to aircraft/gunsight x-axis at the moment of bomb release. L = longitudinal, or x-axis of the ship at the moment of bomb release).

As a result of the perceptibly increased ship defenses noted as early as late 1940, studies were undertaken on how these defenses could best be dealt with and/or avoided. Heavy bombers were regularly forced to fly through ship defensive fire when making low-level attacks if they hoped to enjoy any measure of success; this led to severe losses and considerable damage to the surviving aircraft. Up to this point aerial torpedoes had only been carried by floatplanes, but now their usage saw a considerable increase even by land-based aircraft as the Battle of Britain continued. Those bomber units tasked with attacking British supply transports were able to remain outside the effective fire zone of ship defenses when using this weapon, and the aerial torpedo increasingly found favor as the ordnance of choice when attacking naval targets.

For the faster fighter-bombers, the *Jabos*, which could operate against shipping near the coast, there were two methods mentioned earlier which had been developed:

The Rechlin Method
Diving from a given altitude at an angle of about 30 to 40°, aiming for a reference point well ahead of the target (approximately 1.5 to 2.5 km). Pullout using the speed buildup to advantage. Approach at 5 to 50 meters altitude at maximum inherent velocity and a very small, almost constant angle (3 to 5°). Best release distance: about 500 to 700 m (five to seven times the length of the ship for craft of 3,000 to 7,000 t). In general, wind and target speed do not need to be compensated for. As the reference point in the dive is so far ahead of the target, the angular velocity is so great that the ship's anti-aircraft defenses have little chance of hitting the attacking aircraft. During the last part of the approach the AAA can be fended off with the aircraft's onboard guns.

The Liesendahl Method
If, for reasons of weather or enemy radar, a *Jabo* is prevented from flying at higher altitudes, it then makes its target run at low altitudes. About 1.5 km before the target he pulls up to about 300 m, followed by a direct dive onto the target at an angle of about 20°. At a distance of about 300 m the bomb is released. The angle of elevation is thus minimal and almost constant (3 to 5°). As a result of the long, straight approach the accuracy is somewhat better than with the Rechlin method. However, the attack run is correspondingly more prone to enemy AAA defenses, which the pilot most likely

cannot fully suppress with his onboard guns. Also, in this case the pilot must almost always overfly the target ship. This latter method therefore had less appeal for the *Jabo* pilot, as he was generally unable to avoid the ship's defenses.

Radio Beam Guidance for Bombing Through Overcast
In poor weather and at night it is almost impossible for a bomber crew to spot the target with the naked eye. To improve target location, during the second half of 1940 the experience gleaned from radio navigation was put to good use in determining the coordinates of a target using electronic radio beams or providing reference points with cross beams. These signals would be acquired by the aircraft and exploited by the bombardier for determining the point at which to release his bombs. Of the many methods, all of which were ultimately based on the same principle, we will look at the *X-Verfahren*, or X-method. Selected aircraft were fitted with this equipment to act as target markers for the bulk of the following bombers.

The *X-Verfahren* employed a narrow band guide beam transmitted across the target. A second beam, a so-called advance signal, was transmitted to intercept the guide beam at a predetermined juncture prior to the target, with a second signal beam—the main signal—transmitted to intercept the guide beam at a point just before the target. Similar to a VHF beacon approach, the pilot was able to determine both via a scope and acoustically if he was flying along the guide beam—a constant tone and a non-deviating visual cue confirmed his course. If the scope showed a deviation left or right, the pilot would strive to realign himself with the guide beam by making the necessary course corrections. The bombardier needed to note the advance and main signals. Ground speed was measured between these two signals, and by taking into account the altitude flown in conjunction with the ground speed it was possible to calculate the bomb release point. A cleverly designed stopwatch with a computing mechanism, called an *X-Uhr* (X-clock), calculated the information automatically. The bombardier only had to start the stopwatch when the aircraft crossed the advance signal point and stop it again as the plane intercepted the main signal. The *X-Uhr* then automatically determined the point of release and dropped the bombs, also automatically, via an impulse signal transmitted to the ordnance release device, the RAB.

When it was introduced, the *X-Verfahren* was extraordinarily successful over England in bad weather and during the incessant night raids on British industrial cities and dock facilities which began on 16 September. However, once the enemy soon recognized the operating characteristics of the device and determined the frequencies it operated on, the British began electronically jamming the *X-Verfahren* so doggedly that, other than in a few instances, by the summer of 1941 it had been rendered ineffective.

Another target location system developed at about the same time was the *Y-Verfahren*, whereby the bomber had a special repeater transmitter on board and received precise course correction and bomb release information from ground radio stations. Yet this system proved even easier to jam, and it soon suffered the same fate as the *X- Verfahren*.

Synopsis

Despite a technologically advanced state, the German air weaponry was inadequate for conducting day and night operations against Great Britain under all weather conditions; it was not all-weather capable, and limited in its range and bomb capacity. Its stellar performance in France as part of the tactical air war, i.e. the direct and indirect support of land-based forces, was not in demand during the Battle of Britain. For the war against Great Britain, especially in its primary goal of "crushing the Royal Air Force" and crippling shipping supply lines and its armament industry, an operative—even strategic—type of air warfare was needed. Germany's air forces were inadequately equipped for such warfare and were constantly subjected to serious losses as a result—unless they were protected under effective fighter cover. This, in turn, drastically restricted the already limited penetration range of the bomber formations.

In spite of now having jump-off bases closer to England, Germany's available air weapons were unable to beat the British air forces, particularly their fighter defenses, nor were they able to establish air superiority. Exactly as had been predicted by *General der Flieger* Felmy back on 22 September 1938, yet again on 13 May 1939, and for a third time in the summer of 1939.

Within the limits of the given technological potential at the time, the bombing methods were generally refined and a recipe for successful missions when employed by well trained crews. However, these methods fell foul of enemy defensive countermeasures and the vagaries of air operations, which was the daily bread of operational crews. As a result, the collective voice of these crewmen reached a crescendo in demanding technological improvements on the basis of their combined experience, which were not always easily addressed or immediately realizable. Furthermore, both the *Technisches Amt* and the aviation industry frequently ran into considerable problems in their efforts to meet these demands, demands which could—and did—often interfere with the orderly pace of production!

13. Final Totals for 1940 Following the Western Campaign and Battle of Britain and their Effect

The campaign in France had been won in six weeks with full *Luftwaffe* support of ground forces. However, the successful British evacuation of its expeditionary forces from the beaches of Dunkirk showed the limits of the *Luftwaffe's* effectiveness for the first time when they encountered enemy fighters on par with Germany's. The British fighter defenses, whose inventory increased by more than 400 modern aircraft per month (while Germany's fighter production of the Me 109 was half that at the time), was first and foremost the reason why the *Luftwaffe* ultimately stood no chance of establishing its air superiority over the Channel and southern England—an indispensable prerequisite for the planned landings in England. To be sure, on October 12th 1940 this had been pushed back, but for all intents and purposes the invasion had to be abandoned. Preparations for Hitler's directed campaign against the Soviet Union had been in the works since the end of August 1940!

In light of the victory in France and the successful defense by the British, what was occurring with air weaponry—especially in the area of bomber and reconnaissance aircraft—in the time period leading up to the outbreak of further fighting in 1941?

In the euphoria over the success in France, and as a result of the *Heer's* needs for the planned Russian campaign, a few days after the cease-fire with the French the *Luftwaffe* was dropped to prioritization class 5 on the emergency list for raw materials. In compliance with Hitler's directive of 11 September 1940—in the midst of the costly daylight bombing raids on London—work was halted on all developmental projects which were not expected to reach operational status within a year.

While the Battle of Britain continued into 1941 with unrelenting tenacity through night raids on industrial cities and attacks on shipping convoys, momentum was increasingly developing behind the plans for Russia. In February 1941 the Ob. d. L.'s intelligence branch published a familiarization pamphlet on the Soviet Union "with particular emphasis on the flying corps...and the aviation armament industry...." In it, the fighting strength of the air arm was categorized as much less than that of the *Luftwaffe*, and the operational strength of the Soviet front-line units was estimated at no more than 50% of their on-hand aircraft due to weaknesses in the ground organization and supply system.

On the other hand, the familiarization booklet conceded that the widely dispersed armament industry and its concentration in well-protected areas hindered any type of large-scale disruption of the economic branch, which would play a crucial role in the overall supply within the Soviet Union. A serious miscalculation was the assumed aircraft strength in early 1941, which for the European portion of the Soviet Union was estimated at 2,100 bombers, 620 reconnaissance aircraft, and just over 3,000 fighters—numbers that undoubtedly were incorporated into German operational planning.

Was the German bomber and reconnaissance aviation, with their aircraft and weaponry, up to the task that was waiting in the wings of 1941? An inventory of the material available in early 1941 seems appropriate at this point.

Long range Bombers

Focke-Wulf Fw 200

At this point the Fw 200 long range airplane, modified into an auxiliary bomber, was still the only strategic bomber and maritime reconnaissance airplane available. The "militarization" of this airplane has already been discussed earlier.

Fw 200 civilian aircraft and Fw 200B military transport conversions, plus a handful of military C-0 series, carried out several resupply flights in support of the Norwegian campaign in conjunction with the Ju 90, also a four-engined type. The true military version of the C-series, the Fw 200C-1, was delivered beginning in mid-1940, and shortly after the end of the Western campaign was assigned to bases along the French Atlantic coast. From these airfields the type flew reconnaissance and bomber missions against shipping bound for Great Britain, sometimes in direct cooperation with German submarines. These missions, which penetrated deep into the Atlantic and west of Ireland, initially enjoyed considerable success. Between 1 August 1940 and 9 February 1941 the Fw 200 destroyed no fewer than 85 Allied ships and approximately 363,000 BRT. Numbered among these losses was the *Empress of Britain*

(42,348 BRT), a Canadian passenger liner pressed into service as a troop transport. On the 26 October 1940 it was badly damaged in a bomb attack by an Fw 200 and caught fire. Two days later, while under tow to Ireland, it was torpedoed and sunk by a U-boat which had been radioed to the site. Cooperation between submarines and the Fw 200 intensified even further. In January 1941 alone the Fw 200s, flying armed reconnaissance patrols in the Atlantic, sank 15 ships totaling 63,175 BRT, and on 9 February 1941 they responded to the radio reports of a submarine which had been stalking a convoy in the waters between the Azores and Portugal. Five armed Fw 200s attacked the convoy over a period of six hours at low levels and sank five freighters for a total of 9,200 BRT. Prosecuted by both submarines and the Fw 200, the Battle of the Atlantic had now become a harsh reality for the Allies.

In 1940 a total of 36 Fw 200C models were delivered, including 10 of the C-0 pre-production batch and 26 Fw 200C-1s. In 1941 this increased to 58 aircraft, mainly the structurally improved C-3 series which appeared beginning in the summer of 1941 with the more powerful BMW Bramo 323R-2 nine-cylinder radial engine rated at 883 kW/1,100 hp. The initial defensive armament, with just three MG 15s, was beefed up as the defensive guns on ships improved. With the C-1, in front-line service since 1940, this armament included a flex-mounted 20 mm MG FF in the forward section and an additional MG 15 in the aft part of the gondola. In place of the forward, hydraulically powered turret on the fuselage was now a fixed cupola with a flex-mounted MG 15 fitted. Considered an interim version, the Fw 200C-2 had the same engines as the C-0 and C-1, but had better streamlined bomb racks and engine nacelles. It was just as vulnerable as its predecessors, though.

It was not until the summer of 1941, with the arrival of the Fw 200C-3, that the operational units saw a real improvement in the type. The sub-variant's structure and engines have already been discussed earlier. Crew seats and gun stations were armored. Crew was increased to six men. In addition to the *D-Stand* with its MG FF and the *C-Stand* with its MG 15, the gondola also housed two bomb magazines with ETC racks holding 500 kg bombs. The outer nacelles were each fitted with a PVC 1006L rack for a maximum of 2 x 1,800 kg bombs or 2 x LMA/LMB. As a long range bomber the C-3 carried 1,230 kg of bombs and a total of 8,060 l of fuel (four cells at 380 liters in the wings = 1,520 l, 4 x 260 l [takeoff tank with C3 fuel] = 1,040 l, also in the wings, as well as 5 x 1,100 l in the fuselage = 5,500 liters). Its takeoff weight as a strategic bomber

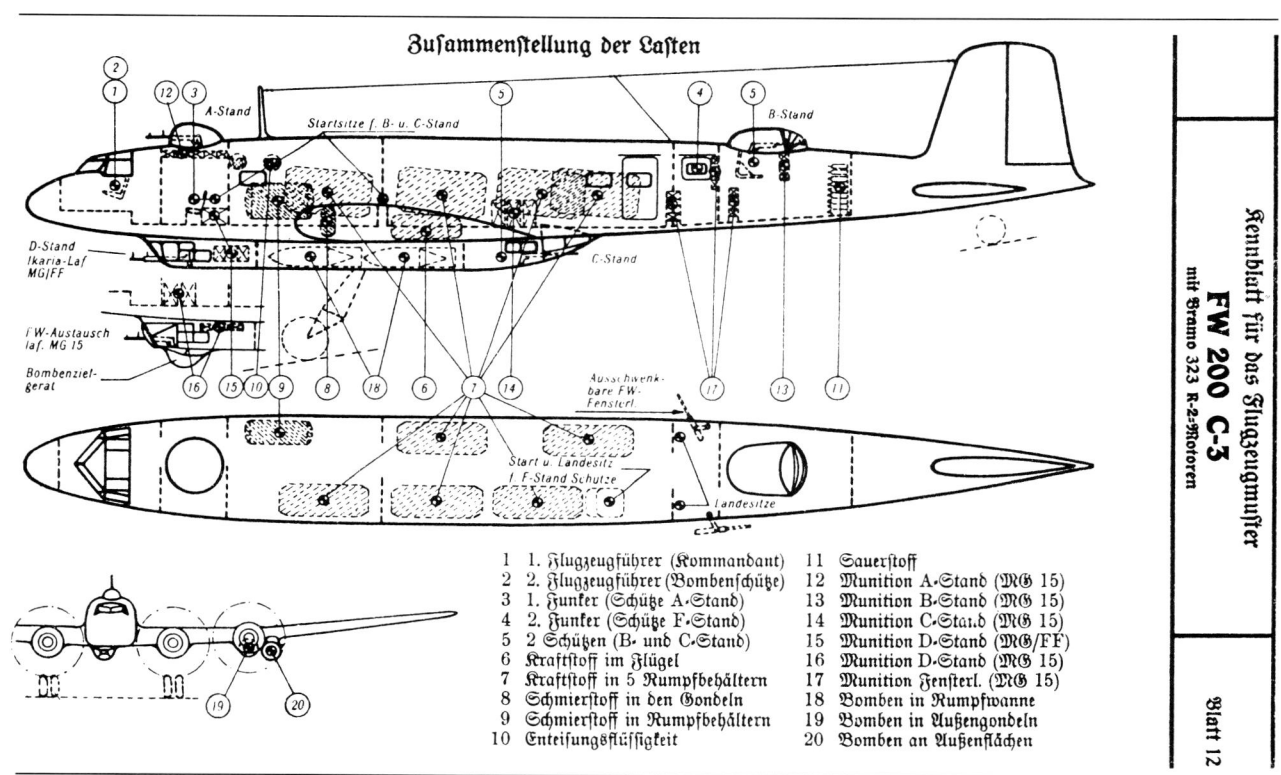

This typesheet for the Fw 200C-3 shows the crew seating, fuel layout, and ordnance carried (1941).

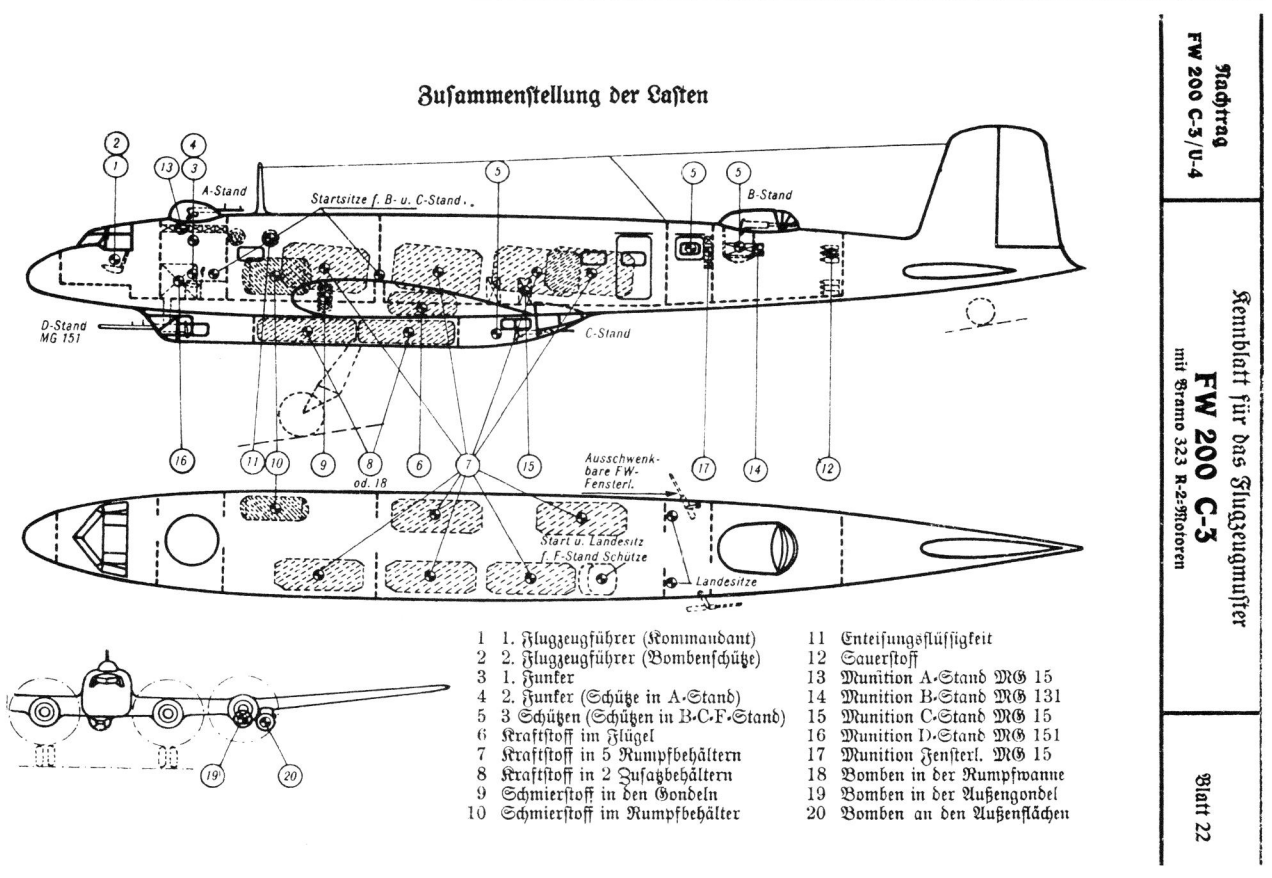

The typesheet for the Focke-Wulf Fw 200C-3/U4 shows the differences with the previous Fw 200C-3 as described in the text.

with a payload of 1,230 was 22,705 kg and 21,451 kg as a strategic reconnaissance platform (without bomb payload). Its greatest operational radius at roughly the same outbound and inbound routes, including climb to 4,000 m and descent (without safety reduction) was 4,490 km for an endurance of 14 hours and 40 minutes.

Beginning with *Werknummer* 0070 the Fw 200C-3 was fitted with better armament, even more range, and a torpedo launch system which supplemented the previously mentioned bomb payload potential. It had a crew of seven and was designated the Fw 200C-3/U-4.

The C-3/U-4's bomb bay in the fuselage was outfitted with two additional fuel tanks at 625 liters each (total of 1,250 l), so that the total fuel capacity had now risen to 9,310 liters, of which 9,030 could be flown. When fully fueled up, the U-4 carried no bombs. However, with reduced capacity, like the C-3 it was capable of carrying a maximum of 540 kg of bombs. In the event the mission called for torpedoes, it could carry two LT S5 on the outer wing stations. In its role as a bomber it weighed in at 22,690 kg with 1,030 kg of bombs, while with 1,500 kg of torpedoes the torpedo bomber weighed 22,717 kg, and the strategic reconnaissance configuration was 22,540 kg. Its maximum radius under the same conditions as that given for the C-3 above was 5,245 km in 17 hours. Improvements to the on-board weaponry extended to the *B-Stand*, which was fitted with a 13 mm MG 141 in place of the MG 15, and the *D-Stand*, which carried the better performing MG 151/20 in place of the 20 mm MG FF.

With this last variant, it seemed that all that could be drawn from the former civilian airliner for military purposes had been. Nevertheless, in spite of the structural improvements to the design, the high operational strain led to several broken fuselages, which clearly highlighted the structural limitations inherent in the basic design of an aircraft that had originally been conceived as a passenger airliner.

Heinkel He 177

Planning began in 1936 on the basis of a military requirement issued on 3 June 1936 for an aircraft with the following performance:

5,000 km range
500 km bomb load
500 km/h minimum cruise speed at engine's rated altitude
and the following deadlines:
1 August 1936 for submission of the preliminary project plans
December 1936 for final mockup inspection
January 1938 flight certification

On 2 June 1937 Ernst Heinkel was awarded the contract for further development of what he designated the Project 1041 bomber design. This had been developed by his technical director at the time, *Prof. Dr.-Ing.* Heinrich Hertel, and was tailored specifically to the above-mentioned military requirement.

On 5 November 1937, nearly a year after the original deadline, the *Technisches Amt* was able to complete the final mockup inspection at the Heinkel Rostock-Marienehe plant; the mockup's overall layout was approved, and at the conclusion of the inspection the type was given the official designation He 177. According to the calculations at the time the He 177 was expected to have a takeoff weight of 25 metric tons, a cruising speed of 500 km/h at 5,500 m altitude, and would even exceed the 5,000 km range called for in the requirement. On 12 November 1938 the Heinkel company was given the nod by the RLM to plan on six prototypes, with this number being raised to 12 on 24 February 1939.

The ideas of Heinkel and his designers with regard to the He 177 focused on a layout using two double engines, each channeling their power to a single airscrew. This long range bomber would have the appearance of a twin-engined heavy aircraft and had a practical precedent that was entirely acceptable to the *Technisches Amt*, as this entity had called for a dive bombing capability for the type, something which would have been a first for an aircraft with four single engines.

Heinkel He 119 Test Prototype

The practical precedent for the Heinkel team's thinking had its foundation in Germany's lagging engine development programs. Due to the lack of more powerful single engines Heinkel, under the strictest secrecy, had used his own resources and gambled on the construction of a prototype series of eight aircraft which had two DB 601 12-cylinder engines side-by-side in the fuselage. Located behind the cockpit, these engines drove a single four-blade propeller via an extended shaft running through the cockpit. With this double engine, designated the DB 606, this streamlined airplane had 1,725 kW/2,350 hp available to it. Conceived by the brothers Siegfried and Walter Günter as an unarmed high speed bomber and strategic reconnaissance aircraft, the Heinkel He 119 (as this experimental craft was designated) had a top speed of 620 km/h. On 22 November 1937, just after approval of the He 177 mockup, the He 119V4 set a world's record for a 1,000 kg payload with Heinkel chief test pilot Gerhard Nitschke at the controls, when the He 119 flew a 1,000 km course at an average speed of 504.998 km/h. For propaganda purposes Heinkel reported the plane to the FAI as the He 606, while it was designated the He 111U in the press, although this single-engined looking plane only resembled the He 111 in the shape of its fuselage and wing design.

Despite what was for its day an impressive performance, the machine never went into production; the RLM did not order it, its newfangled double engine was difficult to access for maintenance and, although minimal in drag, the surface evaporating radiators were not at a production stage and had not yet been thoroughly evaluated. Two of the last prototypes (V7 and V8) were bought by the Japanese, along with a production license, and in 1940 were shipped to Japan on an Italian warship. Even there, the He 119 never went into production following the crash there of one of the two prototypes during testing.

He 119V4 Data:

wingspan	15.9 m
length	14.8 m
height	5.4 m
wing area 50.0 m^2	
takeoff weight (V4)	7160 kg
wing loading	143 kg/m^2
power loading	3.05 kg/hp
service ceiling	9300 m
maximum speed	620 km/h
cruising speed at	
60% rated power at 4,500 m	550 km/h
optimum range	1,320 km

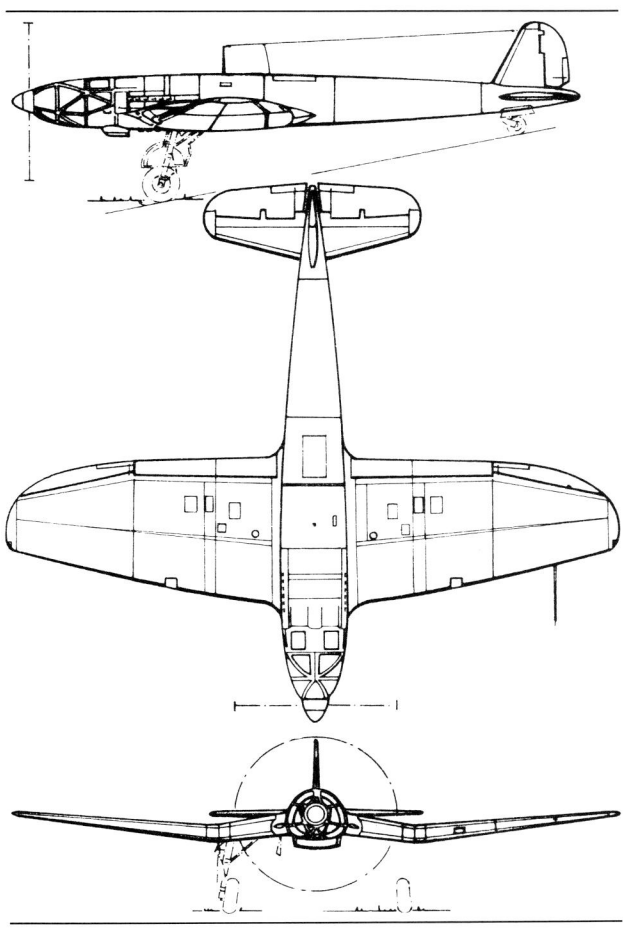

Heinkel He 119 - an ideal fuselage shape was achieved by burying the dual engines behind the extensively glazed canopy designed without any protrusions into the airstream.

Heinkel He 119 - the relatively high telescoping undercarriage was necessitated by the 4.30 diameter four-bladed propeller.

In closing, it is worth noting that the remaining six V-types primarily saw use as testbeds for the improved DB 606 engine, and the 200 kW/270 hp more powerful DB 610 (consisting of two DB 605s). The He 119V3 was fitted with floats and served as a test prototype floatplane for maritime reconnaissance and anti-shipping. Nevertheless, the *Technisches Amt* remained disinterested even in these He 119 versions; the end of a brilliant attempt, even in 1937, to break a new speed barrier!

He 177 Prototypes

Building on the design knowledge gained from the He 119 and under the supervision of *Prof.* Heinrich Hertel, *Dipl.- Ing.* Siegfried Günter drew up the He 177 as a long range bomber powered by the DB 606A-1, now rated at 2,000 kW/2,700 hp. Structurally, it was laid out for bombing from a shallow glide.

In March 1939 Hertel left the Heinkel company and took up the posts of chief of development and technical director at Junkers, a step which did not immediately affect the continuity of the He 177's development. On 19 November 1939 the He 177V1 succeeded in making its first flight at the hands of *Dipl.-Ing.* Carl Francke, the director of the flight testing department at the *E-Stelle* Rechlin. At the time, Heinkel estimated that the new heavy bomber would be ready for full-scale production within two years, with initial deliveries being made as early as late 1940/early 1941. But it was not to be so. Even though the maiden flight was "considered good in every respect,"* Francke was forced to break it off after just 20 minutes because the oil temperature had risen dramatically and threatened to damage the engine. A brief historical overview of the first 12 prototypes of what seemingly was a radically advanced heavy aircraft shows that, with the General Staff's renewed requirement of attacks from diving angles up to 60° yet to be surmounted, the type's production maturity was yet a long way off.

He 177V1

But let us return to the V1's first flight on 19 November 1939: the condition of the aircraft was described as follows in the previously mentioned EHF report:

- duration of flight 20 minutes
- airborne weight 16 metric tons
- with large vertical stabilizer and small horizontal stabilizer
- undercarriage extended entire flight
- maximum altitude 2,000 m

---

* EHF memorandum no. 4297 re: meeting with Prof. Heinkel on 11/20/39 concerning the He 177's maiden flight.

Flight and flight characteristics were assessed by *Dipl.-Ing.* Francke as follows:

"- Takeoff: takeoff and climbout after takeoff were considered quite good.
- Elevators: operation, forces, and motion good.
- Rudder: rudder forces were felt to be somewhat to great. As a result, nothing accurate can be said about the range of motion. Rudder effect seemed to be rather limited, but cannot clearly be ascertained due to the rudder forces being too great.
- Ailerons: effect quite good. Forces too limited. Close to overbalanced. Overbalancing increases with pressure increase. For the next flight the trim tab is to be adjusted to a lower setting. With extended Fowlers, the aircraft is tailheavy, i.e. at a trim tab setting of -6°. With Fowlers retracted the machine is somewhat noseheavy at a trim setting of +2°. It might be better if the adjustment were reduced from -6° to - 4° and from +2° to 0°.
- Stability: good around the pitch axis and good stability. Stability slightly weak around the roll axis. In the event that later flights confirm these findings, a larger tailplane must be fitted.
- Rolling moment due to sideslip: slight rolling moments due to sideslip were noticed. A change in this characteristic will be made on the basis of further flights and/or on the basis of the tactical requirements.
- Engine: at 2,000 m altitude gauges: 1.1 *ata* at 2,200 rpm, time 10:30. Water temperature good. Oil temperature 120° (too high, see evaluated film).
- Undercarriage: during takeoff and landing quite bouncy. The shock absorption for this weight is probably not strong enough."

Thus, there was no real glaring problem noted, only the standard shortcomings encountered with virtually every new design. The overall assessment from the reference document is quoted as:

"All in all, the first flight by *Herr* Francke is considered good in every respect. *Herr* Francke was so comfortable with the aircraft on the first flight that he wanted to extend the flight. Nevertheless, as a result of the increasing oil temperature the flight had to be curtailed after 20 minutes."

Although already undergoing flight testing, on 27 February 1940 the RLM awarded the Heinkel company an official construction contract, signed by Udet, for the "delivery of He 177V1/0001, to be fitted with DB 606 double engines at the current price of RM 1,357,000...Aircraft recipient is *E-Stelle* Rechlin."

With a three-man crew, the V1 carried out a series of generally successful test flights, but crashed on 26 April 1940 as a result of a damaged propeller and was written off.

He 177V2 through V5

The V2, test flown by Francke shortly after the V1, was lost in a crash two days before the V1 on 24 April 1940 during the first dive tests with *E-Stelle* Rechlin's test pilot Rickert at the controls.

The V3 flew beginning in mid-February 1940 at Rechlin for engine testing, but had considerable aerodynamic problems.

The V4 remained with the Heinkel company, where another Rechlin test pilot by the name of Ursinus flew stability trials. During one of these test flights the V4 could not recover from a straight dive and crashed into the Baltic near Ribnitz.

These first four prototypes were externally virtually the same, though V2 through V4 were designed to accommodate a crew of four.

The subsequent He 177V5 had a DB 606A-1 74 kW/100 hp more powerful than earlier versions, providing a maximum permissible cruise rating of 1,765 kW/2,360 hp at 5,500 m. Weapons testing was the primary focus of this prototype at Rechlin. During a simulated low-level attack in early 1941, the engines caught fire and the plane crashed and exploded.

Protokoll Nr. 5935, the minutes from the final investigative discussion on 30 April 1940 on Rickert's accident with the He 177V2, give a glimpse into at least some of the initial teething troubles facing the He 177. An excerpt from the results of the investigation showed, among other things, that:

"...the pilot trimmed the aircraft tailheavy before the dive in order to make pullout easier at higher speeds. There is the probability that the tailheavy trim was set too great, and investigation by the EHF shows that in this case the manual forces being applied no longer translated into the direction of 'push,' but became greater as speed built up (in the direction of 'pull'). This change in behavior can also be aggravated by the change in elevator setting. If the pilot were distracted into releasing the control column for whatever other reason, e.g. canopy pane flying off, momentary vibrations in the aircraft, part of the propeller blade breaking off, etc. the aircraft would pull out of its own volition. The available power control in the airplane has the feature of having the previously deflected trim tab suddenly jump to a neutral position, and under certain conditions even beyond the neutral position. To be sure, there exists oil damping between the trim tab and the main elevator, but in this case it was set to minimum effectiveness. It is thought that the rapid return of the elevator initiated a pullout which was too sharp, so that the resulting aerodynamic forces led to the destruction of the airplane. The EHF has conducted a detailed theoretical investigation of just such a pullout scenario under various presumed conditions and with regard to inertial forces. Events may have continued with the pilot attempting to counter the sharp pullout by pushing forcefully downward. In doing so, he may have been unable to prevent the pullout due to the aircraft's inertia, but the load on the control surfaces had increased so dramatically that it would explain the upward break of the elevator.... The location of the badly damaged propeller parts leads to the conclusion that the breakage of the propeller occurred simultaneously with the initial damage to the airframe itself...."

Regarding further testing of the He 177, the RLM and Heinkel recommended in the same minutes that:

"based on the test results* there is no thought of abandoning further testing of the He 177. All test flights not requiring high speeds may be carried out without problem. Diving at high speeds and dive brake flight tests should be conducted by gradually increasing the rate of speed. These should be deferred until the results of the wind tunnel testing are available."

From this it is clearly apparent that the He 177 had a long way to go before it would have the capability of diving at 60 degree angles. Also obvious is the fact that the V2's crash was not due to propeller damage, as has been sometimes reported in aviation literature, but was caused by too great a load being placed on the control surfaces when pulling out of a dive at high speed, something which led to a break in the elevator. With regard to the propeller breakage, it was assumed that "the breakage of the propeller occurred simultaneously with the initial damage to the airframe itself...."; in other words, it was considered a secondary cause of the accident, and that conclusive wind tunnel tests for dives had not yet been accomplished, which in turn resulted in flight operations paying a heavy price.

Despite what was turning into a series of accidents, on 27 April 1940 work began on the jigs and acquiring the accessory parts for the long runup to full-scale production of the He 177. However, in the latter half of 1940 the *Generalluftzeugmeister* directed that production be limited to five aircraft per month until the He 177's teething troubles had been resolved satisfactorily.

Thus, as 1940 turned into 1941 the He 177 was still not available for front-line operational use. A small start in that direction was made in late 1940 with the beginning of construction of the He 177A-0 pre-production/null series, 15 examples being built in Marienehe, 14 at the Heinkel works in Oranienburg, and five at Arado's Warnemünde facilities.

Medium Bombers

By the time 1940 had turned into 1941 production output had fully compensated for the bomber losses incurred during the campaigns of 1940. In 1940 the numbers of bombers produced were:

1,816 Junkers Ju 88
756 Heinkel He 111
260 Dornier Do 17 and
20 Dornier Do 217, i.e. a total of 2,852 medium bombers.

This was nearly four times the amount produced in 1939, which saw 737 bombers built. Type composition in comparison to the time at the war's outbreak had changed by a marked increase in Ju 88s and the Ju 86's phase-out as a bomber, as well as the start of re-equipping the units with the new Do 217. Due to the lower prioritization class for air armament, the growth expected to have taken place throughout 1941—despite preparations for the Russian campaign—only occurred on a modest scale. Details of the types in production at the time are as follows:

Junkers Ju 88

With the Ju 88 being a "*Quantité négligiable*" at the outset of the war—in 1939 just 69 Ju 88s rolled off the production line, with the last being the A-1 type—by 1940 it dominated the bomber sector with a total of 1,816 aircraft built. The Ju 88 comprised 64% of all bombers built in 1940. As such, it had advanced to the standard bomber of the *Luftwaffe*. Its life and technological development up to early 1941 are described thusly:

Ju 88A-1 through A-8 (minus A-4)

Details of the Ju 88A-1, particularly its armament and equipment, has already been described in part in the chapter entitled "Offensive Weapons and Reconnaissance Package Platforms Through the End of 1939."

Takeoff weight of the A-1 had risen to a maximum of 12,300 kg in overloaded configuration; airframe strengthening, including that of the landing gear, had made an appreciable difference in weight. This led to some problems with the design, typified by ten Ju 88A-1s built at Arado-Brandenburg suffering undercarriage damage to the main gear's struts. Another problem was that when the dive brakes were extended the airframe was subjected to additional stresses, causing limitations to be placed on certain maneuvers at higher speeds and a restriction on any type of aerobatics whatsoever. Stop-gap improvements made during the A-1 production run were carried out in conjunction with the new Jumo 211G-1 engine series (with the same performance as the Jumo 211B-1, but with stronger crankcase housing and shaft, larger main bearings, and reinforced exhaust vents, making the G-1 more robust all around), as well as installation of fittings for RATO pods for takeoffs under overloaded conditions. All these changes evolved into the Ju 88A-2, under which designation the series continued.

---

* This may have been a reference to "accident investigation results."

In the meantime, a conversion trainer designated the Ju 88A-3 was derived directly from the A-1 model, this having dual control columns, dual throttle layout, and some duplication of instrumentation and lacking armament or bomb racks. It served as a follow-on trainer for pilots completing their C-class and IFR training.

In early 1940 development began of a generally reworked version of the Ju 88, which was designated the Ju 88A-4 and was to have had more powerful engines, increased wingspan, and a strengthened undercarriage. A more powerful version of the Jumo 211, the J model, was to have replaced the B and G series, but delays in its development led to the Ju 88s coming off the assembly lines being fitted with current engines. This new interim variant, designated the Ju 88A-5, did indeed have the larger wings. These were increased in span from 18.26 m by almost two meters to 20.08 m and had an area of 54.7 m^2. The A-5 was delivered to the front-line units armed with four MG 15 machine guns.

These variants, the Ju 88A-2 (derived from the A-1) and the Ju 88A-5 with its increased span, were the ones which saw action with bomber units in Norway, France, and in the Battle of Britain. Despite its dive bombing capability, which would generally allow the type to escape British fighters by diving away from them in sticky situations, even the Ju 88 losses were heavy; the type's armor and armament proved to be too weak.

Technical Data for Ju 88A-5

The Ju 88A-5 was a four-man bomber conceived as a twin-engined cantilever low wing design of all-metal construction and metal-skinned.

Two Junkers Ju 88A-5s, loaded with four 250 kg bombs carried externally, seen here taxiing out for takeoff. For night operations over England (during the winter of 1940/1941) the usual light blue undersides were painted a sooty black.

Airframe
Fuselage: monocoque with for longitudinal spars and bulkheads perpendicular to the longitudinal axis. Stressed metal skin fitted with countersunk rivets.
Undercarriage: single strut landing gear for dual braking wheels which were hydraulically retracted to lie flat under the wings. Could be hydraulically extended manually using a hand pump in emergency situations. Retractable tailwheel, pivoting, and capable of being hydraulically locked in place from the cockpit.
Control surfaces:
a) ailerons: extending across approximately 2/5 of the wing, metal-skinned with trim tabs and counterbalance.
b) flaps: extending over about 3/5 of the wing, with flap locks, metal-skinned, and with gap coverings.
c) horizontal stabilizer: cantilever, tailfin hydraulically linked to flaps, metal-skinned, elevators with counterbalance. Trim tab effect supported by protrusion of the tailfin, automatically initiates dive and pullout, hydraulically operated, spring adjusted.
d) vertical stabilizer: metal-skinned, rudder with trim tab, spring controlled and counterbalanced.
Wings: two-spar, metal-skinned
Dive brakes and automatic recovery system: brakes situated on the outer wing undersides, hydraulically driven. Automatic dive recovery system hydraulically driven, coupled to ordnance release system, may be activated with our without dive brakes.
Heating and de-icing: safety heating for crew compartment. Wing leading edge heated by warm air. Horizontal stabilizer fitted with rubber de-icer boots, propeller blades with liquid de-icing system. Windscreen equipped with electric de-icer.
Hydraulic system: following were hydraulically driven: undercarriage and tailwheel, landing flaps and tailfin, dive brakes and automatic dive recovery system.

Powerplant
Engines: Jumo 211B or G with mixture control, 1:1.68 reduction. Takeoff power 2x 883 kW/1,200hp = 1,766 kW/2,400 hp at 2,400 min^{-1} at 1.35 *ata* (= 1.324 bar). Radiator: laminated annular in front of each engine.
Propellers: three-bladed (some four-bladed) VDM variable-pitch metal airscrews, 3.60 m diameter. Position gauge 1 deg. increments = 10'. Base position: 25 deg. with propeller position at 12:20. Starting position: 27 deg. = 12 o'clock.
Fuel tanks and fuel: fully self-sealing as follows:

1x tank in forward bomb bay	1,220 l
1x tank in aft bomb bay	680 l
2x tanks in outer wings	850 l
2x tanks in inner wings	830 l
total	3,580 l

Non-useable reserves total 45 l. Fuel dump for both fuselage tanks. Fuel used: standard A2 (87 octane rating).

Oil tanks, oil, and oil radiators: two self-sealing tanks in the wings, each with a capacity of 136 l; contents 2x ¹25 l = 250 l quantity. Additional unprotected tank in the port wing with 105 l capacity, for a total of 355 l of oil. Rotring or Aero-Shell oil. Oil radiator: air-tube type in upper section of annular radiator ahead of each engine.

Armament
A-Stand: semi-fixed MG 15 with 450 rounds
B-Stand: 2x MG 15 in bubble mount with 1,125 rounds
C-Stand: MG 15 in small bubble mount with 600 rounds
Total # of rounds: 2,175

Bomb racks and bombsights
External wing-mounted racks, either 4x ETC 500/IXb or 2x *Träger 1000* for carrying 4x 500 or 4x 250 kg bombs, or 1x 1,800 kg or 2x 1,000 or 2x 500 kg bombs.
Bombsights:
- for level bombing: Lotfe 7a or Lotfe 7b or BZG 2L or E or GV 219d
- for dive bombing: BZA-1, Suti 5, Revi C 12c with swivel base

Radio/Navigation equipment
radio   FuG X
direction finding   PeilG 5
instrument landing   FuBl 1
internal comms   EiV

Dimensions, weight and performance
dimensions and weight:
wingspan   20.08 m (A-5, some also 19.95 m)
height (with antenna)   4.85 m (without = 4.12 m)
length   14.36 m
wheel track   5.80 m
wing area   54.7 m²
takeoff weight   13,000 kg
max. permissible landing weight
(no bombs, fuselage fuel tanks empty)   10,800 kg
max. permissible descent and dive weight
(with wing bombs and empty fuselage fuel tanks) 10,700 kg
max. wing loading   240 kg/m²
Performance:
Max. speed
- outbound with 13 mt weight (with 2x SD 1000 on outboard racks)
at 5,500 m altitude   376 km/h
- return with 10.4 mt (no bombs) at 5,500 m altitude   440 km/h

Max permissible speeds
- at 50 deg. dive angle (with brakes) at all altitudes
(registered airspeed)   575 km/h
- at 20 to 30 deg. descent angle (no brakes)
   - at 0 to 2,000 m altitude   675 km/h
   - over 2,000 m altitude   600 km/h
- at low level under minimum visibility conditions   400 km/h
- with undercarriage extended   250 km/h

Other details
- K4ü automatic pilot with additional switching for single engine flight
- Power boost system (RATO takeoff for overloaded conditions)
- Single engine flight: just possible with no bombs, empty fuselage tanks, and jettisoned bomb racks (= 10 mt flying weight) at cruise power (2,100 min⁻¹ at 1.10 *ata* setting for low-level turbocharger)

A heavily laden Ju 88A-5 using RATO packs to shorten the takeoff run (a so-called Bläserstart, lit. "blower start").

A Walter RATO pod ready to be fitted to a Ju 88.

A RATO pod seen fitted underneath the outer wing section of a Ju 88.

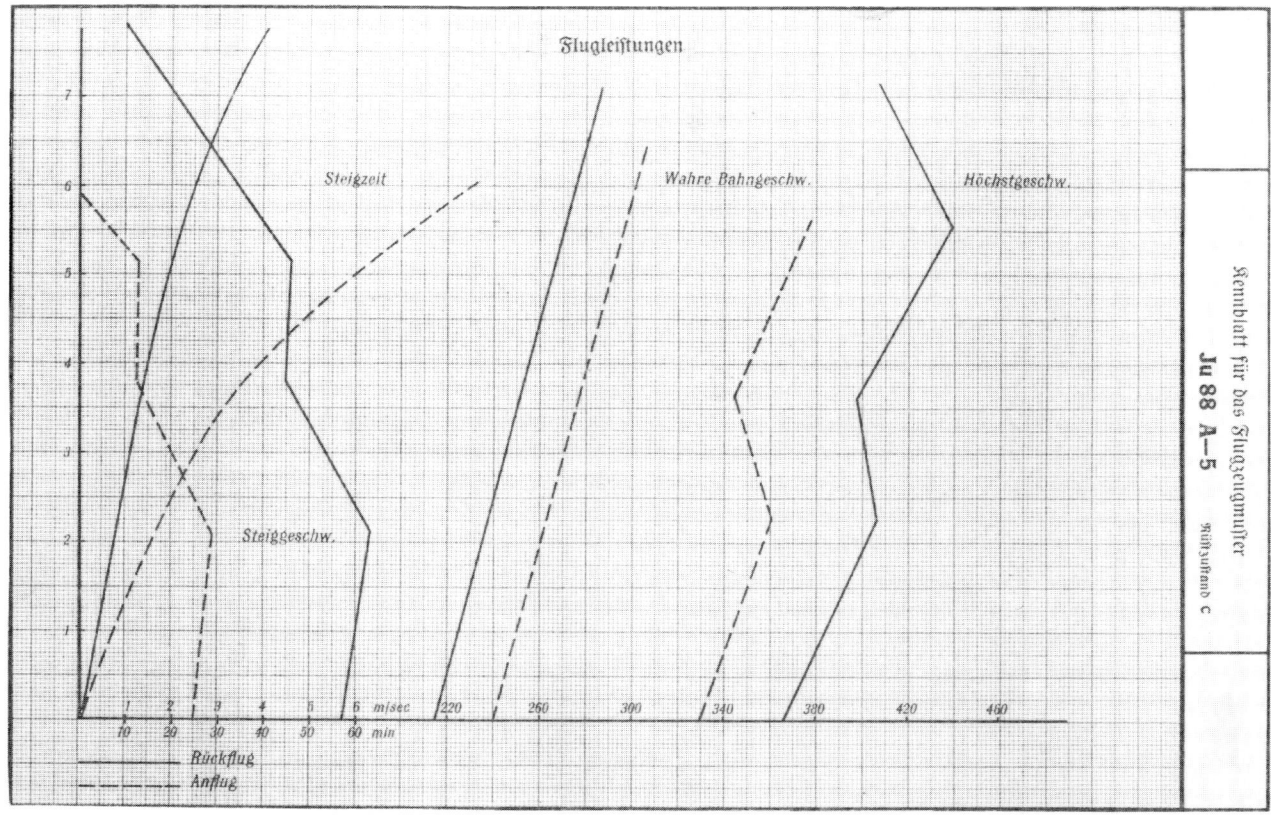

The actual performance figures for the Ju 88A-5 can be seen here on this chart.

Summary and Other A-5 Derivatives

The preceding technical data for the Ju 88A-5 reveals just how versatile and capable, yet—in comparison with older types such as the Ju 86—how highly refined the instruments of war had become with the Ju 88. With its layout as both a horizontal and glide bomber (the term "dive" was apparently deliberately avoided in the official literature of the *Technisches Amt*, since diving angles of greater than 50 deg. were extremely rare in practice, and the absolute dive capability the General Staff had stipulated for the Ju 88 had certain technical and practical limitations in actual front-line service), by late 1940/early 1941 the Ju 88 had matured into the most flexible bomber in the inventory even before it had achieved maturity in its design and as a weapons platform. It had been tailored to military requirements from the outset. The Ju 88 hovered not only near the top of the quantitative list, but also that of the qualitative one when it came to German combat aircraft of the day.

A few other interesting variants of the Ju 88A-5 which appeared simultaneously are worth mentioning.

The variant of the A-5 designated the Ju 88A-6 was fitted with balloon cable deflectors. This was a frame which projected out from ahead of the canopy and the propellers, joining the airframe at the wingtips where a cable cutter would cut the cables of tethered balloons the British stationed as obstacles around major potential target areas.

This equipment made the Ju 88 quite nose-heavy, and to offset this a 60 kg counterweight was fitted in the tail section. The total installation weighed 381 kg and dropped the maximum cruise speed of the A-6 by 30 km/h. In this configuration the Ju 88A-6 was particularly susceptible to fighters, and in any event only enjoyed limited success in its cable cutting role; the bomber formations following behind the "cable cutters" generally operated at altitudes well above the balloon obstacles. The A-6 was therefore relieved of these duties a short time later and, after the cable cutting apparatus was removed, was assigned as a standard bomber. Later a few of the A-

History of German Aviation: Bombers and Reconnaissance Aircraft

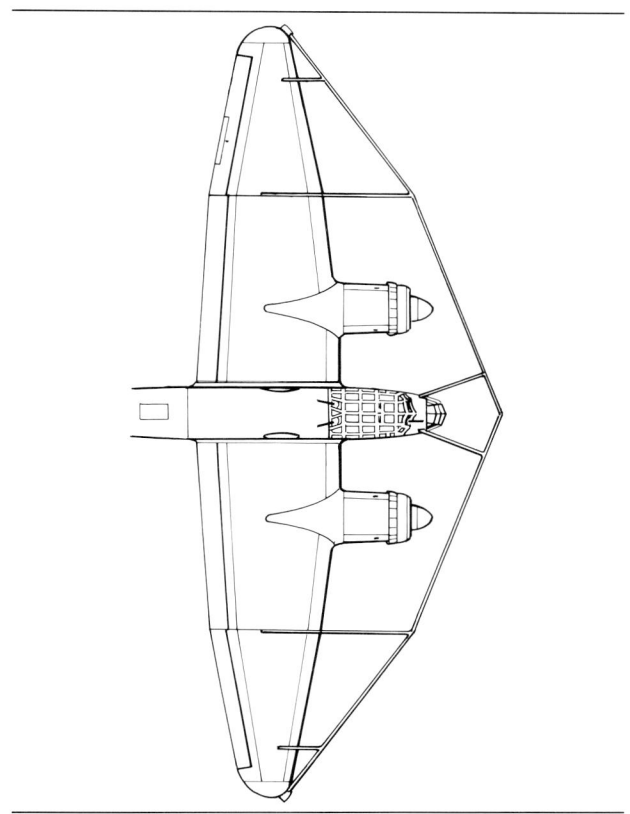

Junkers Ju 88A-6 with balloon cutting attachment.

6 variants appeared in the maritime reconnaissance role as the Ju 88A-6/U, a sub-type which will be reviewed later on.

Here is a good point to mention that a small number of Ju 88A-5 airframes were fitted with dual controls and used in the pilot conversion training role as the Ju 88A-7, powered by the Jumo 21H-1.

So as not to completely be without the ability to cut balloon cables—such methodology was quite handy for low-level raids on point targets or convoys protected by barrage balloons—there was a small number of Ju 88A-8s built interspersed among the ongoing A-4 production which had begun in 1941. The three-man A-8 had a so-called *Kuto-Nase* fitted into the wing leading edge and nose canopy. The premise was that the balloon cable would be pushed outward along the razor-sharp leading edge and thus be cut in two. A frame attached around the engine housing and propellers ensured that the cable would be guided to the wing leading edge and its cutting surface. In comparison to the similar, but much less effective system on the A-6, the A-8's device only minimally affected the aircraft's performance—not least of which was due to the fact that the latter variant was powered by the better-performing Jumo 211F-1 (990 kW/1,350 hp).

The Indispensable Heinkel He 111

The former numerical backbone of the bomber units, the He 111, had shown itself to be a reliable workhorse in the campaigns up to the end of 1940 and, despite its average flight performance, proved itself just able to meet the demands of the air war as 1940 turned into 1941—thanks to its robustness and ability to absorb punishment under enemy fire. With a production of 452 He 111s in 1939 this type was easily at the top of the bomber list, but despite wartime production increasing to 756 examples in 1940 the He 111 was overshadowed by the Ju 88's figure of 1,816 built. Originally, production of the He 111 was to have been entirely abandoned in favor of the Ju 88, but this idea was scrapped in view of the impending onset of the Russian campaign, since production quantity now took absolute priority. Switching construction from the He 111 to large-scale production of the Ju 88 at this point would have meant—initially at least—a perceptible drop in bomber output. Thus, the He 111 remained a permanent fixture of Germany's air arsenal well beyond 1940.

End of the P-Series

Under the guise of establishing commonality, production of the DB 601 powered He 111 P-series ceased in the summer of 1940. Up to that point there had indeed been some further developments, described as follows:

The He 111P-2 available to the bomber units at the war's outbreak soon gave way to the He 111P-4 with increased armor, two additional MG 15s, and two PVC external racks; powerplant for all P-variants up to and including the P-4 was the DB 601Aa or A-1 rated at 810 kW/1,100 hp. Between these two sub-variants came the He 111P-3, a bomber trainer with dual controls produced in small numbers from modified P-0 and P-1 airframes starting in the autumn of 1939.

The last of the P-series, also built in limited quantity, was the He 111P-6. This type was powered by the DB 601N with 865 kW/1,175 hp and made use of the commonly fitted eight ESAC bomb racks for 8 × 250 kg bombs in the fuselage. When this sub-variant was replaced in the units by the H-series models, a number of these found continued employment as glider tow aircraft under the designation He 111P-6/R2, and a further ten aircraft were passed on to the Hungarian Air Force in 1942 as strategic reconnaissance aircraft. In conclusion, with regard to halting P-series production, it should be noted that this measure had been prompted by the engine situation as part of the type standardization program, as Daimler-Benz was expected to provide engines almost exclusively for fighter production, while the Junkers would ensure that sufficient quantities of its Jumo 211 would be available for continued He 111 production.

He 111H-2 through H-5

After the experience of the Polish campaign the armament was beefed up by two additional MG 15s in the side windows of the fuselage. Although initially fitted to the P-4, it was also included on some He 111H-2 versions, which began appearing on the front lines as early as October 1939. Whereas the H-2 was a relatively small production run, the He 111H-3 that followed beginning in November 1939 was put into large-scale production. Instead of the Jumo 211A-3 the H-3 was powered by the Jumo 211D-1 at 880 kW/1,200 hp, some 74 kW/100 hp more powerful than the earlier engine and thus able to alleviate some of the criticism of the He 111's single-engine flying ability. Standard payload called for eight 250 kg bombs. In some instances a 20 mm MG-FF cannon was also carried in place of the MG 15 in the nose canopy. The H-3 achieved notable success during and after the Norwegian campaign in the anti-shipping role, bombing from both low levels and at altitude. However, it was discovered that its armor for protection of the crew

Loading a Heinkel He 111H-4 with a single SC 1800 bomb by means of a special block and tackle system (Flaz 2000). The number of men needed is remarkable!

was still inadequate—chiefly during the subsequent daylight raids on London.

In late 1940/early 1941 small numbers of the He 111H-4 were produced, also with the Jumo 211D-1 and later with the H-1. Payload in the H-4's fuselage was limited to four 250 kg bombs, with the optional ETC external racks being replaced by the stronger PVC racks and enabling the H-4 to carry the heaviest bombs up to 1,800 kg.

Also built in relatively small numbers was the He 111H-5, which followed the H-4 in February 1941. This sub-variant remained only a few months in service with a handful of units. It was also powered by the Jumo 211D and H, but was quite interesting as a weapons platform. First, it was given significantly more armor in the cockpit and gun stations.

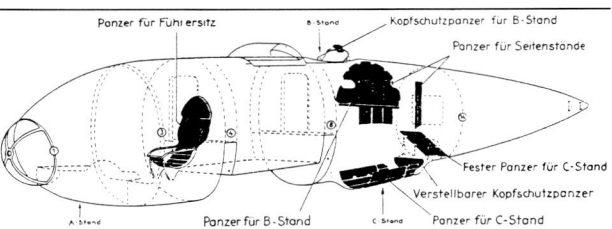

Heinkel He 111H-5 - armor for the pilot and gun positions.

For the He 111H-5 there were four main operational possibilities as a bomber, specifically:
- bomber for heaviest payloads: 2x PVC 1006L under the fuselage (4 vestigial ESAC magazines remained within the bomb bay)
- bomber for heavy and light payloads: 1x PVC 1006L fitted to underfuselage port side and four ESAC 250/IX on starboard side of bomb bay
- strategic and nighttime reconnaissance (primarily for the *Kampfgeschwader* HQ squadrons): six 50 ETC externally under port fuselage side and 2x RMK 50/30 and RMK 20/30 cameras carried internally in starboard bomb bay section.
- bomber with rack for carrying parachute retarded flares: six ETC 50 under port fuselage side and 1x PVC 1006L under starboard fuselage side, or 4x ESAC 250/IX fitted internally in starboard bomb bay section.

Each PVC was capable of carrying: 1x SC 1800 or 1x SD 1000 or 1x SC 500 or 1x BSB 700 (incendiary bomb canisters) or 1x LMA 3 or 1x LMB 3 or 1x S 300 (RATO pack for overloaded take-offs under short-field conditions)

The asymmetrical loading of a single SC 1800 had a minimal effect on the aircraft's directional stability, or stability along its yaw axis, and could easily be compensated for with proper trim. Fitted

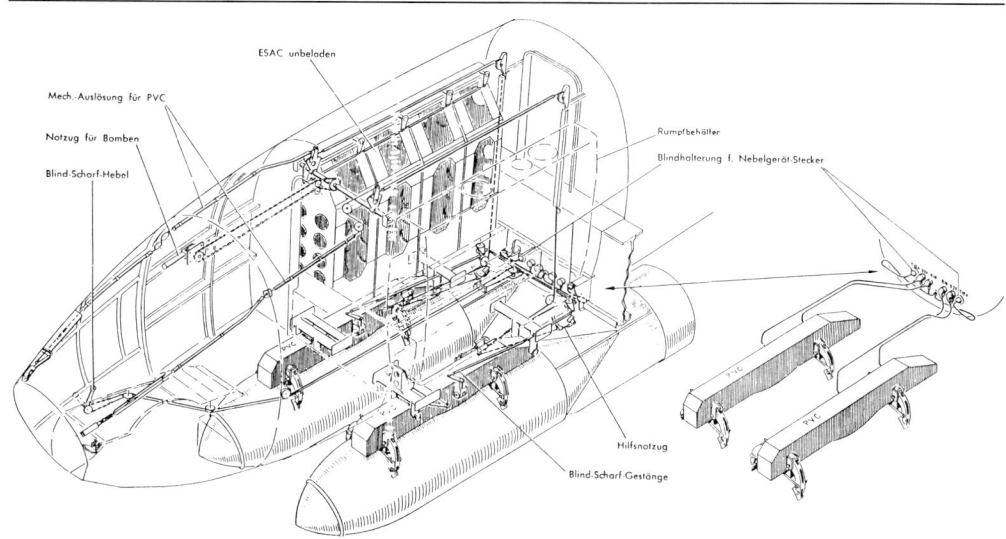

He 111H-5 ordnance option A - PVC is the acronym for a single rack for heavy external loads, with the shackles being released by an electrically fired cartridge vice a coil magnet, as in the case of the ETC external rack (see Appendix 6).

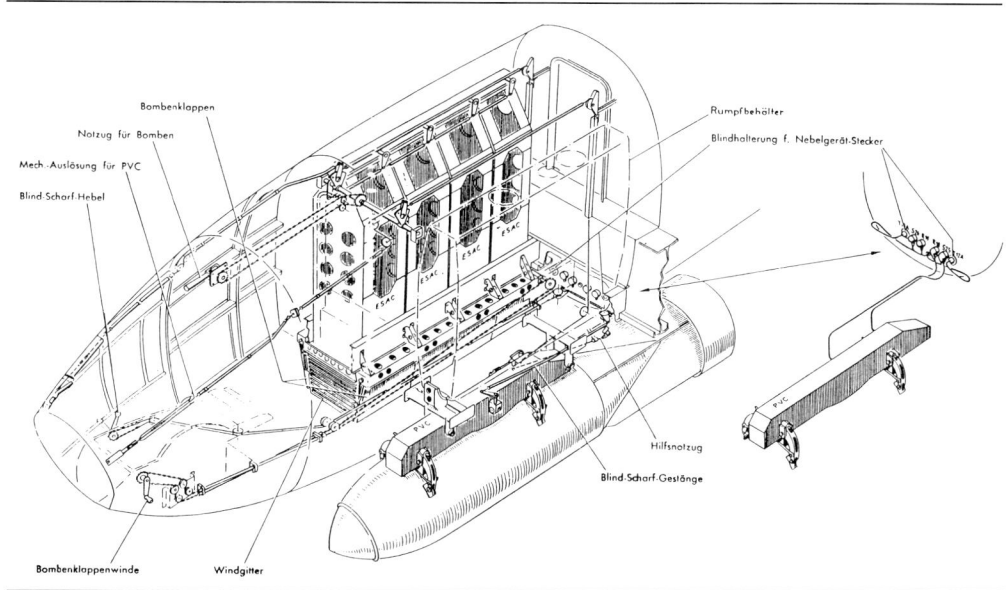

He 111H-5 ordnance option B - ESAC is the acronym for internal bomb magazines in which the bombs—here 250 kg—are carried vertically with the nose pointed downward and released electrically. A mechanical emergency release mechanism is also provided for the ESAC magazines.

with two PVCs for heavy loads, the increased frontal drag of the PVCs in conjunction with the additional weight of the armor made single engine flying much more difficult in the He 111H-5 than in the H-3. Field shops rectified this somewhat by redesigning the armor so that it could be unbolted and jettisoned in case of emergency. As a strategic and nighttime reconnaissance platform, in addition to the previously mentioned photo apparatus the He 111H-5 also carried internally either sic LC 50C parachute retarded flares or six 50 kg standard bombs on the ETCs attached externally to the port fuselage underside. With a field conversion, it could also be fitted with a single PVC rack, even as a strategic reconnaissance plane, and carry a bomb payload of up to 1,800 kg by reducing its fuel capacity correspondingly. The special load plan (p. 143) under the payload specifications for the He 111H-5 reveals just how much ordnance this aircraft was permitted to take off with, although this was only possible when specifically authorized.

He 111H-5 Special Load Plan

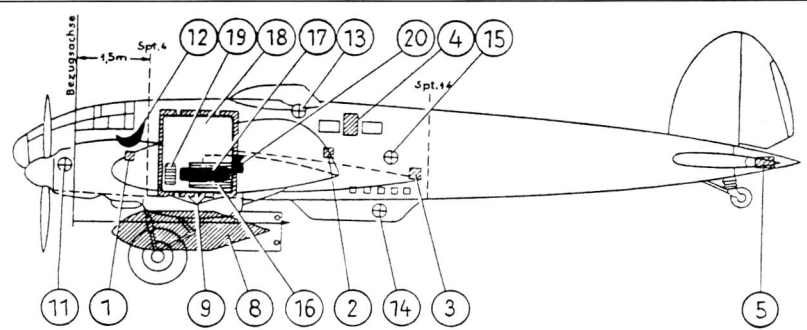

Pos.-	Bezeichnung:	Bomber 2 × 1800 kg Lasten kg	Bomber 1 × 1800 kg Last 4 × 250 kg Lasten kg	Fernerkunder 1 × 1800 kg Last kg
	Leergewicht (höchstzulässige Toleranz ± 1,5%)	7215	7215	7215
	Zusätzliche Ausrüstung	1025	985	1065
	Rüstgewicht	8240	8200	8280
1	Munition im A-Stand	0— 30	0— 30	0— 30
2	Munition im B-Stand	0— 30	0— 30	0— 30
3	Munition im C-Stand	0— 45	0— 45	0— 45
4	Munition MG Rumpf, seitlich	0— 20	0— 20	0— 20
5	Munition Störkörper	0— 15	0— 15	0— 15
8	1 bzw. 2 × 1800 kg Lasten an PVC	0—3600*)	0—1800*)	0—1800
9	4 × 250 kg Lasten in ESAC	—	0—1000	—
11	Mann in Kanzel	70— 100	70— 100	70— 100
12	Führer	70— 100	70— 100	70— 100
13	Mann im B-Stand	70— 100	70— 100	70— 100
14	Mann im C-Stand	70— 100	70— 100	70— 100
15	2. Mann im C-Stand	70— 100	70— 100	70—100
16	Kraftstoff in Flächenmittelstückbehältern 2 × 700 l	0—1040	0—1040	0—1040
17	Kraftstoff in Außenflächenbehältern 2 × 1025 l	0— 790	0—1515	0—1515
18	Kraftstoff im Rumpfbehälter 1 × 835 l	—*)	0— 620*)	0— 620*)
19	Schmierstoff in Flächenbehältern 2 × 105 l	0— 190	0— 190	0— 190
20	Schmierstoff im Laufgangbehälter 1 × 120 l	—*)	—*)	0— 110
	Fluggewicht	14500	14500	14190

*) Werden statt der 1800 kg-Bomben leichtere Lasten geladen, so kann für das Mindergewicht Kraft- und Schmierstoff in den Rumpfbehältern mitgenommen werden.
   Bei der Landung mit der 1800 kg-Bombe an der linken Seite muß der Rumpfkraftstoffbehälter entleert sein.

One interesting item on the list is Position 5, "*Störkörper*" (harassment/jamming ordnance), material which could be deployed from the tail section—the gunners' dead zone—into the path of attacking fighters. In practice, little use was made of harassment ordnance as it did not prove to be particularly effective. The all-up weight of 14,500 kg for overloaded takeoffs shown in the special load plan was undoubtedly the structural limitation of this aircraft, and even with RATO packs must have been a challenge for even the most experienced pilots.

Dornier Do 17Z and Do 217A, C, and E

Production of the Do 17Z ceased in the early summer of 1940. As the successor to the Do 17, Dornier had planned on the Do 217, a type the company had been working on for years, and differed from the Do 17 as follows: more powerful engines, increased take-off weight, new detailed design features, automatic dive recovery with different dive brakes, increased wing area (from 55 to 73 m²), larger fuselage, reinforced undercarriage, warm-air de-icing, hy-

History of German Aviation: Bombers and Reconnaissance Aircraft

draulic system replaced by electrically driven servos, and rudder trim tabs for trimming for single engine flight. In place of the four main components for the Do 17, the Do 217 follow-on type was broken down into a total of six:
- forward fuselage section
- combined fuselage and wing center section
- both outer wing sections
- fuselage mid-section
- fuselage tail section.

First flight of the Do 217V1 had taken place back on 4 October 1938; a total of nine prototypes were built, and these were used to carry out a comprehensive test program using various powerplants. The V7 and V8 had their aft dive brakes replaced by an end cap. The Do 217 V9 was the prototype for the later Do 217E-series, the first large-scale production version of the Do 217. Series production was initiated with the Do 217E-0 and E-1 beginning in 1940, both of which were powered by the BMW 801A-1 (1,147 kW/1,560 hp).

Prior to the E-model, the Do 217A-0 also appeared in 1940, although only eight of these were constructed. These eight did indeed have a bomb bay, but were laid out as reconnaissance aircraft and fitted with cameras (2x RB 50/30 and 1x RB 20/30 aerial cameras) in an extended ventral gondola and flown with a three-man crew. These machines reached a maximum speed of 475 km/h with their two DB 601R engines (each rated at 1,073 kW/1,410 hp), had a service ceiling of 7,000 m, and a range of 1,500 km with a takeoff weight of 10,845 kg. The first A-0 variants entered service with the *Aufklärungsgruppe Ob. d. L.* in the spring of 1940, and in the early winter of 1940/1941 these aircraft, together with other strategic reconnaissance planes in the unit, flew secret photo-reconnaissance missions over Russia in preparation for Germany's invasion of the Soviet Union.

That same year also saw construction of nine pre-production Do 217C-0 level bombers. These primarily served as testbeds for the different engines (Jumo 211B and DB 601), different armament configurations, bombsights, bomb racks (capable of carrying a maximum payload of 3,000 kg), as well as various dive brake designs including pivoting perforated dive brakes located between the fuselage and engine nacelles.

As a result of all this, 1940 can be defined as the year of the beginning of Dornier's bomber modernization. Gradually, the new Do 217E replaced the Do 17Z in front-line service and, although not making a numerical impact—just 20 Do 217s were delivered to the *Luftwaffe* in 1940—the Do 217 did provide a not insignificant boost in quality for Dornier's bombers and reconnaissance aircraft. The Do 217E and its follow-on developments will be examined in greater detail in a subsequent chapter.

Dive Bombers

Junkers Ju 87

In this category ruled the Ju 87, a pure dive bomber (or *Stuka*), with the Henschel Hs 123 being represented at the war's outbreak with just 40 remaining examples serving in the strike aircraft role.

In 1939 577 Ju 87s were built; in 1940 production increased to 769, so that the operational losses suffered during the first 1 1/2 years of fighting were fully offset. It had been planned to taper Ju 87 production off as early as late 1939, but the type's success in the Polish campaign led to entirely the opposite; the decision to step up production was one which paid for itself in operations against Norway and France—until the heavy losses in August 1940 over England exposed the weaknesses of a machine designed exclusively for dive bombing. Nevertheless, the Ju 87B continued to be built under the planning assumption that the Messerschmitt Me 210 would be replacing it in front-line service sometime in the spring of 1941.

In the meantime, the Ju 87 underwent a few minor improvements. The B-1 series, ending in late 1939, was replaced by the Ju 87B-2 built at Berlin-Tempelhof by Weser-Flugzeugbau. Compared to its predecessor the B-2 boasted several refinements, including hydraulically activated radiator gills, improved propeller with wider blades (VS 5, later VS 11), and an improved exhaust system with exhaust pipes. Normally flown as a two-seater, as a single-seater the B-2 could also carry a 1,000 kg bomb, the PC 1000 specially developed for hitting ships. Other improvements included the Ju 87B-2/U2 with a better radio navigation system, the B-2/U3 with supplemental armor, and the Ju 87B-2/U4 with skis in place of the wheeled undercarriage.

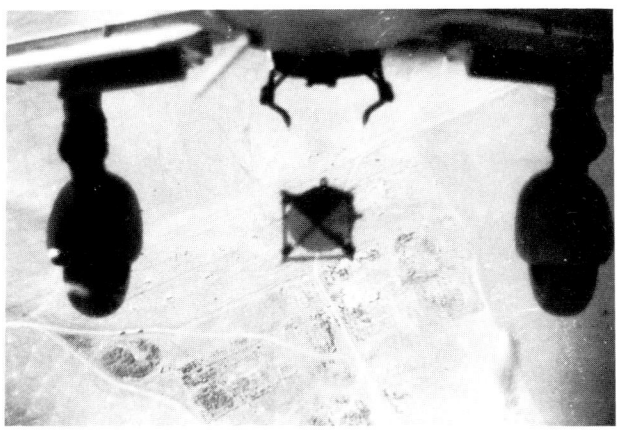

The Junkers Ju 87 in a dive. The 250 kg bomb, which the pilot basically aims along the aircraft's lateral axis, has just cleared the propeller arc and is plunging earthward on a near vertical trajectory.

Junkers Ju 87B-2/U4 with ski undercarriage (from 1940 onward) during technical maintenance.

Another version of the B-2 was the Ju 87B-2/Trop, which together with the regular Ju 87B-2 version was supplied to the Italian Air Force beginning in the summer of 1940.

Parallel to the B-2 series was an R-series (with R indicating *Reichweite*, or range), with two 295 l drop tanks being carried beneath the outer wings in place of bomb pylons. In this configuration, however, its payload was limited to just a single 250 kg bomb. Otherwise the Ju 87R-1 was identical to the Ju 87B-1. The Ju 87R-2 was a derivative of the B-2, and there was a Ju 87R-2/Trop with drop tanks carried beneath the outer wings and sporting a camouflage scheme typical for operations in southern climes. The Ju 87R-3 and R-4 differed only in a few minor equipment changes, mainly with regard to their radio navigation systems. From early 1940 to mid-May these long range Ju 87s initially went to a 40-strong *Stuka* group, which took part in the Norwegian campaign and was used against both fortifications and shipping targets.

The *Luftwaffe* entered 1941 with this minimally improved dive bomber that had not been able to meet the demands of the air war in the West as early as August 1940. Nevertheless, in awareness of the type's even more limited usefulness, in the spring of 1940 work had begun on a new version, officially designated the Ju 87D in May of that year. The Ju 87D was to have flown in December 1940, but the more powerful engine was not yet available by then; also, the designer of the Ju 87, *Dipl.-Ing.* Hermann Pohlmann, had left Junkers that year to take up a position with the Blohm & Voss company in Hamburg. At this point in time, the Ju 87 seems to have generally exhausted its developmental potential, and the intended

Junkers Ju 87R-2/Trop, loaded with auxiliary fuel tanks on the outer wings and a 250 kg bomb beneath the fuselage. The bomb trapeze can clearly be seen, and the typical camouflage pattern for southern operations is also shown to good effect.

MISSING CAPTION

History of German Aviation: Bombers and Reconnaissance Aircraft

Junkers Ju 87 with its sirens blocked in cruise flight. The two sirens, each with a diameter of 0.7 m, can be made out on the upper part of the undercarriage, as can the four 50 kg and one 250 kg bombs carried by this aircraft on what appears to be an operational sortie.

D-series were viewed only as a somewhat better, albeit interim solution until the availability of a truly improved dive bomber arrived on the scene.

Despite these circumstances the Ju 87's reputation as a unique instrument of war was far better than its actual combat capability—except for its accuracy when the defensive situation in the target area permitted safe diving. This reputation was enhanced primarily by two features:

To reinforce the morale effect of dive bombing ground troops, whistles were fitted to the bombs between the guide vanes, with the penetrating whistling sound upon release of these weapons earning them the nickname "Jericho Trumpets." Somewhat later, mainly during the French campaign, wind-driven sirens were also fitted, which generated an ear-splitting wailing sound in a dive. In normal flight these could be turned off and the noise-generating propellers locked by hydraulic pressure, but this led to a slight loss in speed for the Ju 87's already slow cruising speed and were therefore later removed.

The "Retailored" Fighters and *Zerstörer*

General Thoughts

The Ju 87's slow speed and consequent vulnerability to fighter attack made it highly advisable to operate the type only under a protective fighter umbrella, particularly in the area of the south coast of England and the Channel in 1940. Using *Stuka* units with full fighter protection was compounded by problems with forming up and regulating the different speeds. Ideally, fighters were a high altitude component, designed to secure the airspace above the *Stuka* formation, and to force them to provide close-cover escort and ward off enemy fighters breaking through was a strength-diluting approach.

What then, was better than taking a markedly faster aircraft, such as a fighter, and hanging bombs underneath it? Then, after dropping its load it could use its speed and maneuverability to independently operate as a fighter and even jump into aerial combat. At the very least, the cruise speed problem was minimized for the outbound portion to the target, since the bomb-carrying fighters, or fighter-bombers, would still be much faster than the cumbersome dive bombers—and the return flight would be no more difficult for the fighter-bombers (now in a fighter role) than for the escorting fighters.

This was the line of thinking which spawned a new type of development in 1940, one which was not really so new after all. This development will be examined in closer detail in the section entitled "1940/1941 Closing Balance."

Messerschmitt Me 109 and 110 As Fighter-Bombers

Review of Germany's air weaponry and methods following the Battle of Britain provided solid factual and numerical evidence that

Messerschmitt Me 110C-4/B of a Zerstörergeschwader sometime in 1940, with its ETC 500 rack being loaded with a 250 kg bomb by means of a hydraulic bomb trolley.

the Ju 87 was a markedly inferior aircraft when used as a dive bomber in the skies over southern England. The same could be said for the Me 110 as an escort fighter for bomber and dive bomber formations when operating in the arena dominated by British fighter defenses. The Me 109E fighter-bomber evaluation group mentioned earlier, along with the partial reequipping of Me 109 fighter units for fighter-bomber operations, had given rise to a new weapons category, the hybrid fighter-bomber; a hybrid, because the *Jabo* was unable to carry the same load as a true bomber on the one hand, and on the other because after releasing its bomb load its pylons still made it somewhat slower than a pure fighter. In the fighter units, on 28 September 1940 the 138 Me 109s converted to the *Jabo* role already comprised a third of the total strength on hand of 414 airplanes. In the meantime, the Me 110 had also been converted to a bomb platform, and as such entered operations beginning in July 1940. Because of the parallel development of the Me 110C-4/B fighter-bomber with that of the Me 110C-5 reconnaissance aircraft, the combat potential of both types will be discussed in the section entitled "Tactical Reconnaissance Aircraft" below.

Strike Aircraft

Henschel Hs 123 and Messerschmitt Me 109

With only 40 examples of the Hs 123A-1 at the outbreak of the war, the strike aircraft achieved notable success in providing direct support to spearhead movements of friendly forces during the Polish campaign; their pilots flew up to ten sorties a day in their robust sesquiplanes. The intent to fully replace the Hs 123 with the Ju 87 was therefore abandoned for the time being. In the French campaign these numerically limited planes—with numbers increased to just 50 aircraft—then acquitted themselves well yet again in all flashpoints of the ground war. They proved to be relatively impervious to ground fire. One example of the type's effectiveness was at Cambrai, where it beat back an attempted breakthrough by enemy armor toward the south and, flying non-stop and working with friendly AAA units, destroyed 40 enemy tanks just a few kilometers north of their own staging area. More powerful than its weaponry, however, was the psychological impact of the Hs 123's engine noise on ground troops; at a certain rpm setting the BMW 132 sounded like a rapid-fire heavy machine gun. This phenomenon occurred at about 1,800 min^{-1}, a relatively low rpm, and a rate at which the firing of the two synchronized MG 17s through the propeller arc was not recommended due to the potential damage to the propeller blades.

Despite its proven effectiveness and ruggedness in operations, the slow Hs 123 found itself replaced by the much faster Me 109E following the cessation of hostilities with France. The Me 109E began operations from the Channel coast in September 1940, striking at coastal airfields in southern England with bombs and guns. Having fully converted to the fighter-bomber role in the interim, Me 109s hit oil storage facilities, docks, railway junctions, and maritime targets in the area of Ramsgate/London. To this end the strike fighter units were equipped with the Me 109E-7, capable of carrying either 4 x 50 kg or 1 x 250 kg of bombs.

The parts of those Me 109 fighter units operating as fighter-bombers were at first equipped with refitted E-1 series fighters, which as the Me 109E-1/B were initially able to carry only a single

Messerschmitt Me 109E-4/B fighter-bomber, which carried the same bomb load (a single 250 kg bomb) as did the Me 109E-7 strike fighter (1940).

SC 50 bomb underneath the fuselage. Soon, however, there followed aircraft from the Me 109E-4 assembly line with production fitted bomb racks. These fighter-bombers, designated Me 109E-4/B, were able to carry the same payload as the Me 109E-7, the dedicated strike fighter variant.

It is here that the Achilles' Heel of the strike fighter program can be seen; the Hs 123's range and speed were too little for the air war over England, and its successor, the Me 109, offered not much better in the way of range plus had to be diverted from what was already insufficient fighter production.

Foundations for the Henschel Hs 129

Viewed in such a light, a realistic strike fighter planning program which met wartime needs would not seem to have existed, otherwise it would not have been necessary to bridge the first 1 1/2 years or so with aircraft types considered unsuited to the role. This was not entirely accurate, for as early as the spring of 1937 the *Technisches Amt* had been directed to publish specifications for a strike aircraft which was to be relatively small, heavily armed, armored, and specially tailored to direct support of ground forces. As a result the *Technisches Amt* issued a requirement for an aircraft: with minimal dimensions to provide the smallest target possible for ground defenses; two light engines; armored glass for the cockpit; armor protection for crew and engines; and at least two fixed 20 mm MG-FF, plus additional machine guns. Otherwise, the designers were generally given free rein. The request for tender was issued to the companies of Hamburger Flugzeugbau, Focke-Wulf, Gotha, and Henschel. By 1 October 1937 the four companies had submitted their proposals to the *Technisches Amt* for review. The designs of Focke-Wulf and Henschel were accepted, the two others rejected. Henschel and Focke-Wulf were awarded developmental contracts, and by the spring of 1938 the two mockups were ready for inspection.

At this time Focke-Wulf had a design for a tactical reconnaissance aircraft—the Fw 189V1—that was so far advanced, with production ready to start, that the *Technisches Amt* agreed with *Dipl.-Ing.* Kurt Tank's recommendation to build this for the ground support role, i.e. as a strike plane following its initial flight testing. The attractive part seemed to be that all components of the aircraft with the exception of different cockpits for the reconnaissance and strike versions were identical—as a reconnaissance version, however, the Fw 189 had been conceived as a two- or three-seater.

The mockup of Henschel's single-seat design, now designated by the RLM as Hs 129, came up short in several areas and was therefore rebuilt from scratch. After two more viewings and even more changes the design was eventually accepted. In mid-1938 Henschel was awarded a contract for building three prototypes. The first of these, Hs 129V1, was completed and test flown in the spring of 1939.

During this period Focke-Wulf was test flying its own strike plane version, the Fw 189V1b; however, because of poor flight handling characteristics it required a thorough revamping in order to be able to have a comparative fly-off with the Hs 129V1.

In spite of these setbacks Focke-Wulf received a follow-on contract for modifying another prototype, the Fw 189V6, into a strike aircraft from the reconnaissance series. The V6 thus converted had an armored cockpit for two crewmen, four MG 17s, two MG 151s, and the twin-firing MG 81Z in the rear. Its equipped weight now stood at 4,610 kg; the two air-cooled twelve-cylinder inverted inline engines built by Argus, the As 410A-1, each rated at 342 kW/465 hp, proved to be quite underpowered for the V6 to be able to achieve any kind of acceptable flight characteristics.

Henschel on the other hand had taken on a contract for a further eight pre-production Hs 129A-0s in addition to the three prototypes then under construction. With this turn of events the Fw 189 strike version dropped behind the Hs 129, designed from the outset as a strike fighter, and became seen solely as a backup project in the event that the Hs 129's performance figures turned out to be wholly unsatisfactory. In any event, this fear proved to be well founded, for the Hs 129V1 exhibited flying qualities little better than those of the Fw 189 during flight testing beginning in the spring of 1939. A fly-off between the two types at the Rechlin test center resulted in no clear verdict favoring one or the other aircraft—both held little attraction from a flying perspective. Despite this, work continued on the eight Hs 129A-0s, as these machines were more economical to build than the Fw 189.

Following the Hs 129V1's flight testing starting in the spring of 1939 were the V2 and V3 that summer, while the first Hs 129A-0 was finished by the end of that year. The subsequent seven examples of the null series were available for testing by the end of the first month of the year 1940. Powered by the same Argus As 410A-1 as the Fw 189, the Hs 129 proved to be much too underpowered during flight testing; its heavy handling, inadequate acceleration, average maneuverability, and poor visibility from the cockpit found few admirers. Nevertheless, several Hs 129A-0s began combat trials in the late fall of 1940. Not entirely unexpected, their poor showing led to the *Luftwaffe* declining the planned A-1 series, which was to have incorporated minor changes in comparison with the pre-production batch.

It was this unacceptable situation that prompted the designers of the Hs 129 under *Dipl.-Ing.* Friedrich Nicolaus to draw up a somewhat larger model to be powered by the air-cooled 14-cylinder radial Gnôme Rhône, now available with the capitulation of France and rated at 515 kW/700 hp. Under the project designation P 76, this version was to have rectified several of the main shortcomings of the Hs 129A-0.

However, jig construction for the Hs 129 at Henschel was so far advanced that switching to what was practically a new design would have led to delays which the RLM was not in a position to accept. Instead, *Dipl.-Ing.* Nicolaus was given a contract for studying the possibility of fitting the Gnôme Rhône engine to the Hs 129 airframes on hand. Two of the Hs 129s accordingly were returned to Henschel for a test fit of the larger Gnôme Rhône 14M, an engine almost twice as heavy and big as the As 410.

This, then, is a rough overview of the rather unpleasant state of affairs regarding strike aircraft as 1940 turned into 1941.

Long range Reconnaissance Aircraft

Focke-Wulf Fw 200

Subsequent development and total numbers of the Fw 200 combined role strategic maritime reconnaissance aircraft and auxiliary bomber has been discussed in detail up to mid-1941 in a previous section dealing with the Fw 200 long range bomber. At 5,245 km,

History of German Aviation: Bombers and Reconnaissance Aircraft

The Focke-Wulf Fw 200, which found itself increasingly hard-pressed to fulfill its role as a naval bomber from late 1941 onwards thanks to improved shipborne defenses (notice the four ETC bomb racks on the underside of the wings outboard of the outer engines).

the maximum range of the Fw 200C-3/U4 armed recce variant appears impressive, but when reduced to its theoretical penetration range this figure drops to just around 2,600 km from its takeoff point. This does not take into account reserve fuel, which is fundamental to mission planning for ranges and was generally considered to be 10% of the total quantity. Depending on weather conditions, additional time to a diversionary field also had to be factored in, as well as loiter time in the reconnaissance area, so that the grandly sounding 5,245 km in practice was reduced to less than a 2,000 km radius of action.

AAA from ships and convoys, as well as safety provided by warships and, in some cases, shipborne aircraft had increased so perceptibly in 1940/1941 that the role of the vulnerable Fw 200 increasingly shifted from anti-shipping (which dropped considerably in the late summer of 1941) to maritime reconnaissance and the shadowing of convoys. Attacks against ships on the high seas were now almost exclusively the domain of a strengthened U-Boat arm, sometimes in close cooperation with maritime reconnaissance platforms. Since early acquisition of convoys is an absolute necessity for timely interception by submarines, demands for extending the reconnaissance range of the Fw 200 as well as structural improvements increased in tempo. Focke-Wulf continued to intensively work on these issues.

Junkers Ju 86 High altitude Recon Plane

Limited to roughly 40 examples, the Ju 86 was designed to operate at high altitudes above the effective range of enemy air defenses. This was most alarming to the British Air Staff, and in late 1940 specified a requirement for a high altitude fighter with pressurized cockpit. The initial stipulation, on 29 April 1941, was a ceiling of 12,500 m/41,000 ft, but this was soon amended to 12,725 m/45,000 ft. In August 1941 the British high altitude fighter project got a boost of urgency, not so much as a result of the German high altitude reconnaissance planes operating with impunity over England, but because of the reported inferiority of the Spitfire V compared to the Me 109F above 9,500 m/31,000 ft, both in speed and climb rate. Rolls-Royce began working at a fever pitch in a bid to make the Merlin engine and its two-stage supercharger suitable for the specified altitudes; the same went for the Supermarine company of Vickers Armstrong Aircraft Ltd. in their efforts to design a pressurized cockpit for the Spitfire. A contract was issued for 300 such Spitfire cockpits—yet in 1941 Germany's Ju 86P-2 and P-3 were unimpeded masters of the upper skies over England.

Dornier Do 215B

After the bulk of the strategic reconnaissance units entered the war with the Do 17P, these were soon being equipped with so-called reconnaissance bombers, built under the designation Do 215A. These were an "unrefined" Do 17Z originally intended for export. However, with the war's outbreak they were taken into the *Luftwaffe* as the Do 215B and replaced the Do 17 in several reconnaissance squadrons, where many of them served for a number of years. In 1939 three Do 215Bs were delivered, followed by 92 in 1940 and another six in 1941, when production ceased on the type. Externally the Do 215 was identical to the Do 17Z, described in detail earlier in this book.

Junkers Ju 88D

The remaining long range reconnaissance units began converting to the better-performing Ju 88 reconnaissance version *sans* dive brakes from 1940 onward. 1940 saw 330 examples produced, with a much larger number expected for 1941. The long range reconnaissance version was designated the Ju 88D, with the first of these

History of German Aviation: Bombers and Reconnaissance Aircraft

being derived from the Ju 88A-5 bomber version as the Ju 88D-2 with a wingspan increased to 20.08 m. The more powerful engine planned for the D-0 and D-1 series, the Jumo 211J-1, was not yet available in quantity at this time. The Ju 88D-2 retained the Jumo 211B-1, the G-1, and some even with the H-1 rated at 883 kW/1,200 hp, and some examples were fitted with the four side-by-side ordnance pylons of the A-5 series, with each pair capable of carrying a jettisonable fuel tank for increasing range. Other D-2s were delivered without the pylons and thus were suited to reduced range operations. Dive bombing equipment was dispensed with in its entirety for the D-series. The version carried two to three remotely operated aerial cameras of varying focal lengths and a heating system. These were located inside the fuselage aft of bulkhead 15. Standard camera configuration was a long-focus Rb 50/30 for photo recce flights up to about 8,500 m and an Rb 20/30, normally only used under 2,000 m. For operations in tropical climates the Ju 88D-2/Trop reconnaissance version, later designated the Ju 88D-4, was built in 1941.

Tactical Reconnaissance Aircraft

Henschel Hs 126

First flown in the fall of 1936, the Hs 126 had advanced to become the standard tactical reconnaissance platform by the outbreak of the war, with 275 examples serving in front-line units, and had replaced the antiquated 14 He 45s and 67 He 46s still operating in short order. Yearly production in 1939 was 137. Production of 368 Hs 126s in 1940 was enough to offset the tactical reconnaissance aircraft losses in Poland, Norway, and the West, and to supply those units established before the Western campaign. In 1941 just five machines rolled off the assembly line in January, as this workhorse had been undergoing replacement by the more modern Focke-Wulf Fw 189 since late 1940. At the war's outbreak the He 45 and He 46 had mainly been assigned to surveillance duties on the Westwall, with a handful used as battlefield reconnaissance platforms in Poland; beginning in October 1939 these were replaced entirely by the Hs 126, which was fully saddled with tactical reconnaissance responsibilities during the Western campaign. On 11 May 1940, at the beginning of the Western campaign, the Hs 126 was represented by 277 aircraft in front-line units, of which 234, or 85%, were serviceable. An impressive state of readiness, despite the fact that the Hs 126 had suffered painful losses at the hands of enemy fighters as early as the beginning of 1940. During the Battle of Britain several Hs 126s were kitted out as smoke generating aircraft in the event of an invasion, but had little opportunity to serve in this capacity. Some, however, served along the Channel in the fall of 1940 in the search and rescue capacity.

Messerschmitt Me 110

Although maneuverable and rugged, the Hs 126's vulnerability necessitated finding a faster platform for reconnaissance sensors from current suitable types already in production; in certain situations the Hs 126 was no longer able to penetrate enemy territory without sacrificing the crews. "A dead recce pilot is a bad recce pilot"—a rather morbidly humorous saying in which was couched a pragmatic recognition of reconnaissance flying.

In the meantime, the Me 110 had been tailored to a fighter-bomber role in addition to its primary function. As a *Jabo*, it normally was capable of carrying two 250 kg bombs (or heavier, with accordingly reduced range). Driven by the more powerful DB 601N (880 kW/1,200 hp) in place of the standard DB 601A-1 (770 kW/1,050 hp), the fighter-bomber bore the designation Me 110C-4/B. It was successfully employed against shipping in the English Channel, sometimes operating at low level and sometimes by dive bombing at angles of up to 45°. With a maximum permissible diving speed of 650 km/h and good low level handling for flying under enemy radar, the Me 110 had considerable operational potential—which would now be exploited in the reconnaissance role.

Thus was born a parallel development to this *Jabo*, a special reconnaissance version designated the Me 110C-5. Externally, it was virtually identical to the C-4, but had only four fixed MG 17s in the nose and a flexible MG 15 in the *B-Stand*, plus it was initially powered by the less powerful DB 601A-1. It carried an Rb 50/30 aerial camera installed in the cockpit flooring. This two-seat reconnaissance variant became available in the spring of 1940, with some being assigned to Do 17P and Do 17Z units. It therefore also flew strategic reconnaissance and target surveillance missions within the limits of its range.

At its maximum cruise speed at low levels, the Me 110C-5's range without external tanks was 775 km. However, an altitude of a good 4,000 m was needed in order to effectively use the photo equipment. At this altitude, and by economical use of the engines, the Me 110C-5 could notch up nearly 1,100 km. Not counting a safety reserve, this equated to a penetration radius of about 500 km. A somewhat better-performing recce-110 was fitted with the previously mentioned DB 601N and was designated the Me 110C-5/N. Altogether, 75 Me 110 *Jabos* and reconnaissance versions were built in 1940, with a further 190 following in 1941. With these numbers, the Me 110 placed second numerically on the list of tactical reconnaissance aircraft in 1940, just beneath the Hs 126.

Focke-Wulf Fw 189

The Fw 189 was mentioned in connection with the Henschel Hs 129 as a potential strike fighter option, but was rejected for this role; it did, however, become a remarkable tactical reconnaissance aircraft.

Ar 198 and Fw 189 Prototypes and Selection
As early as February 1937, as the Hs 126 was beginning its flight testing phase, the *Technisches Amt* had issued new specifications for the next generation of tactical reconnaissance aircraft. These went well beyond the requirements which had spawned the Hs 126. The new tactical recce plane was to have a 360 degree field of defensive fire, fly with a three-man crew, and have considerably better flight handling characteristics than aircraft hitherto serving in this capacity had possessed.

The Focke-Wulf designed project, drawn up by *Dipl.-Ing.* Erwin Kosel under the direction of *Dipl.-Ing.* Kurt Tank, called for a twin-engined aircraft with a twin-boom configuration, retractable undercarriage, and a cockpit/crew compartment situated between the booms. This design proposal initially caused a bit of a surprise at the *Technisches Amt* which, although it had not specified the number of engines for the prescribed power output, had expected that any new tactical reconnaissance plane would be a single-engined design like its predecessors. After studying the design more closely, however, the proposal gradually won hearts over. This was not least due to Kurt Tank's accompanying suggestions showing how such an aircraft, by retaining all major components with the exception of a swapped out cockpit, could be used in a variety of roles, e.g. as a strike fighter, a pilot trainer, or for crew training. However, critics feared that twin-boom designs were heavier from the outset than a fuselage of standard design, and the open frame formed by the two booms would warp under the stress of hard maneuvering.

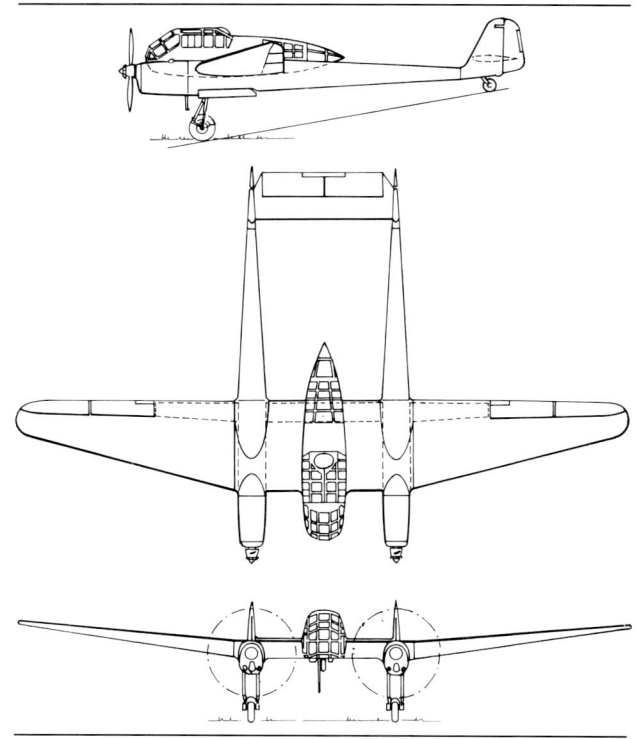

Focke-Wulf Fw 189A - its rather odd design is shown in this three-view from late 1937. The RLM envisioned the Fw 189 as a replacement for the Henschel Hs 126 beginning in February 1937. The Fw 189V1 completed its maiden flight in July 1938.

At the end of April 1937 both the Arado and Focke-Wulf companies were given contracts for building three prototypes each. The Arado design was of a standard layout and was designated the Ar 198, while Focke-Wulf's was designated the Fw 189.

Ar 198 Prototypes, the Fw 189 and the Decision
The Ar 198 was conceived as a single-engined, cantilever, high-wing aircraft with fixed landing gear and powered by the BMW Bramo 323A Fafnir engine (662 kW/900 hp). The forward fuselage section was designed with a bulging lower area to give the observer a full 360 degree view. In the spring of 1938 the Ar 198V1 flew without armament and, because of its extensive fuselage glazing, was nicknamed "The Flying Aquarium." However, from the outset of its testing it was clear that its overall flight performance was well under the required specifications.

The Ar 198V2 was to have been fitted with full armament, which called for two fixed forward-firing MG 17s and two flexible aft-mounted MG 15s, one firing upward and the other downward, plus wing racks for carrying four 50 kg bombs. But due to the poor showing of the Ar 198V1 further flight testing of the V2 was canceled, and the incomplete V2 and V3 were scrapped.

Prototype assembly at Focke-Wulf also came under the jurisdiction of *Dipl.-Ing.* Erwin Kosel, who handed over the Fw 189V1 in July 1938 some fifteen months after the contract had been awarded and work had begun. The type was powered by two Argus As 410A-0 engines, and chief design engineer Kurt Tank himself took it up on its maiden flight. In August 1938 there followed the Fw 189V2 and the Fw 189V3 in September. In all, once a few teething troubles such as separation of airflow between the fuselage booms and the control surfaces had been rectified, flight testing went satisfactorily. After the poor flight test results of the Ar 198V1 the decision was clearly in favor of the Fw 189.

Additional Fw 189 Prototypes

Focke-Wulf, after the first three prototypes, was contracted to build an additional four. The Fw 189V4 was finished in late 1939 and was planned as the pre-production pattern for the A-series tactical reconnaissance version. It also served as a testbed for various specialized equipment, for example the S 125 smoke generating system or a spray system for chemical weapons which could be hung from the outer wings in place of the ETC racks. The anticipated start of A-series production in the summer of 1939 never took place, however, as the ongoing rearmament of the tactical reconnaissance units with the Hs 126 and its performance was proving satisfactory for the time being.

The Fw 189V5 was seen as the pre-production aircraft for the B-series trainer and began flying in early 1939. There followed a contract for three Fw 189B-0 null series and ten Fw 189B-1 production trainers, with the three B-0s and three of the B-1s being built before the year's end. The remaining seven of these dual-control five-seat versions followed in January and February 1940 and were handed over to the *Luftwaffe* for field testing.

The Fw 189V6, with the Argus As 410A-1 powerplant, was intended as the prototype for the Fw 189C strike fighter version. The V6 was finished in early 1940, at which point it began its initial flight trials. This version's fate has already been discussed in connection with the Hs 129's early history. Heavy armament and an armored cockpit made for a compact, albeit relatively heavy machine whose flight performance was no better than that of the negatively viewed Fw 189V1b, the first prototype for the strike role converted from the Fw 189V1 reconnaissance version. Despite the equally poor flight results of the Hs 129—during field tests the machine was unanimously rejected by its pilots—the RLM nevertheless decided in favor of producing the Hs 129. Several promised improvements to the Hs 129 and a cost barely over two-thirds of the unit price for the Fw 189 were the deciding factors. Thus died the idea of production the Fw 189C strike variant.

This, then, brings us to the last of the seven prototypes, the Fw 189V7. As a floatplane, it was to be the initial pattern for a subsequent Fw 189D series. It was under construction in late 1938 with an airframe roughly similar to the V5, but was stricken from the program due to a change in the acquisition planning for maritime aircraft. The semi-completed airframe was eventually completed as one of the previously mentioned three Fw 189B-0 trainers in late 1939.

Fw 189 A-Series

In the spring of 1940 Focke-Wulf was finally awarded the production contract for building the Fw 189A reconnaissance plane, to include a pre-production batch of ten Fw 189A-0s as well as the true production Fw 189A-1s, of which 20 had been built at the Bremen plant by the end of 1940. At this time began field trials with the type in various reconnaissance squadrons, where the Fw 189 acquitted itself well. This favorable showing, coupled with the high losses suffered by the Hs 126 in the Western campaign and the older type's increasingly apparent inferiority in the tactical reconnaissance role, led to a high priority being placed on Fw 189 production. Where just 38 Fw 189s in total were produced in 1940, in 1941 this figure jumped to 250 Fw 189As built both in Bremen and at the newly opened assembly line in Prague. Bremen's output that year was just 99, as the plant converted over to production of the new Fw 190 fighter, with the remaining 151 machines coming from Prague.

Beginning in mid-1941 defensive armament was beefed up by a 7.9 mm MG 81Z twin-firing machine gun in place of the two MG 15s. Thus equipped, the type was designated the Fw 189A-2. Parallel to this was the Fw 189A-3, a conversion trainer built in small numbers with dual controls. This trainer series was enhanced by a handful from the A-0 and A-1 series which were also refitted with dual controls.

Around the end of 1941 even more production facilities were moved abroad or expanded, and the necessary jigs for building the Fw 189 were moved from Bremen to the Atlantic coast in the area of Bordeaux, where final assembly of the plane was planned to take place at Bordeaux-Mérignac.

Thus concludes a brief summary of the Fw 189's development period through to its full operational maturity in 1941.

History of German Aviation: Bombers and Reconnaissance Aircraft

Focke-Wulf Fw 189B-1 - unarmed trainer (1940).

Technical Data for the Fw 189A-1

The Fw 189A-1 was an army reconnaissance aircraft for battlefield reconnaissance, artillery spotting, and smoke laying operations, as well as a light ground support aircraft. It had a three-man crew, and was a twin-engined low-wing design with twin booms and of all-metal construction with retractable landing gear.

Airframe
Fuselage: all-metal crew compartment, extensively glazed, located between the two booms. All-metal booms, interchangeable engine nacelles.
Undercarriage: main wheels hydraulically retractable aft into the booms, dual struts with rocker arms. Tailwheel under the tailplane also hydraulically retractable laterally into well in the tailplane.
Control surfaces:
a) ailerons: two-piece, fabric covered.
b) flaps: electrically powered split flaps between aileron and boom, as well as along entire centerwing underside.
c) horizontal stabilizer: all-metal tailplane at the extreme end of booms. Metal-frame elevator covered in linen, balanced and fitted with electrically driven adjustable trim tab.
d) vertical stabilizers: all-metal tailfins at the extreme end of both booms. Rudders same as elevator.
Wings: three-section, triple-spar all-metal wings. Rectangular center section between the two booms supports crew compartment. Wing outer sections with removable leading and trailing edges.

Powerplant
Engines: 2x Argus As 410A-1, each rated at 340 kW/465 hp, an air-cooled twelve-cylinder inverted V-type (two banks of six cylinders) turbocharged engine
Propellers: automatic two-blade Argus variable-pitch with two rpm settings for takeoff and climb/cruise.
Fuel tanks: full self-sealing with 450 l capacity.

Armament: two fixed MG 17s in wing roots, two flexible MG 15s in the *B-Stand* and in an Ikaria tail turret at the extreme aft end of crew compartment (MG 15s replaced by an MG 81Z beginning with the Fw 189A-2). ETC 50/VIIId pylons underneath the wings for carrying four 50 kg bombs/illumination flares, or two S 125 smoke generating systems.

Camera equipment: an Rb 20/30 as standard aerial camera with sequence and overlap controls, or RB 50/30, Rb 21/18, or Rb 15/18 also possible. Electric blower drive for maintaining continuity of film, with extendable pitot tube in the event of electrical failure. HK 12.5 or HK 19 were also normally carried.

Crew: responsibilities for the three-man crew were as follows:
1. Pilot/gunner for fixed armament
2. Observer/bombardier and radio operator, plus *B-Stand* gunner
3. Engineer, tail gunner

Performance, dimensions, and weights
Performance:
maximum speed at 2,400 m altitude	350 km/h
cruise speed at 2,400 m altitude	325 km/h
maximum speed in a dive	502 km/h
service ceiling	7,300 m
standard range	670 km
endurance	2 hrs 10 mins

Dimensions and Weights:
wingspan	18.40 m
height	3.10 m
length	12.03 m
wheel track	4.28 m
wing area	38.00 m^2
empty weight	2,830 kg
equipped weight	3,245 kg
takeoff weight (no bombs)	3,950 kg
takeoff weight (w/ bombs)	4,170 kg

Features: Prototypes and the Fw 189A-0 had a single-strut undercarriage, both guiding the wheels and absorbing the shock. In action, these proved to be too susceptible to problems, and therefore on the A-1 these were replaced by dual-strut landing gear and VDM shock absorbers.

Assessment

The Fw 189 Focke-Wulf succeeded in creating what was undoubtedly an orthodox design, albeit a design which was particularly well-suited for the tactical reconnaissance role when it was introduced into service in 1941. Its significantly improved all-around visibility in comparison with the Henschel Hs 126, the better armament (aft gunner's station on the Fw 189A-2 fitted with the 7.9 mm MG 81Z twin-firing gun), a roughly 100 km faster top speed, better survivability thanks to its two engines, and full cold-weather capability thanks to its enclosed cockpit—all these contributed to making this type an enormous advance in the reconnaissance aircraft class. The nearly 2,000 m lower service ceiling of the Fw 189 does not enter the equation, for neither its 7,300 m nor the 9,000 m of the Hs 126 are of value for the operational roles of these aircraft. Their operating altitudes were at low or, relatively seldom, at medium altitudes.

The Me 109E-5 generally corresponded to the pure E-4 fighter version and was produced parallel to the fighter. It was powered by the same aero engine, the DB 601A (770 kW/1,050 hp), and inside the fuselage behind the pilot's seat it normally carried an Rb 21/18 or Rb 50/30 camera. The reconnaissance version had only two fuselage-mounted MG 17 machine guns in the nose; the two MG-FF of the E-4 series in the wings were removed on the E-5.

The recce fighter version fitted with the somewhat more powerful DB 601N (880 kW/1,200 hp, with one minute 935 kW/1,270 hp emergency boost at 5,000 m) was designated the Me 109E-6 and generally carried an automatic hand camera, as well. Then there was the Me 109E-9, also appearing in the fall of 1940, which in addition to the two MG 17s now retained the two MG FFs, and as a rule carried the Rb 50/30 in the fuselage. The E-9, however, was powered by the DB 601E (995 kW/1,350 hp) and had improved

Rear guns of a Focke-Wulf Fw 189A-2 showing its 7.9 mm MG 81Z twin barreled machine gun.

Messerschmitt Me 109

The limited operational capability of the Henschel Hs 126, noted as early as the Western campaign but especially felt in the Channel theater, led to acquiring the faster Messerschmitt Me 110C-5 reconnaissance type. Yet an even faster tactical reconnaissance aircraft was needed, one which would have little or no fear of enemy fighter defenses. To this end was the Me 109 made available, of which a total of 26 Me 109E-5, E-6, and E-9 recce fighters were provided in 1940. There are no records indicating these were supplemented by additional aircraft in 1941.

Messerschmitt Me 109E-9 (from late 1940) following its return from a reconnaissance flight. Note the 300 liter drop tank for increased range. Photo technicians are removing the film cassettes from the cameras while the pilot uses a map to show the photo interpreter the target areas which were photographed.

backplate armor for the pilot. For increased range it could carry a 300 liter drop tank, as well.

Measured against the overall total of 507 tactical reconnaissance aircraft produced in 1940, the 26 Me 109 reconnaissance fighters comprised just 5% of that figure. Yet they offered the potential of returning home with photos even from areas threatened by enemy fighters, something that even the Me 110 reconnaissance platforms were no longer up to, not to mention the bulk of the much slower original tactical reconnaissance planes. However, it is deceptive to assume that the faster a reconnaissance plane was, the better it was. To be sure, this was true at higher altitudes where the Rb 50/30's long focal length was still capable of providing good quality pictures. But at lower altitudes the human eye first had to identify the target or target area to be photographed before acquiring much reduced area through the viewfinder. The lower the altitude and the higher the aircraft's speed, the more difficult it is to get a good picture.

1940/1941 Closing Balance

The inventory of what the *Luftwaffe* had at its disposal in early 1941 does not seem to be altogether bad at first glance, yet there were many weaknesses hidden within those numbers. With a view toward the East, it can be assumed that these shortcomings would hardly be noticed in light of the average quality and quantity of the Soviet Air Force assumed from intelligence reports. Even the damper cast by the British in the Battle of Britain did little to change the German feeling of superiority over the Soviets, a feeling enhanced by a series of successful campaigns and one which prompted the highest German leadership to envision total military subjugation of European Russia by the end of 1941.

But the war's events in late 1940/early 1941 took a different course than planned.

An analysis of Germany's air war potential shows that the air attack units—when expressed in numbers—had reached an absolute low in December 1940. On 21 December 1940 the number of operational aircraft in the units was:

1,271 bombers
436 dive bombers
24 strike aircraft, i.e. a total of 1,701 bomb platforms.

Shortly before the Western campaign, on 4 May 1940 there were:
1,706 bombers
425 dive bombers
49 strike aircraft, i.e. 2,180 bomb platforms.

With dive and strike bombers, the differences of +11 and -25 for what was in any case related roles was not so glaring, but the absence of 435 bombers gives pause for serious thought. Especially if the operational figure of 1,271 bombers at the end of 1940 is contrasted with the 2,852 bombers produced in 1940. These numbers are a good indicator of the bloodletting suffered by the bomber forces, which by the end of 1940 still had a long way to go before being offset. A painful lesson from the daylight raids on England in August and September 1940 was that the classic bomber was extremely vulnerable when operating without friendly fighter protection. Even so, fighters and heavy fighters suffered even heavier losses. The numbers below speak for themselves:

Operational	On 5/4/1940	On 12/12/1940	Losses
Fighters	1,330	859	-471
Hvy. Ftrs.	427	218	-209

These 680 not yet complete aircraft losses were only somewhat cushioned by night fighters, which on 4 May 1940 had not yet been included in the operational category for front-line units, but on 21 December that year had reached an operational figure of 168 aircraft. The net substance loss to the *Jagdwaffe* was still 512 aircraft, or approximately 30%! It wasn not until early January 1941 that a general period of rest and recovery set it, particularly in the fighter area. Despite the unexpected developments of the war beginning in early 1941, by 21 June 1941 (the fateful date when Germany attacked the Soviet Union), this recuperative period had led to the following operational strength:

		Compared to 5/4/1940
bombers	1,511	-195 aircraft
dive bombers	424	-1 aircraft
strike aircraft	51	+ 2 aircraft
fighters	1,414	+ 84 aircraft
heavy fighters	188	-239 aircraft
night fighters	263	+263 aircraft

As usual, the bomber units at this time were facing some new, difficult challenges in a weakened state, while the loss of heavy fighters was more than offset by a plus of day and, primarily, night fighters.

However, by 21 June 1941 much had happened in the air war which again took its toll on all front-line units, particularly the air attack units. Beginning in January 1941 the *Luftwaffe* not only had to deal with the continuing war against Great Britain, but was now faced with a strength-sapping "second front" in the air over the Mediterranean.

14. Developments in the Mediterranean and Southeast

First Operations in Expanding the War

The intentions and activities of the Italian allies led to them declaring the North African coast and the Mediterranean an operational theater on 20 August 1940.

In September 1940 Italian troops pushed out from Italy's colony in Libya some 100 km into Egyptian territory, where the problems of resupply soon caused the offensive to come to a standstill.

On 28 October the Italians launched an attack against Greece from Albania, which they had occupied in April 1939. This in turn led to British air and ground forces on 29 October landing in Crete from their staging positions in Egypt as a result of a British assistance pact signed with Greece on 13 April 1939. In November 1940 Greek units, operating with indigenous air support to include previously supplied Heinkel He 70s (which acquitted themselves rather well, it must be said), threw the Italians back over the Albanian border. The first British units landed near Athens that same month. In December the Greeks conquered about a third of Albania, receiving some British air support operating out of airfields around Athens.

It was also in December that the British launched a successful counterattack from Egypt into Cyrenaica and began their advance toward Libya.

Escalation in the South

The rapid sequence of events building up in the Balkans and the Mediterranean in late 1940/early 1941—to the detriment of the Italian challengers—are briefly summarized below as follows:

- Due to concerns that the approach avenues for the upcoming campaign in the East would be threatened, and that the British would be able to carry out air raids on the Romanian oil fields, on 12 November 1940 Hitler signed his Directive No. 18 for initiating preparations for a relief attack against Greece "so that, if necessary," the prerequisites could be established for German flying unit operations against targets in the eastern Mediterranean, "especially against those British airfields which threaten the Romanian oil fields."
- On 10 December 1940 there followed an OKW directive deploying German air attack units to southern Italy.
- On 13 December Hitler signed Directive No. 20, which established the attack on Greece (Operation Marita).
- On 14 December the intended overall strength of the German *Luftwaffe* in southern Italy was established as:
  - 59 Ju 88s for the bomber role
  - 36 He 111s, primarily for torpedo strikes, but also for bombing missions, as well
  - 10 He 111s, primarily for aerial minelaying, but also for bombing missions
  - 78 Ju 87s for dive bombing, especially against shipping targets
  - 11 Ju 88 strategic reconnaissance aircraft
  - 34 Me 110C heavy fighters, plus a number of transports

  In all, a rather impressive air force totaling 226 operational aircraft, placed under a German command headquarters to be based out of Sicily.
- On 18 December 1940 followed Directive No. 21, stating that the *Wehrmacht* must be ready "to cast down the USSR even before the end of the war against England (Operation Barbarossa)." D-Day was set for 15 May 1941.
- On 19 December 1940 the Italians sent out an urgent request for military support in North Africa, followed by another on the 28th of December for their defensive operations in Albania.
- On 9 January 1941 156 of the ordered 226 aircraft were actually on Italian soil, of which 124 of these were in Sicily. These planes had been pulled in the middle of winter from Norway, western France, and Belgium, and some of the airfields they were to be assigned to in southern Italy and Sicily had to be first built from scratch, not to mention the multitude of logistical and other problems which first had to be overcome. The most prominent absence at the time was mainly the Ju 88, of which just four had arrived. Nevertheless, the framework for the planned "naval war from the air" was generally in place.

Naval War From the Air

An Aircraft Carrier's Achilles' Heel
On 10 January 1941, precisely one month after the OKW's deployment directive, 60 He 111s and Ju 87s went into action for the first time in the Mediterranean and attacked British protective forces providing cover for a convoy sailing from Gibraltar to Malta and

Greece. A battleship was struck by one bomb and an aircraft carrier by six, forcing the carrier to pull out of the convoy and seek shelter in La Valetta on Malta. Setting out the next day from Malta toward Alexandria, the main base for the British Mediterranean fleet in the eastern Mediterranean, the cruisers were flushed out once again by Stukas and suffered the loss of two of their number. There followed several bombing raids on Malta, resulting in bitter air battles with the approximately 60 British fighters based on the island. On 16 January 1941 the Ju 88 went into action for the first time, now that 54 machines were based at Catania on Sicily. 17 Ju 88s, escorted by 20 Me 110s formed the first wave, followed by a good 40 Ju 87s escorted by Italian fighters in the second wave. Both groups hit the docks and harbor facilities of La Valetta, dropped even more bombs on the wounded aircraft carrier, and damaged a large merchant ship and a dock. One Ju 88 was lost in this raid. Two days later the airfields of Malta were struck, as was the carrier yet again. Nevertheless, during the subsequent four days of bad weather the ship was made seaworthy enough so that it slipped out of the harbor on 23 January and, after surviving further attacks, safely dropped anchor in Alexandria on 25 January 1941. It was subsequently out of action for several months and had to undergo lengthy repairs in the U.S. This somewhat detailed example showed that a modern aircraft carrier—it was the *Illustrious*, built just the previous year—of the 23,000 ton class was not so easy to destroy with the ordnance available at the time (which in 1941 included an armor piercing 1,000 kg bomb, the PC 1000), despite being hit several times. It could, however, be knocked out of operations for a time. From this point on the German bomber pilots' daily bread in the Mediterranean became attacking British supply convoys and their protective forces.

Problems of Range
But Greece and Malta were not the only neuralgic points; also of concern were the two eyes of the needle—Gilbraltar at the western entrance to the Mediterranean and the Suez Canal at the eastern access. During the night of 17/18 January 1941 eight He 111s took off from Bengazi airstrip to attack a large convoy sailing from Aden to Suez. An agent had originally reported the movement, and the ships had been sighted and confirmed in the Suez Roads the previous day by a reconnaissance plane. Even with extra tanks, the 1,200 km to Suez with a 2,000 kg payload was at the extreme limit of the He 111's range and left virtually no reserve fuel. The convoy was no longer outside of Suez, and only one crew found it in the Great Bitter Lake—too late for a combined attack. Because of unfavorable winds three crews ran out of fuel and were forced to make emergency landings behind British lines and were taken prisoner. Only one crew succeeded in returning to Bengazi, the remaining four being forced to land in Cyrenaica where they were rescued.

This complete failure of an operation designed to interrupt British supply lines by attacking shipping in the vicinity of the Suez Canal clearly showed the limits to which bombers could be pressed without subjecting them to senseless risks. Seven of eight aircraft and three crews had been lost. In any event, the Suez Canal remained taboo for the time being—German bases were simply too far away from the target. Despite this, in the naval war the He 111s, when equipped with torpedoes and operating at low levels, proved to adequately complement the mostly dive bombing Ju 88s and Ju 87s.

North Africa

On 11 January 1941 Hitler ordered Directive No. 22, the deployment of German troops to North Africa in support of the threatened Italians there. As a result, on 6 February 1941 the *Deutsche Afrikakorps* under *General* Erwin Rommel was formed (Operation Sonnenblume). From then on problems not only increased and became aggravated with regard to the deployment, but primarily with supplying the *Afrikakorps* in its increasingly intensified operations against the British. The dive bombers and a third of the heavy fighters were deployed from Sicily to Africa on 13 February 1941 in an attempt to stabilize the front against the advancing British Army, and these planes began operating in support of the ground forces by the next day. It was not until 22 February that additional aircraft began arriving in the Mediterranean. These planes, pulled from units in the West, included:

9 Ju 88 strategic reconnaissance planes
32 Ju 88 bombers
74 Ju 87 dive bombers
39 Me 109 day fighters

From this point onward this "Second Front" under tropical conditions became a front which tied up and drained away the strength of the Luftwaffe at an alarming rate.

The Balkans and Crete

Beginning on 4 March 1941 additional British troops were shipped to Greece—the number totaled 58,000 by 24 April of that year. On 25 March 1941 Yugoslavia joined the Tripartite Agreement, but three days later the signing government was toppled by a national strike. Operating under the assumption that the new leaders in power would be supported by England and possibly the USSR, that same day Hitler decided to occupy Yugoslavia in addition to the planned attack on Greece. On the evening of 27 March 1941 he signed Directive No. 25 "*Blitzkrieg* against Yugoslavia in conjunction with the attack on Greece," which as a result of this was postponed from 1 April to 6 April 1941.

Yugoslavia and Greece

Without any declaration of war, the German invasion of Yugoslavia and Greece began at 0515 hrs on 6 April 1941. German troops advanced through Hungary, Romania, and Bulgaria. The Luftwaffe bombed Belgrade heavily and attacked the Royal Yugoslav Air Force in central and southern Yugoslavia. There was no air defense of any worth, for the Yugoslav aircraft inventory did not include any fighters on par with those of the invaders. Belgrade was taken by German forces on 14 April. On 17 April 1941 the Yugoslav Army capitulated.

At the same time, Stukas were bombing fortifications along the Metaxos Line, where the Greeks were putting up solid organized resistance. The line was broken and Saloniki fell into German hands. On 21 April 1941 the Greek Army commander in the north was compelled to surrender. Following additional fighting with British forces, these began boarding ships on 26 April bound for Crete and Egypt. On 27 April Athens fell, followed by the occupation of the Peloponnesus and the Greek islands.

Crete

Debate and Decision

Or would it have been better to eliminate Malta with a landing operation? Certainly the better option would have been both Malta, whose air and naval bases posed a constant threat to the supply lines into North Africa, as well as Crete, from whence it was possible to control the eastern Mediterranean and the Aegean. But the urgency lay with the invasion plans in the East, which had already fallen victim to delays and logistical problems due to the military developments in North Africa and the Balkans. Two such island assaults were simply not feasible. The decision fell upon Crete, since the occupation of this large island was both the key to the Aegean and would at the same time eliminate the possibility of the British again trying to push their way into southeastern Europe. Furthermore, with Alexandria and the Suez Canal just 700 km or so away, it would be easier to conduct offensive operations against these two targets and undermine supply efforts for Malta from Egypt, or have a major impact on planned enemy operations in the central Mediterranean. Following this decision, the action against Crete was handed over to the Luftwaffe *in toto*, which drew up its plans for the invasion in short order. Operation Merkur called for a large-scale airborne assault followed by using air and naval transports to bring troops in to combat the roughly 30,000-strong British Expeditionary Corps and the remainder of the Greek Army.

Execution of the Plan

In addition to the Mediterranean units operating in support of the fighting in North Africa, even more forces were provided for the invasion of Crete. These were deployed to areas around Attica, the Peloponnesus, and the Italian island of Skarpanto in the Straits of Kaso. By 17 May 1941 the following were ready for action:

47 strategic reconnaissance aircraft
209 bombers   ⎫
                    ⎬ = 341 bomb platforms
132 dive bombers   ⎭
110 heavy fighters ⎫ = 222 for low level attacks, and,
112 fighters   ⎬   in part, fighter-bomber operations
                    ⎭

as well as the real backbone of the entire operation in the form of 542 transport aircraft.

Non-stop attacks on British airfields along the north coast of Crete forced the British to pull their last Hurricane and Gladiator fighters back to Egypt on 19 May to prevent their certain destruction. By the day prior to the airborne operation the Luftwaffe had complete air superiority over Crete, and had thus obtained the most critical prerequisite for the success of Merkur.

On 20 May 1941 at 0715 hrs German airborne troops began landing on Crete. However, German aerial reconnaissance had not been able to provide a complete picture of the defensive measures of the enemy, which in part included doing an excellent job of concealing heavy weapons and infantry forces. From their positions in the terraced hillsides overlooking the airfields and drop zones, the British brought their weapons to bear on the transports and troops using these areas, causing unusually high casualties among the attacking forces and preventing the invaders from reaching a majority of their objectives on schedule. This was not only compounded by the fact that the island defenses had been in a high state of alert since the 17th of May—meaning a surprise attack had been out of the question—but also that the entire British Mediterranean fleet, with 45 warships and numerous other smaller vessels, had been at sea since the 15th of May and was now located in four groups in the approaches to the Aegean Sea. The British had assumed that the bulk of German strength and logistical support would arrive by sea, an assumption which proved only partially true and was based on an underestimation of Germany's air transport capabilities. Thanks to uninterrupted air support the German attackers gradually began gaining the upper hand (after a few isolated setbacks and with heavy losses) against an unusually tenacious and numerically superior foe. Supply ships, bridgeheads, and airfields became the targets of British warships. This resulted in the focus of bomber and dive bomber operations increasingly shifting to attacks on the British fleet. The fleet was now subject to incessant attacks from the air and forced to absorb considerable punishment without the luxury of fighter protection—during this critical time the only aircraft carrier in the Mediterranean fleet was still in port awaiting its aircraft. On 22 May two British cruisers and a destroyer were sunk, and another two battleships and two destroyers were damaged. Of three destroyers which shelled the airfield at Malemes during the night of 23 May, two of them were sunk by Stukas early the next morning as they were pulling back. On that day, too, fighter-bombers sank a further five motor torpedo boats in Souda Bay, so that total British losses in the space of just 24 hours amounted to two cruisers, three destroyers, and five torpedo boats. During these anti-shipping operations two Ju 88s, seven Ju 87s, three Me 110s, and an Me 109 were shot down, for a total of 13 aircraft lost.

The commander of the British fleet was compelled to radio London that the navy was no longer able to operate by day in the Aegean or off Crete in light of these attacks. In spite of this, higher authority intervened and the British continued their shelling, although now mainly at night. On 26 May 1941 they even attacked the island of Skarpanto, where Ju 87s were based, with air support; the aircraft carrier *Formidable* (23,000 t) had finally joined the fray with 12 operational aircraft. A sustained victory was not in the cards for the British—the Ju 87s continued their unceasing bombing activity. Under pressure from the attackers, the British began evacuating their troops on 28 May 1941, bringing them to collection points at harbors along Crete's northern coast. Two additional destroyers succumbed to bombs from Ju 88s and Ju 87s during the attempt, so that the remainder of the evacuation took place from the island's south coast and entirely at night. Over a period of four nights, beginning on 28/29 May approximately 17,000 men were evacuated from the beaches by British warships, with German mountain troops hard on their heels, in what amounted to just a few hours. Three cruisers were hit by bombs, and on the last day of the evacuation, 1 June 1941, an AA cruiser was found and sunk by two Ju 88s about 100 nautical miles north of Alexandria. After almost 12 days of heavy fighting, the island was now in German hands.

Results and Lessons

With losses totaling nearly 50% dead, wounded, or missing in action of the 8,000 paratroopers involved, this German victory was bought at a heavy price. Of the approximately 6,500 mountain troops participating in Merkur, about 18% became casualties. 100 German operational aircraft were lost over the 12 day period: 26 bombers, 18 Stukas, 52 fighters and heavy fighters, and four reconnaissance aircraft. Hardest hit were the transports, which lost 136 of their number out of the 542 involved in the operation (=25%)—another bloodletting, when one considers that these aircraft would also no longer be available for pilot training, either.

The enemy's loss in personnel—dead, wounded, and missing—totaled 17,750 men, more than three times the German casualty figure. Nine warships had been sunk, with numerous others damaged and put out of commission for several weeks or even months. The massive use of offensive air power had ultimately, despite several initial setbacks and the tough fight put up by the British and their Mediterranean fleet, been the key to capturing Crete. Indeed, the fleet's role as masters of the sea had soon been relegated by Ju 87 and Ju 88 dive bombing to operating only at night and, even then, under restricted conditions.

In capturing Crete, the Germans had undoubtedly won a position of great strategic importance. The question remained as to whether Germany's leaders would be able to make sufficient combat resources available to effectively utilize this strategic position and thus play an influential role in the battles of North Africa, where Rommel's counterattacks from 24 March to 15 April 1941 had thrown the British back to the Egyptian border where, for logistical reasons, the *Afrikakorps* was forced to halt. Once air elements had been diverted for the Eastern campaign on 21 June 1941, there remained for Germany's air war in the eastern and central Mediterranean, as well as in North Africa:

19 Ju 88 and Me 110 strategic reconnaissance aircraft
76 Ju 88 bombers
28 He 111 bombers (torpedoes and mines)
49 Ju 87R dive bombers

= 172 operational aircraft for offensive operations

and

25 Me 110 heavy fighters
12 Me 110 night fighters
49 Me 109 day fighters

= 86 operational aircraft for defensive operations, some of which could be used in the fighter-bomber role

i.e. a total of 258 operational aircraft, which at the beginning of the invasion of Russia were tied down in the South by this "second front," were also not available for the air war over England.

15. The Campaign in the East and its Effects on Air Armament

Course of the War on the Eastern Front to Winter 1941/1942

On 10 June 1941, ten days after the conclusion of operations in Crete, the buildup of air assets began for the campaign in the East. On 17 June Hitler gave the order to begin Operation Barbarossa on 22 June 1941. At 0315 hrs on that day the Germans launched their great offensive from the Baltic to the Carpathians with the goal of using advance armor wedges to cut off and destroy the Red Army, and prevent their retreat into the vast expanse of Russia. From a line running between Archangelsk, the Volga, and Astrakhan, the Luftwaffe would then eliminate the surviving industrial region in the Urals. The Red Army, initially comprised of some 2.5 million troops along the USSR's western frontier, was mostly caught by surprise by the roughly 3.25 million attacking Germans, although Stalin had been warned many times about a possible German preemptive invasion.

German bombers on the first day flew massive raids against about 60 Soviet airfields and various cities, during which (according to Soviet data) they destroyed over 1,200 enemy aircraft. In what had become a routine procedure, over the next days and weeks the attack spearheads were supported by dive bombers, strike planes, and fighter-bombers in rolling operations. This was according to instructions given to the Luftwaffe to make as much of its strength available for the invasion so that ground operations could be concluded quickly, without hindering the war against England—particularly the efforts to isolate the island nation.

The Luftwaffe was thus faced with the fateful situation of fighting a three-front war, one which over time would lead to its personnel strength—already suffering some wear and tear—being almost entirely consumed. Ignoring this, the ground operations proceeded apace with the full air support of close combat units, to include the total involvement of all operative bomber units. The intended surrounding of Soviet troops led to the pocket battles of Minsk and Bialystok, ending on 9 July 1941, with Smolensk falling on 5 August, Uman on 10 August, and Kiev on 26 September 1941, whereby more than 1.4 million Russians were taken prisoner.

Just three weeks after the invasion had begun, on 14 July 1941 a Hitler certain of victory ordered Germany's armament industry to again focus on construction for the air force and navy—to the detriment of the army; like the chief of the army's general staff, Hitler thought that the war in the East had already been won. On 19 July 1941 he issued Directive No. 33 for continuing the advance toward Moscow. The first German bombing raid on Moscow took place during the night of 21/22 July 1941 by 195 bombers, an event which was to be repeated over subsequent nights. On 2 October 1941 German troops launched their attack on Moscow (Operation Taifun). On 7 October 1941 Hitler forbade the acceptance of Moscow's tentative offer to surrender, and on 14 October the OKH ordered Moscow surrounded. This prompted Josef W. Stalin, chairman of the People's Commissars and commander-in-chief of the Soviet armed forces, to declare a state of siege on 19 October 1941. In October the first snows were already beginning to fall, and the muddy season was bogging the advance of ground forces. Helped by a light frost, it was not until 15 November 1941 that the second phase of the assault on Moscow could begin. On 28 November 1941 Germany's Moscow offensive reached the Volga-Moscow Canal and formed a temporary bridgehead at Dmitrov. On 5 December 1941 began the Soviet counteroffensive, launched from Moscow with the intent of encircling the German forces and destroying them. The Germans, within sight of Moscow, were forced to withdraw, as they were wholly underequipped for fighting a winter war of such Russian intensity. On 16 December 1941 Hitler forbade any type of operational retreat. He ordered the Army of the East to defend itself to the last man on 28 December 1941. By this time German troops had lost 25% of their original strength, while on the other hand a total of 3.35 million Soviet soldiers had become Germany's prisoners of war. But the USSR's sheer vastness of its human reservoir was just beginning to mobilize for the "Great Patriotic War of the Soviet Union," as the Central Committee of the CPSU had declared on 29 June 1941.

Setbacks in North Africa

The winds of change were blowing in North Africa, as well, this time against the German/Italian Axis powers. On 18 November 1941 the British launched a counterattack from along the Libyan/Egyptian border, and within seven weeks had advanced as far as El Agheila, southwest of Bengazi, which the British took on 16 December 1941. Their victory had been made possible in large part due to the fact that, with over 400 aircraft in the theater, they enjoyed a nearly two to one advantage over their German and Italian foes. The Germans had just roughly 50 fighters and heavy fighters, plus there were another 155 Italian fighters based in Libya. These superior numbers were only marginally offset by the arrival of the Me 109F-4 (Trop) in September, a variant superior to the then-current British fighters in nearly every respect, and the Italian MC 202 fighter, quite a good aircraft in its own right. Here, too, the establishment of air superiority played a crucial role in the outcome of ground battles. To make matters worse, Malta had taken delivery of 48 Hurricane fighters, flown off of two British carriers, on 21 May 1941, and from that point on increasingly evolved into the main source of nuisance for the German/Italian supply lines to North Africa. The situation is exemplified by the fact that, from October to December 1941, of the 123,000 BRT leaving German and Italian ports, only 44,500 BRT made it to Tripoli unscathed.

Japan and the United States

Hitler compounded what had become a most precarious situation when on 5 December 1941 he offered assistance to the Japanese in the event of war and signed a Japanese-proposed military agreement preventing a separate peace by either of the partners. With their backs thus covered and without any prior declaration of war, on 7 December 1941 the Japanese attacked Pearl Harbor, the main base of the U.S. Pacific Fleet in Hawaii, using carrier-based aircraft. They knocked 19 heavy warships out of action and destroyed 188 American airplanes for a loss of 29 of their own number.

On 11 December 1941 Germany also declared war on the United States, and the three powers of Germany, Italy, and Japan signed an agreement on a cooperative prosecution of the war, obligating each country "to not negotiate a cease fire or peace with the U.S. or England without full prior mutual agreement."

This, then, was the situation in late 1941, a situation into which Germany had maneuvered itself under a miscalculating leadership out of touch with reality.

Status of German Air Arms in Late 1941/Early 1942

Framework

The 11 September 1940 declaration mentioned earlier, which stated that developmental work on all projects not expected to be operational within a year was to be halted, led to the aviation industry being strictly prohibited from undertaking any kind of independent developmental work. The RLM effectively throttled development of new aircraft types and considered them of secondary importance, that once the war was over such development would not be productive. Ten months later, on 14 July 1941, the *Luftwaffe's* priorities had changed with the following directive (quoted here verbatim):

"Military dominance of European territory following the overthrow of Russia makes it possible to significantly reduce the size of the Army for the time being.... Armament focus is transferred to the Air Force, which is to be strengthened on a large scale."

But just two months later, in September of 1941—even before the *Führer's* decision had an effect on aerial rearmament—this prioritization directive became watered down by a new order. As a rule, the three branches of service were to pass their requests for development and acquisition to the chief of the OKW who, together with the recently appointed minister for armament and munitions *Dr.-Ing.* Fritz Todt, would review the requests for feasibility and, acting under Hitler's authority, decide whether and to what extent a contract would be awarded. However, the expanded aircraft acquisition program remained one of the focal points of this new directive.

These premises formed the framework in which the aviation industry's otherwise rocky development and output moved up to late 1941, a framework which was subject to a constant tug-of-war of overlapping and contradictory instructions and directives.

The "Göring Program" and the Technical Leadership Apparatus

The contractor, and therefore the guiding force behind the aviation industry in wartime, was the *Generalluftzeugmeister*. His of-

---

* Der Führer und Oberbefehlshaber der Wehrmacht, OKW WFSt/Abt. L (II Org.) Nr. 441219/41 gKdos vom 14. Juli 1941

** Der Führer und oberste Befehlshaber der Wehrmacht, Chef OKW, Nr. 340/41 gKdos vom 11 September 1941

fice, with 26 departments directly subordinate to him, had been responsible for issuing no less than 16 aircraft acquisition programs from the time the war began until mid-1941. Each of these generally took six to seven weeks before succumbing to the facts, i.e. the sights were initially set high before reality set in and they were lowered. Problems built up, and instead of the needed radical increase in air armament fully able to meet the needs of a three-front war, the output of the aircraft factories in 1941 either rose only slightly over that of the previous year or, in some cases, slipped backward. To be sure, in 1941 3,373 bombers were built (1940 = 2,852), which equated to an increase in production of about 18%. Yet with a 1941 output of 507 production of strike aircraft and dive bombers was almost 100 (16%) less than in 1940, while reconnaissance aircraft showed an increase of 108 (11%) over 1940 with 1,079 being manufactured. Not a particularly bright picture for offensive air power, which bore a heavy burden at this point and whose losses in 1941 were a major impact, one which could not be offset with such production numbers. On the other hand, fighter and heavy fighter production in 1941 jumped by 36% to 3,744, 998 more than had been produced the previous year. It should be noted that some of these fighters were configured as fighter-bombers and were used to strengthen close air support operations in support of the army, filling the gap left by a lack of dedicated strike planes. The campaign on the Eastern Front, which now could no longer be viewed as a *Blitzkrieg*, naturally required greater numbers of aircraft than before if the consolidated resistance and counterattack capability of the Red Army were to be broken after the winter of 1941 in the spring of 1942. Even before the *Luftwaffe's* official restructuring of priorities in July 1941 mentioned earlier, Göring had called for quadrupling the *Luftwaffe's* front-line strength and given a special assignment to his state secretary and *Generalinspekteur der Luftwaffe*, *Generalfeldmarschall* Milch, to investigate the feasibility of such a program. Together with *Generalluftzeugmeister* Udet, he was to explore what industrial capacity was available and what additional capacity would need to be created. *Generaloberst* Udet, who in 1941 was also Inspector of Night Fighters, saw no potential for a notable increase in aircraft production based on the current air armament framework. As a result, at the end of June 1941 Milch effected special authority from Göring, who authorized him to close down or impound factories, have additional facilities built for aircraft construction, and forcibly transfer workers. In addition, managing personnel in the industry could also be replaced or removed with no regard for legal private work contracts, all in order to bring about this so-called "Göring Program." Milch went to work immediately. He contracted to have three factories, each the size of the Volkswagen plant, to be built in Brünn, Graz, and Vienna for the manufacture of aircraft engines, the main bottleneck in aircraft production. These were "supplemental" factories, and were to be torn down after the war. He had the aluminum requirements of the aviation industry reassessed, and through strict controls was able to prevent the wasteful and non-associated use of aluminum, as well as gain access to the reserve supplies within the companies. Göring had given general director *Dr.* Koppenberg from the Junkers company special authority to double Norwegian aluminum production by mid-1942, and this authority was extended to include acquiring bauxite from other countries, as well. Milch also became involved in the matter of the so-called *Bomber "B,"* the replacement for the He 111 and Ju 88, and demanded a definitive decision. This Udet was not yet able to give, since the Ju 288 was the favorite choice for the program and was running into serious developmental problems—which will be discussed later. Into this problematic situation, on 18 August 1941 Milch presented an outline of his ideas to the industrial board concerning an increase in production of offensive aircraft as follows: monthly production must be increased from the current levels to:

300 Ju 88s, from 220 currently (the production goal which *Dr.* Koppenberg had been given back in 1939)
160 He 111s, from 100 currently
100 Do 217s, from 24 currently
156 Ju 87s, from 55 currently
140 Me 210s, from 90 currently, plus bring production of the He 177, just beginning, up to a monthly output of 120 aircraft.

This new program applied to two years, 1942 and 1943. During these years there was to be no new aircraft type going into production in any quantity, with the focus instead being on large-scale production of the types already proven. This direct involvement in the responsibilities of the *Generalluftzeugmeister* included Milch proposing organizational and personnel changes within that office, which Milch felt was not being led properly at the time. Compounding these problems was Messerschmitt's inability to develop the obsolescent Me 110 into the better performing Me 210 by the established deadlines—despite an intensive test program the Me 210 was not suited for production as a dive bomber and reconnaissance aircraft in addition to its original role as a heavy fighter, despite Messerschmitt having anticipated the type being available for front-

---

* "Der Industrierat des Reichsmarschalls für die Fertigung von Luftfahrtgerät" (Reichmarschall Industrial Council for Aircraft Production) was established by Göring on 14 May 1941. It was comprised of six reputable business leaders from the aviation industry, whose experience would be used for a planned increase in military aircraft production. The chairman was the Generallu*ftzeugmeister*, at the time *Generaloberst* Udet, with whose offices the industrial council was expected to work.

line service by mid-1941. All these issues led to additional pressing problems for the *Generalluftzeugmeister*, who saw his difficulties piling higher and higher. Another heated argument followed on 12 November 1941 between Milch, Udet, Messerschmitt, and other senior personnel regarding the ratio of Me 109 to Fw 190 output within the fighter production program. This was the last straw, and given his already tenuous state of mental health Ernst Udet soon felt himself unable to fulfill his duties any longer; he resigned from his post, and on 17 November 1941 committed suicide.

The era of Udet was at an end—but not the accumulating problems!

The Era of Milch

It is not within the scope of this book to examine and evaluate the details of personnel changes in senior positions, but within this context is it necessary to put the situation at the time into perspective, to mention the background and personalities who had a major influence on the development of the aircraft discussed within these pages, and who used their technology or made decisions based on that technology.

One of these key figures was undoubtedly *Generalfeldmarschall* Milch, who assumed Ernst Udet's responsibilities following the latter's death.

During the last months of Udet's time in office, Milch—under orders and with full authority—had become heavily involved in the organization and assignment of personnel to senior positions within the office of the *Generalluftzeugmeister*, something which Udet protested loudly against at first. To make the technical aspects of air armament easier to manage, Milch reintroduced interim review boards within the office and had them staffed by individuals he selected. When Milch became responsible for the office of *Generalluftzeugmeister* it was in addition to his position as State Secretary for Aviation and General Inspector of the *Luftwaffe*. On 8 January 1942 Göring lifted the OKW's restrictions from October 1941 and empowered Milch "to restrict all the *Luftwaffe's* acquisition plans to the main focus programs and align them with the requirements of the *Luftwaffe* General Staff. Submission of developmental plans to the OKW has been done away with."

In addition, Milch was tasked to review the *Luftwaffe's* development plans in light of their feasibility with regard to raw material and industrial situation at the time.

On 8 February 1942 *Dr.* Todt, the Minister for Armament and Munitions, was killed in a plane crash. With his replacement, it was expected that the rivalry of the three branches of service for prioritization of raw materials and industrial potential would vanish. Göring had hoped to use his "Four-Year-Plan" office for this, or possibly appoint Milch in Todt's stead. Hitler, however, had already decided upon the 36-year old *Dipl.-Ing.* Albert Speer, who became Minister for Armament and War Production. Milch soon became quite loyal to Speer and generally succeeded in establishing a good working relationship with him.

1941/1942 Aircraft Production and Subsequent Models

a) Bombers

Bomber "B" Prototypes

The Bomber "B" project was to be the follow-on to the Ju 88, and specifications for it were issued in July 1939, with the companies of Arado, Dornier, Focke-Wulf, and Junkers competing for the contract. From the outset, the RLM concentrated on the work of the Junkers company, which was developing a design for a twin-engined high altitude bomber with pressurized cockpit and the new liquid-cooled 24 cylinder Jumo 222 rated at 1,840 kW/2,500 hp. Several preliminary designs were drawn up by the design department under the direction of *Dipl.-Ing.* Heinrich Hertel, who had transferred in May 1939 from Heinkel and was now Junker's technical director and development manager. Finally, in December 1939 a showing of a pressurized crew compartment mockup took place, one which had undergone several changes in order to more closely match the ideas of the *Technisches Amt*. As a result of this, the RLM approved construction of a complete fuselage mockup with full instrumentation, control systems, armor, and periscope sights for the remote-controlled gun stations. Inspection of the mockup occurred on 29 May 1940, and a few days later Junkers was given an initial contract for building three prototypes under the designation Ju 288.

Junkers Ju 288

Fully expecting the RLM to provide the funding for the construction of the prototype Ju 288, in February 1940 (i.e. almost four months before the RLM contract was formally awarded) the Junkers company had already begun work on the prototype. With the intent of gaining as much flight experience as early as possible, even before the Ju 288's maiden flight, Junkers brought the Ju 88V2 and V5 into the test program. These were used to evaluate certain design features and had been kitted out with two complete Ju 288 cockpits and the dual rudders planned for the Ju 288. The engines, however, were still the Jumo 211 powerplants. Flight testing of these two modified Ju 88 prototypes began in the late spring of 1940 under pilots Holzbauer, Joop, and Preuschen, the latter two being company test pilots. Despite being underpowered, the results were generally acceptable. Another Ju 88 test airframe underwent static tests in the fall of 1940, until it was intentionally destroyed as part of the stress testing. Following rigorous ground trials, the Ju 288V1 took to the air for the first time in late January 1941. Since the Jumo 222 was not yet certified for standard flight operations—this was not expected until 1942 now—the Ju 288V1 flew with the 14 cylinder BMW 801 radial engine, which had a slightly larger frontal diameter than the Jumo 222, and at 1,174 kW/1,600 hp provided 662 kW/900 hp less power than planned for the production aircraft.

In early spring there followed the Ju 288V2, which had the hydraulic dive brakes layered onto the landing flaps, while the V1

Junkers Ju 288V2, spring 1941. The "swallow's nest" for the remotely operated gun barbette on the cockpit sidewall has not yet been fitted with a gun. Neither the V1 nor the V2 turned in satisfactory flight handling characteristics, forcing the designers to lengthen the fuselage by about one meter. Both prototypes suffered accidents due to collapsed landing gear, as the V2 shows in the lower photo.

Junkers Ju 288V1, provisionally fitted with the BMW 801, since the more powerful Jumo 222 had not yet been certified for flight operations.

and the Ju 288V3 appearing in the early summer of 1941 both still retained the split-type dive brakes. Since these three prototypes flew without armament or payload, the underpowered engines were adequate for evaluating the general flight characteristics during the early stages of the test program.

During this time additional test prototypes were approved, the first of which was the Ju 288V4 powered by the BMW 801MA but lacking dive brakes. On one of the first test flights the port engine caught fire on final approach to the runway, and the flames spread so rapidly that as the airplane was taxiing to a stop the remaining fuel in the forward tank ignited, breaking the cockpit section away from the fuselage. Despite the severe damage the V4 was completely repaired and resumed flight testing in late November 1941.

In mid-July 1941 work slowed down on construction of the Ju 288V5 in anticipation of the availability of the first Jumo 222A/B engines. During the interim the payload capacity specifications for the aircraft had been increased, leading to an increase in the wingspan and wing area. The Ju 288 was planned to have a span of 22.0 m and an area of 60 m², with the A-1 sub-variant having large cheek gun stations without periscope sights, while the A-2 sub-variant would have flatter profile gun positions remotely controlled with periscope sights. Both versions, the Ju 288A-1 and the Ju 288A-2, were designed for a three-man crew and were to have a maximum takeoff weight of 18.5 metric tons with a bomb load of 5.0 metric tons. This would have meant a significant jump in bomb capacity compared to the medium bombers of the day. Three prototypes were started for the intended A series, the Ju 288V6, V7, and V8. New ailerons were designed, along with the increased span wings, and these were tested in wind tunnels in March 1941. It was planned that Junkers would eventually reach a production output of 80 Ju 288s per month; the Arado, ATG, Dornier, Heinkel, Henschel, and Siebel companies would convert their production lines to the type and would have an additional combined output of 300 per month. But these plans fell victim to a reassessment by the General Staff and the *Technisches Amt*, which now called for a four-man crew. As this necessitated considerable redesign work, plans for the construction of the Ju 288A were abandoned in favor of a Ju 288B with a larger, four-seat cockpit and fully redesigned wings of even greater span and area.

The 1 1/2 year plus delay with the delivery of the first Jumo 222 certified for flight meant that the Ju 288V5 did not complete its maiden flight until 8 October 1941, with the Ju 288V6 (also powered by the new engine) following within the month. Whereas the V5 still had the original wing shape of the V1, the V6 was given the new larger wings with a span of 22.67 m and an area of 64.7 m².

Junkers Ju 288 V-series with Jumo 222A/B, which was made available for the four-bladed VS 7 propeller used on the V5, V6, V8, V12, and V14. The Ju 288V5 completed its first flight on 8 October 1941. This engine layout permitted the cooling air to flow through to the annular alloy radiators via the large tunnel-like propeller spinners—so-called double spinners—giving the engine nacelles a cleaner design shape. The only one of these prototypes to make use of a standard cowling was the Ju 288V9, also fitted with the Jumo 222A/B. Remaining prototypes in the V-series received the BMW 801 or 801J with exhaust turbocharger, or the DB 606A/B. Planned for the A-1 series, the cheek barbette (again without its gun) is clearly visible.

The Ju 288V7, which was not available for trials until the spring of 1942, was virtually identical to the V6, but due to a shortage of additional Jumo 222s was fitted with the BMW 801C engines and accordingly suffered somewhat in performance. In addition, it had a serious impact on flight testing when an engine fire forced it to be temporarily pulled from the program. During repairs it was given a larger tail similar to the one fitted to the Ju 288V8, which was again powered by the Jumo 222 and had begun flight testing shortly after the V7. It was planned as the prototype for the Ju 288B, on which Junkers had planned starting production of as early as September of 1941.

The specifications prepared by Junkers and accepted by the *Technisches Amt* for the Ju 288 called for: aft and center fuselage sections as for the A-series, joined to a somewhat bulbous-looking new forward section, and the reworked and enlarged wings as tested on the Ju 288V6. In order to keep the normal and maximum weights to 17.7 and 18.6 metric tons, respectively, despite the marked increase in the aircraft's empty and ground weight, the maximum bomb load was limited to 3,600 kg. For armament, it was planned to make use of remote-controlled gun stations sighted with periscopes located on the dorsal and ventral fuselage areas. These included:

- the *A-Stand* under the tip of the cockpit, also called the chin position.
- the *B-Stand* just behind the pressurized cockpit. Both stations were to have two 13 mm MG 131 twin guns each.
- the rear position, also with a remotely-operated gun, either
  - an MG 131 twin-barreled machine gun, standard, or
  - an MG 151 (15 mm), or
  - an MG 151/20 (20 mm)

For this version the RLM awarded Junkers with a contract to begin production of the Ju 288B starting in March 1942. However, shortly afterward this agreement was modified so that production would initially be limited to 35 aircraft per month until the matter with the engine had been solved; the optimistic acceptance of production for the Jumo 222 had again proven to be unrealistic.

Another prototype for the B-series was the Ju 288V9, which flew (also with the Jumo 222, but lacking the ducted propellers) in May 1942. This type had the gun stations, but was not fitted with the guns themselves. The next prototype was the Ju 288V11, which was test fitted with the liquid-cooled Daimler-Benz DB 606A/B 24-cylinder engine. Despite its higher weight and the larger dimensions of the DB 606 it was considered a viable alternative to the Jumo 222, since the DB was able to offset these disadvantages with its greater power (1,986 kW/2,700 hp on takeoff and 1,949 kW/2,650 hp at 4,800 m.); as a comparison, the Jumo 222 only provided a maximum of 1,618 kW/2,200 hp at 5,000 m. The Ju 288V11 began its ground testing in May 1942 after Junkers had been ordered the previous month to begin production of the Ju 288 within ten months, in February 1943. But the type was to be built with the DB 606 engine fitted or, if ready in time, with the DB 610 double engine (2,170 kW/2,950 hp), since all hope for the Jumo 222 entering production had effectively been abandoned by this time. Even before the RLM had given Junkers this directive, Prof. Heinrich Hertel had already carried out major changes to the airframe intended to take the DB engine. These changes included stretching the cockpit by 33 cm and improving its instrumentation, as well as structural strengthening and adding a *C-Stand* behind the bomb bay. Thus modified, the Ju 288 was given the designation Ju 288C, so that the intended B-series suffered the same fate as the A-series and never went into production.

The first prototype of the planned C-series was the Ju 288V101, completed with the DB 606A/B in August 1942. The Ju 188V102 followed just a few weeks later and was powered by the same engine.

This somewhat detailed history of the Ju 288 highlights the difficulties under which the aviation industry had to struggle. On the one hand there were the inconsistencies on the part of the contractor (the RLM in this case), and on the other hand the imponderability with regard to the development of new, high-performance engines. Also, it seemed that the way to developing an optimally designed airframe in the hope of having the powerplant available in time was not always the smoothest path to achieving production maturity for an urgently needed operational type.

It should be mentioned here that there were four companies competing with Junkers for the most suitable Bomber "B": Arado, Dornier, Focke-Wulf, and later—due to experience with building pressurized cockpits—Henschel. After submitting their proposals in July 1940 the *Technisches Amt* eliminated Arado's Ar 340 from the competition, while it issued contracts to Dornier for the Do 317, to Focke-Wulf for the Fw 191, and to Henschel for the Hs 130, in addition to Junkers for its Ju 288.

Dornier Do 317

Six prototypes of the Dornier Do 317 were contracted for, with the Do 317V1 flying for the first time on 8 September 1943—some two and a half years after the Ju 288V1 had begun its flight test program. Compared to the Do 217, the Do 317's dimensions were somewhat larger, and it had a more spacious fuselage. It was powered by the DB 603 engine (delivering 1,287 kW/1,750 hp on take-off) driving a four-bladed VDM variable pitch propeller with a diameter of 4.3 m. The twin vertical stabilizers were triangular in shape.

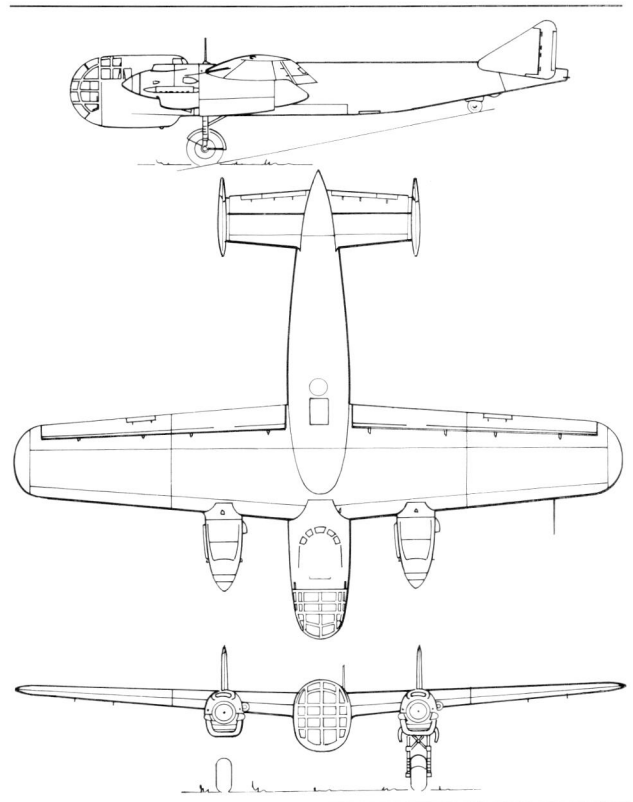

Dornier Do 317V1 (1943).

This prototype was to have been the pattern for the Do 317A series, and was to have been followed by a prototype for the Do 317B with pressurized cockpit, the DB 610 double engine, and a wingspan increased to 26 m. But when it was discovered that the Do 317's flight performance was little better than that of the Do 217P-0 (examined later on in the book), the whole program was canceled in favor of the Ju 288. The remaining five Do 317 prototypes were built without pressurized cockpit, specially modified for using guided weapons against shipping targets and delivered to the front lines under the designation Do 217R.

The Focke-Wulf Fw 191

The Focke-Wulf Fw 191 was a serious competitor to the Ju 288, and initially two prototypes were built under the direction of *Dipl.-*

Ing. Kosel. The first of these, Fw 191V1, entered flight testing in early 1942—roughly one year after the Ju 288V1—under the supervision of *Dipl.-Ing.* Mehlhorn, and the Fw 191V2 followed a short time later.

In this case, too, the intended Jumo 222, or Daimler-Benz DB 604 (1,835 kW/2,500 hp) powerplants were not yet available, and both the V1 and V2 were fitted with the much less powerful BMW 801MA (1,015 kW/1,380 hp) engines. The unusual feature of the Fw 191 was its complete reliance on electrical circuitry. At the behest of the RLM's research department, all systems or components normally driven hydraulically or mechanically were operated by electro-servomotors, so that the Fw 191's nickname of "The Flying Powerplant" seems not altogether off the mark. However, electrical short circuits soon became so common that the test program had to be broken off after just ten flying hours on both airframes. The electrical problems were compounded by difficulties with the four-piece, combined landing flap/dive brake design, the so-called Multhopp flaps, which was the source of major flutter problems when extended. Further prototype construction was halted until the reliability of the electrical system had been proven, the flutter problem eliminated, and the more powerful engines became available. At this time there were three additional prototypes in an advanced stage of construction. These were to have been given the Jumo 222, as well, but because of the problems with supply mentioned earlier, in late 1941 it was decided to venture using the DB 606 or DB 610 double engines in place of the Jumo 222. These powerplants were made up of two DB 601 or DB 605 engines side-by-side driving a single propeller via a single reduction gear. The problems associated with such designs have already been addressed in the section dealing with the He 177.

Eventually Focke-Wulf was able to obtain a pair of pre-production Jumo 222s for testing the Fw 191. After repeated demonstrations by *Dipl.-Ing.* Kosel showing that the problematic parts of the electronic on-board systems could be replaced by standard hydraulic systems, the RLM finally gave its approval for the changes in late 1942. A fourth prototype, the V6, which had not yet been built with the electrical system installed, was fitted with both the hydraulic system, as well as the available Jumo 222s. The Fw 191V6 successfully completed its first flight in the spring of 1943 with Focke-Wulf's chief test pilot, Hans Sander, at the controls. Subsequent flight testing at Delmenhorst, however, was not entirely satisfactory. The V6 was to have spawned the Fw 191B series, as well as an Fw 191C with four single engines and lacking pressurized cockpit and remotely-operated gun stations, but both of these designs never left the drawing board. In June 1943 increasing material shortages and the priority of the fighter program not only led to cancellation of further development of the Fw 191, but also of the entire Bomber "B" program.

The fate of the longest running and therefore most promising Bomber "B" project, the Ju 288, will be examined again briefly in another section. Junkers continued working with the Ju 288 after the cancellation of the Bomber "B" program in order to gather data for more advanced developments in the areas of high altitude bomber and reconnaissance aircraft.

Henschel Hs 130

As early as 1940 Henschel had been test flying a high altitude reconnaissance aircraft design (favored by Rowehl), a design which was in turn based on the quite successful Hs 128 high altitude research plane, of which two examples had been built in 1939. This reconnaissance platform was the Hs 130A, and served as the basis for Henschel's proposal. Yet, like the Hs 130A the Henschel project for the Bomber "B" never made it to production, despite three Bomber "B" Hs 130C prototypes having entered flight testing in late 1942/early 1943, and giving good reason to hope that this design would make an effective high altitude bomber. As late as the summer of 1943 there was talk of initially producing 100 to 130 examples of the Hs 130C, but with the ultimate abandonment of the entire Bomber "B" concept that same year the writing on the wall came too late even for this development.

The so-called *"Höhenzentrale,"* or high altitude turbocharger system (HZ), fitted to the markedly improved Hs 130E testbed, will be looked at more closely in the section dealing with the Do 217P. The HZ system installed in the first three prototypes of this three-seater provided the Hs 130E with excellent high altitude per-

Henschel Hs 130E-0, three-seat high-altitude reconnaissance platform. First flight May 1943.

formance, at altitudes of 14,500 m, and led to a contract for four pre-production versions of the Hs 130E-0. The first of these pre-production high altitude reconnaissance platforms took to the air with the HZ system beginning in May of 1943. As the high altitude bomber variant was capable of carrying two 1,800 kg bombs on external racks in place of the drop tanks, the RLM even envisioned the construction of 100 Hs 130E-1s—for the other Bomber "B" competitors were having their problems, as well!

However, constant difficulties with the HZ system—not yet fully developed—led to the Henschel contract being reduced in stages, so that by the end of 1943 the number had dropped to just 30 aircraft. When further development of the problematic HZ system stopped in 1944, the Henschel high altitude reconnaissance program also fell by the wayside, so that not a single one of the promising Hs 130 prototype and pre-production aircraft ever went on to full-scale production.

Old Workhorses and New Disappointments with the Heinkel He 177

Among the front-line bombers the Ju 88 led the pack, with a total of 2,146 examples of the bomber version alone being built in 1941. This was 330 more than had been produced in 1940, while 950 He 111s were built in 1941, as well, a rise of 196 over the production output of 1940. At 277, the Do 217's numbers remained roughly the same as the 260 Do 17s built at the end of their production cycle, plus the 20 Do 217s replacing them in 1940. The Fw 200, used as a long range bomber and maritime reconnaissance platform, totaled 58 for 1941, 22 more than in 1940. When the problems with the Bomber "B" pushed the modernization of the medium bomber fleet well into the future, Milch's emphasis on those front-line aircraft which he had postulated before the industrial council on 18 August 1941 generally held for the manufacturing practices of 1941. Only the He 177 lagged well behind its expectations. A brief report bringing the reader up to date following the earlier discussion of the He 177 V1 through V5 prototypes is therefore appropriate at this point.

The He 177V6 and V7 began combat trials on 2 August in the role of anti-shipping, being assigned to the same unit operating the Fw 200 on the French Atlantic coast. However, continuing technical problems, breakdowns, and modification work meant that these two aircraft seldom saw action.

In September 1941 the He 177V8 was ready for testing, but was assigned to the Rechlin Test Center for engine trials. Other evaluation priorities caused the aircraft to be returned to Heinkel after just forty days; it did not resume its interrupted engine testing at Rechlin until February 1942. The V8 was the last of the original He 177 prototype series, with other prototypes being pulled as needed from the null-series or pre-production series, which had begun in late 1940.

The first of the null-series was the He 177A-01, flying for the first time in November 1941 at Rostock-Marienehe, where it conducted weapons testing—albeit with unsatisfactory results. Before modifications could be made to the inadequate defensive armament, both of the airplane's engines caught fire on takeoff, making it a total write-off.

The second null-series aircraft, the He 177A-02, began its flight testing on 8 February 1942, but in May of that year an engine caught fire and the plane crash-landed. The crew was able to escape just seconds before the aircraft blew up. Thus, even in 1942 it was impossible to predict when or whether this heavy bomber would be ready for combat.

After the sobering experiences with both the Bomber "B" and the Heinkel He 177, it became necessary to return to those bombers either entering or still in production. A performance boost of the available air attack potential could only be achieved gradually by the use of more powerful engines, something which had achieved limited success in 1941. The Junkers Flugmotorenwerke succeeded in taking its proven Jumo 211 and building an F-series with an output of 990 kW/1,350 hp. Further development led to its J-series, rated at 1,040 kW/1,420 hp. These engines were prioritized for the Heinkel He 111 and the Junkers Ju 88, and resulted in measurable performance improvements for these two workhorses, mainly with the Ju 88's weak area of single engine flight—something which often determined whether a crew would return or be lost after being hit by AAA or attacked by enemy fighters in combat. The first bombers powered by these engines were the He 111H-6 with the Jumo 211F and the Ju 88A-4 with the Jumo 211J, both of which entered front-line service in 1941.

Heinkel He 111H-6

In addition to fitting the more powerful Jumo 211F-1, whose 956kW/ 1,300 hp drove a fully automatic Junkers VS 11 variable-pitch three-bladed wooden propeller, the defensive armament of the H-6 was also improved. This version was produced in large numbers for a crew of five, and had five MG 15s, as well as a 20 mm MG-FF in its A-Stand. Some even had an MG 17 firing from the base of the tail, remotely controlled by the radio operator from the B-Stand. It was designed to cover the dead zone between the firing field of the radio operator and that of the rear gunner in the ventral gondola. In place of the MG 17 several H-6 aircraft carried a *Störkörperausbringungsvorrichtung* (SKAV), which would deploy tiny charges to prevent an enemy fighter from lining up in a firing position. These nuisance devices (Störkörper, or SK) were deployed on small parachutes and had a timer, which would detonate a charge weighing 0.3 kg after a period of about two seconds. However, they were of limited success, as they turned out to be little more than a harmless scare tactic, and the majority of front-line units considered them of little value. Normally the H-6 was equipped with external racks for carrying a 2,000 kg bomb load or two LT 5b aerial torpedoes (each weighing 765 kg), or two LMA or LMB aerial mines (500 kg each). Field conversion kits, or *Rüstsätze*, enabled several other weapons/payload combinations, and the field units could normally fit these kits with little difficulty. The performance and flight characteristics of the He 111H-6 in single engine flight finally showed a high degree of flight safety, something which crews up to that point were generally not able to rely upon. The 14,500 kg maximum takeoff weight using RATO packs continued to be the upper limit for the He 111H-6; its improved performance with the more powerful engines—the Jumo 211F-2 rated at 990 kW/1,350 hp would appear somewhat later—mainly revealed itself in improved speed and climb rates, as well as its service ceiling.

The performance values for an He 111H-6 powered by the Jumo 211F-1 follow. Note that the speed values refer to the aircraft in an unloaded state with half-fuel, while those in parentheses are for an aircraft at maximum load.

After the He 111H-3 the H-6 series was the second real large-scale production variant of this remarkably robust and reliable bomber; all other interim series before and after these two standard versions were generally small production batches for special roles.

Heinkel He 111H-6 (from 1941). This variant had a 20 mm MG-FF in the C1-*Stand* (lower forward gun position, just behind the SC 1000 bomb).

Performance Data for the He 111H-6

Maximum speed (km/h)
- at sea level   365 (349)
- at 2,000 m   399 (379)
- at 4,000 m   410 (389)
- at 6,000 m   434 (405)

Range (with maximum load)
- at sea level   @ 329 km/h = 1,950 km
- at 4,000 m   @ 370 km/h = 1,930 km
- at 5,000 m   @ 384 km/h = 2,060 km

Time to climb (with maximum load)
- to 2,000 m   8.5 min
- to 4,000 m   23.5 min
- to 6,000 m   42.0 min

Service ceiling
- with maximum bomb load   6,700 m
- unloaded with half-fuel   8,500 m

Weights
- empty (with permanent and removable equipment)   8,680 kg
- maximum takeoff weight   14,000 kg
- maximum takeoff weight with RATO   14,500 kg

The He 111H-6 gradually began stocking front-line units starting in May 1941, and over time replaced the older models.

Junkers Ju 88A-4

Driven by the more powerful Jumo 211J, the Ju 88A-4 was the Ju 88 variant produced in greatest numbers, and was also assigned to front-line units beginning in 1941. Data for the Ju 88A-5 has already been provided earlier in the book, and the following data for the A-4 is listed only in those instances where it differs from the A-5.

Airframe
Fuselage: as A-5
Undercarriage: as A-5, but with retractable shock-absorbed tailwheel, free swiveling non-locking
Control surfaces
a) split ailerons, trim tab on left aileron
b) flaps as on A-5
c) horizontal stabilizer as on A-5
d) vertical stabilizer as on A-5

Wings: as on A-5
Dive brakes and automatic recovery system: as on A-5
Heating and de-icing: as on A-5
Hydraulic system: as on A-5

Powerplant
Engine: two Jumo 211J with rich-lean mixture controls, 1: 1.833 reduction, takeoff rating for each was 986 kW/1,340 hp at 2,600 rpm (maximum 2,666 rpm) at 1.40 *ata* (= 1.37 bar). Automatic switching of turbocharger from low to high altitude at 3,000 m. Radiator: heavy metal, later alloy, annular radiator in front of each engine.
Propellers: fully automatic, hydraulic three-blade Junkers VS 11 wooden type. 3.6 m diameter. Selectable blade position via rpm selection between 1,800 and 2,600 rpm using throttle. Standard position 25°. Feathered position selected electro-hydraulically.
Fuel tanks and fuel: same as on A-5
Oil tanks, oil and oil cooler: same as on A-5, but heavy metal cooler

Armament
A-Stand: semi-flexible MG 81 with 750 rounds
B-Stand: 2x MG 81i in bubble mounts with 1,800 rounds
C-Stand: MG 81Z (*Zwilling*) in a small bubble mount with 1,800 rounds
Total rounds carried: 4,350.

Bomb Racks and Sighting Systems
2x ETC 500/IXb(M 14) external racks under the wings plus 2x Träger 1000 (M 3) in addition to the potential for fitting two more ETC (500) racks in the vicinity of the ailerons, for an optional load of:
- six 500 or six 250 kg bombs
- 1x 1,800 or 2x 1,400 or 2x 1,000 or 2x 500 kg bombs
Sighting system: same as on the A-5

Radio Systems
Same as on A-5, but with additional FuG 16 on-board radio system (linked to FuG 10)
FuG 25 IFF system

Other Features
Automatic pilot: same as A-5
RATO system: same as on A-5, with additional option for smoke generating system
Single engine flight: as for A-5, but possible with a flying weight of 10.5 metric tons at combat power setting (2,400 rpm at 1.25 *ata* for the low-altitude turbocharger setting) with the right engine running, and up to 10.3 metric tons flying weight with the left engine running. In both cases the radiator gills on the damaged engine must be closed. In order to achieve these critical weight numbers the bombs must be jettisoned, as with all bomb racks, both fuselage tanks drained, and the sight as well as all non-essential armor, ammunition, and armament thrown out. Indicated speed when flying on one engine: 240 km/h.

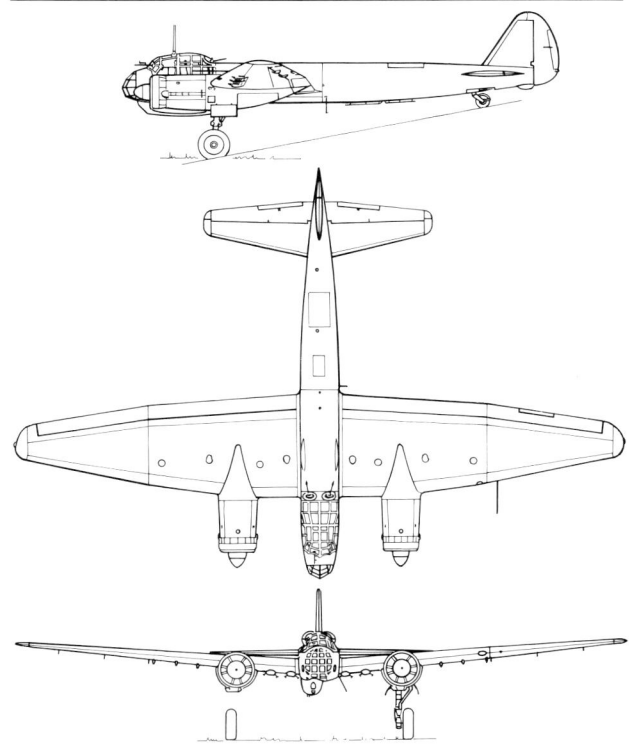

Junkers Ju 88 A-4 (ab 1941).
Junkers Ju 88A-4 (from 1941).

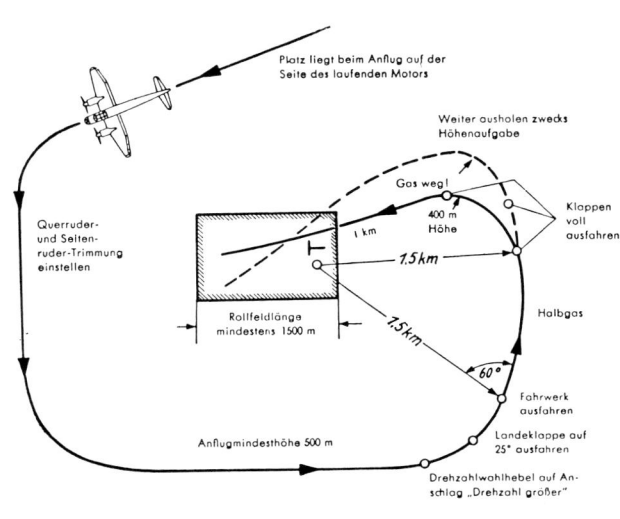

Junkers Ju 88A-4 diagram showing the proper way to make a single-engine landing.

The pilot's handbook for the Ju 88 reveals that flying on a single engine required not only concentration and good piloting skills, but that landing on a single engine called for an ingrained mastery of landing procedures. In the Ju 88A-4, there was no opportunity for a missed approach go-around when flying on one engine with the landing gear down or with full flaps. Making a safe single-engine landing rested solely with the pilot and his ability to summon up his experience and control the airplane.

Dimensions, Weights, and Performance	
Wingspan	20.0 m
Other values as the A-5, but takeoff weight	13.75 metric tons
Maximum permissible landing weight (no bombs, fuselage tanks empty)	
- normal	11.0 metric tons
- overloaded	12.0 metric tons
Maximum permissible glide and dive descent (with wing bombs and empty fuselage tanks)	11.7 metric tons
Maximum wing loading	250 kg/m²
Maximum speeds	
- outbound with initial weight of 13.75 metric tons (with 2x SD 1000 on outboard wing sections) at 5,000 m altitude	408 km/h
- return with initial weight of 10.8 metric tons (no bombs) at 5,500 m	453 km/h
- at 60° dive angle (with dive brakes deployed)	575 km/h
- other speed limitations: as for A-5	

Ranges

The Ju 88A-4's maximum range in standard configuration with four external racks was 2,030 km with 3,100 liters usable fuel* at an altitude of 4,000 m, with the turbocharger set at the low-altitude stage. This assumed an approach to the target, including an initial climb and dive, with an average weight of 13 metric tons (with two SD 1000 bombs) and an outbound leg of 1,020 km at 330 km/h true airspeed. The 1,010 km return leg involved a second climb out and descent at an average weight of 10 metric tons at a true airspeed of 350 km/h. For all other altitude and performance variables the potential ranges and times were in many cases much less.

---

*   Of the total 3,580 liters carried by the Ju 88A-4, 480 liters were reserved for other purposes: 290 liters for maintaining trim, 50 liters for warm-up and taxiing, approx. 20 liters for landing approach from 4,000 m (giving 2 1/2 minutes' flight time), 60 liters for a one-time missed approach, and 60 liters minimum remainder.

Opposite: The RLM type sheet for the Junkers Ju 88A-4 dated 1 January 1942 shows the differences between the Ju 88A-5 and A-4 series in the lower section, and points out the planned follow-on developments for the Ju 88E variant with its new cockpit.

Below: Junkers Ju 88A-4 chart showing flight performance.

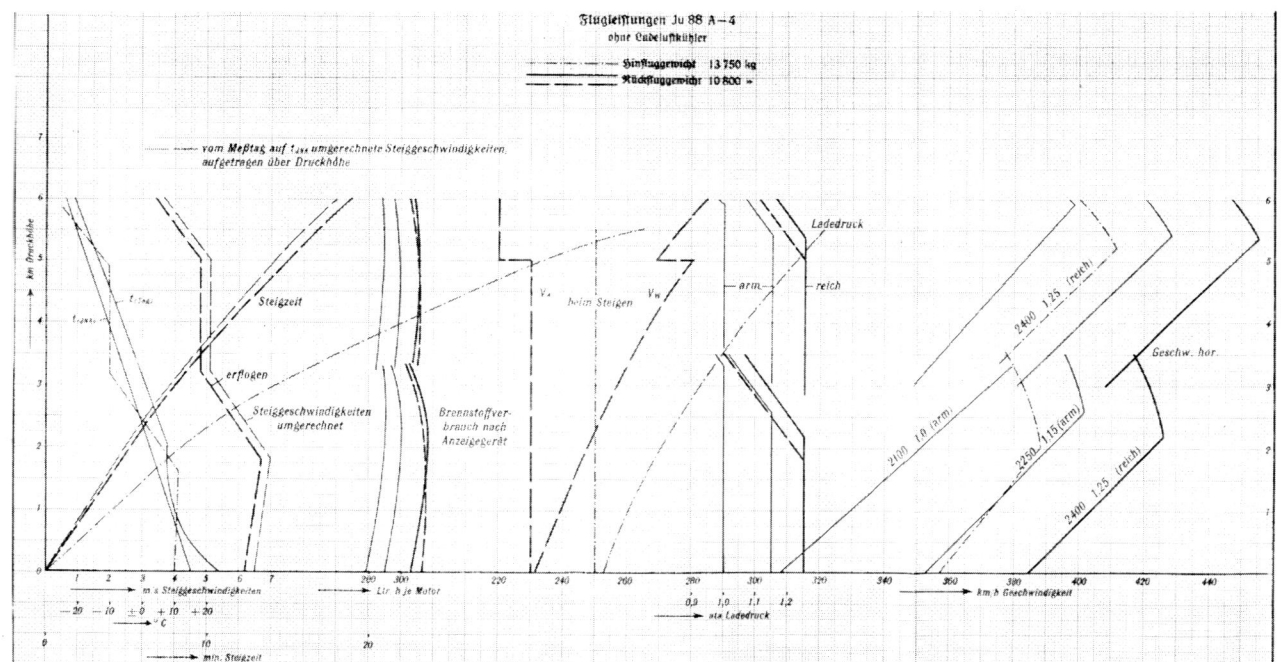

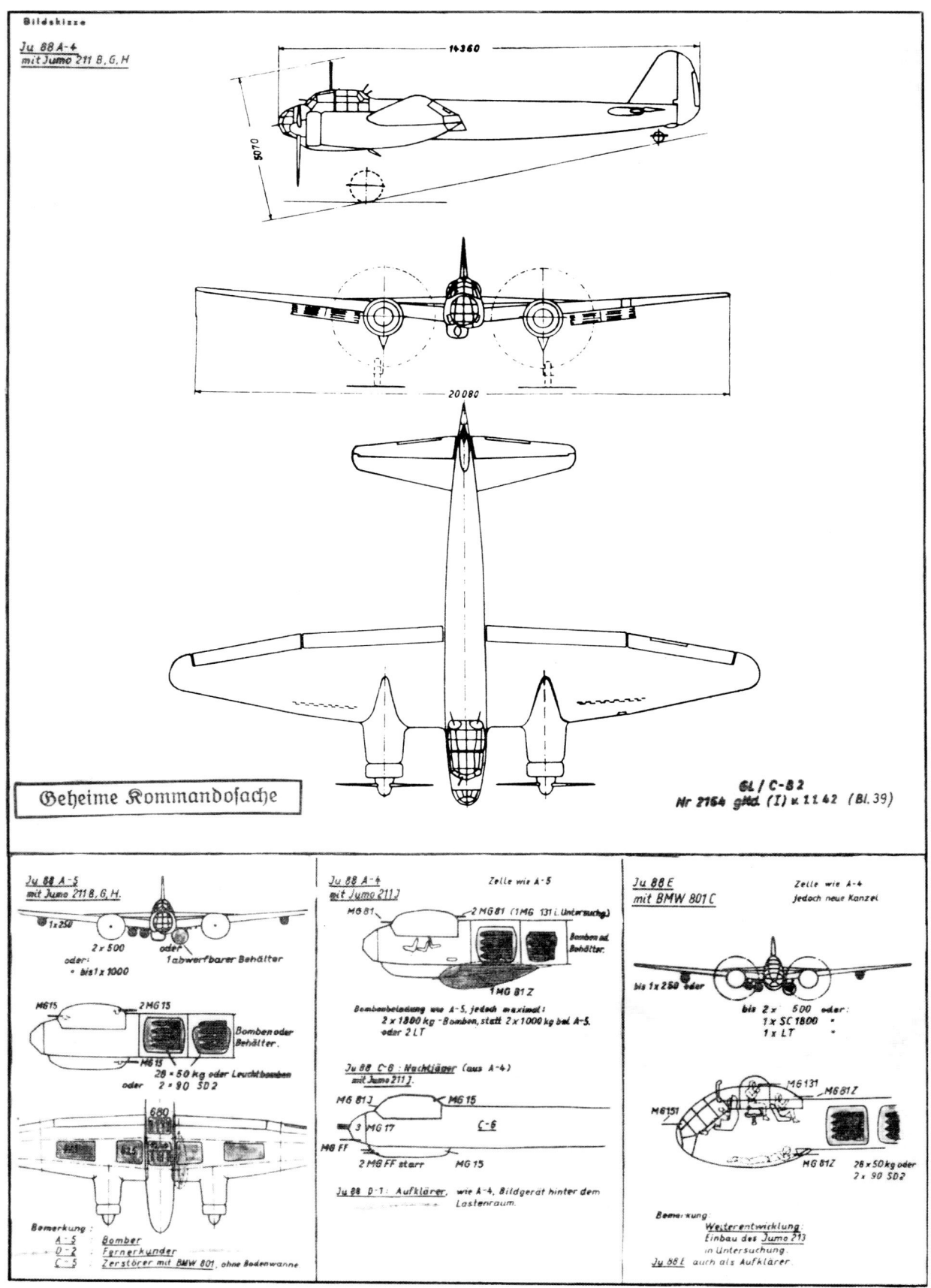

If the A-4 had just one external bomb rack instead of the standard four, it could attain a range of 2,525 km carrying one SD 1000 with the same mission profile with full tanks. This meant that with half the bomb load it was able to fly almost 500 km more. When loaded with just one 500 kg bomb and an additional drop tank, a total fuel capacity of 4,480 liters gave the A-4 a maximum range of 3,150 km (in this case the outward leg would be flown at an altitude of 2,500 m, the return leg at 4,000 m).

Assessment

With the A-4, the Ju 88 had reached a mature state in its developmental cycle. Powered by the Jumo 211J it was noticeably better performance-wise than the A-5, i.e. it was faster and, even more importantly, was easier to fly. To be sure, single engine flight continued to pose difficulties, but it was much less problematic than with the A-5. Armament had been significantly improved with the rapid-firing MG 81 (1,600 rounds per minute, or 3,200 for the MG 81Z), and on-board ammunition had been doubled. By having the option of replacing the standard four bomb racks with up to six, its bomb capacity had been increased to a theoretical maximum of 5,000 kg (in this operationally unrealistic case it was only possible by reducing fuel capacity to a minimum, since the 13.75 metric ton takeoff weight was not to be exceeded, i.e. the end result was a significantly reduced range). Two 1,800 kg bombs, the heaviest in Germany's arsenal, could also be carried by reducing the fuel capacity to 1,000 liters. Compared to the He 111H-6 the Ju 88 was undoubtedly the better performer, although this was not as obvious as is generally assumed. The A-4's trump card, however, was its dive bombing capability, something the He 111 lacked. Despite these different applications, both aircraft fell under acquisition group H 3, which had only been expanded for the Ju 88's dive bombing capability with the inclusion of the caveat "with special ordnance carrying capabilities for glide bombers." This was a cautious way of avoiding reference to the Ju 88 dive bomber, instead calling it a "glide bomber," which in practice was optimized for a diving angle of 60°.

b) From the Messerschmitt Me 210 High Speed Bomber and Recce Plane to the Me 410

The Me 210V1 completed its first flight on 2 September 1939 with *Dr.-Ing.* Hermann Wurster at the controls. The Me 210 was seen as an improved successor to the Me 110. However, in line with the earlier mentioned—and abandoned—idea of the "*Kampfzerstörer,*" it would assume the roles of a high-speed reconnaissance aircraft and a high-speed dive bomber and strike plane

Messerschmitt Me 210 of *Versuchsstaffel 210*, where front-line testing of this aircraft took place. Intended to be a jack-of-all-trades, it was expected to be produced as a high-speed bomber capable of making dive-bombing attacks, as well as a strike plane and a reconnaissance aircraft. In any event, only one *Zerstörergruppe* was fitted out with the Me 210, and this only briefly in 1942 in the Mediterranean theater.

in addition to its main role as a heavy fighter, thus making it heir to the Ju 87 as well.

From the outset, the Me 210 proved to be highly unstable in both the yaw and roll axes, with a fatal tendency to spin, leading to numerous control surface and other modifications. Its evaluation and developmental cycle as a heavy fighter is described elsewhere, and we will only look at the type insofar as it relates to the areas of bombers and reconnaissance aircraft.

37 aircraft were used during the two and a half years the Me 210 was in development and testing. 16 of these were prototypes, eight came from the A-0 series, and 13 from the A-1 series. A special commission recommended in January 1942 that the type be dropped from production and manufacture continue with the less powerful but untroublesome Me 110. Yet on 14 March 1942 Messerschmitt finally succeeded in demonstrating an aircraft that showed acceptable flight handling in further testing when the company rolled out the Me 210V17 with a new aft fuselage section. Göring had officially ordered all construction to be stopped on the Me 210 on 14 April 1942, and this went into effect on 25 April of that year, so that all work, including that of the sub-contractors, came to a standstill on that date. However, the order was eventually lifted, and—now powered by the better performing DB 603 engine (1,200 kW/1,625 hp on takeoff)—the improved Me 210 now be-

came known as the Me 410. This was primarily considered a high-speed bomber with horizontal bomb racks in the fuselage and began its flight testing with the Me 410V1 in the fall of 1942.

Hungarian Interest and Realization-

On the basis of an arms assistance agreement signed with Germany in June 1941, the Hungarians were able to obtain a manufacturing license for the Me 210. They built a new factory in Horthyliget, the Donau Flugzeugwerk, which the Messerschmitt company provided with the necessary jigs and machine equipment for building the Me 210. At the time Me 210 production was halted in Germany, preparations for license building in Hungary were at such an advanced stage that switching to a new aircraft type would have incurred severe delays. The Hungarians therefore decided to stick to producing the Me 210, but with more powerful engines and the structural improvements of the Me 210V17. They equipped their Me 210s with the DB 605B (1,085 kW/1,475 hp), built under license by the Weiß company in Budapest. With Messerschmitt's approval the new type was designated the Me 210C and had the new aft fuselage section of the Me 210V17, along with its leading edge slats. It was to be built in two versions: the Me 210Ca-1, a combined heavy fighter and dive bomber, and the Me 210C-1, with a primary role of long range reconnaissance and a secondary role as heavy fighter.

According to the original production contract, two thirds of the Me 210Cs built in Hungary were to be supplied to the *Luftwaffe*, with the remaining third going to the Hungarian Air Force. Production started slowly, and by the time it had ceased in March 1944 when the Donau Flugzeugwerke shifted its production over to the Messerschmitt Me 109G, a total of 267 Me 210Cs had been built. Of these, 108 were delivered to the *Luftwaffe* starting in April 1943, where they served mainly as heavy fighters with a secondary role as reconnaissance aircraft until being replaced by the Me 410. As we have now gone far beyond the late 1941/early 1942 time period, the ongoing role of the Hungarian Me 210 should be mentioned as well insofar that starting in early 1944 the Hungarian Air Force was able to equip several high speed bomber squadrons with the Me 210Ca-1, which saw considerable action against the Soviets.

The German Me 210

Of the Me 210B—originally conceived purely as a high-speed reconnaissance aircraft—only two Me 210B-0s and two Me 210B-1s had been completed before production halted in early 1942. These were not followed by any further Me 210Bs when production resumed. An Me 210D-1 planned by the Germans, which would have had similar performance characteristics to the Hungarian Me 210C, was never realized, as it was made superfluous by the Me 410. Seven of the Me 210A-1s, along with one Hungarian Me 210Ca-1 were modified by Blohm & Voss with dual controls for pilot conversion training. Another A-1 became a testbed for maritime reconnaissance systems and was fitted with the FuG 200 Hohentwiel surface search radar at the DLH facilities in Staaken. This Me 210A-1 flew for the first time with this radar system on 7 May 1943.

Me 210Ca-1 and Me 210C-1

In effect, the most "usable" version of the Me 210, whose life cycle from 1939 onwards can generally be considered as the failure of a project filled with potential, was the Me 210C. To be sure, with just 267 examples of the C variant built the production run cannot be considered overly impressive, but with only 352 Me 210s manufactured altogether in both Germany and Hungary (including the four Me 210B reconnaissance aircraft), this was still 76% of the total. Overall production of the 352 Me 210s is broken down by year as follows:

1941 = 94
1942 = 95
1943 = 89
1944 = 74, the year that Me 210 production ceased in Hungary, as well.

The most important performance data for the most produced Me 210Ca-1s and C-1s (reconnaissance) is worth noting, if for nothing more than to compare it to other air assets of the day.

The Me 210Ca-1 was a two-seat heavy fighter/*Zerstörer* and dive bomber, while the Me 210C-1 was a reconnaissance aircraft and heavy fighter (data for the C-1 below is in parentheses).

Engines: 2x liquid cooled Daimler-Benz DB 605B 12-cylinder inverted V-type, each delivering 1085 kW/1475 hp on takeoff and 982 kW/1,335 hp at 5,700 m, plus GM 1.

Armament: 2x 20 mm MG 151/20 each with 350 rounds of ammunition plus 2x 7.9 mm MG 17 each with 1,000 rounds. Both pairs of guns were fixed to fire forward. 2x 13 mm MG 131 in remotely controlled FDL 131 barbettes firing rearward plus either 8x 50 kg or 2x 500 kg or 2x 250 kg bombs in an internal bomb bay.

Flight performance:
Maximum speed at a weight of 6,500 kg (7,200 kg)
at sea level	478 km/h (450 km/h)
at 6,500 (6,300) m	578 km/h (554 km/h)
dive speed	up to 750 km/h
maximum range	1,730 km (1,368 km)
service ceiling	8,900 m

```
Weights:
  empty equipped           7,283 kg
  takeoff                  9,744 kg (10,705 kg)

Dimensions:
  wingspan                 16.34 m
  length                   12.13 m
  height                   4.28 m
  wing area                36.2 m²
```

* GM 1 = designation for nitrous oxide fuel boost, which gave a brief increase in performance above the maximum pressure altitude

A Focke-Wulf Fw 200C-4 (from 1941), with its large HDL 151 turret above and behind the cockpit, seen cleared for takeoff in central Norway.

The reconnaissance version carried two aerial cameras inside the bomb bay as standard equipment, to include the necessary technological equipment as has already been discussed in the section dealing with long range reconnaissance aircraft and systems.

In all, the Messerschmitt Me 210 can be described as a multi-role aircraft which only partially met the high expectations placed on it, and even then only after a long, drawn out developmental period.

c) Heavy Bombers

Fw 200 and the Henschel Hs 293 Glide Bomb

Yearly production of just 58 Fw 200s in 1941 can be attributed in part to bomb damage to the Bremen factory, the necessary transfer of production to Blohm & Voss, as well as the required construction of a second assembly line in Cottbus. Only the Fw 200C, which began arriving at front-line units in mid-1941, is worth examining for the fact that in addition to several minor improvements, there were four quite major enhancements to the variant.

The first of these measures was an improvement of the reconnaissance capabilities at night and in poor weather through the installation of radar search systems. The initial system fitted was the Rostock system manufactured by the Gema company, which had a range of about 15 km against ships and required supplemental antennas on the forward fuselage and above and below the outer wing sections. The Rostock search system, with its limited search range, was soon replaced by the FuG 200 Hohentwiel made by the Lorenz company. Depending on the altitude, it provided the ability to acquire a medium size ship at ranges up to 80 kilometers. Since Hohentwiel could be switched to close range acquisition (10-15 km), where it offered an accuracy of +/- 50 m, it could also be used to bomb through overcast. To be sure, the additional antennas protruding from the airplane were not altogether beneficial for flight performance, but the important thing was that it was now possible to conduct systematic searches of specific areas of ocean by using the quasi-optical transmission of search pulses and their reflections in all weather. This Fw 200, with a strengthened fuselage, became a long range bomber and was designated the Fw 200C-4. The large majority of C-4s operated with the Hohentwiel system. The second major improvement was the large HDL 151 gun turret, which was fitted with either the 13 mm MG 131 or the 20 mm MG 151/20 and was also found on the earlier Fw 200C-3/U1.

With the FuG 200 the Fw 200C-4 could engage ships from beyond about 1 kilometer distance. At closer distances the target's radar echo blended in to the reflection from the water surface, the so-called "sea serpent" effect, which with the Rostock system occurred at about 5 km. This effectively rendered Rostock unsuitable for bombing through overcast and was one of the reasons why it was replaced by the FuG 200.

The third improvement was the introduction of the Lotfe 7D bombsight for high altitude level-flight bombing, which with a well-trained crew provided an accuracy of 20 to 30 meters from an altitude of 3-4,000 m. Those aircraft equipped with the Lotfe were from the Fw 200C-3/U2 and U3 series, the latter also with improved armament and the Atlas-Echolot for precision altitude reckoning over water, plus the U4 with its seven-man crew for operating all the guns. The fourth improvement was made to the Fw 200C-5,

---

* Details of the Hs 293 can be found in volume 10 of this series, "Flugkörper und Lenkraketen," ed. Benecke.

which could engage ships with stand-off weapons, and to this end was fitted with pylons for two so-called Henschel Hs 293* glide bombs carried under the fuselage. These glide bombs received their control commands via signals from an onboard FuG 203a control transmitter, which in turn was given inputs by a small control stick operated by the bombardier/gunner. The glide bomb would be dropped 3.5 to 5 km from the target at an altitude of 4,000 m and up to 16 km from the target at an altitude of 8,000 m. A short-endurance rocket engine would burn for eight seconds, accelerating the Hs 293 to about 0.8 Mach. The bombardier would control the Hs 293 in such a manner that a light source in the tail of the bomb would be lined up and held with the intended target until, after flying for 100-120 seconds, the Hs 293 impacted the ship. The Henschel Hs 293 had a wingspan of 2.9 m, a length of 3.4 m, weighed 791 kg, and had a warhead of 500 kg.

However, at barely 6,000 m the Fw 200's service ceiling was far below that for optimally using this new stand-off weapon. Prerequisite for a hit was not only the precise manual control of the missile, which responded to inputs like an airplane (i.e. if the bombardier failed to control it properly it could pitch up and stall out), but also a straight and level flight without any change in speed on the part of the mother aircraft. Improvement of attack tactics was undoubtedly attained against weakly armored or unarmored ships, since the Fw 200 remained outside ship defenses as it approached the target with its Hs 293 load, even at medium altitudes. However, this type of attack was only possible in those areas free from enemy fighter defenses. During the approach to the target, the mother plane was not able to carry out defensive maneuvers or engage in aerial combat, unless it first broke off the target approach and thus abandoned control of the missile.

The Fw 200s modified with the FuG 200 search radar and capable of carrying the Hs 293 glide bombs were given the interim designation Fw 200C-3/U1 and U2, which ultimately became the Fw 200C-6, and as such was practically the pinnacle of what this multi-faceted aircraft had to offer. The C-6 was produced with the FuG 203b Kehl transmitter as standard equipment. Kehl transmitted the signals to the Hs 293's FuG 230b Straßburg receiver, which moved the aileron and elevator controls accordingly.

Then there was another series, the Fw 200C-8, which in its long range bomber version resembled the C-4, but with glide bomb pylons and equipment for the Hs 293 was designated the Fw 200C-8/010. It differed from other glide bomb capable versions by having its ventral gondola extended further forward to give the bombardier a better view for controlling the Hs 293.

In all, beginning in 1943 only about 20 of the above-mentioned Fw 200s were built for operating the Hs 293 remote-controlled glide bomb.

A brief synopsis of the aircraft's further development: in 1942 the Fw 200 reached its production peak with 84 examples being built. In 1943 this number was still relatively high at 76, but with a dramatic increase in convoy defenses, which in some instances now even had catapult-launched fighters, the Fw 200's days were numbered. By 1944 only eight of these elegant looking aircraft were built, the last version being the Fw 200C-8. These were delivered in January and February, just before its military production—comprising some 263 aircraft of all variants—closed down for good. The Heinkel He 177 was expected to replace it as a maritime bomber.

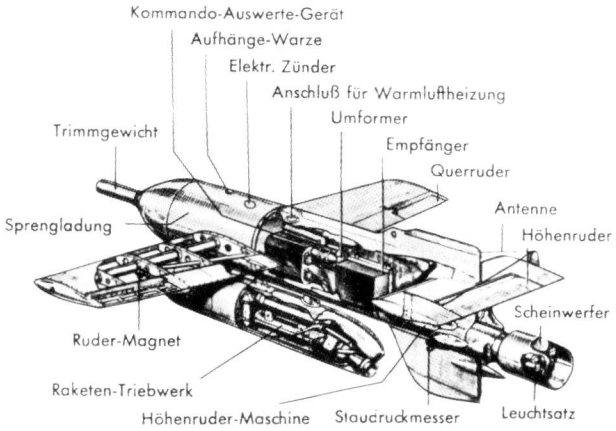

The components of the Henschel Hs 293 guided bomb.

Henschel Hs 293 guided bomb, in production from 1943 onward and used chiefly against naval targets. Of the approximately 500 maritime missiles built, about 100 hit their ship targets. A better preserved example than that shown here can be found in the Deutsches Museum in Munich.

Fw 200F

In conclusion, a particularly interesting project variant worth mentioning is the Fw 200F. This was to have been developed from the C-6 as a long range reconnaissance aircraft with increased range. Up to then, 19° West Longitude was considered the western limit of armed reconnaissance, yet the F variant would fly well beyond this line in search of Allied supply shipping. Only in extreme instances had missions been flown out to 25° W, something which was not exactly an optimal range for bringing submarines in to destroy acquired convoys in a timely manner.

On the Fw 200, fuel capacity would have been increased by 2,600 liters to 12,000 liters by replacing the standard fuselage fuel system with eight new fuel tanks, each with a capacity of 1,100 liters, plus fuel tanks in the aft part of the ventral gondola. The oxygen system and all bomb pylons were to have been removed, although all defensive armament was retained. In order to increase takeoff power to 883 kW/1,200 hp per engine for this overly heavy machine (25.26 metric ton flying weight), it was planned to make use of a water-methanol fuel injection system (MW 50) to provide a brief performance boost for the Bramo 323R-2 engines, which had proven themselves in the Fw 200C-3/U4 series earlier. In this configuration the Fw 200F was to have had a range of 6,600 km—of seven options examined for increasing range, option six was the one developed as the best with regard to range.

This project never left the drawing board, and the role of strategic reconnaissance at extreme ranges was later assumed by the Junkers 290—also a four-engine type—which would operate from southern France on its long range patrols.

Synopsis and Performance of the Fw 200C-3/U4

With all the military demands placed on it, the Fw 200 had overstepped the limits of its developmental potential, which in some instances was made at the expense of structural safety. For the most advanced anti-shipping weapon of the day, the Hs 293 glide bomb,

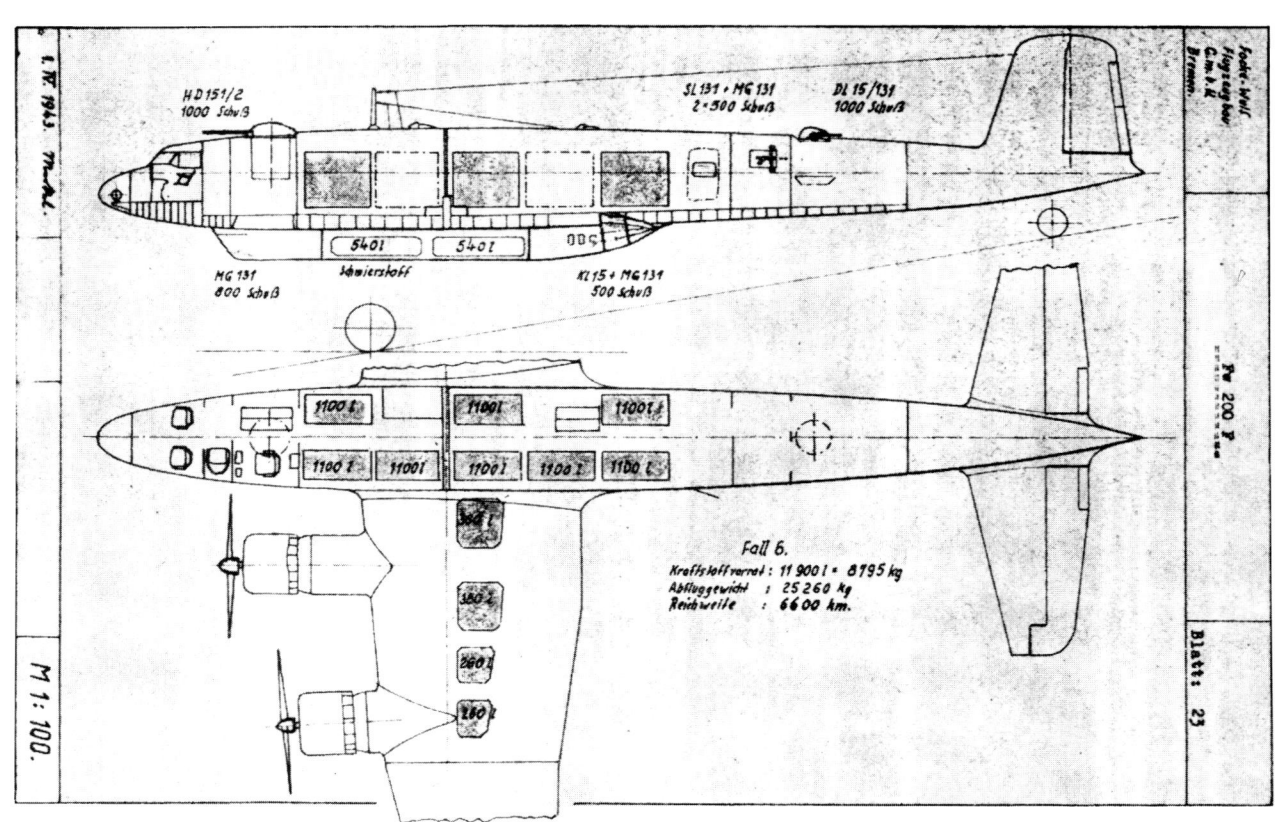

Focke-Wulf Fw 200F showing the optimal fuel layout for increasing range; this variant never went into production, however.

it was too slow and too unmaneuverable a platform to be able to provide consistently effective results. In addition, the number of Fw 200s built for carrying this kind of weapon was too small for a continuous naval war from the air with the Hs 293. The Fw 200 was only employed against shipping targets in very limited numbers, striking from altitudes above 2,700 m. Generally it was used in the armed reconnaissance role, primarily in the North Sea operating out of Norway. Using its Hohentwiel radar system, pilots developed the tactic of making so-called search circuits, whereby the aircraft would periodically pop up from low altitude to 500 meters and make a full 360° circle. In this manner relatively large bodies of water, e.g. between Iceland, Greenland, Jan Mayen, and the ice packs, could be monitored by single aircraft practically without any gaps in coverage—even during the polar night periods.

The Focke-Wulf Fw 200C-3/U4 was a seven seat long range maritime reconnaissance aircraft and bomber.

The 20 mm MG 151/20 located in the Focke-Wulf Fw 200C-3/U4's *D-Stand*, showing its excellent forward field of fire.

Focke-Wulf Fw 200 C-series. The three FuG 200 Hohentwiel antennas on the nose were used to assist the type in its role as an armed maritime reconnaissance platform (notice the 250 kg bombs being manhandled to the plane). Attacks were made from high altitude (above 2,700 m) on targets of opportunity (from 1942).

Engines: 4x air-cooled BMW Bramo 323R-2 Fafner nine-cylinder radials, each rated at 880 kW/1,200 hp on takeoff, with water-methanol fuel injection (MW 50), 736 kW/1,000 hp at sea level and 690 kW/940 hp at 4,000 m.

Armament: 1x 15 mm MG 151 with 1,000 rounds in hydraulically operated upper turret, 1x 13 mm flex-mounted MG 131 with 1,000 rounds in upper aft turret, 2x 13 mm MG 131 each with 500 rounds in the waist positions, 1x 20 mm flex-mounted MG 151/20 with 500 rounds in the forward gondola (*D-Stand*), 1x 7.9 mm flex-mounted MG 15 in the aft gondola section. Maximum bomb load: 2,100 kg, consisting of 2x 500 kg plus 2x 250 kg plus 12x 50 kg bombs. At maximum fuel load of 9,310 liters bomb payload was limited to 1,230 kg.

Flight performance:

Maximum speed	
- at sea level	305 km/h
- at 4,800 m	360 km/h
Maximum cruise speed	
- at sea level	227 km/h
- at 4,000 m	335 km/h
Economy cruise speed	254 km/h
Ranges at economy cruise speed	
- with 8,060 liters fuel (standard)	3,550 km
- with 9,950 liters fuel (with external tanks)	4,440 km
Service ceiling	5,800 m

Weights:

empty	12,950 kg
full	22,700 kg

Dimensions:

wingspan	32.88 m
length	23.47 m
height	6.30 m
wing area	119.84 m^2

Heinkel He 177 - In Production At Last!

Many hopes, expectations, and skeptical optimism rested on the He 177. Its absence from an operative prosecution of the air war was sorely felt, not to mention a strategic one—like the Allies were increasingly carrying out against Germany—necessary for interrupting and destroying the enemy's power bases. The bulk of the medium bomber fleet, Ju 88s and He 111s, were more and more frequently needed for direct and indirect cooperation with the ground forces—a trend which became even more pronounced during the campaign on the Eastern Front. Just where was the He 177 during that critical year of 1942?

The destruction of He 177A-02 in a crash landing in May 1942 has already been mentioned briefly. During the three months of trials before the crash, the only recommendations for alleviating the numerous engine fires on the type came in the form of minor changes: the engine mounts should be extended forward by about 20 cm so that the fuel and oil lines could be better laid out; a firewall should be fitted; the oil reservoir should be moved to an area less susceptible to fire, plus the exhaust system should be entirely reworked. These improvements were only gradually introduced, beginning with the He 177A-1 series then in production. For the time being, it sufficed just to move the oil reservoir, since this would not mean any delays in production already well underway at Arado in Warnemünde. He 177 production at the Heinkel plant in Oranienburg had been reserved exclusively for the pre-production series up to this point, so that Arado alone was fully responsible for manufacturing the He 177A-1. To this end, the company received complete empennages and several fuselage sections from Mielec in Poland. Somewhat later there went out a call for flame dampers for the He 177's night operations, and the manufacturer took this opportunity to rework the exhaust system. It was not until the He 177A-3 began production in tandem with the A-0 at Oranienburg that the engines were shifted the recommended 20 cm forward. A 1.60 m fuselage plug was added behind the bomb bay as a counterbalance to maintain the center of gravity. The first production example of the He 177A-3/R1 rolled off the assembly line at Oranienburg in the fall of 1942, followed by about a dozen more A-3/R1s before the end of the year. It was expected that 70 aircraft per month would be produced, but constant modifications led to an output of little more than five per month up to early 1943.

Pre-production Series and Optimistic Appraisals

In this section, we will briefly discuss the pre-production series and look further at events surrounding the He 177 around the end of 1942. 35 examples of the He 177A-0 pre-production series were produced, predominantly at the Oranienburg factory, but also at Warnemünde, as well. The A-0s mainly saw service with the *Luftwaffe* as crew training aircraft. Several of these pre-production machines, however, were used as testbeds, such as the He 177A-05 as He 177V9, and the 06 and 07 as the He 177V10 and V11. Shallow gliding tests were conducted with the first of five He 177A-0s built at Warnemünde, with speeds reaching 713 km/h, but the g forces on pull-out proved too great for the airframe and wings—these became deformed under the enormous stress. Then, on 9 October 1942, the *E-Stelle* Rechlin released an evaluation report with test results showing that the strength of the wing was 1/3 less than had been estimated by Heinkel. The reason was cited as the unequal stiffness of the individual components, with resulting warpage under stress. Heinkel had not recognized this condition in time, and structural tests had been carried out too late to affect the program.

Since Hitler had placed the highest priority on the problem-fraught He 177 program, on 28 October 1942 *Generaloberst* Jeschonnek wrote the *Generalluftzeugmeister*, *Generalfeldmarschall* Milch, that "the Führer" would like to see the He 177—even in its most basic version—on the Eastern Front for high altitude night raids outside the range of other bombers, as well as in the Atlantic to provide top cover for submarines and blockade runners. For East Front operations a unit of group strength (approximately 30 aircraft) was to be established and maintained, with remaining production being directed toward maritime roles.

On 11 November 1942 Milch responded by stating that the He 177 in its current state was able to drop all "big fish" from level flight, including the Hs 293 glide bomb and the SD 1400X Fritz X free-fall bomb, which could be controlled from the bombsight. In view of its speed and good defensive armament it was possible—in Milch's opinion—to use the He 177 against distant targets behind the Eastern Front during daylight hours, as well. Thus equipped, maritime operations would pose no problems and, with the necessary minor changes for carrying the "big fish," a group would be ready for action in the East by the end of January 1943.

Tough demands and optimistic expectations on the one hand and the malady of testing and manufacturing problems on the other led to *Prof.* Heinrich Hertel being appointed to assist with the project in November 1942. Up to then Hertel had been working for Junkers, and with the move was given special authority from the RLM to reorganize the He 177 production program.

He 177A-1 and A-3

In the meantime Arado had been building the He 177A-1 at Warnemünde, with 130 examples eventually rolling off the assembly line. Parallel to this was production of the He 177A-3 in Oranienburg. 19 aircraft from the A-1 series had already been writ-

History of German Aviation: Bombers and Reconnaissance Aircraft

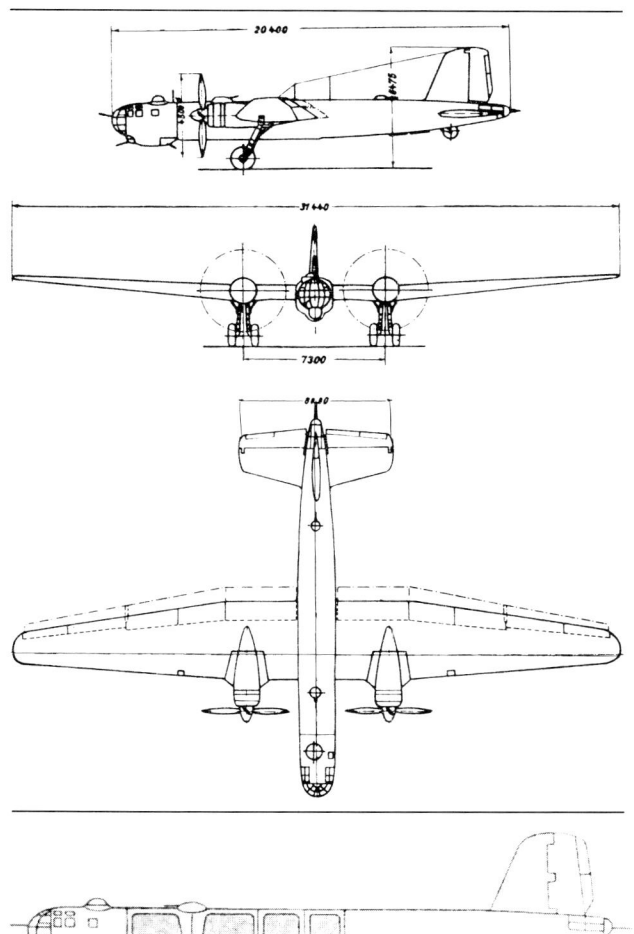

The above drawings of the Heinkel He 177A come from the type sheet for the He 177 published by the RLM, GL/C-B2, Nr. 2154 gKd., dated 1/1/1942, as does the layout of the fuel cells in the fuselage and ordnance spacing options for three profiles: short range/tactical bomber, medium range bomber, and long range bomber. The sketches show the elegant lines of this unusually heavy bomber to good effect.

ten off as total losses, with only five aircraft undergoing combat testing at the time. Two of these were with a front-line unit operating over the Atlantic, where their multifarious technical shortcomings caused their crews no end of frustration. The other three were taking part in special missions for long range reconnaissance with the reconnaissance element of the Ob. d. L.'s test unit, led by Theodor Rowehl. The remainder were scattered among the various test centers and conversion training units.

Prototypes for the improved He 177A-3 were the He 177V15 and V16, which did not have the lengthened fuselage of the A-3 but were fitted with the more powerful DB 610 double engine consisting of a pair of DB 605s. This double engine provided 2,170 kW/2,950 hp on takeoff and 2,280 kW/3,100 hp at an altitude of 2,000 m. All A-3 aircraft were to have received this more powerful engine, as well, but delays in delivery and other priorities forced the less powerful DB 606 (1,986 kW/2,700 hp) to be retained. A previous section has already dealt with the laborious start of He 177A-3 output in late 1942/early 1943. Total yearly output for the above mentioned variants in 1942 amounted to 166 aircraft. Production continued, if only slowly, but a true front-line maturity was still not in sight, despite the expectations of the *Generalflugzeugmeister*.

Interim Balance and Performance for the He 177A-1

The mess with the He 177's development led to serious conflicts between Ernst Heinkel and Milch, who was considering replacing him as company director. Göring stepped in, and in September 1942 lifted what was for him the absurd requirement by the General Staff to make the He 177 capable of dive bombing. A bomber unit especially established for operating this heavy bomber in the East, after evaluating the He 177, was extremely skeptical about the aircraft's technical unreliability, although reporting that from a purely flying standpoint, its qualities were good. It remained to be seen whether *Prof.* Hertel, specially commissioned by the RLM and given responsibility for Heinkel's He 177, would be able to make the sweeping improvements the He 177 situation called for.

What follows is performance data for the He 177A-1/R1 for the time period in question, providing the reader an opportunity to assess its potential operational capability for himself.

The He 177A-1/R1 was a five seat heavy bomber.

Engines: 2x liquid cooled Daimler-Benz DB 606 24-cylinder double engine, each with 1,986 kW/2,700 hp on takeoff and 1,763 kW/2,360 hp at 5,800 m; four-blade counter-rotating VDM variable-pitch propellers with a diameter of 4.5 m.

Armament:
1x 7.9 mm flex-mounted MG 81i with 2,000 rounds in the glazed nose
1x 20 mm flex-mounted MG-FF with 300 rounds in the forward ventral gondola
2x 7.9 mm flex-mounted MG 81i with 2,000 rounds each in the aft ventral gondola
1x 13 mm MG 131 with 750 rounds in remote controlled electrically operated forward fuselage turret
1x 13 mm flex-mounted MG 131 with 1,500 rounds in the tail

Maximum bomb loads:
Short range bomber (Configuration A):
Either 48x 50 kg plus 12x 250 kg plus six 500 kg (=8,400 kg) bomb load, or
4x 1,000 kg plus six 500 kg (=7,000 kg) load, or
six 1,000 kg (=6,000 kg) load, or
2x 1,000 kg plus 2x 1,800 kg (=5,600 kg) load, or
2x LMA aerial mines (each weighing 500 kg) plus 2x 1,800 kg (=4,600 kg) combined payload.
Medium range bomber (Configuration B):
16x 50 kg plus 4x 250 kg plus 2x 500 kg (=2,800 kg) load, or
2x 1,000 kg (=2,000 kg) load.

Flight performance:
Maximum speed at 5,800 m altitude	510 km/h
Cruising speed at 5,500 m altitude	430 km/h
Ranges (with maximum bomb payload):	
- with 8,920 liters fuel	1,200 km
- with 10,415 liters fuel	3,200 km
- with 12,820 liters fuel	5,600 km
Service ceiling	7,000 m

Weights:
empty	16,100 kg
empty equipped	18,040 kg
takeoff	30,000 kg

Dimensions:
wingspan	31.44 m
length	21.90 m
height	6.7 m
wing area	102.00 m²

Theoretically, the He 177's operational potential was an enormous improvement over those of the Fw 200, the Do 217, and the medium bombers with regard to its bomb capacity, range, speed, and armament. In practice, however, the number of prevailing technical problems significantly impaired the type's usefulness and contributed to its low state of readiness.

Dornier Do 217 E-Series

Following the 20 examples built in 1940, 1941 saw a jump to 277 Do 217s, while this number increased significantly in 1942 to 564 produced.

Models Do 217E-0 through E-5

The Do 217V9 prototype had been flying since early 1940, and became the precursor to the Do 217E. Its powerplants were the air-cooled BMW 801MA 14-cylinder double radial engines, each rated at 1,160 kW/1,580 hp on takeoff and driving a three-bladed wooden propeller manufactured by Schwarz. The characteristic feature of the V9 was its fuselage, increased in size along its entire underbelly. The bomb bay was braced more solidly to accommodate the heavier bomb load, with the bombs being carried horizontally in the bay's lower section. On-board systems, including the landing gear and the opening and closing of the three-compartment bomb bay doors, were electrically driven.

Parallel to the Do 217V9's flight testing were the preparations for series production of the Do 217E, with the first pre-production series aircraft, the Do 217E-0, rolling off the assembly line in the fall of 1940 and completing its maiden flight on 1 October of that year. The first production aircraft was the Do 217E-1, and followed the E-0 even before the year was out. It, too, was virtually identical to the V9 and was laid out as a level bomber and anti-shipping aircraft without dive brakes. Its bomb bay could hold 8x 250 kg, 4x 500 kg, or 2x 1,000 kg bombs. For low-level attacks it had a pilot-operated fixed 15 mm MG 151 with 250 rounds installed in the lower left section of the nose glazing. The E-1's defensive weaponry included 5x 7.9 mm flex-mounted MG 15s. With photography equipment installed in the bomb bay, an initial batch of ten Do 217E-0s and E-1s were sent to a long range reconnaissance unit deployed to Romania in early 1941, from where they flew photo reconnaissance missions over the Russian border areas in anticipation of the Eastern campaign. Other Do 217E-1s served in the anti-shipping role in the Atlantic and the North Sea beginning in March 1941. Although initially enjoying some success, the type soon revealed certain weaknesses with regard to its armament and crew protection from enemy fire. Measures correcting these deficiencies and a few changes to improve the type's versatility in combat led to the Do 217E-3 series. To suppress enemy fire from ships, this model was armed with an additional flex-mounted 20 mm MG-FF in the lower right section of the nose glazing, plus two further flex-mounted 7.9 mm MG 15s on either side of the crew compartment just behind the pilot. Although the E-3 now had seven light on-board guns, five of these alone had to be operated by the radioman. Nevertheless, he was able to better cover the fields of fire with this armament suite than had been the case with the E-1. Crew seats were given adequate protection in the form of 5 mm and 8.5 mm armored plating.

Dornier Do 217E-2 with a new *B-Stand* containing a 13 mm MG 131, the side-firing 7.9 mm MG 15 behind the pilot's seat, and the retrofitted 20 mm MG-FF in the nose *A-Stand*.

Parabrake variations for the Dornier Do 217E; above is the opened parabrake in flight, while below is the so-called *Rüstsatz 25* dive brake being tested on the ground.

In addition, the type's operational palette was considerably expanded by a whole number of standardized field conversion kits, many of which were used on later versions of the Do 217. Despite headaches with the dive brakes mounted in the extreme tail, the RLM insisted that the Do 217 would have full dive bombing capability.

In October 1940 the Do 217V11 prototype began dive bombing trials for the Do 217E-2, which unlike the E-1 was expected to be dive capable, and ultimately was given higher priority and went into production alongside the E-1 series then underway. The most significant change for the E-2 was an electrically-powered *B-Stand* with a 13 mm MG 131 and 500 rounds of ammunition.

At the same time the small caliber MG 15 in the aft section of the gondola was replaced by a 13 mm MG 131 with 1,000 rounds, and the E-2 was also fitted with the two side-firing MG 15s of the E-3, its port gun position and that of the MG-FF in the *A-Stand* easily seen in the photo. In addition, a further additional MG 15 was installed in the forward-most location in the nose glazing for the bombardier, since the MG-FF with its heavier caliber, slower rate of fire, and limited ammunition was not always the best suited weapon for the task at hand. The E-2 was powered by two BMW 801ML engines having the same performance as the BMW 801MA, but driving different propeller types. In place of the wooden propeller were three-blade VDM variable-pitch metal airscrews with a diameter of 3.9 m.

In the early summer of 1941 one of the first Do 217E-2s was assigned to a dive bomber unit operating the Ju 87 for the purpose of carrying out combat trials. These tests gave unsatisfactory results, as the location of the dive brakes at the extreme aft end of the fuselage led to the fuselage warping under too much stress when the brakes were fully extended. Despite this poor showing the *Technisches Amt* insisted on retaining the dive bombing capability in order to better strike point targets. As a result, on a trial basis Dornier fitted the 36th E-2 series production aircraft with new dive brakes designed as split flaps which could be extended to 90°, and located between the fuselage and the engine nacelles. During one of the diving trials near Friedrichshafen, however, the retracting brakes jammed the elevator trim tab linked to them and the plane crashed. In the late summer of 1941 the RLM finally dispensed with the dive requirement for the Do 217, since the Ju 88 would be

Dornier Do 217E-4s flying in a *Kette* formation (1942).

built in large numbers and was far better at diving than the Dornier plane, thus giving the *Luftwaffe* sufficient dive bombing potential. The remaining Dornier Do 217E-2s and E-3s produced in 1941 were intended for operations against England and anti-shipping missions in the North Sea.

Around the end of 1941 the assembly lines switched over from the Do 217E-2 to the Do 217E-4, which began making its way to front-line units starting in early 1942. It, too, was fitted with the BMW engines with the same performance ratings as the previous variants, but this was not a complete engine assembly like the BMW 801MA and ML were, rather a single engine designated the BMW 801C. One novelty on the E-4 was the fitting of a *Kuto-Nase* for cutting balloon cable, which the reader has already become familiar with in the sections on the He 111 and Ju 88. A small number of this variant were modified on the assembly line as mother ships for the Hs 293 glide bomb. The modifications included two ETC 2000/XII external racks under the wings, plus the Telefunken FuG 203b Kehl III transmitter, a mini-joystick with impulse transmitter for the bombardier to control the glide bomb after it had dropped, as well as heating ducts in the wings to keep the glide bomb at a constant internal temperature. In this modified state the aircraft was designated the Do 217E-5. At first, the E-5 aircraft were used for thorough testing of the Hs 293 and its FuG 230b Straßburg receiver, as well as the SD 1400X free-fall bomb controlled via the Lotfe 7D sight. Also known as the Fritz X, or simply FX, these bombs were dropped from high altitudes in the vicinity of the Baltic Sea, and were tested not only to give the aircrew an opportunity to try out the new weapon, but also to familiarize the ground crews with the bomb. However, it wouldn not be until 25 August 1943 until this promising but troublesome weapon would be dropped from a Do 217 in combat—against Allied naval forces in the Bay of Biscay.

Basic Data for Do 217E-2 and Sub-Variants

The Do 217E-2 was a four seat heavy bomber.

Engines: 2x liquid cooled BMW 801ML 14-cylinder double radial engine, each with 1,162 kW/1,580 hp on takeoff and 1,015 kW/1,380 hp at 4,600 m; three-blade counter-rotating VDM variable-pitch propellers with a diameter of 3.9 m.

Armament:
1x 20 mm fixed MG 151 forward firing with 250 rounds in the lower port section of the glazed nose
1x 13 mm MG 131 with 250 rounds in remote controlled electrically operated forward fuselage turret (*B-Stand*)
1x 13 mm flex-mounted MG 131 with 1,000 rounds in the aft gondola
1x 7.9 mm flex-mounted MG 15 in glazed nose
2x 7.9 mm flex-mounted MG 15s on either side of the cockpit
The Do 217E-2/R19 also had
1x 7.9 mm flex-mounted remote controlled MG 81Z (*Zwilling*) in the tail cone

Maximum bomb loads: 4,000 kg (of which up to 2,500 kg could be carried inside the fuselage bomb bay)

Flight performance:
Maximum speed
- at sea level   440 km/h
- at 5,200 m   515 km/h
Cruising speed (with maximum internal payload)
at 5,200 m altitude   415 km/h
Economy cruising speed   394 km/h
Ranges
- with normal fuel capacity   2,300 km
- with supplemental fuel tanks 2,800 km
Service ceiling (with maximum internal bomb payload)   7,000 m

Weights:
empty   8,909 kg
empty equipped   10,535 kg
takeoff   15,000 kg
max. overloaded   16,465 kg

Dimensions:
wingspan   19.00 m
length   17.30 m (minus tail dive brakes)
height   5.00 m
wing area   57.00 m^2

Field Conversion Sets (*Rüstsätze*):
Of particular interest are the standardized field conversion kits introduced on the E-series. These carried over into other variants, and the more important ones are highlighted here with the applicable types (some of which have not yet been discussed) following in parentheses. Conversion sets for the Do 217 night fighter versions are not included here.

History of German Aviation: Bombers and Reconnaissance Aircraft

R1 1x *Sonderträger* 1800 bomb pylon for 1x SC 1800 bomb with ring-shaped stabilizer (Do 217E-2 and E-3)
R2 2x external bomb racks under outer wing panels for 2x SC 250 bombs (Do 217E-2/E-3)
R4 1x PVC 1006 external rack for 1x L5 aerial torpedo (Do 217E-1/E-2/E-3/E-4 and K-1)
R5 1x 30 mm MK 101 cannon fixed in the left lower nose glazing (Do 217E-2/E-3)
R6 camera package in bomb bay (Do 217E-1/E2/E-4/K-1 and M-1)
R7 1x 4-man rubber dinghy stored above the bomb bay behind the wings (Do 217E-1/E-2/E-4 and K-1)
R8 1x additional fuel tank (750 l) fitted in forward part of bomb bay (Do 217E-1)
R9 as R8, but in aft section of bomb bay (Do 217E-1)
R10 2x ETC 2000/XII external racks for the Hs 293A glide bomb beneath outer wing panels (Do 217E-2/E-4 and K-1)
R12 1x PVC 1006B in bomb bay for 1x SC 1800 bomb (Do 217E series, K-1 and M-1)
R13 alternative tank to R8 with 750 l (Do 217E-2/E-4 and K-1)
R14 alternative tank to R9 (DO 217E-2/E-4 and K-1)
R15 2x ETC 2000/XII external racks for HS 293A glide bomb underneath wing center sections between fuselage and engine nacelle (DO 217E-4 and K-2)
R16 2x ETC 2000/XII D between fuselage and engine nacelles (Do 217M-1)
R17 1 supplemental fuel tank (1,160 l) in forward bomb bay (Do 217E-4 and K-2)
R19 1x MG 81Z (twin-firing MG 7.9 mm) in tail cone (Do 217E-2/E-4/K-1/K-2/M-1 and M-11)
R21 underwing jettisonable fuel tanks (2x 900 l, wooden) (Do 217E-3/E-4 and K-1)
R25 tail parabrake (Do 217E-4/K-1/M-11 and P)

Junkers Ju 87D being loaded with a PC 1000 (on a sled, with pylon already attached) by means of a block and tackle (1942).

Interim Balance and Viewpoint

The Do 217 stemmed from the medium bomber category, where the Do 17Z and Do 215 were classed. As a heavy bomber, its payload capacity was much greater than those of the He 111 and Ju 88, and it was able to perform on par with the "big fish." The Do 217 had evolved into an impressively performing, versatile, and reliable aircraft. Its crews enjoyed flying the type, even if the horizontally configured controls in place of the standard joystick took some getting used to for the pilot, and the two-step canopy did not exactly offer the crew the best view. Thus, in 1942 there was still every opportunity to optimize what was indeed a highly capable airframe even further.

d) Dive Bombers and Strike Aircraft

Junkers Ju 87D, F, G, H and Ju 187

The Ju 87D mentioned earlier went into production at the Weser Flugzeugbau in Bremen in the late spring of 1941 and gradually replaced the Ju 87B-2. Although the Ju 87D-1 was a marked improvement over the Stukas then in operation, altogether the 1,000+ aircraft produced in 1941 did not amount to much more than that of the previous year. This was undoubtedly in anticipation of a better performing aircraft type, such as the Me 210 intended for this role.

In the meantime the Russian campaign was beginning to take its toll, and still nothing newer or better was in sight since the Me 210 had not lived up to expectations. As a result, in November 1941 the Ju 87 production program was once again placed high on the prioritization list as a sort of transitional emergency measure, so that despite a temporary cutback in production 1942 also saw impressive numbers of the type produced in Bremen and Tempelhof, which led to a yearly output of 967 examples.

The Ju 87D-1 appeared in the spring of 1942 in units on the Eastern Front, where it slowly replaced the Ju 87B-2. At almost the same time it entered combat in North Africa as the Ju 87D-1/Trop. Included in the D-1 assembly lines were a series of aircraft with reinforced aft fuselage sections and tailwheel, fitted with a tow system for glider operations and supplied mainly to North Africa and the Mediterranean area under the designation Ju 87D-2.

In late 1942 D-1 and D-2 production was replaced by the Ju 87D-3, which was primarily conceived as a strike aircraft with im-

Junkers Ju 87D-2s towing DFS 230 transport gliders. Beginning in 1937, approximately 1,600 of these assault gliders were built. They were also called dive assault gliders because of their steep approach angle at speeds of 250 km/h using expanded braking chutes.

proved armor for the crew, the engine, and the radiator. Otherwise, it was similar to the D-1 and had dive brakes. A small number of D-1 and D-2 variants were modified as torpedo bombers. Designated the Ju 87D-4, these never served in their intended role and were retro-fitted as D-3s.

Finally, in early 1943 there appeared the Ju 87D-5. With demands for greater payloads becoming more vocal, the D-5's outer wings were extended to absorb the increased wing loading by 1.20 m, giving the type a span of 15.00 m. The D-5 had jettisonable wheel spats like those of the C-series carrier version in the event of having to ditch. In addition, the design soon dispensed with the need for dive brakes, as this variant was used almost exclusively in the strike role. Beginning in 1943 the Hungarian and Romanian Air Forces also took deliveries of the Ju 87D-5. In any event, it soon became apparent that the Ju 87's operational potential was extremely limited, even on the Eastern Front; by the spring of 1943 Soviet fighter defense had become so strong that the Ju 87 was only able to operate in localized areas of air superiority, with fighter protection, or at night.

The Ju 87D-7 was especially tailored for night operations and was derived from the D-1 series. The D-7 had exhaust pipes stretching backwards over the wings, full night instrumentation, the more powerful Jumo 211P (rated on takeoff at 1,103 kW/1,500 hp and at 4,300 m 1,037 kW/1,410 hp), plus improved forward firing armament in the form of two 20 mm MG 151/20s in place of the 7.9 mm MG 17s. It, too, dispensed with the dive brakes, although the D-7 retained the D-5's jettisonable undercarriage. A small interim variant is worth mentioning, the Ju 87D-6, which was similar to the D-5.

The last of the D-series was the Ju 87D-8, built alongside the D-7. The D-8 stemmed from the D-3 but had the 15 mm longer wings of the D-5, making it similar to the D-7. Unlike the D-7, however, it was not fitted with the long exhaust tubes and night instrumentation, thus making it primarily suited for daylight strike operations.

Basic Data for the Ju 87D-1 and D-7

Ju 87D-1	Ju 87D-7
two-seat dive bomber	two-seat strike aircraft
Engine:	
liquid-cooled Jumo 211J-1 12-cylinder inverted V with intercooler liquid-cooled Jumo 211P on takeoff	similar type with intercooler 1,103 kW/1,500 hp 1,044 kW/1,420 hp on takeoff 1,037 kW/1,410 hp at 4,300 m

Rear armament of the Junkers Ju 87D showing the 7.9 mm MG 81Z twin barreled machine gun.

Armament:
2x 7.9 mm MG 17 fixed forward firing
2x 20 mm MG 151/20 fixed forward firing
1x 7.9 mm MG 81Z (*Zwilling*) flex-mounted rear firing

External Load
1x 1,800 kg bomb (short range overloaded configuration) or
1x 1,000 kg plus (only with maximum underfuselage load of 500 kg)
4x 50 kg or
2x 250 kg bombs under the wings or
2x canister bombs each containing 92 SD 2 fragmentation bomblets against soft targets or
2x weapons pods each with 6 7.9 mm MG 81s or 2x 20 mm MG-FF

Flight Performance
Maximum Speed

with a takeoff weight of 5,700 kg		with a takeoff weight of 5,800 kg
409 km/h at 4,100 m altitude		399 km/h at 4,700 m altitude
Normal Cruise Speed at 72% Setting at 5,100 m altitude		
319 km/h		301 km/h
Economy cruising speed with 1,800 kg payload		185 km/h
Normal range at 4,100 m altitude		820 km
Maximum range at 5,100 m altitude		1,535 km
Time to climb to 5,000 m altitude		19.8 min
Service ceiling at maximum takeoff weight		4,730 m
at 5,715 kg takeoff weight		7,290 m
Weights:		
3,900 kg	empty equipped	3,940 kg
5,840 kg	standard takeoff	5,840 kg
6,600 kg	maximum takeoff	6,610 kg
Dimensions:		
13.80 m	wingspan	15.00 m
11.50 m	length	11.13 m
31.90 m²	wing area	33.7 m²
	height	
4.27 m to top of antenna or 4.24 m to tip of rudder		

Ju 87G Tankbuster

The later reputation of the Ju 87G, derived from the D-3 and D-5 variants, as a "tankbuster" can be traced back to an instance of questionable veracity from the first weeks of the campaign against Russia. In this case, an entire *Stuka* unit allegedly dive bombed a massed group of Soviet tanks and armored vehicles, only to discover later that just one of the tanks had been destroyed. Even if the truthfulness of this unconfirmed report can be called into question, it does illustrate that this type of attack method was no longer the optimal way of hitting tanks on the move.

This reason prompted research and testing of tank busting weaponry designed to hit the vulnerable parts of a tank better than could be accomplished with a standard iron bomb from a dive. The results of this research eventually showed that the Flak 18, a medium 3.7 cm anti-aircraft gun which had also proven its effectiveness at destroying armor, could be installed in a cannon pod and mounted externally beneath the wings of the Ju 87 just outside the landing gear struts. These cannon pods could be alternated with the standard bomb racks, giving this special anti-tank aircraft a certain operational flexibility. In the summer of 1942 a Ju 87D-3 was accordingly modified and used to carry out the first combat trials with this new weapon, now designated the BK 3.7. It proved accurate enough, and beginning in early 1943 the BK 3.7 equipped Ju 87G-1 began appearing in front-line units. To be sure, this aircraft was extremely slow and somewhat clumsy in the air, but it made for a good firing platform, and many experienced pilots achieved outstanding results against Soviet armor. With increasing Soviet fighter defenses and Germany's own flying unit strength being sapped away, the Ju 87's days on the Eastern Front were numbered; beginning in October 1943 their *Geschwader* were redesignated as strike wings and converted over to the Focke-Wulf Fw 190. For a time, these units retained a tank busting squadron equipped with the Ju 87G-1, a plane much feared by the enemy but also extremely vulnerable to him.

With regard to the cannons, it should be noted that they were sighted for 400 m—the optimal firing distance for this caliber against armored targets—and carried 12 rounds per weapon in each magazine clip. The braking oil for absorbing the weapon's recoil was kept in a liquid state with hot air from the engine, an absolute necessity for winter operations.

It should also be mentioned in this context that a Ju 87G-2 version was delivered to combat units, as well, with the only difference being that the G-2 had the lengthened 15 m wings of the D-5 and was thus somewhat easier to fly.

Sighting in the two BK 3.7 guns of a Junkers Ju 87G-1.

Each of the BK 3.7 cannons carried two clips of six rounds each, giving the Junkers Ju 87G-1/G-2 a total of 24 shells per operational sortie.

Air-to-air shot of a Junkers Ju 87G-1 showing the two BK 3.7 cannons. The wheel spats were generally removed in the field because they tended to accumulate a buildup of mud.

Junkers Ju 187

To round out the Ju 87's developmental life, we should perhaps look at the project for the Ju 87F, a radically improved version of the Ju 87D designed to accommodate the Jumo 213 engine (rated at 1,369 kW/1,776 hp). However, the Junkers proposals submitted to the *Technisches Amt* in the spring of 1941 were rejected, since the expected boost in performance over that of the D-series was considered too low.

At this time and because of this, Junkers undertook a thorough reworking of the Ju 87, with special emphasis being placed on an aerodynamically clean design. To this end, the wings were refined and a retracting undercarriage was planned, among other features. With these changes having markedly altered the appearance compared to previous versions of the Ju 87, the RLM designated the project as the Ju 187, and was given the design's final layout in early 1943. Even the rear-firing gun was to have been improved by fitting a remote controlled gun platform on the dorsal fuselage firing a 13 mm MG 131 and a 20 mm MG 151/20 in place of the MG 81Z. Maximum bomb load would be retained at 2,000 kg, broken down to 1x 1,000 kg bomb under the fuselage and 4x 250 kg bombs beneath the wings. With this load, the top speed of the aircraft would have been little over 400 km/h even under the best of conditions, and it was for this reason that the Ju 187 project was finally canceled in the fall of 1943.

To complete the picture, brief mention must be made of the Ju 87H series, a conversion trainer with dual controls and only some with armament. It was built in limited numbers alongside the D-series variants as the H-1, H-3, H-5, H-7, and Ju 87H-8.

End of the Ju 87

In conclusion, we can see that the Ju 87's weapons arsenal had grown enormously from that of 1939, with its payload potential almost quadrupling and the plane's firepower increasing markedly. Despite a commensurate increase in the aircraft's engine power, a prerequisite for improving the Ju 87's combat value, the Junkers dive bomber remained a slow and cumbersome weapon of war in its combat configuration. Its original, initially quite successful role as a dive bomber practically made it of limited usefulness even before the first one rolled off the assembly line, and its use as a strike aircraft can be considered more a successful stop-gap measure than anything else. Despite its robustness and the ability to operate from improvised airstrips, it should have been replaced years earlier—its loss rate to enemy aircraft and AA fire, particularly with regard to its flying crews, simply could no longer be offset as the war progressed.

Ju 87 production peaked in 1943 with a total of 1,629 examples built, followed by 771 in 1944 before production ceased at the Bremen-Lemwerder plant in May of that year and at Berlin-Tempelhof in August of 1944.

This, then, was the end of the era of the classic Ju 87 dive bomber, which turned the name "*Stuka*" into a legend.

Henschel Hs 129B and C

Under pressure from the demands of the Russian campaign, in mid-1941 it was decided to place greater importance on the role of the strike fighters, which had acquitted themselves so well in the Polish and French campaigns. As a result, work on the Hs 129 was not only given a long absent emphasis, but at the same time was assigned the highest priority. A trial fitting of the French Gnôme-Rhône 14M 4/5, rated at 515 kW/700 hp on takeoff and 485 kW/660 hp at 4,000 m, on two Hs 129A-0s at the Henschel company's Berlin-Schönefeld plant was cleared for series production following a very brief test period at the factory and at the *E-Stelle* Rechlin. Production immediately got underway at Schönefeld with the greatest sense of urgency. This variant, designated the Hs 129B, had somewhat better visibility from what was still an extremely small single-seat cockpit. Its control input forces were somewhat better balanced than its predecessors, and in place of the Hs 129A-0's two MG-FFs was armed with two 20 mm MG 151/20s with 125 rounds each. The pilot's sight, the Revi C 12/C reflective sight, was mounted outside the cockpit in front of the armored windscreen since there was no room inside the cockpit. For the same reason several engine instruments were located on the inside part of the nacelles. The control stick was still unusually short, forcing the pilot to make his control inputs carefully and with great attentiveness. With these aesthetic shortcomings the first of ten contracted pre-production Hs 129B-0s rolled off the assembly line in late December 1941, with three more following in both January and February 1942. The final three from the initial B-0 pre-production batch came in March, along with the first three of the Hs 129B-1 variant. The B-1 differed little from the B-0, and by April 1942 31 Hs 129s had already been delivered. But initial necessary design alterations and delays in getting individual parts caused the monthly output to drop, and it was not until October 1942 that it yet again climbed to 33.

The first strike unit was formed with Hs 129B-0s and B-1s, and on 10 May 1942 set out from northwestern Germany for the Eastern Front. There it did not take long to realize just how sensitive the engine was to even the slightest damage from enemy fire, and to the sand and dust blowing off the steppes of Russia. It had a tendency to seize up without warning, and the unit's readiness state

Henschel Hs 129B showing the gunsight located ahead of the cockpit's armored glass frontplate, as well as descent/dive marker aids on the side windows of the canopy.

dropped catastrophically. The mechanics worked feverishly, trying to iron out the engine's teething troubles, as well as those on the aircraft itself. It took five months before the most critical problems had been corrected and further deliveries could be made to front-line units.

Fate dealt an even worse hand to an Hs 129 equipped strike unit in Poland which transferred to North Africa on 10 November 1942. Four aircraft malfunctioned during the ferry flight and had to turn back, and of the eight remaining only four were operationally fit by the time they arrived in North Africa. Yet again, desert sand was public enemy #1. After two or three missions two of the Hs 129s were lost over enemy territory when their engines gave out. The rest deployed to Tripoli for a thorough technical overhaul, but had to be blown up when the airfield was vacated in the face of advancing British troops. A truly promising prelude for the first use of an urgently needed strike plane!

Despite this, by the end of 1942 a total of 219 Hs 129B-0s and B-1s had been manufactured at Schönefeld. It should not be overlooked that the assembly line was affected by constant requests for modifications and delays in the shipment of engines and components, so that it was not until June 1943 that monthly output reached 40 aircraft—a number that was maintained with considerable difficulty until November of that year. Total yearly output for 1943, at 414 Hs 129Bs, was about half of the planned production output. A further 225 aircraft followed in 1944, but the entire Hs 129 program was canceled in September of that year in favor of the fighter production program, with many Hs 129s being replaced in their strike role by the Fw 190.

In spite of these numerous handicaps the Hs 129 soon developed an outstanding reputation, particularly as a tank buster. In early 1943 these were organizationally combined on the Eastern Front and played a pivotal role in many difficult phases of Germany's defensive actions, as well as enjoying spectacular successes during counterattacks. For example, in two instances in July 1943 they succeeded—without the involvement of friendly ground troops—in halting Soviet tank forces up to brigade strength that had broken through the defenses.

Starting in 1944 the Fw 190 increasingly became the standard plane for strike units, whose inventory of Ju 87s and Hs 129s were subject to wartime attrition without being replaced by new aircraft of the same types, as the assembly lines had closed down.

858 Hs 129 operational types were built. With the three prototypes and eight A-0 pre-production aircraft added, the total number manufactured was 869, and we should take a closer look at the development of this single seat strike plane under the combat conditions prevailing at the time.

The standard armament of the Hs 129B-1 consisted of four fixed machine guns mounted in the nose: 2x 7.9 mm MG 17s each with 500 rounds and 2x 20 mm MG 151/20s each with 125 rounds. Field conversion kits enabled front-line units to install:

- as *Rüstsatz* R2: 30 mm MK 101 with 30 rounds in a weapons pod beneath the fuselage, or
- as *Rüstsatz* R3: 4x 7.9 mm MG 17 exposed, also under the fuselage with ammunition totaling 1,000 rounds carried in the left fuselage, or
- as *Rüstsatz* R4: racks beneath the fuselage for 4x 50 kg bombs or 4x AB 24 bomb pylons for carrying 24 SD 2 fragmentation bombs against soft targets or 1x 250 kg bomb, or
- as *Rüstsatz* R5: a Rb 20/30 aerial camera installed in the fuselage for battlefield reconnaissance.

The 30 mm MK 101 machine cannon, located under the fuselage of the Henschel Hs 129B, is being loaded with a 30 round ammunition belt.

Henschel Hs 129B-2/Wa having its MK 103 loaded.

The sub-variant specially fitted with the tank busting MK 101 30 mm cannon was the Hs 129B-1/R2, but after initial successes against lesser armored targets it was discovered that the MK 101's rounds were not able to penetrate the 45 mm armor of the Soviet T-34 tank. An alternative weapon for engaging tanks was found in the form of the SD 4 hollow charge bomb, which were used until a more suitable means could be found for engaging armored ground targets from the air.

In 1943 the Hs 129B-2 went into production. This variant had all the operationally dictated modifications of the B-1 and could be fitted with all the field kits of its predecessors, but because of the situation in the East the factory soon began delivering it exclusively as a tank busting weapons platform under the designation Hs 129B-2/Wa. Initially, it was fitted with the 30 mm MK 103 in an underfuselage pod. The MK 103 had a higher muzzle velocity and a flatter trajectory than that of the MK 101. However, since the caliber was still not big enough, a handful of Hs 129s were fitted with the same BK 3.7 gun carried by the Ju 87G. Fitting the gun involved removing the two MG 17s located in the lower fuselage in order to make room for the 3.7 mm magazine. Both the MK 103 and the BK 3.7 proved effective weapons when attacking Soviet tanks from the side or the back.

Additional testing in the search to find more effective tank busting weapons and integrate them into this aircraft design led to the installation of various rocket systems and even a flame thrower. The so-called "*Förstersonde*," the SG 113A, was installed on a trial basis in three Hs 129Bs. The SG 113A involved six mortars, each containing one 77 mm shell with a 45 mm anti-tank core, which would fire vertically downward automatically when the electromagnetic field generated by a tank triggered a photoelectric cell in the fuselage nose. A turret-like housing protruding from the Hs 129's fuselage was necessary to house the mortar tubes and absorb the recoil. But the tests in Tarnewitz never reached the stage where the weapon could be considered reliable enough for front-line operations, and the SG 113A was never operationally introduced.

Ultimately, a successful installation occurred with the proven 75 mm PaK 40 anti-tank gun, which was placed in a large underfuselage pod and designated the PaK 40L. Following good test results at Travemünde and Tarnewitz, the gun was produced as the BK 7.5 and installed in the HS 129B-3/Wa. The PaK 40L was fitted with a large muzzle brake and operated electro-pneumatically instead of mechanically; thus configured, it was designated the BK 7.5. It had a firing rate of 40 rounds per minute and carried 12 shells, each weighing 11.8 kg. When opening fire at a range of 500 m, this allowed four rounds per pass, so that a single aircraft could theoretically destroy three tanks on one sortie. Exhaust gases were directed aft, and the entire, somewhat capacious BK 7.5 device could be jettisoned in case of emergency, as single engine flying with the BK 7.5 was barely possible. Only about 25 of these Hs 129B-3/Wa strike planes left the Schönefeld plant in 1944 before all Hs 129 production ceased in September of that year.

Technical Data for the Hs 129B

Role: the single-seat Hs 129B-1 and B-2 aircraft were strike planes of all-metal construction, twin-engined low-wing designs with retractable landing gear.

Airframe:
Fuselage: all-metal, monocoque, three section, almost rectangular in shape. It comprised the forward section (armored cockpit), the middle section, and the aft section.
Undercarriage: separate components. Each strut retracted into the wheel well by means of a hydraulic cylinder. The wheel wells were partially covered by automatically operating doors. Hydraulic struts absorbed the landing shock. A 360° rotatable tailwheel was located at the extreme aft end of the fuselage and was spring-centered. Landing shock was absorbed by ring springs.
Control surfaces: cantilever empennage at the aft end of the fuselage. Ailerons located on the wings. All control surfaces counterbalanced. Trim tabs located on elevators, rudder, and ailerons.
Control inputs: elevator and ailerons controlled by joystick. Adjustable pedals controlled the rudder. Trim tabs adjusted electrically. Landing flap extension and retraction powered hydraulically.
Wings: cantilever, consisting of a center section joined to the fuselage by rivets and two outer sections.

Powerplant
Engines: The two Gnôme-Rhône 14M, model 4/5, were air-cooled four-stroke 14-cylinder double radial engines with reduction gear, boost carburetor, and turbocharger. The two engines rotated counter to each other to offset the counter rotation forces caused by the propellers, so that no special trimming was needed. Airscrew shaft: crankshaft reduction ratio = 1:1.417.

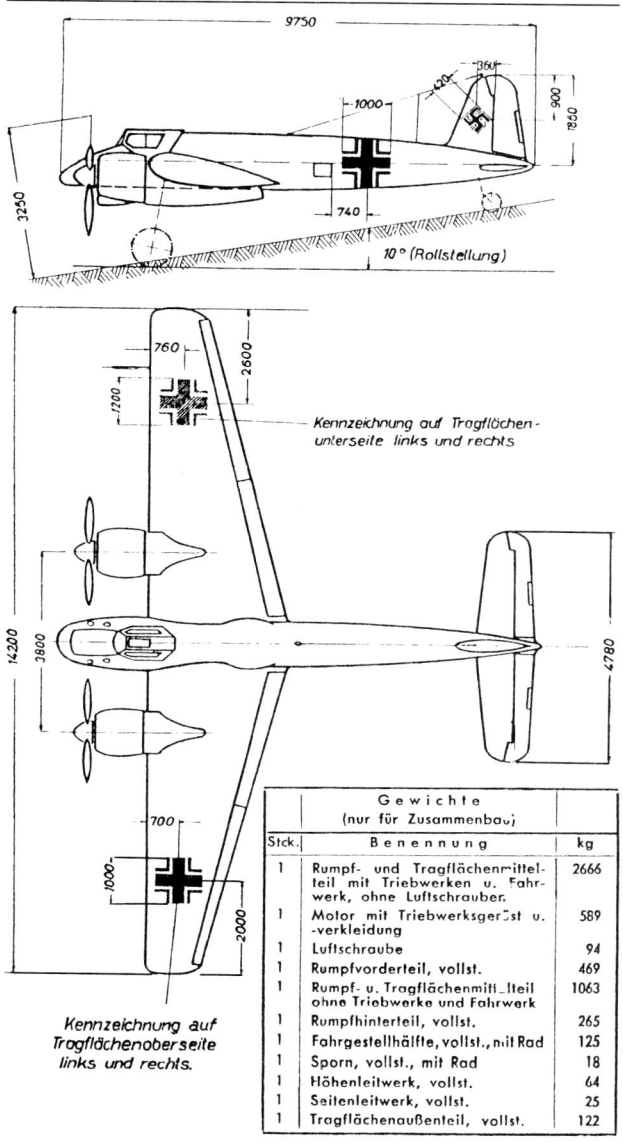

Main dimensions and weights for the Henschel Hs 129B.

Propellers: three-bladed metal Ratier types, with 2.6 m diameter. Manually adjustable electric pitch, as well as automatic. Settings: minimum = 26°, maximum = 50°.

Tanks:
- two wing tanks each with 205 liters + 6% reserve
- one fuselage tank with 200 liters + 6% reserve
- two fuel injection tanks in the wheel wells, each with 3 liters
- two oil reservoirs with cold starting system in the wings, each with 35 liters

Performance

Structural: aircraft met stress group H4 specifications at a maximum takeoff weight of 5,000 kg and—with corresponding identification = H5 up to 5,250 kg takeoff weight. Load factor on pull-out at H4 = 5, at H5 = 6.

Aerobatics: at its maximum weight of 5,250 kg the aircraft was fully aerobatic.

Flight performance (with 3/4 fuel and full ammunition plus ETCs minus bombs):

Maximum speed:
- at sea level ..... 343 km/h
- at 3,000 m ..... 382 km/h

Range and endurance
- at 1.0 *ata* and 2050 rpm
- at sea level ..... 690 km ..... 2 hr 26 min
- at 3,000 m ..... 650 km ..... 2 hr 5 min
- at 0.9 *ata* and 1900 rpm
- at sea level ..... 735 km ..... 3 hr 2 min
- at 3,000 m ..... 680 km ..... 2 hr 29 min

Flight with flaps at full setting ..... max. 240 km/h
Maximum authorized speed ..... 600 km/h
Landing speed ..... 145 km/h

Head-on view of the Henschel Hs 129B, showing the trapezoid shape of the fuselage cross-section to good effect.

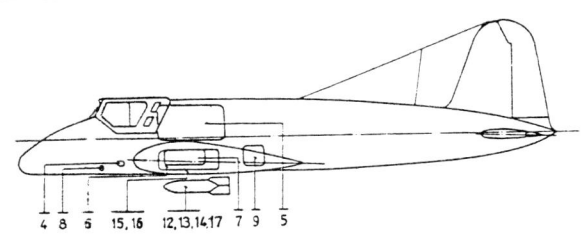

Ladeplan für Schlachtflugzeug Hs 129 B-1
Gültig für Werk Nr. 0151-0200

Dieser Plan hat nur Gültigkeit für das seinem Verwendungszweck entsprechend vollständig ausgerüstete Flugzeug (siehe Beladevorschrift). Fehlende Teile sind durch gleich schweren Ballast am gleichen Platz zu ersetzen.

Nr.	Bezeichnung	I kg	II kg	III kg	IV kg	V kg	VI kg	VII kg	VIII*) kg	IX*) kg
1	Leergewicht mit Gnôme-Rhône 14 M und Ratier-Verstelluftschraube	3661	3661	3661	3661	3661	3661	3661	3661	3661
2	Zusätzliche Ausrüstung	396	422	296	322	211	215	288	245	271
3	Rüstgewicht	4057	4083	3957	3983	3872	3876	3949	3906	3932
4	Führer mit Fallschirm und Schlauchboot	100	100	100	100	100	100	100	100	100
5	Kraftstoff im Rumpf 200 Ltr., Y = 0,74	148	148	148	148	148	148	148	148	148
6	Kraftstoff im Flügel 2×205 Ltr., Y = 0,74	303	303	303	303	303	303	303	303	303
7	Schmierstoff im Flügel 2×35 Ltr., Y = 0,9	63	63	63	63	63	63	63	63	63
8	Munition MG 17	53	53	53	53	53	53	53	53	53
9	Munition MG 151	86	86	86	86	86	86	86	86	86
10	Bomben 2×50 kg	100	—	100	—	100	100	—	100	—
11	Bomben 48×2 kg	—	96	—	96	—	—	96	—	96
12	Bomben 1×250 kg	—	—	—	—	—	250	—	—	—
13	Bomben 4×50 kg	—	—	—	—	200	—	—	—	—
14	Bomben 96×2 kg	—	—	—	—	—	—	—	192	—
15	Munition für MK 101	33	33	—	—	—	—	—	—	—
16	Munition für 4 MG 17	—	—	106	106	—	—	—	—	—
17	Fluggewicht	4943	4965	4916	4938	4925	4979	4990	4759	4781
18	Abgerundetes Fluggewicht	4940	4970	4920	4940	4930	4980	4990	4760	4780
19	Höchstzulässiges Fluggewicht	5000 kg (H 4)								

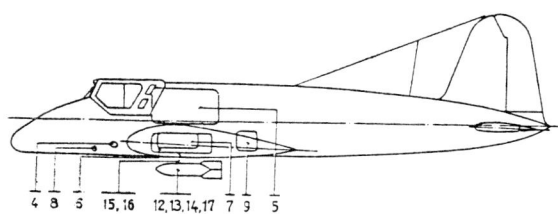

Ladeplan für Schlachtflugzeug Hs 129 B-2

Dieser Plan hat nur Gültigkeit für das seinem Verwendungszweck entsprechend vollständig ausgerüstete Flugzeug (siehe Beladevorschrift). Fehlende Teile sind durch gleich schweren Ballast am gleichen Platz zu ersetzen.

Nr.	Bezeichnung	I kg	II kg	III kg	IV kg	V kg	VI kg	VII kg
1	Leergewicht mit Gnôme-Rhône 14 M und Ratier-Verstelluftschraube	3810	3810	3810	3810	3810	3810	3810
2	Zusätzliche Ausrüstung	387	387	291	291	209	209	214
3	Rüstgewicht	4197	4197	4101	4101	4019	4019	4024
4	Führer mit Fallschirm und Schlauchboot	100	100	100	100	100	100	100
5	Kraftstoff im Rumpf 200 Ltr., Y = 0,74	148	148	148	148	148	148	148
6	Kraftstoff im Flügel 2×205 Ltr., Y = 0,74	303	303	303	303	303	303	303
7	Schmierstoff im Flügel 2×35 Ltr., Y = 0,9	74	74	74	74	74	74	74
8	Munition MG 17	53	53	53	53	53	53	53
9	Munition MG 151	97	97	97	97	97	97	97
10	Bomben 2×50 kg	100	—	100	—	100	—	100
11	2 Bombenbehälter AB 24 t SD 2	—	113	—	113	—	113	—
12	Bomben 1×250 kg	—	—	—	—	—	—	250
13	Bomben 4×50 kg	—	—	—	—	200	—	—
14	4 Bombenbehälter AB 24 t SD 2	—	—	—	—	—	226	—
15	Munition für MK 101	30	30	—	—	—	—	—
16	Munition für 4 MG 17	—	—	106	106	—	—	—
17	Fluggewicht	5102	5115	5082	5095	5094	5133	5149
18	Abgerundetes Fluggewicht	5100	5120	5080	5100	5090	5130	5150
19	Höchstzulässiges Fluggewicht	5250 (H 4 und H 5)						

Final Tally and the Hs 129C

In spite of its numerous initial problems, the Hs 129 evolved into an outstanding close support strike aircraft, especially when it came to the tank busting role. Its Achilles' Heel were its engines, which remained quite sensitive despite a series of improvements, so that the strike units equipped with the type chronically suffered from a low operational readiness rate most of the time. The Henschel company's developmental team under *Dipl.-Ing.* Nicolaus attempted to find a better alternative to the problematic Gnôme-Rhône and thus initiate a new variant under the designation Hs 129C, which would also have had two MK 103 mounted side by side beneath the fuselage with limited lateral movement. In the end, however, this project failed because the intended engine, the more powerful Isotta-Fraschini Delta RC 16/48 12-cylinder inverted-V motor (rated at 618 kW/840 hp at 5,300 m), was delayed in development and did not reach production in time to be fitted to the Hs 129 before production was halted.

Here, too, the Fw 190 was diverted from fighter production to assume the role of the standard strike fighter, a positive development when viewed in light of commonality and logistics, and a development which perhaps should have happened even sooner than it did!

e) Reconnaissance Aircraft

Long range Reconnaissance Aircraft

Junkers Ju 88D: Development in 1941/1942 can be summed up here in just a few words, since the Ju 88 was the primary type serving in the strategic reconnaissance role during this period. 568 Ju 88s were delivered in 1941, and a nearly identical number (567) rolled off the assembly lines in 1942. With the availability of the Jumo 211J improving the powerplant situation somewhat, this engine was installed in the Ju 88D-1, and its tropicalized brother, the Ju 88D-3, while parallel to the D-1 was produced as the Ju 88D-5—standardized to three camera types, including the Rb 75/30 with its even greater focal length. Ju 88 reconnaissance aircraft operated out of Norway, flying over the northern Atlantic as far as Iceland. They also flew from Sicily over the Mediterranean and North Africa, and over England and the eastern Atlantic from Brittany. But their main theater of operations was in support of the Russian campaign, for which the Romanian and Hungarian Air Forces were also later equipped with the Ju 88D-1 long range reconnaissance aircraft.

Dornier Do 215B: The final six Do 215 strategic reconnaissance platforms were assigned to their units in 1941, following which production came to a halt. However, a Do 215 testbed for the FuG 202 Lichtenstein BC radar system developed for night fighting scored its first kills against enemy bombers with the system in September and October 1941, although the additional antennas caused a loss in speed of 30 to 40 km/h.

Focke-Wulf Fw 200C: This type has already been discussed in the section dealing with the Fw 200C heavy bomber, wherein it was mentioned that the reconnaissance potential of this aircraft, constantly flying the mixed "armed reconnaissance" role, had been given a decisive new dimension—namely the ability to conduct all-weather reconnaissance over the ocean—with the fitting of the FuG 200 Hohentwiel surface search radar. Also addressed in this context was the recce tactic developed with this search radar by using a search circuit pattern to pick up shipping over large areas of ocean.

Junkers Ju 86P: In January 1941 the *"Versuchsstelle für Höhenflüge"* (Research Center for High Altitude Flight) continued with their reconnaissance and harassment bombing of the British Isles using a number of Ju 86P-1s and P-2s operating out of northern Germany, flying at altitudes where the enemy could not reach them. The same research center also flew photo reconnaissance missions—also untouchable at their high operating altitudes—from Bucharest and Krakow over Russia in preparation for Germany's attack on the Soviet Union, as well as to determine possible Soviet troop movements.

In January 1942 this research center was incorporated into the *"Versuchsband Ob. d. L."* (Research Unit of the Luftwaffe High Command), under whose authority it continued flying high altitude missions in the West. In May 1942 a handful of Ju 86P-2s were sent to Crete, detached to a strategic reconnaissance unit there, and flew high altitude recce missions over Egypt. Here, too, they remained unscathed—until 24 August 1942 when a specially modified Spitfire Mk. V succeeded in intercepting a Ju 86P-2 north of Cairo at an altitude of about 11,300 m. After pursuing its quarry over a long period of time, climbing ever higher, at an altitude of 12,800 m the Spitfire fired a burst which set the starboard engine on fire; the Ju 86 crashed shortly afterward. Two further Ju 86P-2s were shot down by Spitfire Vs, which had been configured to be especially light and were operating from Aboukir, so that by the time the reconnaissance unit transferred to Greece in August 1943 they only had two P-2s, neither of which were operational.

Opposite: This load plan for the Henschel Hs 129 shows the armament and equipment layout for the B-1 variant up to a maximum of 5,000 kg (in accordance with stress category H4) and for the B-2 up to 5,250 kg (in categories H4 and H5). The profusion of potential combinations for this strike plane is clearly manifest here.

Ju 86R: In anticipation of this disastrous development for high altitude reconnaissance, even prior to the debacle Junkers had already been working on increasing the operational altitude of the Ju 86. This was to be achieved with the Ju 86R-1, which as a high altitude reconnaissance platform had the same camera systems as the P-2, but its wings were extended by 3.20 m each for a total span of 32.02 m, and it had the Jumo 207 diesel engine. Beginning in the spring of 1942 several Ju 86P-2s were converted into this new reconnaissance variant, with a few Ju 86P-1s being converted at the same time into the Ju 86R-2 high altitude bomber. During trials the R-1 and R-2 reached altitudes up to a maximum of 14,800 m. The six-cylinder Jumo 207B-2, later the B-3, diesel aircraft engine with exhaust turbocharger provided 735 kW/1,000 hp on takeoff and 552 kW/750 hp at 12,000 maximum pressure altitude. It had a GM 1 system, which offered a significant performance increase by boosting the intake pressure and increasing the heat exchange in the cylinders by enriching the intake manifold with a liquid oxidant (N_2O nitrous oxide). Due to temperature reasons, however, the GM 1 system could only be activated above 13,000 m altitude. As a result of these technological-physical limitations the operational Achilles Heel of this improved high altitude aircraft lay between 12,000 and 13,000 m, in which combat with enemy high altitude fighters was to be avoided if possible, since engine performance dropped steadily above the 12,000 m maximum pressure altitude. It was not until the aircraft was above 13,000 m that the GM 1 injection system kicked in and briefly provided a 150 to 220 kW/200 to 300 hp boost in engine performance.

A small number of these Ju 86R-1s flew special missions as part of the "*Versuchsband Ob. d. L.*," operating at extremely high altitudes. In February 1944 the unit still had four of these high altitude reconnaissance aircraft with their almost grotesquely large wingspan, but by the end of July 1944 these were no longer listed on the inventory and had probably been phased out. Earlier, in September 1942, two Ju 86R-2s had appeared serving with a bomber unit in Holland, but the unit was disbanded just a month later; here, too, there was no further trace of the fate of these high altitude harassment bombers. From an operational viewpoint, the Ju 86 era can be said to have finally ended with the Ju 86R.

It is, however, worth mentioning that even before the Ju 86R-1 entered service Junkers had drawn up plans for a Ju 86R-3, which was to have been powered by a significantly improved Jumo 208 diesel engine offering 1,103 kW/1,500 hp on takeoff and 809 kW/1,100 hp at about 13,500 m, with a service ceiling of 15,800 m. But the Jumo 208 never made it beyond a prototype stage, and construction of the R-3 was abandoned. The same fate befell a parallel project for a Ju 186, which had been planned as a high altitude testbed by using a Ju 86P fuselage with either four Jumo 208s or two Jumo 218s (double engine made from two Jumo 208s) and a fixed undercarriage.

This was the ultimate end of the Junkers Ju 86 program, a long-obsolete design, but one with an astoundingly long life and one which proved to be most versatile.

Tactical Reconnaissance Aircraft

Henschel Hs 126: This was the main workhorse in the East, as well, with all tactical reconnaissance units operating the type—except for a single unit in North Africa—supporting the Army of the East. The last five Hs 126s rolled off the assembly lines in January 1941, and since Fw 189 production was fully underway by this time, by the spring of 1942 the Hs 126 had almost completely been replaced in the units by the Fw 189A and other tactical reconnaissance platforms.

Messerschmitt Me 110 and Me 109: Me 110 recce planes operated from Norway, observing the ongoing Russo-Finnish War, as well as during the Battle of Britain, where among other things they helped sight heavy coastal artillery guns against targets along England's south coast. They also flew daylight reconnaissance missions over the island. During the Balkans campaign they flew in support of the Army, and with the start of the Eastern campaign were predominantly active in the south and central sectors of the Eastern Front. In August 1943 a portion of these were pulled back for homeland defense in the *Zerstörer* role, while other Me 110 reconnaissance units converted over to the faster Me 109. The Me 110 also was operational with reconnaissance units in the Mediterranean theater beginning in late 1941/early 1942.

Production of reconnaissance Me 110 peaked in 1941 at 190, then dropped by over 50% in 1942 to just 79 before reflecting the

Messerschmitt Me 110E-3 long range reconnaissance plane, photographed during maintenance work in North Africa (1942). The object in the foreground is the Rb 50/30 camera and its film cassette.

History of German Aviation: Bombers and Reconnaissance Aircraft

Focke-Wulf Fw 189A on a forward airstrip somewhere in the East (1942).

great demand in 1943 by rising to 150. This was the last year for production, however, since there had been a fundamental conceptual change in favor of the single seat, single engine, and primarily much faster Me 109. To be sure, no specialized reconnaissance version was built in 1941, but the successes enjoyed by the Me 109E-5, E-6, and E-9 recce fighters over England in 1941 and 1942 led to more of these same high speed tactical reconnaissance version being produced. By the end of 1942 an initial batch of eight had rolled off the reopened assembly lines, but in 1943 141 examples were delivered to the tactical reconnaissance units.

Focke-Wulf 189: Beginning in 1941 the Fw 189—already discussed in detail in a previous section—started arriving at front-line units, where it replaced the Hs 126. In 1941, the year Hs 126 production ceased, the last five examples of these planes were delivered—totally eclipsed by the production of 250 Fw 189s that same year. Fw 189 production increased to 327 in 1942 and dropped back to 208 in 1943. The following year production terminated at 17 aircraft, since tactical reconnaissance units in the meantime had been almost completely reequipping with the Me 109. The Me 109's reconnaissance variants were included as part of the rapid step-up in the tempo of fighter production, with the fighter program taking top priority at this stage of the war, and became the most numerous tactical reconnaissance aircraft built.

Recce Summation and the Outsider

For the strategic reconnaissance aircraft it was only the role of the Ju 86 as a high altitude reconnaissance platform that was worthy of note. However, the tactical reconnaissance role ran the gamut from the specially designed single-engined high wing He 46 and Hs 126, continuing with the twin-engined Fw 189 and its specially suited fully glazed cockpit and ending with the high-speed multi-role aircraft. In this latter category fell the Me 110 first, which played a relatively flexible role to include that of a long range reconnaissance aircraft. Quite soon, however, it had been overtaken in the pure tactical reconnaissance role by the faster Me 109. By 1943 the dedicated tactical reconnaissance design had virtually died out; it simply could no longer survive in a battlefield environment.

We should perhaps nevertheless briefly mention a specialized tactical reconnaissance development and look at it in more detail—not because of its operational potential that this type would never have been able to provide in front-line service—but because of its

Blohm & Voss BV 141, the so-called "flying asymmetry," tested as a tactical reconnaissance platform from 1938 to 1943. However, it was never expected to go into production, not to mention into active service. This photo of the BV 141V12 with its BMW 801A-0 (1,147 kW/1,560 hp) engine emphasizes the unique nature of its design, for which *Dr.-Ing.* Vogt was awarded a patent by the German patent office. The design was not so unusual, however, that the Gotha Werk had not attempted something similar back in 1918.

unusual design, which made this aircraft, to put it bluntly, a definite outsider.

Blohm & Voss BV 141

The first machine had been developed at Hamburger Flugzeugbau as the Ha 141-0 and flew for the first time on 25 February 1938, while the last, the BV 141B-B2 (V13), was not finished until 15 May 1943, nearly a six year history of what was a remarkable aircraft, if ultimately a developmental failure.

"The Flying Asymmetry," as this design by *Dr.-Ing.* Richard Vogt was sometimes called, was a mid-wing layout with a fuselage section and engine offset to port from the wing centerline and a fully glazed crew compartment enveloping the wing offset to the right of centerline. During its tedious developmental cycle the BV 141 failed to materialize as a serious competitor to the Fw 189. Although the BV 141 was adjudged generally favorable at the conclusion of the official flight trials at the Rechlin test center, the *Luftwaffe* leadership was never able to embrace this unorthodox airplane. Instead, it preferred to cancel the intended series production on 4 April 1940, in part justifying it by the fact that engine performance was too weak. *Dr.-Ing.* Vogt had recognized this earlier, and since early 1939 had been working on an improved version, the BV 141B. The RLM inspected a mockup of the B-variant on 14 February 1940. This BV 141B was powered by the BMW 801A-0, with a takeoff rating of 1,147 kW/1,560 hp, nearly twice the power of its predecessor. For the B version the company was awarded a fixed contract for five BV 141B-0s, the first of which began its flight testing on 9 January 1941 as the BV 141B-0/V9. Even during static testing, however, structural deficiencies and problems with the hydraulic system were found, so that the aircraft was limited to a maximum speed of 450 km/h in flight. The subsequent V10 through V12 revealed a multitude of smaller problems, with the latter in particular suffering difficulties with its weapons during firing trials at Tarnewitz. The test program was constantly interrupted as a result, and the last of these aircraft, the BV 141V13, ultimately was not delivered until 15 May 1943.

However, this particular reconnaissance project had been dropped back in the spring of 1942, since the Fw 189—already in production and proven in service—was fully up to the task of the BV 141's intended role of battlefield reconnaissance. This was the last straw for this interesting tactical reconnaissance aircraft which, in spite of considerable developmental investment, never entered service. Normally, this book would not deal with such a design, which neither reached front-line maturity nor entered into production, but as an outsider the BV 141 falls outside the normal framework. The following table contains the most important reference data for the BV 141B, specifically for the BV 141B-02 (BV 141V10).

The BV 141B was an asymmetrically designed, single-engined three-seat tactical reconnaissance aircraft for Army support.

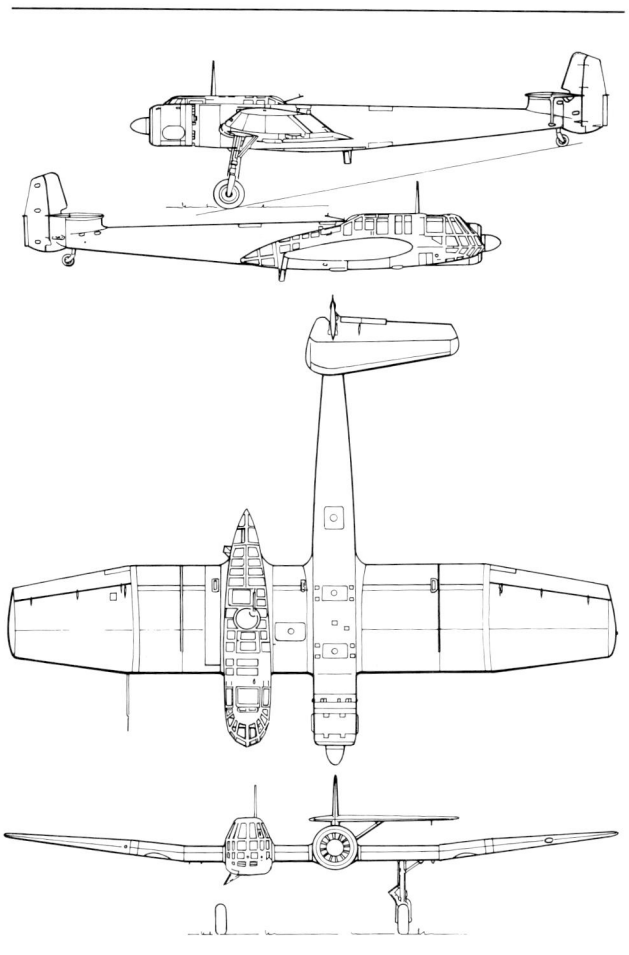

Blohm & Voss BV 141B-0 (1941).

Engine: air-cooled BMW 801A-0 14-cylinder radial engine rated at 1,147 kW/1,560 on takeoff.

Armament: 2x 7.9 mm MG 17 forward-firing fixed machine guns, 2x MG 15 rear-firing turret-mounted machine guns, 4x 50 kg bombs on external underwing racks.

Flight Performance:
Maximum speed
- at sea level   369 km/h
- at 3,000 m   438 km/h
Standard range   1,200 km
Maximum range   1,900 km
Service ceiling   10,000 m

Weights:
Empty equipped   4,700 kg
Normal takeoff   5,700 kg
Max. takeoff   6,100 kg

Dimensions:
Wingspan   17.46 m
Length   13.95 m
Height   3.60 m
Wing area   53.00 m^2

16. Course of the War from 1942 Onward

In order to provide a continuous picture of bomber and reconnaissance aircraft types, we have already looked at certain obsolescent models which were no longer up to their operational roles during this time period. The events of the war, where obvious, which dictated the tactical-technological demands on aircraft are chronicled in brief over the next few pages in order to better understand the role of the bomber and reconnaissance forces and their airplanes.

In the East

On 10 January 1942 Hitler revised his earlier directive from 14 July 1941 calling for the armament industry to focus on building up the *Luftwaffe*; instead, Germany's industrial resources efforts would again be directed toward the *Heer*. Factors which undoubtedly contributed to this reversal included Japan's entry into the war on 7 December 1941, Germany's presumptuous declaration of war against the United States on 11 December 1941, and the failure of the *Blitzkrieg* tactics of Operation Barbarossa in the eastern winter of December 1941, where Hitler personally assumed command authority on 19 December 1941. The Army' struggle in the East—despite the general order forbidding any type of retreat—now seemed as though it would be lasting much longer than any previous *Blitzkrieg*.

Following the bitter defensive fighting during the winter, Germany's offensive was given new life in April 1942. German attacks on Leningrad were stepped up—primarily those involving massed dive and level bomber raids on anti-aircraft positions, port facilities, and shipping. These attacks caused substantial damage without achieving any decisive victories.

In May 1942 the Kertch peninsula was retaken after having been lost six months earlier, and the Crimean fortress of Sevastopol fell to German forces after being subjected to withering aerial bombardment. A Soviet large-scale offensive along the "Southwest Front" begun near Kharkov at the beginning of May concluded at the end of the month with a German counterattack south of the city, which encircled the Soviet forces and resulted in a quarter of a million Russian prisoners. This was to be the last successful German encircling operation. In late June 1942 a large-scale German summer offensive began against Stalingrad and the Caucasus. On 19 August 1942, after German troops had reached the oil fields in the northwestern region of the Caucasus in early August, *General* Friedrich Paulus ordered his 6th Army to attack Stalingrad itself. By October of that year German units had taken 90% of the city, but the offensive bogged down in house-to-house fighting and with the onset of winter.

On 19 November 1942 the Soviets launched a counteroffensive near Stalingrad, and after breaking through Romanian lines the prongs of the Soviet attack striking out northwest and south of the city met behind the 6th Army. In so doing, 284,000 men found themselves cut off in the Stalingrad area between the Volga and the Don rivers. On 23 November Hitler rejected *General* Paulus' request to break out to the west and rejoin German forces after Göring guaranteed that the *Luftwaffe* would supply the trapped soldiers from the air—as had been done successfully in the pockets at Demyansk and around Kholm. Hitler ordered: "The 6th Army is to dig itself in and await relief from the outside."

On 25 November 1942 aerial resupply into the Stalingrad pocket began with a daily average of 95 metric tons of goods instead of the required and intended 300 tons! A relief effort failed just before Christmas 1942. Soon the airfields in the ever-shrinking pocket could no longer be used. Even though every bomber and transport plane that could be found was thrown together and used for dropping supplies to the 6th Army, the resupply effort remained entirely insufficient. The inordinately harsh winter and the much too inadequate air force resources needed to sustain an air bridge of this scale under icy weather conditions, coupled with inadequate ground personnel to handle the supplies and Germany's losses in general, forced *General* Paulus to capitulate the southern pocket after fierce house-to-house fighting on 31 January 1942. The north pocket of the split 6th Army followed suit on 2 February 1942. Approximately 90,000 men became Soviet prisoners-of-war.

Aircraft losses suffered by those flying supply missions into Stalingrad were enormous, especially for the bombers: 169 He 111s, 42 Ju 86s, 9 Fw 200s, 5 He 177s, and 1 Ju 290 were lost. In addition to these 226 machines, most of which fell into the offensive air sector, there were also 269 Ju 52s lost (primarily falling under the category of pilot training) together for the most part with their crews. All these in the space of about two months!

From this point on, virtually all along the Eastern Front German forces began their irreversible retreat under constant pressure from the Red Army. After 900 days of besieging Leningrad the German blockade was finally broken, and the last major German offensives in the central sector, the battles at Kursk and Orel (Operation Zitadelle) failed in July and August 1943. The offensive strength of Germany's armed forces and her allies was broken.

North Africa

Fortunes in Africa soon changed, as well, where *Generalfeldmarschall* Erwin Rommel and his *Afrikakorps*, together with Germany's Italian allies, had been fighting the British with mixed success since the spring of 1941. Rommel had succeeded with a counter thrust which threw the British back to west of Tobruk, but then his offensive came to a standstill on 7 February 1942. Sometime within the first six months of 1942 Malta, a main thorn for resupplying North Africa, was to have been bombed in preparation for an airborne assault. This plan, Operation Herkules, was postponed indefinitely by Hitler on 21 May 1942. In June 1942 German/Italian troops, with massive air support, succeeded in capturing Tobruk and advanced as far as El Alamein, which they had reached with their weak tank forces and reconnaissance vehicles by the end of the month, and where they remained. British counterattacks were half-hearted affairs and initially met with only limited success.

The British lost many ships in the waters off Tobruk through air attacks. In October 1942 Malta once again was subjected to concentrated German/Italian air raids of about 200 to 270 aircraft daily. On 23 October, however, General Montgomery began his decisive series of counteroffensives with over 1,000 tanks against a German/Italian army of just 500 tanks, and Malta fell into the background. Two weeks after the start of Montgomery's offensive, American and British troops under the command of General Eisenhower made a surprise landing in Morocco and Algeria under weak counterfire from the French troops stationed there, who gave up the fight after just a few days. On 17 November 1942, ten days after their landing, Eisenhower's troops moved from Algeria into Tunisia where they engaged German forces for the first time about 50 km west of Bizerta and Tunis. In the east, Tobruk had already fallen on 13 November under the pressure of superior British forces, and on 20 November 1942 Benghazi followed, and with it all of Cyrenaica.

From then on the German/Italian forces in North Africa were involved in a hopeless threat from both the east and the west.

The conference at Casablanca from 14 to 26 January 1943 was to have many consequences, one of the most significant being Churchill's concept of "unconditional surrender" he announced to de Gaulle and Roosevelt, and which Germany, Italy, and Japan would be forced to accede to. It was agreed that the short term objective would be the capture of Tunisia and an Allied landing in Sicily, plus the bombing campaign against Germany would be intensified. In February and early March 1943 Rommel launched a series of failed thrusts in the direction of eastern Algeria in an attempt to interrupt the Allies' advance. Rommel expressed his doubts that the "the Tunisian Bridgehead" could be held for long, whereupon Hitler called him home and replaced him with *Generaloberst* von Arnim. In late March 1943 Montgomery's 8th Army launched its major offensive against the Italians, who were compelled to fall back. In April 1943 the Americans and British closed the ring around von Arnim's *Heeresgruppe Afrika* situated in the northeastern part of Tunisia. On 7 May 1943 Bizerta and Tunis fell, and on 12/13 May 1943 *Heeresgruppe Afrika* was forced to surrender. 130,000 German and nearly 120,000 Italian soldiers became British and American prisoners-of-war. The North African theater had ceased to exist. Instead, it became a springboard for further operations by the Allies in the Mediterranean area.

In the West, Over the Reich and the *Luftwaffe* Leadership Crisis

Germany's weaknesses in the air became even more clear when in 1941 it was unable to force England to its knees. Aggravating the matter even further was the need to scatter the affected flying units in the south, southeast, and Russia, as well. To be sure, the air war against the British Isles continued on, but the relatively weak bomber force only managed to drop 3,600 metric tons of bombs in all of 1942, while that same year the Allies succeeded in unloading 53,000 metric tons of bombs over the Reich and the occupied West—almost 15 times as much ordnance. The protection given to German cities and armament factories was wholly inadequate in the face of Allied air raids. In order to coordinate their attacks, the Americans formed the U.S. Army Bomber Command on 22 February 1942 in Great Britain, and on 23 February 1942 Air Marshal Sir Arthur Harris was given command of the RAF's Bomber Command. He would go down in the history of strategic air warfare as "Bomber Harris."

Following several smaller day and night raids, British Bomber Command carried out its first area bombing of a larger German city on 28/29 March 1942, destroying the city center of Lübeck. After several more bombing raids on numerous cities, on 30/31 May 1942 the first "Thousand Bomber Raid" targeted Cologne, onto which was dropped almost 1,500 metric tons of bombs within the space of just 90 minutes. The war of destruction from the air took on a new dimension with this attack, one which made everything before it seem like nothing more than a gruesome child's game.

History of German Aviation: Bombers and Reconnaissance Aircraft

Compared to this monstrous concentration of strategic destructive power, German bombing raids in April and May 1942 were typically flown with between 25 and 90 aircraft per mission against targets in southern England. These raids caused no more than a fragment of the destruction the Allied bomber fleets were unleashing against the German hinterland.

The large-scale nighttime raids carried out by British bombers against the German Reich were supplemented by the USAAF's (United States Army Air Forces) first daylight mission flown from Great Britain on 27 January 1943. From this point on, their bomber formations—albeit somewhat sporadic at first—could be seen in the skies over Germany. The quantity of bombs and their intended targets began to broaden at an ever greater rate. The arsenal of the Allies seemed incomprehensible, and losses seemed to be replaced without effort. In early March 1943 Göring had designated a special "*Angriffsführer England*" (lit. "Assault Leader England"), and in May of that year he attacked targets in Great Britain with a total of 237 bombers. In June about 100 German fighter-bombers attacked coastal targets—mere pinpricks in comparison with the Allied strikes taking place on almost a daily basis. From 24 to 30 July 1943 Hamburg fell victim to a large-scale series of raids as part of the Allies' Combined Bomber Offensive. About 3,000 heavy British and American bombers dropped approximately 9,000 metric tons of bombs day and night (i.e. around 3,000 kg per plane) mainly on the city center, where a firestorm broke out. As a result of disabling German radar systems by dropping chaff, the Allies suffered relatively minimal losses. Raids against industrial cities, aircraft works, ball bearing factories, test centers, etc. continued systematically over the following weeks. In contrast, Germany carried out only minor air raids against targets in southern and central England. This is the depressing scenario that was playing itself out by the summer of 1943.

But it was an attack on the Peenemünde test center that was to have even more far-reaching consequences and become a trigger for change. After a series of mistakes and failures by the *Luftwaffe*, especially that of its anti-aircraft, the position of the *Luftwaffe*'s chief of staff, held by *Generaloberst* Jeschonnek, became increasingly tenuous. Göring was already looking for his replacement on the sly. Hitler made the gravest accusations against Jeschonnek personally after the disasters of Schweinfurt (ball bearing plant) and Regensburg (Messerschmitt). The subsequent RAF attack on the Peenemünde rocket test center sparked total confusion in giving orders for deploying night fighters using ground controlled interception techniques. The same went for the "*Wilde Sau*" method introduced by *Major* Hajo Herrmann in the face of much prejudice. In the main, these were single-engined fighters adapted to the night fighting role, and on the night of the Peenemünde raid about 200 aircraft had formed up in the skies over Berlin. Mistaken for a smaller formation of British Mosquitoes, these night fighters came under heavy fire from their own flak batteries. Milch, who was watching the operation, was asked to order cease fire himself or to instruct that the order be given, which he did after consulting with Göring, who was in Berchtesgaden. But from East Prussia, Jeschonnek declared the order given to the AAA units to cease fire when friendly fighters were in the area to be invalid. Shortly afterward, there followed a heated exchange between Göring and Jeschonnek over the telephone, which apparently was the final straw as far as Jeschonnek was concerned. On 18 August 1943 he committed suicide, leaving behind a note saying he could no longer work with the *Reichsmarschall*. On 20 August 1943 Göring introduced the new Chief of Staff for the *Luftwaffe* to Hitler; it was *Generaloberst* Günter Korten. However, Korten was able to execute his responsibilities for less than a year; on 22 July 1944 he died from wounds suffered in the attempted bombing assassination of Hitler in his headquarters on 20 July.

Many far-reaching decisions were made during the period of Korten's responsibility for defining the operational concept of the *Luftwaffe*, one of these being the role of jet powered aircraft. The operational deployment of these new weapons was just over the horizon following the successful development of production-ready jet engines by both BMW and Junkers and, in the summer of 1942, the installation of a Jumo jet engine in a Messerschmitt Me 262. The Me 262 had been undergoing flight trials since 18 April 1941 with the piston Jumo 211G, but on 18 July 1942 Fritz Wendel successfully piloted the Me 262V3 on the first jet powered flight of what had been conceived as a twin-jet single-seat fighter. Although the V3 was lost in a takeoff crash in August 1942, development of this jet fighter continued on, and in 1943 a handful of experienced fighter pilots were able to fly the Me 262. One of these was the *General der Jagdflieger* (General of Fighters), *Generalleutnant* Adolf Galland. "It is as though an angel was pushing me" were the famous words of Galland, when he was later asked to describe the experience. There is more than enough authentic material penned by authoritative figures regarding Hitler's belated decision in December 1943, following a flight demonstration of the Me 262 at Insterburg in East Prussia, to deploy it not as a fighter, but as a so-called "*Blitzbomber*" against the upcoming Allied invasion.* Nevertheless, in a later section of the book it will be necessary to briefly examine the use of front-line capable jet aircraft in their development for the bomber and reconnaissance roles.

But for now, let us turn our attention to the precarious situation with regard to bombers.

---

*   Galland, Die Ersten und die Letzten

17. Bomber Planning from October 1943

What follows is a brief situational report on considerations in light of the previously mentioned planning.

Comparison of the Bomber Situation in May 1940 with that of May 1943

Despite recurring losses, a snapshot taken on 31 May 1943 of the actual bomber and reconnaissance aircraft on strength shows a rather impressive overall strength which seemed hardly affected by the ongoing war. These numbers, however, belie the fact that the multi-front war had fragmented the *Luftwaffe's* effectiveness to the point that a concentrated application of force along individual fronts was limited, despite the inherent flexibility of the use of air power. This limitation was primarily due to the *Luftwaffe's* resources being assigned to specific "air fleets," or commands, which in practice effectively eliminated any short-term concentration of force for large-scale strikes. On 31 May 1943 the distribution of forces appeared as follows:

	East	West	South & South-west	North	Total
Bombers	580	300	217	11	1,108
Dive Bombers	434	-	-	-	434
Strike & Hvy Ftrs	217	-	174	-	391
Night Strike Acft	220	-	-	-	220
Total Bomb Platforms	1,451	300	391	11	2,153
Tactical Recce Acft	300	31	23	13	367
Strategic Recce Acft	204	100	49	20	373
Total Recce Acft	504	131	72	33	740
Total Offensive Acft	1,955	431	463	44	2,893

The term "night strike aircraft" requires some clarification. These were aircraft assigned to auxiliary bomber elements established on 7 October 1942 for the purpose of harassing the enemy at night by attacking villages close to the front, occupied wooded areas, etc. The hodge-podge inventory comprised trainer aircraft or obsolescent operational types, such as the Ar 66, Ar 96, Cr 42, Do 17, Fw 58, Go 145, He 46, He 50, Hs 126, W 34 and, from 1944 onward, also the Ju 87, which had become available as the strike units converted over to the Fw 190. Only one night strike group, which later saw service in the West, was equipped from the outset with the Fw 190. In the main, the operational value of night strike aircraft was marginal, and the type was primarily considered nothing more than a harassment measure—a supplemental measure born from the similar and rather unpleasant harassment missions flown by the Soviets with their U-2s, also known as "sewing machines."

In the East, the 580 bombers generally served in cooperation with ground forces, i.e. in the tactical role, whereas the 300 in the West flew in the operative role against England and supply shipping convoys, and the 217 in the South and Southwest were also used mainly against shipping targets, as well as the air and naval bases on Malta. In this situation, the *General der Kampfflieger* (Gen. d. K., General of Bombers)* focused primarily on turning into reality one of the original tenets of L.Dv. 16 "*Luftkriegführung*," namely the targeting of the enemy's power bases, striking at the root of his will to fight and resist. To this end, the aviation industry had provided the *Luftwaffe* with the following:

	1942	1943
Ju 88	2,270	2,160
Ju 188	165	301
Ju 388	-	4
He 111	1,337	1,405
Do 217	564	504
He 177	166	415
Total	4,502	4,789

---

* On 14 March 1943 Hitler appointed 29 year-old *Oberstleutnant i. G.* Dietrich Peltz as *Angriffsführer England* (Task Force Commander England) on the recommendation of Göring, to whom he was immediately subordinate. As an *Oberst*, that very year he became *General der Kampfflieger* (General of Bomber Pilots), was promoted to *Generalmajor*, and as commanding general became in charge of the *Fliegerkorps* specially tasked with prosecuting the air war against England.

History of German Aviation: Bombers and Reconnaissance Aircraft

Referring to the status of 31 May 1943, it is worth noting that the 1,108 bombers available for combat in front-line units was barely a fourth of the yearly output for 1942 or 1943. The remaining three-quarters must therefore be considered to have fallen victim to wartime attrition. At the start of the Western Campaign in May 1940 there were 1,120 bombers available for service in front-line units, i.e. just twelve more than three years later, and (discounting the cyclical drops) their operational losses were roughly offset by a yearly production output in 1940 of 2,852 aircraft and 3,373 aircraft in 1941. With the number of available aircraft on 31 May 1943 being virtually the same, it can be ascertained that—referring to 1942 and 1943—approximately 1,500 more aircraft had to be produced per year compared to 1940 and 1941 in order to maintain roughly the same combat readiness levels. As the war dragged on and increased in intensity, so had the "wear and tear" on the *Luftwaffe*—and losses became considerably greater!

Demands of the *General der Kampfflieger*

It was against this background that the Gen. d. K., in his "*Bomberplanung*" paper of 5 October 1943, pushed for a decision regarding the quantity of bombs to be dropped on the enemy per day in the different theaters of conflict, as this number would determine the requisite number of aircraft. Also to be established was the ratio of special units to normal bomber units. With the *Waffengeneral* being given responsibility for the ground support units at about this time, meaning that these units were no longer under the purview of the Gen. d. K., the Gen. d. K. felt a prerequisite for the establishment of operative bomber units was that these would be available exclusively for purely operative roles "and only in cases of emergency would the exception be made for using them to support friendly attack or defensive ground operations." To this end, he calculated that the current monthly output of about 380 various types of bombers be shifted and increased to 600 modern bombers, with the basic models available solely for the bomber sector, i.e. no more diversions for night fighting, heavy fighter, or reconnaissance units. He rejected as being uneconomical the further production of five different medium, twin-engined bomber types, these being the Ju 88/188, Do 217, He 111, Me 410 (generally classed as a fighter/heavy fighter, but also conceived as a high-speed bomber), and He 219 (primarily a night fighter, but also a bomber in its A-3/A-4 variant). Instead, as the best basic type and for purposes of commonality he called for stepped-up continued production of the Ju 88/188, as this type and its follow-on developments offered the best performance—particularly at higher altitudes.

Nevertheless, he had concerns about fitting the Ju 188 with the Jumo 213A in place of the BMW 801E, since the 213 was much more susceptible to combat damage than the 801, and the Ju 188 was expected to be roughly 40 km/h slower with this engine than with the BMW powerplant. As his strategic bombers Peltz called for the He 177 and Ju 290, the latter to equip so-called "long range wings," with two of these wings serving on the Eastern Front and one in the Mediterranean. In his definition, the previously mentioned special wings were particularly suited to engaging shipping targets by using remotely guided missiles and the new generation of aerial torpedoes. These special wings were to "achieve extraordinary success striking against enemy tonnage with relatively standard prerequisites." Peltz felt that using these special wings as normal bomber units was inefficient in the extreme—except in limited cases. He called for the highest priority to be assigned to remotely controlled missiles and their follow-on developments (frequency hopping, wire guidance, camera nose). In Peltz's opinion, four of these special wings and two aerial torpedo wings (*LT Geschwader*) would be needed for prosecuting an effective anti-shipping war in all waters, and that these units would also work in cooperation with the submarine arm. The remainder of his planning outline encompassed 15 standard bombing wings, of which seven would focus on England, six in the East, one in the Southeast, and one in the Mediterranean. These were the minimum number of units required in order to conduct an operatively effective air campaign. In his demands, Peltz did not overlook difficulties in the areas of personnel and training, without which it would be impossible to create the total of 24 bomber wings from the bomber units then on hand. To this end, he made concrete recommendations regarding both the flying and the technical personnel, recommendations which included converting Do 217 units over to the Me 410 and He 177, as well as re-equipping Ju 88 units with the He 177 and Ju 188.

The running monthly requirement for front-line units and replacement training units, where conversion training took place for the crews to familiarize them with the front-line types and the role of their future unit, was calculated by the Gen. d. K. at 390 Ju 88/188. In late 1943/early 1944 the average monthly output was just

---

* Der Oberbefehlshaber der Luftwaffe, General der Kampfflieger, Nr. 13/43 gKdos Chefs, dated 5 October 1943.

about 170 bomber versions. Additionally, in order to bridge this gap the aircraft needed for conversion would have to be available; for example, to convert the four current He 111 wings would require a total of 600 aircraft. In short, then, the monthly output of around 380 bomber aircraft (of all types) at the time would have to be increased to 600 modern bombers of the types listed, which would then need to be available exclusively for the bomber sector.

The Gen. d. K. harbored no illusions, however, and knew that these altogether most realistic needs could in no way be met by current production. He was also aware of the options and difficulties in converting over to the new types—the decision would have to be made on establishing the requisite priorities for the continued prosecution of the war, and that decision was out of his hands.

With the bomber aircraft type development now acting as the backbone of the operative prosecution of the air war, it is worth looking at the course of action the senior leadership chose in reaction to the given wartime situation, insofar as it followed the demands of the *General der Kampfflieger* or cast them aside under the pressure of events, or for other reasons. In this overview, reconnaissance aircraft are only included and/or mentioned in isolated cases, since by this time they were taken exclusively from the ranks of bombers or fighters. The only exception to this was the Ju 290, which evolved from a heavy lift transport through a strategic reconnaissance aircraft before being specially designed as a maritime reconnaissance platform for armed reconnaissance; it was not until its follow-on development as the Ju 390 that it was conceived as a pure long range bomber, albeit never attaining full production status as such.

Bomber Development to 1945

But first let us examine the heavy bombers' follow-on developments, of which two would be selected once performance data had been submitted. The *General der Kampfflieger* noted in his bomber planning paper that the jet bomber should receive special emphasis; thus, although in his planning document the jet bomber was only briefly mentioned as a comment, it nevertheless went without saying that it was expected to reduce the number of piston-engined bombers as the jet types later became available.

Over the next few pages we will take a look at the development and/or follow-on development of those aircraft which the bomber planning paper considered heavy bombers, or strategic bombers, and show their impact on the bomber sector from October 1943 to the end of the war. These bombers encompassed the He 177 and Ju 290, followed by the Ju 88, Ju 188, Ju 288, and Ju 388. Despite the fact that the jet bomber's sporadic appearance in the air war of 1944/1945 partially overlapped the standard bombers, the role of jet aircraft, hitherto only briefly touched upon, is examined in greater detail in the section of this book dealing with jet airplanes.

Heinkel He 177A-3 through A-7

In Oranienburg, from April 1943 onward series production of the He 177A-3/R1 saw the phasing in of the improved He 177A-3/R2, which had a reworked electrical system. In place of its 20 mm MG-FF it carried the MG 151/20 of the same caliber, as well as an improved rear gunner's station where the gunner, formerly compelled to lie prone, could now sit. The rear gunner's position was now fitted with a 20 mm MG 151/20 in place of the 13 mm MG 131.

Another version was the He 177A-3/R3, a variant production-fitted to carry three remote controlled Hs 293 glide bombs—two

Heinkel He 177A-3/R2 rear gunner's station showing the improvement made by fitting a 20 mm MG 151/20 (1943).

beneath the outer wings and one under the fuselage. Beginning in May 1943 the R3 served primarily to provide crews with the necessary training for the Hs 293, since the attack method with the glide bomb involved special tactics.

The subsequent He 177A-3/R4 had its ventral gondola lengthened by 1.20 meters in order to better accommodate the FuG 203b Kehl III command transmission system for the Hs 294 and the associated control stick for the bombardier.

We should perhaps also mention the unforeseen use of the He 177 in support of the Stalingrad airlift, which resulted in the total loss of five of these aircraft. Field units installed a 50 mm BK 5 anti-aircraft gun in the ventral gondola of several He 177s so as to be able to attack hardened ground targets in between supply flights. The armor piercing ammunition for the gun was located in the forward bomb bay. Following this relatively successful experiment a few production machines, designated He 177A-3/R5 and nicknamed "Stalingrad Types," were built with the 75 mm BK 7.5 cannon and more powerful DB 610 engines. However, only five examples were produced, since the cannon installation negatively influenced the flight handling characteristics and firing the gun placed considerable structural stress on the airframe. Without a doubt an airplane of such dimensions was ill-suited to the role of strike aircraft.

German submarines in British coastal waters were having an increasingly difficult time sinking ships thanks to the improved anti-submarine tactics of the English, and it was for this reason that the *Marine* pushed for the use of the He 177 as a torpedo bomber. Following several experiments, as the He 177A-3/R7 the type was fitted with two torpedo racks to carry the Italian L5 torpedoes underneath the wings, since this variant had simpler Fowler flaps which no longer extended the entire length of the wing's trailing edge. Nevertheless, only three examples of the torpedo variant were ever built. These were used to test the new electric LT 50 torpedoes, which could be dropped by parachute from a height of about 250 m. Following the R7 the entire A-3 series was dropped in favor of the He 177A-5 model. Once the last A-3 variant was completed, the Oranienburg and Warnemünde factories focused all their resources on production of the A-5, the first of which rolled off the assembly line in February 1943. Monthly output for both plants reached 12 aircraft by July 1943, and by the end of the year this number had increased to 42 per month. Near the end of 1943, when

Wearing a live jacket, the rear gunner climbs into his compartment on an He 177A-3/R2.

The prominent cockpit of the Heinkel He 177A-3 in profile.

261 A-5s had already been built, an RLM order went into effect calling for the scrapping of all He 177s extant due to the high number of problems. The Heinkel company, however, silently ignored this order!

In the meantime, Professor Heinkel had tasked an action committee to look into the reasons behind the engine fires, and their recommendations led to an He 177 being modified accordingly at the Rechlin test center. The committee had identified 56 potential causes and had proposed a whole series of changes to the airframe. A lengthy test period revealed that this aircraft suffered none of the problems criticized earlier. The *Technisches Amt* was now convinced that the solution for the He 177's engine problems had been found, but was only able to introduce a few of these changes—and only then bit by bit—into the ongoing production series (which continued apace at both Heinkel plants, despite an order calling for production to be halted at the end of 1943). 1944 saw the manufacture of no less than 565 more He 177A-5s until production finally ceased for good in October of that year, a stop ordered (and this time obeyed by Heinkel) in favor of the fighter production program. By this time a total of 1,146 examples of this highly interesting aircraft had been built, despite never really having reached front-line maturity. The continuous technical problems which kept cropping up undoubtedly led to more losses than those caused by the enemy, and all hopes that this would be the bomber which could seriously compete on an even level with those of the Allies were brutally dashed.

Nevertheless, before we offer a final assessment, let's look at a few variants of the He 177A-5, as well as its role in combat, sporadic as it may have been.

The He 177A-5 was primarily laid out for operations with external payloads, such as the remotely guided Hs 293 glide bomb, the remote controlled FX 1400 Fritz X free-fall bomb, and the LT 50 aerial torpedo. It had reinforced wings with the more simplified Fowler flap system. The undercarriage legs were shorter than those of the A-3, and for all sub-variants the A-5 was driven by the more powerful DB 610 engines. Its armament was the same as for the He 177A-3/R2, so that the first aircraft built were designated the He 177A-5/R2. Subsequent A-5 variants generally differed in the area of armament, such as the R5, R6, and R8, while the He 177A-5/R7 was unique in having a pressurized cockpit enabling it to operate at altitudes up to 15,000 m.

With regard to the operational role of the He 177, the first 20 He 177A-1s began carrying out winter trials in the East beginning in late 1942. However, the situation around Stalingrad necessitated these being pressed into service flying resupply missions. On the first mission using seven He 177s the formation leader crashed, and following the loss of four other planes the idea of using He 177s to fly resupply runs was dropped altogether, a decision which was in no small part reinforced by the knowledge that the smaller yet much more reliable He 111 had almost the same capacity as the more labor intensive He 177, which was also highly sensitive to the theater's almost Arctic weather conditions. Instead, the He 177 began flying bombing missions starting in early 1943 in support of army operations; the temporary fitting of the BK 5 anti-tank cannon has already been mentioned in conjunction with these missions. By this time there were 13 remaining aircraft, and despite their good showing with regard to defending themselves against enemy fighters, no less than seven further machines became total write-offs in spite of having notched up quite a successful service record. Most of these losses were attributable to engine fires, without any visible signs of damage due to enemy action. After Stalingrad capitulated, the remaining He 177s transferred back to Germany in February 1943.

Considerable effort went into perfecting this aircraft over the course of 1943, and despite these efforts the complex airframe still came up short in light of its suitability for front-line operations. After several months of trials and conversion training for the anti-shipping role, a group of He 177s was assigned to the Atlantic coast in the late autumn of 1943, from whence they flew their first large-

Four Heinkel He 177A-3s lined up for a training mission lacking external ordnance (1943). The object in front of the cockpit is a battery cart used for starting the engines. On the left is the six-man crew.

scale mission on 21 November 1943. 20 He 177A-5s, armed with Hs 293s and in poor weather, attacked a 66 ship convoy after nearly 30 German submarines had enjoyed little success in harassing the convoy. The British noted that 16 Hs 293s were dropped, some against a straggler (which was sunk). This ship, plus further reported damage to another cargo ship, was the sum total of the mission's success. Three He 177s were lost on the mission. Five days later 14 aircraft from the same He 177 unit, also carrying Hs 293s, attacked another convoy during daylight hours, lost four of their number due to enemy action, and a further three which crashed on landing, so that 50% of the He 177s used on one mission were written off without any notable success—a costly undertaking! The He 177 crews now switched over to night raids on shipping targets. On brightly moonlit nights or by illuminating the targets with 50 kg parachute flares (which caused the silhouettes of the ships to stand out), the He 177s would make a direct approach—considerably simplifying targeting and control—and drop their Hs 293s from a distance of between 10 to 14 kilometers from the target.

Another operational area for the He 177 was over England. Hitler had ordered massive retaliatory strikes on London in response to Allied raids on Berlin, and the *Luftwaffe* high command had all dispensable bombers pulled out of Italy and the East and assigned to the "*Angriffsführer England*" (Attack Leader England). By December 1943 he had no less than 695 bombers at his disposal, a number which also included 35 He 177s. The first of 31 missions flown up to the end of May 1944 took place on 21 January that year, with 1,025 metric tons of bombs falling on London during the night—i.e. approximately one-ninth of the tonnage which had fallen on Hamburg in July 1943. This operation, known officially as Operation Steinbock, but called the Little Blitz or Mini Blitz by the British, achieved little in the way of spectacular success. It did, however, offer the He 177 much valuable experience. In the hands of experienced crews the machines would climb with full payload to 7,000 m while still over friendly territory, then head toward their targets in the southeast and south of England in a shallow glide at

Loading an He 177A-5 for night operations against England in early 1944. All bombs were carried horizontally; these are 500 kg bombs, of which the He 177 could carry up to six.

speeds of over 650 km/h, speeds at which neither night fighters nor AAA could touch them unless by pure chance. By using these tactics, during the weeks of the intensified German bombing campaign against England only four He 177s were lost to enemy action; technical problems, though, remained the order of the day.

One example of the technical troubles plaguing the type can be gleaned from a mission flown during the night of 13 February 1944, which *Generalmajor* Peltz followed personally. On this winter night 13 machines took to the skies—after one had dropped out shortly beforehand due to tire damage. Immediately after takeoff, eight more aircraft were forced to break off the mission because their engines were overheating; some powerplants were even on fire. It was quite apparent that the cold starting method did not work as advertised. Of the remaining five aircraft, four reached London. One was shot down by a night fighter, while the fifth aircraft broke off the flight over Norwich and jettisoned its bombs. For this aircraft, it would be almost impossible to provide any type of accurate ratio of the methods used to the success achieved!

Other He 177s flew missions during the twilight hours of the 23rd and 24th of January 1944. A new trouble spot had flared up with the Allied landings in Italy in September 1943, and on these two evenings He 177s dropped parachute retarded flares and Hs 293s on the Allied landing fleet at anchor off the bridgehead at Nettuno. No significant results were achieved, and following this action the Heinkel bombers were again used for armed reconnaissance over the Atlantic. At this time operational readiness hovered around 35% on average. Ultimately the remaining aircraft, which suffered heavy losses during the invasion in June 1944, transferred from the Atlantic Coast to central Norway where they replaced the Fw 200s based in Trondheim in the armed reconnaissance role; some

This He 177A-5 is camouflaged for night operations such as that carried out over London in January 1944.

of the latter aircraft stayed on in Norway as transports. Shortages of fuel and personnel limited the number of anti-shipping missions. Numerous He 177s lay scattered about various airfields throughout Europe, grounded because of the lack of spare parts—especially engines. In spite of this bleak picture, older types continued to be refitted to carry the Hs 293, and a handful of missions were flown with the glide bomb up until early 1945.

Another intended role for the He 177 in the East, one which would have fallen into the strategic category, involved striking the Russian armament and electricity plants in the Urals, around Moscow, Rybinsk, and Gorki. With approximately 400 aircraft being rounded up starting in December 1943, this air offensive was expected to result in "the planned attack on the Soviet armaments industry" beginning in February, as had been outlined in Göring's order from 26 November 1943. The He 177 was also a participant in the "*Spezialverbände mit besonderer Treffgenauigkeit*" (Specialized High-Accuracy Bombing Units), but the Soviet winter offensive of 1944 forced Germany's leadership to abandon these plans. Instead, the aircraft which had been gathered together found themselves increasingly needed to provide direct and indirect support to the army's ground forces, primarily by attacking the rail transport system behind the front lines. In so doing, circumstances had compelled them to revert to the tactical-operative combat role. A final attempt to breathe life into the operative-strategic air war was Operation *Eisenhammer*, the destruction of Soviet electrical supplies in the area of Moscow. The He 177 was expected to have played a major role, but these plans were scrapped at the end of March 1945 due to the unfavorable overall situation—the staging fields in East Prussia were no longer available!

These notable examples illustrate just how much of a bitter disappointment the He 177's front-line career was to those who had placed high hopes in this weapons platform which had such promising strategic potential. Nevertheless, it is worth taking a final look at the technological follow-on development of this aircraft. The He 177A-6 was developed parallel to the production run of the He 177A-5, starting in early 1944. The first of six examples rolled off the assembly line in May of that year as the He 177A-6/R1, a heavy strategic bomber. This variant was in effect a re-equipped version of the A-5/R6, but with pressurized cockpit and thus lacking the B2 gunner's station in the aft fuselage section, since defensive fire was expected to be adequate with a new, electrically powered Rheinmetall-Borsig rear turret having four 7.9 mm MG 81s. The A-6/R1 was also designed to carry an external payload of up to 2,500 kg, but in addition could also hold 500 kg in the aft bomb bay. This variant had particularly well protected fuel tanks and had a maximum range of 5,800 km. The seventh A-6/R1, as the He 177V22, became the prototype for the He 177A-6/R2 variant, which had an aerodynamically refined canopy, improved armament in the form of a so-called chin turret in the nose, and could carry six different payload combinations, including the Hs 293 and Fritz X. However, following the construction of the V22 the entire R2 program was dropped in favor of the He 277, which will be briefly discussed shortly, although the verdict of "too little, too late" probably applied more to this aircraft than to many other promising aircraft developments of the day.

Six examples of the He 177A-7, with a wingspan increased to 36 meters, were built as high altitude bombers with DB 610 engines. It was intended to later refit the type with the DB 613, comprising two DB 603G engines each, having a combined takeoff rating of 2,648 kW/3,600 hp and 2,117 kW/3,150 hp for climbing and combat, but these powerplants were not available in time for installation in the A-7. Its empty and takeoff weights were 15,383 kg and 34,600 kg, respectively, and at an altitude of 6,000 m it attained a maximum speed of 540 km/h.

Japanese Interest

At quite an early stage the Imperial Japanese Navy expressed interest in the He 177, even going so far as to begin construction of a factory in Chiba so that the Hitachi Company could build the He 177 under license—once the problem with fires in the double engines had been alleviated, of course. Several jigs and special tools were shipped to Japan via submarine as a result of this interest, and the third He 177A-7, completed in May 1944, was to have been supplied to Japan for evaluation and as a pattern airframe. It was fitted with additional fuel tanks and had most of its armor removed during the summer of 1944 in preparation for a non-stop transfer flight to Japan. Heinkel proposed a flight at extremely high altitude over Siberia, while because of their neutrality treaty with the Soviet Union the Japanese insisted on a route via Persia and India, even after the Soviets renounced their non-aggression pact with Japan. Despite all the measures taken to increase its range, the A-7 was still unable to make the longer southern route, and the transfer flight never occurred. Together with the other five He 177A-7s, plans called for these half-dozen high altitude bombers to make one-way missions against targets in the United States, but again, these plans never came to fruition. One of the A-7s nevertheless did indeed make it to the U.S.; it was taken by U.S. forces as a war prize, while the other five A-7s were scrapped. The single He 177A-7 arriving in the States served as a testbed.

The "*Bomber-Zerstörer*"

In conclusion, to illustrate the variable history of the He 177 we should perhaps mention an experiment conducted with a view toward home defense. Three He 177s, which had their bomb bays and forward fuselage tanks replaced by a battery of 33 rocket tubes, were provided to a test fighter group in Pardubice. The tubes were angled at an elevation of 60° and inclined slightly to starboard. Along its upper decking the fuselage had corresponding openings from which the rockets could be fired individually or selectively in groups of 15 or 18, or as a single full salvo. These *"Bomber-Zerstörer"* (bomber-destroyer) He 177s were to fly somewhat to port and about 1,800 m beneath enemy bomber formations and fire their rockets into the mass of bombers. Following initial technical trials in Pardubice, testing continued operationally at Rechlin. A few missions were even carried out during the day, but no contact was made with the enemy. In the meantime, however, American escort P-38 Lightnings, P-47 Thunderbolts, and mainly P-51 Mustangs began appearing in ever increasing numbers in the bomber streams, the first escorts flying over Berlin as early as 6 March 1944. Despite the He 177 being fitted with a welcome sighting system for the rockets, the entire project was abandoned.

Summary Including Data for the He 177A-5

The example of the *Bomber-Zerstörer* further serves to highlight the immense developmental and labor efforts invested in this remarkable bomber. It was an airframe that, despite the ingenious brilliance of its parents Günter and Hertel, had problems which were never fully surmounted by the means available to the aviation industry during the war. As a result, the He 177 is considered to have been a poor investment, and an investment bought at a heavy price. Its most glaring defect can be traced back to the belief, bordering on obsession, of the key decision makers at the time, who felt that even a heavy long range bomber must be capable of dive bombing and thus must only have two engines as part of its configuration. When the dive bombing requirement was dispensed with in September 1942, the path had already been irreversibly laid out with the time and effort involved in building the jigs, and production of this controversial aircraft type was already well underway. Nevertheless, the designers did succeed in improving the combat potential of the He 177 over the course of its life, from its earliest beginnings to the most-built A-5 variant (data for which is provided below).

The He 177A-5/R2 was a six-seat heavy bomber, long range reconnaissance platform, and anti-shipping strike plane.

Powerplant: 2x liquid-cooled Daimler-Benz DB 610 A-1/B-1 (A-1 = port engine, B-1 = starboard engine) 24 cylinder double engines each rated at 2,170 kW/2,950 hp on takeoff and 2,280 kW/3,100 hp at 2,100 m altitude; four-bladed VDM variable-pitch propellers with a diameter of 4.5 m.

Armament:
1x 7.9 mm MG 8li with 2,000 rounds flex-mounted in nose (*A-Stand*)
2x 13 mm MG 131 with 750 rounds each in remotely operated forward station on dorsal fuselage (*B1-Stand*)
1x 13 mm MG 131 with 750 rounds in electrically powered aft gun position on dorsal fuselage (*B2-Stand*)
1x 20 mm MG 151/20 with 300 rounds flex-mounted in the forward underfuselage gondola (*C1-Stand*)
2x MG 8li with 2,000 rounds each flex-mounted in the aft underfuselage gondola (*C2-Stand*)
1x 20 mm MG 151/20 with 300 rounds in the aft fuselage tail.

Payload:
Internal:
16x 50 kg plus 4x 250 kg = 1,800 kg payload, or
2x 500 kg bombs = 1,000 kg payload
External:
2x LMA III aerial mines or
2x LT 50 aerial torpedoes or
2x Hs 293 glide bombs or
2x FX 1400 Fritz X free-fall bombs

Performance:
Max. speed
- with 27,200 kg takeoff weight
  a) at sea level = 400 km/h
  b) at 6,000 m = 490 km/h
- with 31,000 kg takeoff weight at 6,000 m = 440 km/h
Max. cruising speed at 6,000 m = 415 km/h
Economic cruising speed at 6,000 m = 338 km/h
Max. range
- with 2x Hs 293 = 5,500 km
- with 2x Fritz X = 5,000 km
Time to climb rates
- to 3,000 m altitude = 10 min
- to 6,000 m altitude = 39 min
Service ceiling = 8,000 m

Weights:
empty 17,210 kg
empty equipped 18,940 kg
normal takeoff weight (loaded) 27,200 kg
max. takeoff weight 31,000 kg

Dimensions:
wingspan 31.44 m
length 22.02 m
height 6.67 m
wing area 102 m²

Heinkel He 274 and He 277 (a.k.a. "He 177B")

As early as 1940 Heinkel's project team in Wien-Schwechat had come up with a recommendation to the *Technisches Amt* for solving the He 177's double engine problems by fitting four single engines instead. At this time the first prototypes of the He 177 were just beginning their test phase, and the RLM rejected the proposal because of the aerodynamic drawbacks caused by the greater frontal drag and the lesser maneuverability of the four-engined version; furthermore, the RLM felt that the initial engine problems of the He 177 could be easily overcome. The *Technisches Amt* eventually approved the construction of a high altitude bomber with four single engines and a pressurized cockpit in the fall of 1941. As this follow-on development involved a general makeover of the He 177 and would have little in common with the type, the high altitude bomber was given the designation He 274. Further working of the project was assigned to the French company "Société Anonyme des Usines Farman" (SAUF) at Suresnes, near Paris. SAUF was given a construction contract for two prototypes, the He 274V1 and V2, as well as for four pre-production airplanes. Following an overly long preparation period, work finally got underway on building the two prototypes in 1943, but in late 1943 the contract for the four pre-production variants was canceled. In July 1944 the He 274V1 was readied for its first flight in Suresnes, but this was postponed and—like its subsequent transfer to Germany—ultimately fell by the wayside in the face of the advancing Allied invasion forces. The explosive charges placed by the Heinkel personnel for destroying the airframe upon their departure may have indeed destroyed the engines, but the fuselage survived relatively intact. Obtaining a new set of engines, the French flew the He 274V1 (now designated the AAS 01A) in December 1945 and continued to fly the type as a testbed until it was written off in late 1953 at Istres, the French test center. The He 274V2 was never completed by the French.

Now to the He 277: in late 1941 the *Technisches Amt* again rejected the installation of four engines for any follow-on development of the He 177, despite the fact that Heinkel felt this was the best solution for the double engine problems and insisted on forcing the issue. The conflict went so far that Göring forbade the use of the He 277 designation for the four-engined Heinkel project. Heinkel paid little heed to this and simply referred to the type in any official correspondence as the He 177B, but in design work, calculations, etc. for the four-engined heavy bomber he continued referring to it by the designation He 277.

Ultimately, however, the project received an official sanction after Hitler called for a two-role bomber at a meeting of the leading aviation industrialists at the Obersalzberg on 23 May 1943. This bomber would be capable of hitting London both day and night from altitudes above enemy fighter defenses, as well as striking convoys on the open Atlantic. Heinkel gave assurances that the "He 177B" was the perfect bomber for this dual role, and as a result was ordered to continue immediately with the development of the He 177B. To this end, Heinkel took the airframe of an He 177A-3/R2 and redesigned it to accept four single DB603A engines with annular radiators. Under the designation He 277V1 this machine began its flight testing program at Wien-Schwechat in late 1943. With regard to the RLM and Göring, this machine continued to be referenced as the He 177B-0. The He 277V2, designated between Heinkel and the RLM as the He 177B-5/R1, was a reworked He 177A-5/R8, which first took to the skies on 28 February 1944 at Schwechat and transferred to the Rechlin test center in April 1944 for official evaluation.

The He 277V3 was similar to the V2, and like it had initial stability problems around the yaw axis, so that it was fitted with an entirely redesigned twin rudder empennage in place of the earlier single rudder design. These new vertical stabilizers reestablished full directional stability.

During a conference on 25 May 1944 Göring reemphasized that the heavy bomber would continue to be the cornerstone of the air armament program, and ordered the immediate full-scale production of the He 277B. He went on to call for a monthly output of 200 airframes, an entirely unrealistic goal at this late stage in the war.

The first production variant was the He 277B-5/R2, a heavy bomber designed for medium and long range missions with four DB 603A engines outputting 1,287 kW/1,750 hp each, for a total rating of 5,148 kW/7,000 hp. Only eight He 277s were ever built, of which two or three actually entered flight testing before the order of 3 July 1944 suspending all bomber construction programs in favor of the "emergency fighter program." Both those He 277s which had been completed as well as those still under construction were scrapped.

At this time work was well underway on the He 277B-6. This variant had a wingspan increased to 40 m and was powered by four Jumo 213F engines, each rated at 1,515 kW/2,060 hp on takeoff, equating to a total output of 6,060 kW/8,240 hp. Another advanced project was the He 277B-7, a long range reconnaissance platform derived from the He 177A-7. One example of the He 277B-7 was completed with DB 603A engines in place of the intended Jumo 213s, but was destroyed shortly before the arrival of Soviet troops.

With regard to those heavy bombers for the so-called "special wings," the bomber planning of the *General der Kampfflieger* thus had finally been laid to rest for good.

Junkers Ju 290 and the "New York/America Bomber"

The long range bomber the *General der Kampfflieger* (Gen. d. K.) had in mind when he called for a second bomber type in his planning paper of 5 October 1943 was the Ju 290. The Ju 290 had been

History of German Aviation: Bombers and Reconnaissance Aircraft

Junkers Ju 90V7, the four-engined precursor to the Ju 290 developed from the Ju 90V11 and which first flew in July 1942. The cargo ramp designed for the heavy lift transport, the rear gunner's station, and the turret situated on the dorsal forward fuselage all indicate its military applications.

Three-quarter frontal view of the Junkers Ju 90V7.

originally developed from the Ju 90 civilian airliner and transport, and was initially conceived as a large-capacity military cargo transport. Within the framework of the Ju 90's follow-on development the Ju 90V11 became the Ju 290V1, taking to the air on its maiden flight on 16 July 1942 with Junkers test pilot Pancherz at the controls. Flight testing began in August of that year. Used from 10 January 1943 as an armed transport for resupplying the encircled 6th Army at Stalingrad, the Ju 290V1 crashed just three days later, on 13 January 1943, while attempting to take off with wounded on board.

Using the Ju 290V1 and a few other prototypes as a basis, a series of nine Ju 290A-1s was produced, with the last three being kitted out as maritime reconnaissance aircraft. The same role applied to the Ju 290A-2 series, of which three were built starting in the summer of 1943, and the A-3 series (five examples). The latter were fitted with the more powerful BMW 801D/G (1,250 kW/1,700 hp) engines, had a total fuel capacity of 18,450 liters and an endurance of up to 18 hours. Starting on 15 October 1943 they flew from southern France against convoys between Gilbraltar, the northwestern coast of Africa, and out to 30° W west of the Azores, not to mention as far as Iceland, providing reconnaissance data and shadowing the ships on behalf of the *Fliegerführer Atlantik* and the submarine forces commander. Without exception, all airframes flying these long-distance recce missions were equipped with the FuG 200 Hohentwiel search radar.

Five examples of another small batch of Ju 290A-4s were also built, one of which served as the testbed for fitting the Kehl equipment for remote controlling the Hs 293 glide bomb or the remotely piloted Fritz X free-fall bomb. The glazed nose of the A-4 was dropped for the subsequent Ju 290A-5 appearing in late 1943, of which eleven were built with the BMW 801D engine. The A-5 was configured for the armed reconnaissance role from the outset, and in addition to an emergency fuel jettison system had three ETC 2000 external racks for carrying heavy ordnance, plus an additional 20 mm MG 151/20 in its solid nose. Finally, in the spring of 1944 the Ju 290A-7 entered flight testing as a long range bomber/reconnaissance aircraft. It had a reinforced airframe, the glazed nose of the A-4 as standard, the Kehl system with the FuG 203e, as well as cameras for taking offset angle pictures, and was armed with a total of seven 20 mm MG 151/20s and one 13 mm MG 131. By the time production ceased in June 1944, 19 A-7 variants had been manu-

Junkers Ju 290A-7 (1944) carrying a Henschel Hs 293 glide bomb.

factured. Takeoff weight had risen to 46 metric tons, maximum speed was reached at an altitude of 5,000 m at 438 km/h, and the A-7's maximum range extended out to 5,800 km. Parallel to the A-7 was the Ju 290A-8, in production until the end of 1944. Ten examples of the type were planned and laid down, but only two or three were actually completed. The A-8 differed from the A-7 mainly as a result of its more modern and increased defensive firepower, to include two hydraulically powered gun turrets along the dorsal fuselage, as well as a new tail position in which the formerly prone gunner could now sit and avail himself of two 20 mm MG 151/20s. With only a minimal loss in performance, the heavily armed A-8 now had:

    four 20 mm MG 151/20s in four hydraulically driven turrets,
    one 20 mm MG 151/20 in the glazed nose,
    one 20 mm MG 151/20 in the forward section of the fuselage gondola (offset to port),

one 13 mm MG 131 in the aft section of the fuselage gondola, and two 20 mm MG 151/20s in the tail position.

This made for a total of eight 20 mm MG 151/20s and one 13 mm MG 131. The maximum bomb capacity amounted to 3,000 kg, or up to three Hs 293s, Hs 294s, or Fritz X free-fall bombs—an impressive weapons platform indeed!

The final variant worth mentioning is the Ju 290B, a long range high altitude bomber with an eight-man crew, vastly improved defensive armament (pressurized nose and tail stations, each with four 13 mm MG 131s, two pressurized dorsal fuselage turrets, each with two 20 mm MG 151/20s, remotely operated twin MG 151 (15 mm) in a flat profile fuselage gondola in place of the earlier design) totaling 14 machine guns instead of the earlier 9! Offensive load was the same as for the A-7. The prototype Ju 290B-1 carried out its first test flights without the pressurized cockpit and with wooden mockups of the nose and tail gunner positions during the summer of 1944 and was then transferred to Prague. There the B-1 continued flying up to March 1945 as the developmental aircraft for the Ju 290B production aircraft originally scheduled for production starting in mid-1945; however, production of the Ju 290 had effectively come to a standstill by the fall of 1944 as a result of the overall deteriorating situation in Germany.

Summary and the Ju 390

All told, about 50 Ju 290s of all variants were built, but only the A-7 and A-8 came close to fulfilling the idea of the Gen. d. K. for equipping a "long range wing," although numerically their numbers fell far short of what he expected. The Ju 290B-1, as a high altitude bomber optimal for this role, looked as though it had a promising future, but there was no future to be had.

The four-engined Ju 290 also spawned a six-engined version as a pure strategic bomber, the Ju 390 developed by *Dipl.-Ing* Kraft.

Junkers Ju 390V1, a six-engined strategic bomber developed from the Ju 290, entered flight testing in August 1943.

The RLM called it the "New York" bomber, and it had an increased wingspan (50.36 m) and lengthened fuselage (34.20 m), as well as reinforced landing gear (second set under the central engine nacelle, which was only used when heavily loaded). Only two of these were ever built. The Ju 390V1 entered flight testing in August 1943 and was initially envisioned as a strategic transport with 34,000 liters of fuel and a 10,000 kg capacity, with which it covered a distance of 8,000 km at a speed of 330 km/h at 2,000 m altitude. In 1944 it successfully carried out air-to-air refueling of the Ju 290A over Prague-Ruzin. The Ju 390V2 was built at Bernburg and first flew in October 1943. In January 1944 it was assigned for combat trials as a strategic maritime reconnaissance aircraft and served in France. By conserving fuel, it could fly for up to 32 hours and, following a few shorter missions, on one of its ranging missions it managed to reach a point about 20 km off the American coast north

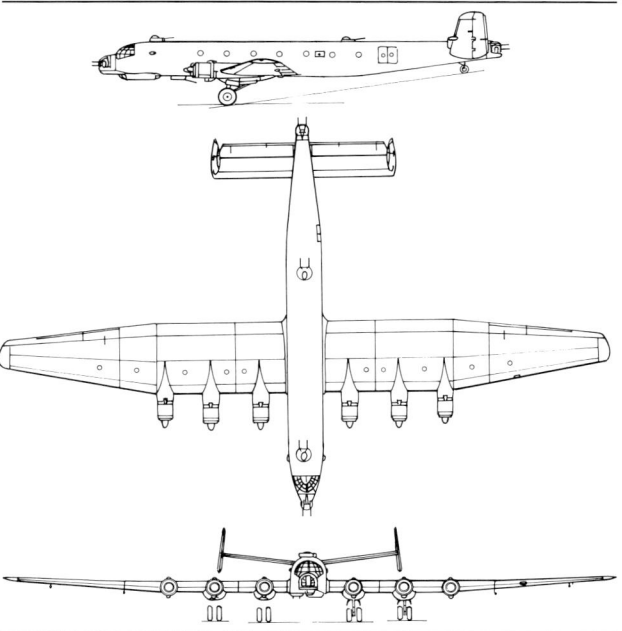

Junkers Ju 390V2, also known as the Ju 390A, offered exceptional performance as a maritime reconnaissance platform during field testing in January 1944, hinting at the type's potential for the New York bomber requirement.

of New York before returning back to its starting point of Mont de Marsan in southern France.

Messerschmitt Me 264

Parallel to the "New York" bomber by the Junkers Company outlined above was a development at Messerschmitt for a future "America Bomber," the four-engined Me 264. In designing this

History of German Aviation: Bombers and Reconnaissance Aircraft

Messerschmitt Me 264, with the V1 completing its maiden flight in December 1942.

bomber Messerschmitt was effectively ignoring his official restriction to produce only fighters. The Me 264V1 flew for the first time in December 1942 without armament and initially powered by four Jumo 211J-1 engines (the same as in the Ju 88A-4), each with a takeoff rating of 986 kW/1,340 hp. Under the direction of Dr. Konrad, the Messerschmitt team had developed the Me 264 as an all-metal, mid-wing design with a large aspect ratio (initial wingspan was 38.9 m). It was a remarkably aerodynamically clean lay-

Messerschmitt Me 264, the sole German four-engined aircraft with a nose gear.

out and was the only German four-engined aircraft design of its time to go into flight testing with a nose gear. In the spring the V1, built without a pressurized cockpit, was refitted with four BMW 801G-2 engines, but this model was destroyed in the fall of 1944 during a bombing raid on Memmingen. The same fate had already befallen the Me 264V2, which was powered by the BMW 801 engine with GM 1 system from the outset. In late 1943 the V2 had been destroyed in a bombing raid on the Neu-Offing factory near Ulm. Work on the Me 264V3, the prototype of an Me 264A series, came to a halt in early 1944 due to other material priorities.

The data calculated for the V3 called for a takeoff weight of approximately 49 metric tons, with a maximum speed using the GM 1 system at 565 km/h at an altitude of 8,000 m with a weight of 34.4 tons. Both the V2, as well as the V3 had wings lengthened to a span of 43 meters, four BMW 801D or G engines with a takeoff rating of 1,250 kW/1,700 hp each and 1,060 kW/1,440 hp at 5,700 m altitude, the GM 1 system, and six RATO packs for shortening the takeoff run at maximum weight—of which two were held in reserve in order to be used to boost climb rate. Plans called for both aircraft, which were expected to have a range of 12,500 km, to be the pattern for the Me 264A production, as well as serve as high altitude testbeds. With a takeoff weight of around 54 metric tons, it was hoped that the Me 264A would have a range of 15,000 km, while the Me 264B—the actual "America Bomber" which, like the Ju 390, would have six engines—never got off the drawing board.

Focke-Wulf Ta 400

Finally, within the scope of the "New York Bomber" project there was the six-engine long range bomber design by Focke-Wulf for a crew of nine. In a brief description from 13 October 1943 the company listed the Ta 400 as a "long range bomber-destroyer and strategic reconnaissance platform" with six BMW 801E engines, a 42 m wingspan, and a takeoff weight of 62.5 metric tons. Like Messerschmitt's competitor to the Ju 390, which never got beyond the prototype stage, Focke-Wulf's design had just as little success.

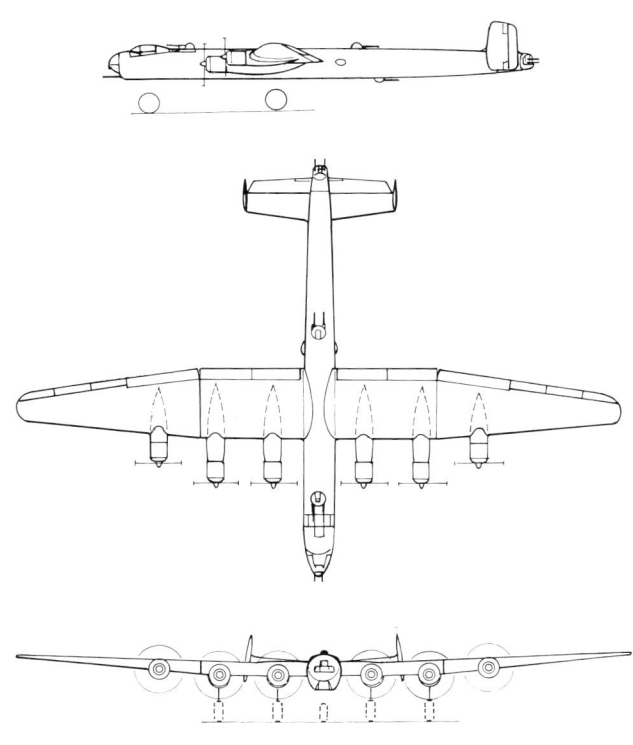

The six-engined Focke-Wulf Ta 400—conceived in 1943 as a "strategic strike bomber and reconnaissance aircraft"—was technologically well ahead of its time. It never advanced beyond the drawing board stage, though.

Since this aircraft was primarily tailored to the maritime reconnaissance and anti-shipping roles it was not designed with a pressurized cabin in mind. It did, however, have an oxygen system capable of supplying the crew with oxygen at altitudes above 4,000 m for up to four hours.

Thoughts on Six Engines
At some future time the six-engined "heavies" were to have been used as harassment bombers against the city of New York, far removed from the events of the war. It would be ridiculous to think of such an undertaking as anything more than a psychological tool, much less an instrument of destruction having a major influence on the war's outcome. In any event, the operation never took place. Nevertheless, by looking at this project we can gain some insight into the state of technology at the time, which can be considered quite impressive overall. Even though the Focke-Wulf Ta 400 was nothing more than a project, its design is indicative of the available technological potential of the day. Had the Ta 400 project been realized, however, it would have been overtaken by the actual flight performance of the Junkers Ju 390 in several areas, since with the same engine output (total of 7,940 kW/10,800 hp) and payload (both 10,000 kg) the Ju 390 carried more fuel and therefore offered greater endurance and range. In spite of this, the takeoff weight was a good 15% higher than that of the Ta 400, and it was about six meters longer (34.20 m) and had a wingspan roughly eight meters greater (50.36 m) than the Ta 400. Its greater wing area (254.4 m^2) gave it less wing loading, but the tradeoff here was a lower maximum speed—although for its intended role this may have been of secondary importance. Comparative assessment of the advantages and disadvantages of this or that type is speculative at best, however, for one never left the drawing board, whereas the other had the advantage of flying under operational conditions, sporadic though they may have been.

The Ju 390 was undoubtedly the most advanced six-engined bomber and reconnaissance aircraft at the time that reached some semblance of operational maturity, even if—other than its enormous range for its day—it was unable to achieve any type of spectacular success.

Ju 88A-4 Follow-on Variants
As was mentioned earlier, cancellation of the "Bomber B" project meant the Ju 288 was never able to go into production, much less attain front-line service. Initially, this only left follow-on developments of the Ju 88 as outlined in the Gen. d. K.'s bomber planning document. We have already looked at the Junkers bomber up to its most-produced variant, the A-4, but now let us examine the A-4's successors.

The Ju 88A-4 was followed by an imposing number of variants, tailored to a specific given role or differing in powerplant/armament. These variants can be found in the listing below, which only provides a brief description insofar as the variants shown were all evolutionary follow-ons to the production series discussed in a previous section.

Ju 88A-6: It was similar to the A-5 with two Jumo 211G-1 (880 kW/1,200 hp) and was permanently fitted with the M9 field conversion set for deflecting balloons. This consisted of thin steel tubing ahead of the canopy and propellers. The deflector tubing ran arrow-like out to the wingtips and guided the balloon cable out to the tips where electric snippers, the so-called "*Klettenmagazin*," would cut it. The weight of this system necessitated a counterbalance of up to 59 kg in the tail, and together these caused a loss in speed of about 35 km/h.

Because of these disadvantages, it was decided to switch over to the well known "*Kuto-Nase*" as used on the He 111, with its steel blade attached beneath the wing leading edge, so that some of these aircraft from this series flew as maritime reconnaissance platforms without the balloon deflectors under the designation of:

Ju 88A-6/U: with the FuG 200 Hohentwiel surface search radar system.

Ju 88A-8: Derived from the A-4 with more powerful Jumo 211F-1 engines (2x 990 kW/1,340 hp), fitted with the *Kuto-Nase*, but only a few examples were built. The Ju 88A-7 which preceded it in the type listing was a trainer stemming from the A-5, with dual controls and larger wings.

Ju 88A-9: Produced with supplemental tropical weather equipment weighing 190 kg and carried in the rubber dinghy compartment, derived from the Ju 88A-1; also designated as Ju 88A-1trop.

Ju 88A-10: A tropicalized version derived from the A-5, also known as the Ju 88A-5trop.

Ju 88A-11: A tropicalized version derived from the A-4, production version of the Ju 88A-4trop.

Ju 88A-13: This version was derived from the Ju 88A-1 with the smaller wing (18.38 m span) and had no dive brakes. It carried 500 kg SD 2 fragmentation bombs, as well as a number of light caliber fixed guns, and was used in the strike role. The Ju 88A-12 listed beforehand is again a trainer derived from the Ju 88A-5 with enlarged cockpit having dual controls and fitted with the larger wings.

Ju 88A-14: This was an improved version of the A-4 series with larger wings and the *Kuto-Nase*, but was no longer equipped for dive bombing operations.

Ju 88A-15: A one-off model, similar to the Ju 88A-4, with a wooden belly enlarged to accommodate 3,000 kg of bombs; three-man crew.

Ju 88A-17: Torpedo bomber derived from the A-4 and A-14 series with two Jumo 211J-1 engines, also larger wings, some lacking the

Junkers Ju 88A-17, loaded with two LT F5b aerial torpedoes. The torpedo's wooden guide vanes were designed to break apart on impact with the water.

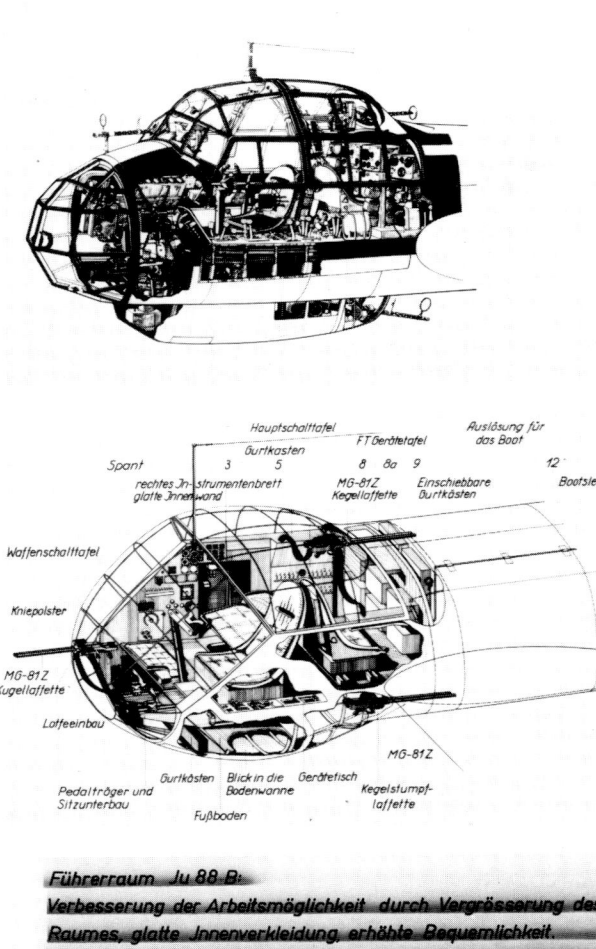

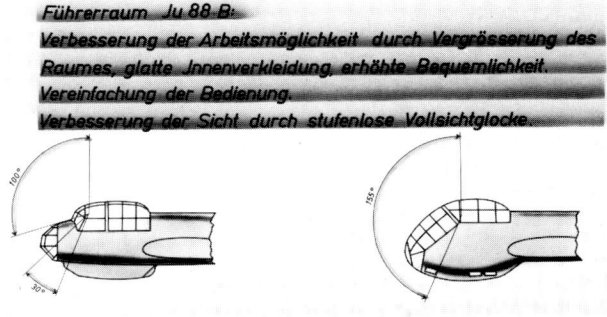

ventral gondola; depth and angle indicator rods on the wings for the torpedo launch system, three-man crew.

The Ju 88A-16 coming before the A-17 was an A-14 modified as a trainer with dual controls and lacking the ventral gondola.

Only five examples of the Ju 88B-0 were built, including the Ju 88V101 and V102 testbeds, which were fitted with two BMW 801MA engines (1,150 kW/1,560 hp) and a more streamlined, spherical glazed canopy. They had the larger wings (20.08 m span) and achieved a maximum speed of 500 km/h.

Based around the same powerplant was the Ju 88E-0, of which a few examples were built in 1942 to include the modified Ju 88V27 and V30. Both were equipped with an MG 131 turret.

Also appearing in 1942 was the Ju 88P-1, a tank buster modified from the A-4 series, meaning that it had the Jumo 211J-1 (1,040 kW/1,420 hp), but lacked dive brakes or ordnance. Instead, in place of the forward ventral gondola section it was fitted with an enormous gun pack below the fuselage into which was installed the heavy 7.5 cm BK derived from the PaK 40. At the aft end of this pack was an MG 81Z (7.9 mm) for providing defensive fire downward and to the rear. The gun pack, weighing some 1,200 kg, could

As early as 1939, the Junkers design department began carrying out follow-on developmental work on the Ju 88A under the direction of Dipl.-Ing. Ernst Zindel. Designated the Ju 88B (later designated the Ju 88E by the RLM) the Junkers drawings below show several interesting details. Among other features, these reveal a newly designed cockpit with better visibility being provided to the pilot.

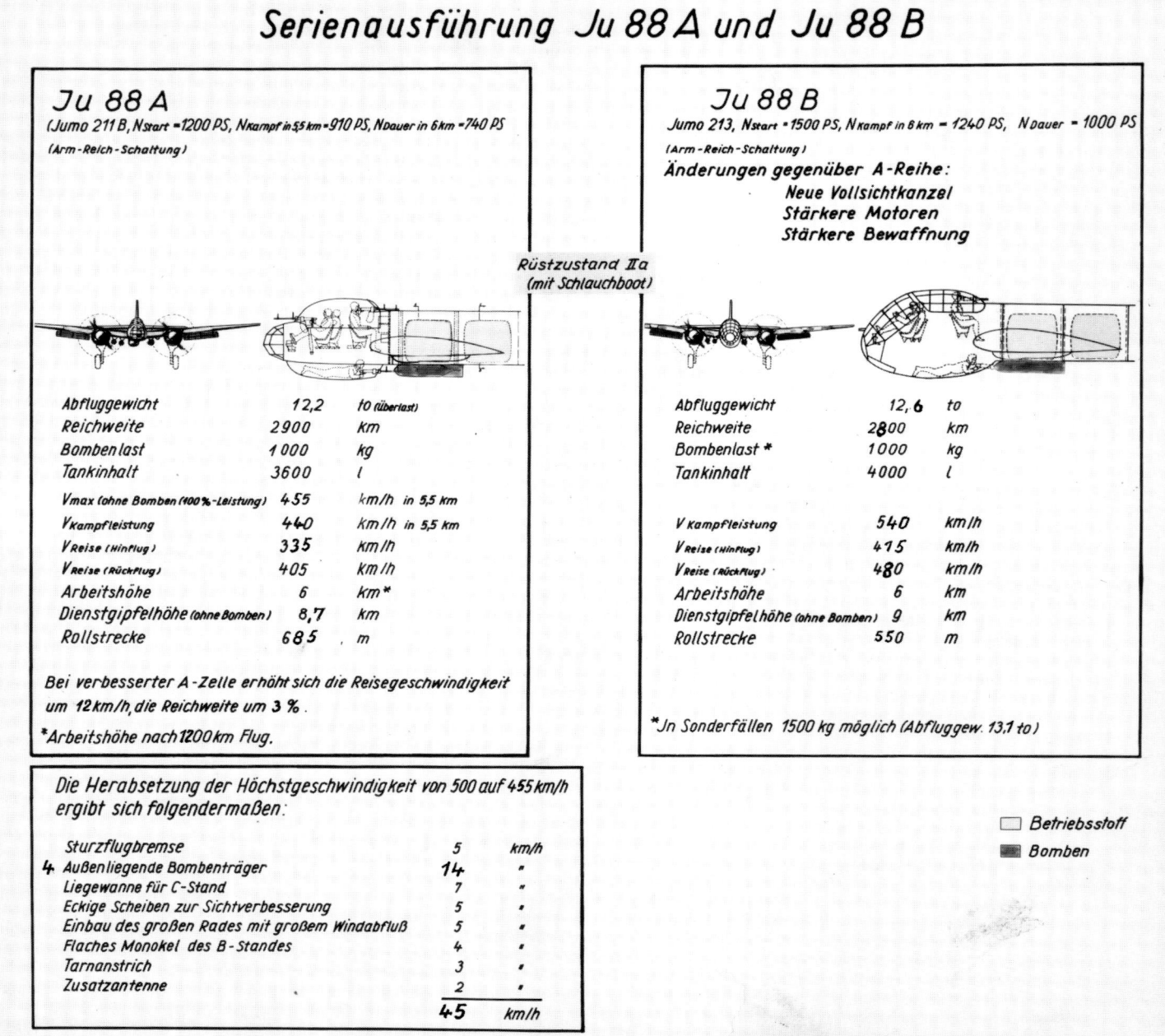

More powerful engines and better speeds for the Ju 88B in all categories:

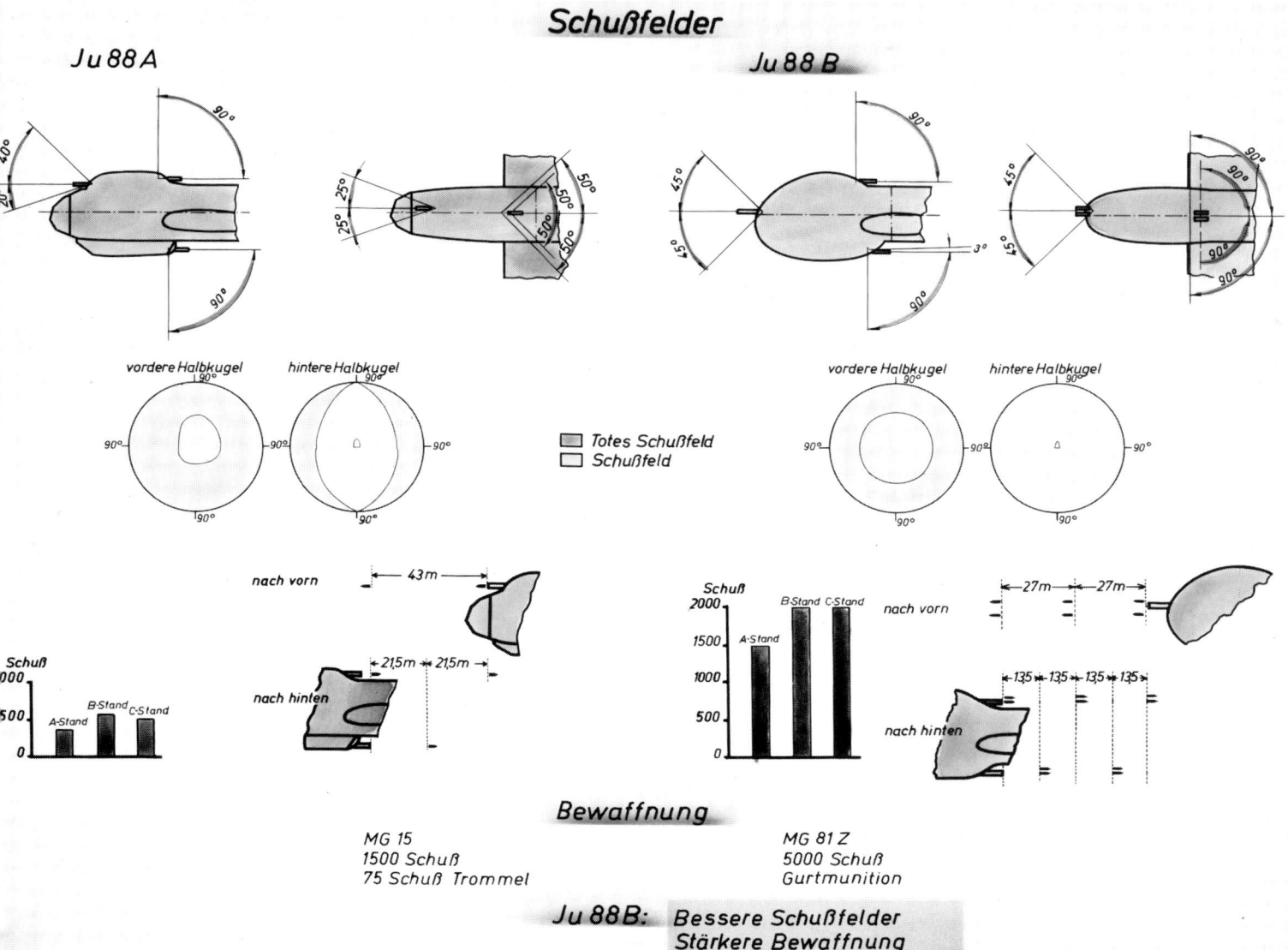

Improved fields of fire, armament, and increased ammunition capacity:

Other applications of the Ju 88B included a *Zerstörer* and a strategic reconnaissance platform.

The aerodynamic improvements of the Ju 88B alone were expected to provide a benefit of 32 km/h in comparison with the A-series, as shown below:

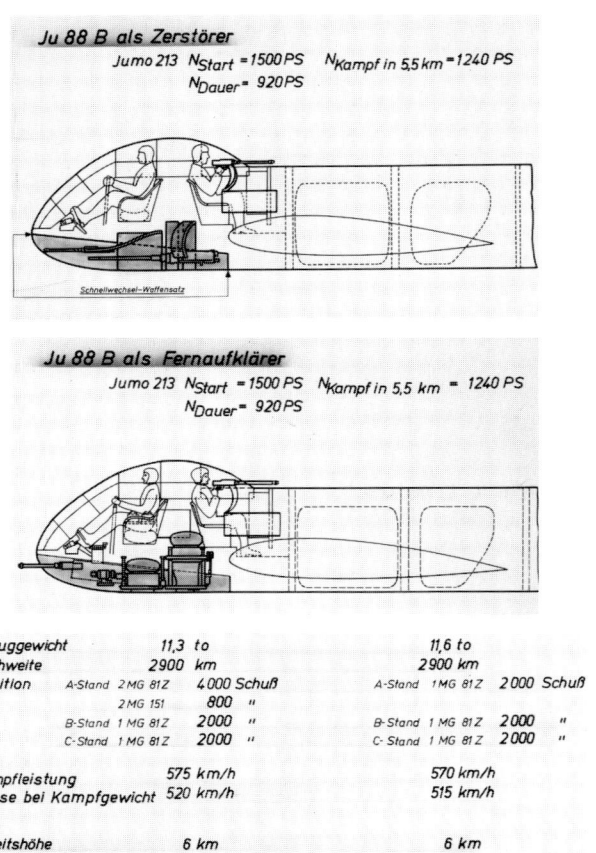

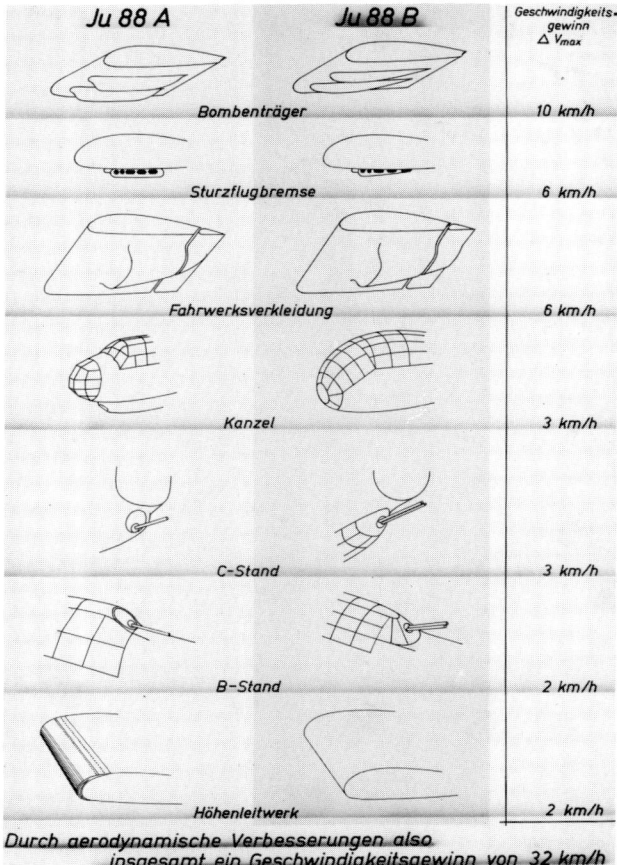

be jettisoned for single engine flight or in the event that this cumbersome and quite slow aircraft had to make a belly landing.

The P-1 had been preceded in the summer of 1942 by an experimental conversion of an A-4 at the Bernburg plant. The resulting aircraft, designated the Ju 88P-0 or Ju 88P V1, exhibited problems with the muzzle brake of the BK 7.5 developed from the PaK 40L. Satisfactory firing results were not achieved until the gun was fitted with a lengthened sieve-type brake as a muzzle brake. Ammunition was fed to the chamber via a nine-round clip loaded from above. The BK 7.5 was the heaviest caliber gun used by the *Luftwaffe* especially for engaging tanks. Only about thirty BK 7.5s were produced capable of firing armor piercing shells each weighing 11.9 kg and having a rate of fire of 30 rounds per minute. At a distance of 1,000 m the round was capable of penetrating armor 130 mm thick at an impact angle of 90°!

History of German Aviation: Bombers and Reconnaissance Aircraft

The following diagram provides a general overview of the Ju 88B, of which only a few prototypes were built as predecessors for the Ju 188. Certain elements and components, however, influenced the Ju 288 and Ju 388.

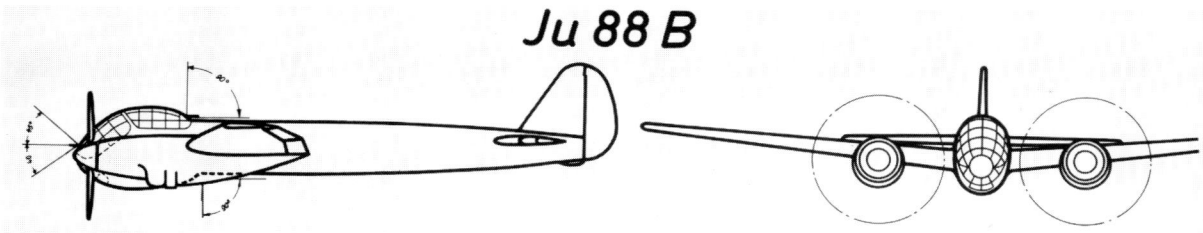

Ju 88 B

KENNZEICHEN:

Neue Vollsichtkanzel
Dadurch:
- Bessere Sicht
- Größerer Innenraum
- Geringerer Luftwiderstand

Stärkere Motoren
JUMO 211 F, später JUMO 213
Günstigere Triebwerkseinbauten
Neue Junkers Luftschraube VS 11
Dadurch: Höhere Geschwindigkeit
Kürzere Startstrecke

Stärkere Bewaffnung
3 Doppel-MG 81 mit Schußfolge 2800/Min
(bisher 3 MG 15 mit Schußfolge 1000/Min
Munition 5000 Schuß
(bisher 2000 ")
Günstigere Schußfelder

Leistungen		JU 88 A mit JUMO 211 B		JU 88 B mit JUMO 211 F		JU 88 B mit JUMO 213	
Startleistung	PS	1220		1350		1500	
Kampfleistung	PS	910		1050		1240	
V max	km/h	440	460	510	525	540	555
V Reise	km/h	370	415	420	445	445	465
Reichweite	km	2900	1420	2800	1400	2800	1400
Bo	kg	1000	1400	1000	1400	1000	1400
Roll-/Startstrecke	m	685/1050		640/960		550/830	

Durch Einbau des JUMO 211 F wird der Serienanlauf der JU 88 B auf Herbst 1940 vorverlegt. Dadurch werden bis 1.4.1941 bereits 75 B Maschinen geliefert.

Opposite: Junkers Ju 88P-1 armed with the heavy BK 7.5 cm cannon. Unlike the Ju 88P-V1 prototype's nose glazing, the P-1 made use of an armored fuselage underside in order to deflect the gas from the BK 7.5's high muzzle velocity to either side without damaging the underside of the cockpit.

Junkers Ju 88P-2/P-3 (1942/1943) with two 3.7 cm cannons carried in a gondola mounted on the underside of the fuselage.

The Ju 88P-2 and P-3 were other tank buster developments from 1942/1943, each equipped with two 3.7 cm BK guns (developed from the FlaK 18 AAA gun) carried in a weapons bay side by side beneath the fuselage. Both were one-off versions built from the A-4 series, with ten P-2s being produced and serving on the Eastern Front with a tank busting operational testing unit.

The experimental use of a few Ju 88P-2s against American bomber formations by day, the role it was originally intended for, proved to be senseless due to the P-2's poor speed and maneuverability, whereas the type, despite having less armor than the P-3, was found to be somewhat effective against armored vehicles. The Ju 88P-4 was a result of the realization that the penetration effectiveness of the P-2's cannon could no longer hold up in the face of the increased armor thickness of Soviet tanks. The P-4 was a batch of 32 A-4 variants converted into tank busters at the Merseburg plant to carry a 5 cm BK in a smaller weapons tray beneath the fuselage. The BK 5 was developed by Rheinmetall-Borsig from the KwK 39 tank gun and adapted for installation in aircraft as an automatic onboard cannon, with the automatic loader and magazine having had to have been developed from scratch. The magazine encompassed the aft end of the gun in a ring fashion and could accommodate an endless supply of 22 round ammunition belts. At a range of 1,000 m and an impact angle of 60°, the armor piercing shells could penetrate armor up to 65 cm thick. An experimental use of the P-4 as a day or night fighter met with as little success as had the P-2, but in late 1944 the variant achieved notable success on the Eastern Front against enemy armor, although the sheer numbers of Soviet armor formations meant that these victories had little impact.

The S-Series
The idea of a high speed bomber again came into vogue in light of the increasing challenges of the air war over England, particularly as a result of the speed need for offensive aircraft to operate in the face of the constantly improving British day and night fighter defenses. By dispensing with the dive capability of the Ju 88A-4, the most effective model up to that point, some of its armor was first removed and the external ETC bomb racks taken off, and in place of the multi-angular canopy design the Ju 88V93 prototype was fitted with a hemispherical all-around view canopy streamlined to cut slip through the wind. Driven by air-cooled BMW 801D engines (1,250 kW/1,700 hp takeoff rating and 1,060 kW/1,440 hp at 5,700 m), the Ju 88SV1 (as it was now called) attained a maximum speed of 535 km/h, well above that of the A-4 variant. However, the *Technisches Amt* considered this to be insufficient for successfully escaping pursuing enemy fighters. In spite of this skepticism, a small pre-production batch of Ju 88S-0s was built based on the SV1, but in the late autumn of 1943 this version was replaced by the markedly improved Ju 88S-1 production model. For its engines, the S-1 was powered by the BMW 801G-2 with GM 1 fuel injection, giving it a performance of 1,272 kW/1,730 at an altitude of 1,500 m and 1,052 kW/1,430 hp at 7,300 m. A brief detailed look at the GM 1 system in the Ju 88S-1 is appropriate at this point: the kit weighed 182 kg; the N_2O in liquid form was carried in the aft bomb bay in either three cylinders well insulated with spun glass, or in a single specialized 415 liter container which, when full, weighed another 410 kg. This system gave a performance boost for a limited

Junkers Ju 88S-1 with the new, spherical glazed nose (1943).

time at two different settings. Compressed air was used to inject the N_2O into the engines, and at "NORMAL" setting gave an operating time of 45 minutes, while "NOT" (EMERGENCY) provided 27 minutes of boost. At an altitude of 8,000 m the Ju 88S-1 without external load and no GM 1 boost had a maximum speed of 548 km/h, but with GM 1 this jumped to 610 km/h! The S-1 could carry two SD 1,000 bombs beneath the wing center sections, while in the forward bomb bay was housed an additional fuel tank with a capacity of 1,220 liters. This aircraft began reaching field units in small numbers starting in early 1944, where it primarily served in the pathfinder role for bombers following behind.

A small batch of Ju 88S-2s followed in the spring of 1944. This variant was powered by the BMW 801TJ engine with turbocharger, making the space- and weight-hogging GM 1 system of little use. The 801TJ offered 1,330 kW/1,810 hp on takeoff and developed 1,100 kW/1,500 hp at an altitude of 12,000 meters. In order to provide more room for the internal fuel tanks the S-2 was fitted with a wooden gondola beneath the fuselage—similar to that tested with the Ju 88A-15—capable of carrying 3,000 kg of bombs. In the forward bomb bay located above this gondola was the 1,220 liter fuel tank of the S-1, but in the aft section the N_2O containers were replaced by an additional fuel tank holding 680 liters of fuel. The S-1 had just a single rear-firing 13 mm MG 131 in the way of defensive armament, whereas the S-2 carried two additional fixed forward-firing 7.9 mm MG 81s in the wooden gondola. The S-2, in spite of the large gondola, paid only a small price in speed at all altitudes compared to the S-1.

The final development of the S-series followed in the late summer of 1944 with the production of a small number of Ju 88S-3s, the manufacture of which came to a halt after only a brief period in the face of the higher priority given to other programs. In principle, the S-3 was the same as the S-1, but had the liquid-cooled Jumo 213A with GM 1 as its powerplant. The N2O was housed in a 340 liter container in the aft bomb bay. With the GM 1 set to "NORMAL," takeoff performance jumped from 1,306 kW/1,776 hp to 1,563 kW/2,125 hp, and when set to "NOT" this increased to 1,692 kW/2,300 hp. Maximum takeoff weight of the Ju 88S-3 totaled 13,790 kg. At an altitude of 5,500 m the standard output of 1,434 kW/1,950 hp and emergency power of 1,560 kW/2,120 hp could be boosted by 155 kp of exhaust thrust. At a normal flying weight of around 10,500 kg, i.e. after the bombs had been dropped, the S-3 with GM 1 had a maximum speed of an impressive 615 km/h at an altitude of 8,500 m, slightly more than the rather quite fast S-1.

Summary

The S-series was virtually the end of the line with regard to development of the original Ju 88 concept. Further performance increase for the basic model would be difficult to achieve in 1944. With the elimination of the Ju 288 from the production program, it was left to the Ju 188—the follow-on development of the Ju 88—and Ju 388 to carry the banners of hope for the *General der Kampflieger's* October 1943 call for high performance bombers. What had been happening with these two aircraft types in the meantime?

Junkers Ju 188 - Prototypes
Development potential of the original Ju 88 had lain practically dormant for over three years, limited to numerous, operationally driven modifications and engine performance improvements. The Ju 288 was expected to have resulted in a notable performance boost within the framework of the "Bomber-B" project. By late autumn of 1942, it was apparent to the RLM's Tec*hnisches Amt* that the Ju 288's availability for front-line service might be delayed considerably due to the planned Jumo 222 engines having suffered setbacks in their development, setbacks that would take some time to correct.

Fortunately, Junkers—as sort of a fallback position for the "Bomber-B"—had continued developing (albeit at a lower priority) the Ju 88E-0 briefly mentioned earlier. In 1942 the company had built a Ju 88V44, similar in configuration to the Ju 88V27, which had a modified and enlarged rudder resulting from instability problems with both the Ju 88E-0 and the V27.

In October 1942 the Ju 288's situation prompted a decision to bump the Ju 88 up to a higher developmental priority, and the Ju 88V44 became the prototype model for development of the Ju 188. Retroactively designated the Ju 188V1, the V44 was followed by a contract for a Ju 188V2, which in January 1943 also entered flight testing. Both Ju 188 prototypes were laid out for both level and dive bombing with the same dive brakes and automatic recovery system that had proven so effective with the Ju 88.

Junkers Ju 188V2, which entered flight testing in January 1943.

While jig construction got underway for series production of the Ju 188 at Bernburg, in light of the engine situation the RLM authorized the option of fitting the Ju 188 with either the Jumo 213 or the BMW 801, with the only stipulation being that the engine mounts could not be altered.

Ju 188A, E, and G Bomber Variants
The first version to go into production was the BMW 801 powered E series. The pre-production variant, the Ju 188E-0 fitted with two BMW 801ML (each rated at 1,177 kW/1,600 hp on takeoff), entered flight testing in the spring of 1943, was fully dive capable, and had the same offensive potential as the A-series, with the only change being a 13 mm MG 131 in the fuselage turret instead of the Ju 188A's 20 mm weapon.

Junkers Ju 188E-0 with BMW 801 and MG 131 in the fuselage turret (1943).

The subsequent Ju 188E-1 initially had the same powerplant as the E-0, but this was soon replaced with the more powerful BMW 801D-2 or G-2 with 1,250 kW/1,700 hp output. This version and the A-series, which differed only in engine details, formed the bulk of Ju 188 bombers built; just over 500 were manufactured by the time production ceased in June 1944. What follows is a somewhat detailed description of these aircraft, whose performance and serviceability in front-line operations were assessed to be quite positive.

Ju 188A and E General Description

The Jumo 213 powered A-series and the BMW 801 powered E-series were four-man bomber aircraft with a spacious cockpit whose walls had been pushed outward (compared to the original Ju 88) and designed with symmetrical curved glazing. The wings were pointed at the tips and had a much greater negative decalage, with the outer section of the two-part aileron having a noticeable bend. The empennage was much larger; the hydraulically powered elevator position setting was linked to the flap setting. At higher stick force, rudder and aileron trim tabs acted as auxiliary controls by means of compressed control rods. The undercarriage included wide-rim wheels and hydraulic dual brakes. The antenna mast behind the upper aft *B1* gun position could be electrically lowered out of the field of fire. The aircraft was armed with a 20 mm MG 151/20 having a limited field of fire in the nose, which could be locked at a depression of 4% 30' for making high speed firing runs using the cockpit gunsight. The *B2-Stand* above the cockpit was a turret for a 13 mm MG 131 or 20 mm MG 151/20 (Ju 188A), and the *B1-Stand* behind it had a 13 mm MG 131. The *C-Stand*, the lower aft gun position beneath the cockpit, had a 7.9 mm MG 81Z with speed-controlled reflective gunsight. Maximum bomb capacity was 3,000 kg, and the aircraft had two bombsights—the Lotfe 7D for level bombing and the BZA 1 with Stuvi 5 for dive bombing.

At a later date the dive system ceased being fitted to the aircraft, since in practice the dive bombing role was seldom used. Also dropped was the so-called "dive activation" setting on the engine control system, which adjusted the propeller pitch during a dive.

Weights and Dimensions:

wingspan	22.00 m	empty equipped weight	9,860 kg
wing area	56.00 m²		
length	15.01 m	takeoff weight	14,500 kg
wheel track	5.77 m	max. speed	500 km/h
width of empennage	8.00 m	range	2,600 km

It should be noted that the Jumo 213 equipped A-series came after the E-series in order of production, but its pre-production version, the Ju 188A-0, was completed at Bernburg just shortly after the E-0 in the early summer of 1943. The Ju 188A-0 was designed exclusively for the level bombing role and carried a maximum of 16 SD 65 bombs in the fuselage (= 1,040 kg), plus two 1,000 kg or four 500 kg bombs on external ETC racks beneath the center wing section.

The dive bombing components were to have been included on the subsequent Ju 188A-1 variant, but when this operational role was dropped, the A-1 series also fell under the axe in favor of the Ju 188A-2. This was a pure level bomber which began rolling off the assembly line in early 1944. The Ju 188A-2 had the same engine as the A-0, the Jumo 213A-1 rated at 1,306 kW/1,776 hp on takeoff and 1,177 kW/1,600 hp at an altitude of 5,500 m. In addition, it had

Junkers Ju 188E-1 (1943), an early production aircraft with dive brakes mounted underneath the wings. The wings (span: 22.0 m compared to the Ju 88A-1's 18.25 and the A-4's 20.08 m), which tapered to a point, and the "bent" ailerons can be seen to good effect in this air-to-air photo.

Junkers Ju 188A-2 being loaded with a 1,000 kg bomb. Using a hoisting device, a human chain lifts the bomb off the improvised bomb cart. The 20 mm MG 151/20 can be seen protruding from the A-*Stand*.

the MW 50 water methanol boost system, which boosted the takeoff power to 1,648 kW/2,240 and gave it a maximum performance in flight of 1,383 kW/1,880 hp at 4,800 m altitude. The A-2's maximum speed was barely higher than the E-series, i.e. a good 500 km/h. Torpedo bombers were built in small numbers on the basis of both the A-series and the E-series, these being the Ju 188A-3 and Ju 188E-2, respectively. The two outer ETC racks were removed from beneath the center wing sections, and two torpedoes, either the 800 kg LT 1B or the 765 kg LT F5b, could then be carried on the remaining two racks. The torpedo bomber variant had a long extension along the starboard side of the lower nose housing the adjusting equipment for the torpedo control mechanism. These bombers were equipped with the FuG 200 Hohentwiel surface search radar in the nose, and a few examples were supplied to the field lacking the *B1-Stand*.

Like the original Ju 88, operations showed that all Ju 188 aircraft suffered from a vulnerable spot with regard to their defensive armament—the dead zone just behind the rudder. To compensate for this somewhat, it was planned to fit the type with a remotely operated FA 15 tail turret with two 13 mm MG 131 guns, one above the other. Problems with the targeting and control systems, however, caused so much difficulty that it was decided to pursue a manned rear position with the same armament. Such a turret was fitted to a Ju 88A-4 and evaluated. Like a Ju 188C-0 predecessor, the Ju 188V2 was also modified accordingly and, with an enlarged and strengthened rear fuselage, became a testbed for manual and remotely controlled rear turret designs. Fitted with the manned turret, the V2 was designated the Ju 188G-0. The dimensions of this turret, however, were so small that its range of movement was severely restricted and only gunners of extremely small stature could fit inside it. Additionally, in their tiring, crouched positions they barely had a chance to bail out in the event of an emergency. As a result the L*uftwaffe* rejected this solution, instead pressing for further development of the FA 15 turret for installation in the planned Ju 188G-2 series. The G-2 design dispensed with the external racks altogether, instead carrying 3,300 kg of bombs in a wooden "bathtub" under the fuselage; thus configured, it was expected to have had a range of 2,400 km and achieve a speed of 540 km/h at 6,000 m. However, the Ju 188G-2 never entered testing, as it was overtaken by development of the Ju 388.

Ju 188D, F, and H Reconnaissance Variants

Parallel to the bomber versions described above were the Ju 188D-1 and D-2 Jumo 213 powered variants based on the A-series. These lacked the nose armament (*A-Stand*) and flew with just a three-man crew, their layout focused exclusively with an eye toward range

Junkers Ju 188F-1 strategic reconnaissance plane carrying two 300 liter drop tanks for increasing range.

and speed. With full internal tanks and two external 300 liter drop tanks beneath the center wing sections, the Ju 188D had a takeoff weight of 15,196 kg, a maximum range of 3,400 km at 6,000 m. and a cruising speed of 480 km/h. Maximum speed at the same altitude was 540 km/h. Camera equipment could be varied depending on the specific mission; for daylight reconnaissance the suite included two Rb 50/30 or 75/30, and for night photography two NRb 40/25 or 50/25. The Ju 188D-2 was mainly designed for the maritime reconnaissance role and also carried the FuG 200 Hohentwiel radar system.

Junkers Ju 188F-1 strategic reconnaissance plane. When fitted with the FuG 200 Hohentwiel surface search radar it was designated the Ju 188F-2.

The Ju 188F-1 and F-2 were the BMW 801D-2 or G-2 equipped counterparts to the D-1 and D-2, with the Ju 188F-2 also being equipped with the FuG 200 Hohentwiel system for maritime reconnaissance missions.

Finally, there was the reconnaissance counterpart to the uncompleted Ju 188G-2, the Ju 188H-2. This also was to have had a rear turret with two remotely controlled 13 mm MG 131 machine guns, but as a result of the stepped up development of the Ju 388 it, too, never entered flight testing. Of the total 1,076 Ju 188s the Lu*ftwaffe* received, about 570 examples were D and F series reconnaissance variants.

Ju 188 through Ju 388 Follow-on Developments and the Ju 488

Fate of the Ju 288 and the Ultimate Sacrifice of the Ju 188

We have already taken a look at the life of the Ju 288V101 and V102, both available for testing in August 1942, in conjunction with the planned Ju 288C production series. Despite the return of D*ipl.-Ing.* Hertel, who had been responsible for design supervision of the Ju 288, to the Heinkel company with special tasking for the He 177 in November 1942, production preparations for the Ju 288C proceeded at Junkers with utmost urgency. Shortcomings discovered during flight testing of the V101 and V102 were taken into account when building the Ju 288V3, and in preparations for full scale production. Among other things, the V103 had reinforced dive brakes, four remotely operated gun turrets, and the more powerful DB 610A-1/B-1 engines (each rated at 2,170 kW/2,950 hp on takeoff and an impressive 2,280 kW/3,100 hp at 2,000 m). The V103, planned as the prototype for the Ju 288C-1, began its flight testing in the spring of 1943, followed by the Ju 288V104 and V105 in May and the Ju 288V106 in June of that year, all fitted with the DB 610 powerplant.

There were three bomber versions envisioned at this point in time, all of which differed practically only in their weapons layout. The Ju 288C-1 was to have had forward firing remotely controlled 13 mm MG 131 twin guns in the chin position, and in the upper and lower fuselage a 15 mm MG 151 each, plus one 20 mm MG 151/20 in a rear turret—also remotely controlled. The Ju 288C-2 was to have been heavily armed with 15 mm MG 151 twin guns in place of the MG 131s or MG 151s in the three gun positions, as well as two 13 mm MG 131s in the aft position, or four of these if the manned rear turret then in development became available. Finally, there was the Ju 288C-3, conceived as a pure night bomber with only an MG 131 twin gun in the lower fuselage for defensive armament firing aft.

In June 1943 the T*echnisches Amt* informed Junkers that the entire "Bomber-B" project was to be canceled. Shortages of material, unavailability of the Jumo 222 engine planned for the Ju 288, and the negative effect of starting up a new production program on the current bomber production at a critical phase in the war were behind this unexpected decision. In spite of this, in July 1943 Junkers began flight trials with the Ju 288V107 and V108 (both had been near completion at the time of the Techn*isches Amt's* decision), and flight evaluation for the Ju 288 even continued up until the summer of 1944. By then, out of a total of 22 Ju 288 prototypes 17 had been lost in accidents—a high price to pay for development! When flight testing stopped, some of the leftover B and C-series aircraft were fitted with a receptacle for a 50 mm BK 5 cannon and handed over to the L*uftwaffe*, where they were used individually during the last stages of the war. These modified prototypes have sometimes inaccurately been called Ju 288Es or Ju 288Gs.

In all, over 1,000 variants of the Ju 188 series were built, including the following (available late 1943/early 1944):
- a small production batch of Ju 188S-1s as a high-speed bomber without any defensive armament, otherwise basically similar to the A-2
- a small production batch of Ju 188S-1/U1s as a tank buster with the 5 cm BK 5 cannon.
- Ju 188T-1s, also in small numbers, as a high-speed reconnaissance platform, unarmed like the S-1.

The fiasco with the Ju 288 and the cessation of the Ju 188's production, however, meant that all the *General der Kampfflieger's* hopes for a high altitude, high-speed bomber now rested with the Ju 388.

The "Hubertus" Program

Begun in September 1943, the "Hubertus" program (effectively follow-on development of the Ju 188 into the Ju 388) called for three basic versions of the Ju 388:

Left: Manned rear gunner's station test fitted with four 13 mm MG 131 planned for the heavily armed Ju 288C-2 (1943).

History of German Aviation: Bombers and Reconnaissance Aircraft

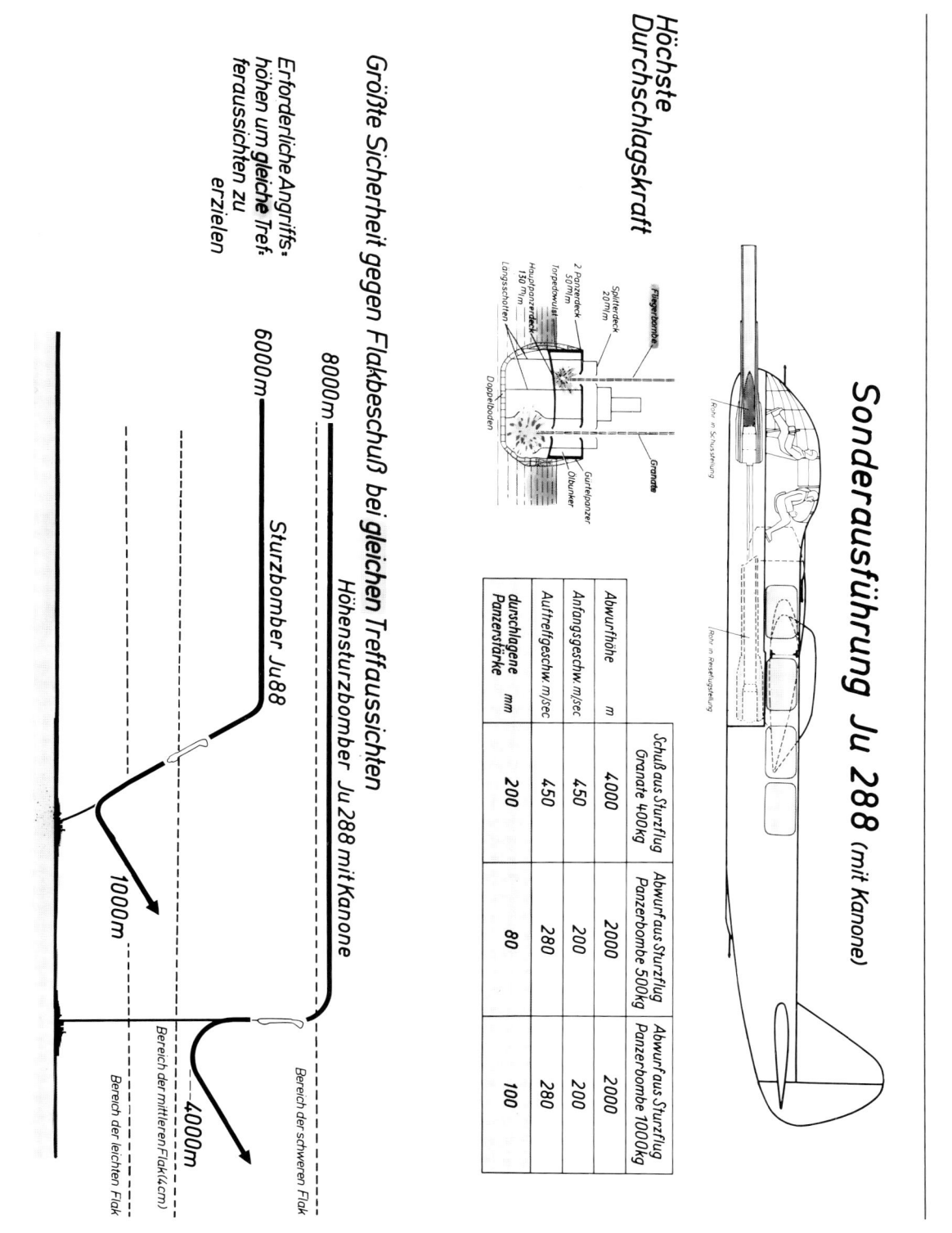

Sonderausführung Ju 288 (mit Kanone)

	Schuß aus Sturzflug Granate 400kg	Abwurf aus Sturzflug Panzerbombe 500kg	Abwurf aus Sturzflug Panzerbombe 1000kg
Abwurfhöhe m	4000	2000	2000
Anfangsgeschw. m/sec	450	200	200
Auftreffgeschw. m/sec	450	280	280
durchschlagene Panzerstärke mm	200	80	100

Höchste Durchschlagskraft

Größte Sicherheit gegen Flakbeschuß bei gleichen Treffaussichten

Höhensturzbomber Ju 288 mit Kanone

Sturzbomber Ju88

Erforderliche Angriffshöhen um gleiche Treffaussichten zu erzielen

8000m
6000m
1000m
4000m

Bereich der schweren Flak
Bereich der mittleren Flak (4cm)
Bereich der leichten Flak

Interesting project study from 1943 for the Junkers Ju 288C as a high altitude dive bomber with "heavy armament." It called for a 400 kg bomb to be dropped in a vertical dive from 8,000 to 6,000 m, enabling it to penetrate ship armor up to 200 mm thick. This method made it 100% more effective than dropping a PC 1000 armor piercing bomb in a dive from about 1,000 m! As the method of hitting a precision target from a distance of 6 to 8 kilometers was crucially dependent upon weather/visibility, this project also remained nothing more than a pipe dream.

- Ju 388J as a high-latitude all-weather fighter
- Ju 388K as a high altitude bomber
- Ju 388L as a high altitude photo-reconnaissance platform

Since is was questionable whether the Jumo 213E-1 (the engine called for in the initial design proposal) would be available in time, it was determined that the first variants of all three types would all be fitted with the BMW 801TJ with turbocharger. The reconnaissance version was to receive the highest priority of the three.

Junkers Ju 388L

The first completed prototype was the long range high altitude Ju 388L V1, completed in Dessau in late 1943. It had been built using the airframe of a Ju 188T-1 which, with its Jumo 213E-1s and GM 1 fuel injection system and just two cameras, had itself attained a top speed of 700 km/h at an altitude of 11,500 m. Following successful flight trials at the Rechlin test center the ATG company (Allgemeine Transportanlagen G.m.b.H.) at Merseburg was given a contract to convert ten Ju 188S-1 airframes into the pre-production Ju 388L-0 reconnaissance version. The first of these modifications was handed over to the *Luftwaffe* in August 1944, by which time preparations for production of the Ju 388L-1 at Merseburg had practically been completed. From the outset all Ju 388 versions were to be fitted with the remote controlled MG 131Z/FA15 rear turret with its two over-and-under 13 mm MG 131 guns, as had been tested for the Ju 288C-2. However, like earlier trials, the arrangement ran into technical problems for the ten pre-production versions, and in the end the only defensive armament carried was a single, rear-firing fixed arrangement of two 7.9 mm MG 81s fitted in a teardrop shaped bubble (called a "*Waffentropfen*" and designated WT 81Z) under the fuselage. The Ju 388's speed and operating altitude, however, were such that this weapon needed to be little more than a scare tactic to stave off any enemy fighters actually able to get that high and maneuver into a firing position.

The Ju 388L-1, built from October 1944 onward at Merseburg and from November 1944 at the company of Weser-Flugzeugbau in Bremen, differed quite markedly from the L-0 pre-production batch. In place of the three-bladed wooden propellers the L-1 had four-bladed VDM Dural props and the FuG27 Neptun rear warning radar. Also, like the Ju 388K bomber version developed alongside the L, it had a large wooden fuselage belly for carrying a jettisonable 900 liter fuel tank and two cameras in different configurations. These included the Rb 20/30, 50/30, the long focal length 75/30 or, for night reconnaissance, the NRb 35/25, 40/25, or 50/25. With the extra fuel, the Ju 388L-1 had a maximum fuel capacity of 4,800 liters and at economic cruise setting at 11,000 m had a range of a good 3,400 km.

By the time the L-1 began deliveries, the technical problems with the FA 15 rear turret had, for the most part, been ironed out,

Junkers Ju 388L-1/b (1944), a four-seat photo reconnaissance aircraft designed with a pressurized cockpit for operating at high altitudes. This photo shows off its remotely operated FA 15 rear gun position with two 15 mm MG 151s.

and this replaced the WT 81Z. However, front-line trials revealed that at high speeds the gun's target angle was not always in alignment with the angle sighted by the gunner using the PVE 11 periscope sight. Because of the potential inaccuracy, units fitted a flex-mounted MG 131 in the upper aft canopy section and increased the crew by one (a gunner) to four. In this configuration the machine was designated the Ju 388L-1/b.

With its BMW 801TJ engine rated at 1,324 kW/1,800 hp on takeoff and an output of 1,037 kW/1,410 hp at 12,300 m, the L-1 had a speed at sea level of 415 km/h and 615 km/h at 12,300 m. Added to that was a service ceiling of 13,500 m, giving the L-1 remarkable high altitude performance indeed, yet ultimately still not as fast as the Ju 188T-1!

There was also a Ju 388L-3 series laid down, with the Jumo 213E-1 engine, MW 50 water-methanol fuel injection, and the four-bladed VS 19 wooden propeller; at a maximum speed of 460 km/h it was faster than the L-1 at sea level, but had a service ceiling of just 12,200 m. On the other hand, with the MW 50 system it had a maximum speed at 9,000 m of 655 km/h. Just one or two examples of this excellent high altitude reconnaissance platform were built at Merseburg. By the end of 1944 Merseburg had supplied 37 Ju 388L reconnaissance aircraft, with Weser-Flugzeugbau adding another ten before both plants ceased production in early 1945—the priority no longer existed by this late stage in the war. Their operational service was limited to a few examples with the *Versuchsverband Ob. d. L.*, while the true long range reconnaissance units were unable to finish their conversion training for this higher performing reconnaissance aircraft because of the progress of the war.

Ju 388J and K

Just three pre-production examples of the Ju 388J high altitude all-weather fighter version entered flight testing in 1944 under the codename "Störtebeker." Despite the urgency given to the program, construction of the Ju 388J-1—scheduled to begin in January 1945—was canceled due to the situation developing the previous month.

On the other hand, the Ju 388K high altitude bomber enjoyed a slightly more fortunate history. A pre-production batch of ten Ju 388K-0s began rolling off the Dessau assembly line starting in July 1944. Precursor to, and prototype for, this high altitude bomber was the Ju 388V3, which had begun its flight testing in January 1944. It had the same wooden belly of the Ju 388L-1, enabling it to carry a maximum 3,000 kg of bombs over shorter distances. Normal payload included two 1,000 kg or one 2,000 kg bomb, with which the Ju 388K had a range of 2,250 km at maximum cruise setting at an altitude of 11,000 m. Neither the V3 nor the ten Ju 388K-0s had the FA 15 rear turret and were thus unarmed. It was not until the Ju 388K-1 series, of which five were built by early 1945, that the rear station was fitted with the two remotely operated MG 131s, each with 600 rounds, as its sole defensive form of armament.

The Ju 388K-2, powered by the Jumo 213E-1 engine, and the Ju 388K-3 with the Jumo 222E/F never made it to the flight testing stage due to production being canceled in early 1945.

The last variant of note was the Ju 388M-1, also uncompleted, with the pre-production version still under construction when production ceased in early 1945. This four-man version lacked the wooden belly and was designed as a platform for the LT 950 aerial torpedo.

Summary and the Ju 488

The Ju 388 was the penultimate bomber stemming from the original Ju 88 high-speed bomber developed for purely military purposes back in 1936. Developed with pressurized cockpit for high altitude operations and impressive performance in spite of numerous wartime annoyances, it was a bomber optimally suited to its intended role. Had it become available earlier, in the numbers called for by the *General der Kampfflieger*, it could have posed serious short-term problems for the Allies, whose air defenses had not yet been tailored to the high speeds and operating altitudes of such an aircraft. However, other than a few operational sorties by the reconnaissance version, the Ju 388 did not involve itself in the war's events.

Interesting, and certainly worthy of note, is a proposal submitted to the T*echnisches Amt* by the Junkers design bureau for taking components of the Ju 188, Ju 288, and Ju 388 and designing a four-engined strategic bomber. This recommendation was obviously made in light of the misery caused by the He 177 debacle. The *Technisches Amt* accepted this hybrid design in principle and gave the project the designation Ju 488. According to the way Junkers envisioned it, it would have incorporated the wings and entire forward fuselage of the Ju 388K/L, the fuselage center and aft sections from the Ju 188E, and the wooden belly of the Ju 88A-15/Ju 388K. The control surfaces from the Ju 288C would, by using additional fuselage components, have been combined with new center wing sections, which would have also accommodated the two additional engines. To relieve some of the workload at Dessau, the former Latécoére factory at Toulouse would have been assigned the production of the entire fuselage and new wing center sections. Work began immediately on construction of the fuselages for the first two prototypes, the Ju 488V401 and the V402, making use of parts supplied from Germany. Further thinking resulted in a complete rework of the fuselage design, however, which was now envisioned as a welded steel tubing construction with alloy metal skinning of the forward and center sections, with the aft fuselage fabric covered. The next four prototypes, Ju 488 V403 through V406, were contracted for using the new fuselage construction in anticipation of the planned Ju 488A production model. The goal was to have the Ju 488A in service by mid-1945. Both the Ju 488V401 and V402 were to have been powered by four BMW 801TJ engines (1,324 kW/1,800 hp takeoff rating, 1,248 kW/1,710 hp at 9,000 m), have had a three-man crew, and been streamlined and unarmed testbeds, capable of carrying a 2,000 kg bomb load over 2,000 km at economic cruise setting. The four single struts of the landing gear retracted into each of the four engine nacelles. The prototypes from

Junkers Ju 388K-0 high altitude bomber (1944). Only ten Ju 388K-0 and five Ju 388K-1 bombers had been built by early 1945. The five K-1s were unarmed with the exception of their two remote controlled 13 mm MG 131 rear guns. The stretched wooden "bathtub" for the type's increased bomb load can be seen to good effect in the upper photo.

the Ju 488V403 onward all had the new fuselage, with only the pressurized cockpit from the Ju 388K/L sharing any kind of commonality with the V401 and V402. Otherwise, this new fuselage design was much wider and some three meters longer than on the first two prototypes. The wooden belly was dropped and the wings were set further back along the fuselage. Instead of the BMW engines the Ju 488V403 incorporated the 24-cylinder liquid-cooled Jumo 222A-3/B-3 rated at 1,840 kW/2,500 hp on takeoff and 1,410 kW/1,920 hp at 7,200 m. In the center of the fuselage spine and in the tail the V403 was fitted with remote controlled gun turrets (FDL) with two 20 mm MG 151/20s in the fuselage and two 13 MG 131s in the tail. Maximum payload had now risen to 5,000 kg, and the bomber could carry 15,000 liters of fuel in a total of 14 tanks. Maximum estimated on economy cruise at 485 km/h and an altitude of 7,200 m was 3,400 km, with a planned long range reconnaissance version even projected to have a range of over 5,000 km. Service ceiling was expected to have been somewhere around 11,500 km.

Yet all these ideas, plans, and calculations were to remain nothing more than pipe dreams, for during the night of 16/17 July 1944 French saboteurs damaged the fuselages and wing center sections to such an extent that they were no longer repairable. The saboteurs struck in Toulouse after the components had been finished and were being loaded onto railcars for shipment to Bernburg for final assembly. Work on the Ju 488V403 through V406 nevertheless continued apace until November 1944, when the RLM stopped the entire Ju 488 program and the completed parts were scrapped.

In January 1945 Japan received authorization for license construction of the Ju 488, but production never got underway there, either. This was the final chapter in the development of an outstanding bomber aircraft. Only one other role for the type remains to be discussed, that of an unmanned Ju 88 with a heavy warhead in the nose being flown by a controller plane against point targets.

Junkers Ju 88-*Mistel*

In early 1943 Junkers factory test pilot Siegfried Holzbauer resubmitted an idea (originally rejected by the RLM in 1941) for a piggyback bomber. This was not altogether an original concept; the British and Americans had explored different piggy-back configurations as early as 1916 and 1918, and the British put these ideas into practice in 1937 with the Short Mayo combination (four-engined flying boat with a four-engined floatplane on top). The Soviets, too, had been playing with the idea since 1931. According to Holzbauer, those Ju 88 airframes not up to the latest standards would be used as a warhead platform in piggy-back arrangement with a fighter plane functioning as the control aircraft. After separating, by using a new three-dimensional autopilot the control plane would guide the unmanned "disposable" bomber onto heavily armored naval or strategically important land targets. This decision had been predated in Germany by successful flight trials by Fritz Stamer and Karl Schieferstein, who in 1942 at the *Deutsche Forschungsanstalt für Segelflug* (DFS) had towed a Klemm Kl 35 and Focke-Wulf Fw 56 to altitude, as well as demonstrated an Me 109E taking off under its own power as a control plane for a DFS 230 glider. Using the DFS designed frame between the *Mistel* (the lower airplane) and the control plane on top, *Dr.-Ing.* Fritz Haber developed the prototype frame for a flying team consisting of a Ju 88A-4 as the *Mistel* and an Me 109 as the controller. A contract was issued for an initial batch of this combination using Ju 88A airframes in July 1943. Known by its nickname of "Father and Son," this now familiar combination was given the covername of "*Beethoven-Gerät*" (Beethoven Device). The Beethoven flight program went smoothly, and in the spring of 1944 Junkers company test pilots began training the first operational crews on this novel and complex weapons system.

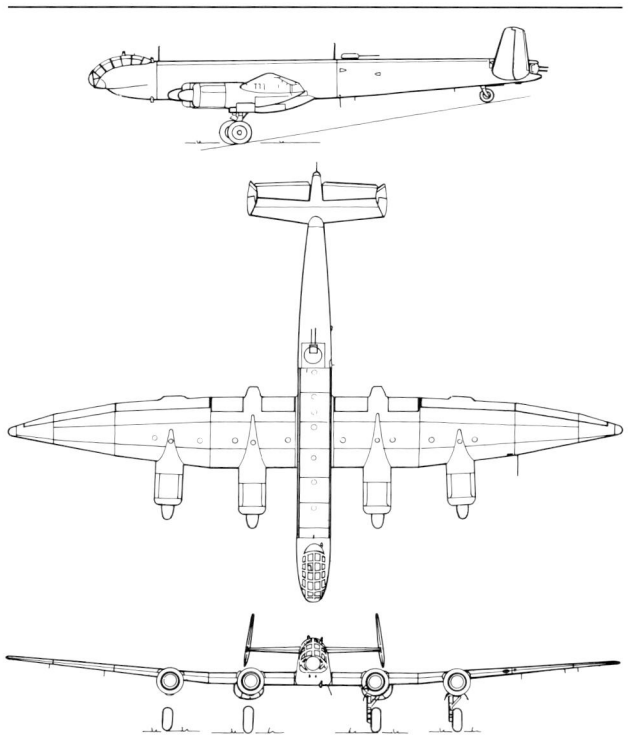

Junkers Ju 488V403 (1944), a prototype which was never completed, with armament and new fuselage. Work on the V403 through V406 was halted in November of 1944.

In late 1943 tests were carried out against an old French battleship in Toulon, as well as against heavy iron/concrete facilities in East Prussia using a hollow charge load in conjunction with the explosive charge's steel core weighing another 1,000 kg, for a total explosive charge of 3,600 kg. These tests were quite successful, and Ju 88 trials shifted to Peenemünde in April/May 1944 where the design was declared ready for combat—despite not being satisfactory in all flight profiles.

With regard to the explosive charge, it should be mentioned that it was attached to the Ju 88 using speed screws in place of the unglazed nose just before an upcoming "live" mission. At least six man-hours were required to switch from the nose to the warhead, with a full workday needed to set up the explosives. Half of the 3,600 kg of explosive material was contained within the warhead, which consisted of 70% hexogen and 30% TNT.

The first series run of this team was designated Mi*stel 1*, consisting of a Ju 88A-4 and Me 109F-4, with the first examples being used as *Mistel S1* trainers to familiarize the crews. The first *Mistel 1* operations were flown from northern France in late June 1944 against Allied ship targets in the mouth of the Seine. 12 Mis*tel* teams took off and hit their targets without a single loss, although it was not possible to determine how many ships had been sunk. Reconnaissance flights showed that six ships had been hit and one cruiser sunk. Subsequent missions in the West were generally not as successful. Nevertheless, champions of this unorthodox concept felt justified by the initial results, and a series of 75 *Mistel 2*s were contracted for using Ju 88G-1 (night fighters), at the Leipzig-Mockau factory for repairs, mated with the Fw 190A-6 or F-8 control plane. The first of these was completed in November 1944. The teams, now weighing 18.9 metric tons on takeoff, were to be used in a decisive strike against the British fleet anchored in Scapa Flow. About 60 aircraft combinations were collected in the northern part of Denmark for the strike, but weather forced the attack to be called off. An alternative plan was drawn up, involving the long anticipated blow against the Soviet armament industry, especially its power plants. Operation Eisenhammer was planned for March 1945.

By this time at least 100 M*istel* combinations would be available, following a contract for a further 50 examples in December 1944. This meant that the "Beethoven" program—despite a number of other priorities—had been accelerated dramatically. Even newly built airframes were joined into teams with Fw 190A-8s. These included the latest Ju 88H-4 and G-10, the latter with the Jumo 213A-12 engines and originally designed as a heavy fighter with extended fuselage for additional fuel and extreme endurance. The Fw 190s also carried extra fuel in streamlined tanks on the wings (so-called "Doppel*reiter*" tanks), and an additional 300 liter tank under the fuselage. Now known as the *Mistel 3*, the Ju 88H-4 combination was designated the *Mistel 3B* and the Ju 88G-10 the *Mistel 3C*; these weighed in at 22 metric tons, a remarkably high weight for the Ju 88's standard landing gear to support! As a result, the *Mistel 3* had a third undercarriage leg beneath the canopy, which was jettisoned after takeoff.

In March 1945, however, the staging airfields intended for use in Operation "Eisenhammer" were overrun by the Red Army—the ten power plants around Moscow/Rybinsk and, most assuredly, the three near Gorky were no longer within range of the Mis*tels*. The *Mistel* combinations, now numbering well over a hundred, could now only be used in small groups as part of operations in support of the "*Brückenbevollmächtigter*" (literally, "bridge authority") established on 1 March 1945. Strikes were made against bridges—mostly hidden by smoke—over the Oder, Neiße, and Weichsel Rivers. On one of the first of these raids on 8 March 1945 the northern and southern Oder bridges near Görlitz suffered near misses and were blocked for a brief period. In mid-March there followed operations

Junkers Ju 88 *Mistel* with a Messerschmitt Me 109G as control plane (1943).

Ju 88 *Mistel* with Focke-Wulf Fw 190A-6/F-8 as control plane (1944).

against the bridge at Remagen on the Rhine River, but with little success; all in all, these were costly, expensive operations in the East and the West which were very weather dependent. One of the last operations involved 12 Mistel teams on 16 April 1945, striking the bridges on the Oder in enemy hands with considerable success, but not even successful raids could significantly slow the advance of the Red Army or the Western Allies.

The awkward Mistel combinations were often shot down by the ever-present enemy fighters shortly after takeoff or destroyed by strafing while still on the ground, forcing them to generally be used at night or in particularly favorable weather conditions.

Under the deceptive assumption that these Mistel operations might somehow influence the course of events, in early 1945 Junkers/Nordhausen drew up a Mistel 4 concept consisting of a Ju 188 with an Me 262 control plane, but it was never tested. Despite the futility of this type of operation, the "Beethoven Program" continued throughout the last two months of the war; a so-called "control plane" was even built—developed from the Mistel 3B with the Fw 190A-8, which could act as its own fighter protection if attacked. This control plane would fly with a three-man crew in the Ju 88's cockpit, which instead of an explosive charge contained a search radar in an extended nose, as well as light defensive armament (one 13 mm MG 131 firing upward and aft). It was to have acted as a long range pathfinder. For increased range, this control plane would have had two 900 liter drop tanks under the wings.

As late as April 1945 there was another Mistel combination tested, this time involving a Ju 88G-7 (advanced Ju 88 night fighter) and a Focke-Wulf Ta 152H (high altitude fighter) as a control plane. Evaluations took place at the Römergraben airfield near Quedlinburg until here, too, the course of the war shut down all industrial and flying activities.

The last hopeless Mistel missions took place as late as 26 and 30 April 1945 against bridges on the Oder—without measurable results!

Summary and the Ta 154

Evidence shows that at least a total of 195 Ju 88 Mistels were built, although unofficially this number fluctuates between 97 to about 250. Following Germany's surrender on 9 May 1945, 50 of these alone fell into the hands of the Allies near Burg/Merseburg, where the bulk of pilot training also took place.

The Mistel solution was an unusual and costly stop-gap solution to correct the Luftwaffe's long-standing lack of effective strategic air power elements, and as such, it was a failure. Not only did

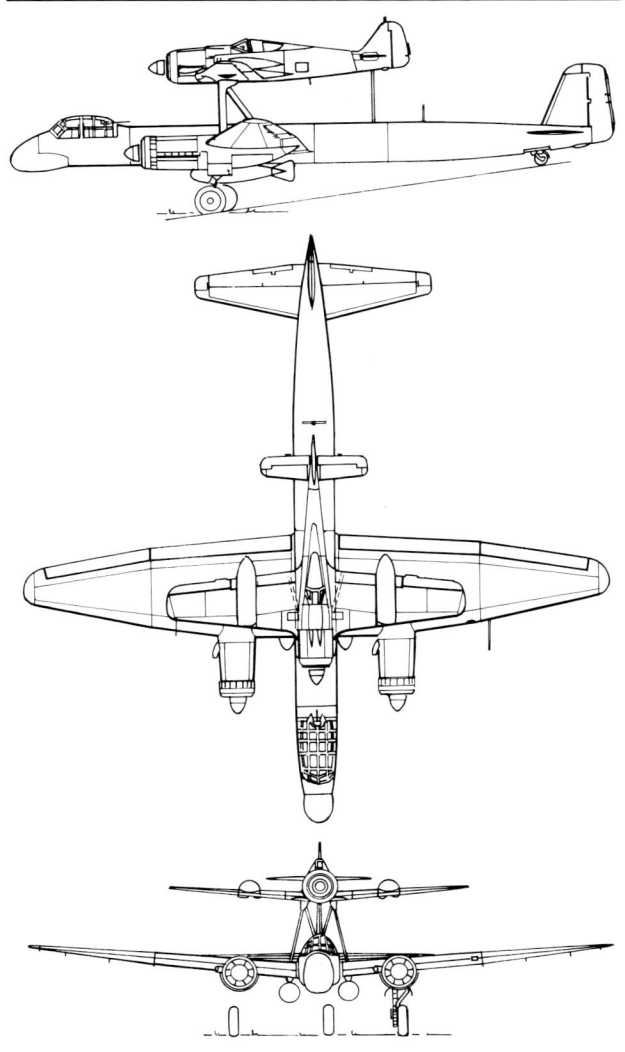

Three view of the Junkers Mistel command and control system consisting of a Junkers Ju 88H-4/Fw 190A-8 combination (1945).

---

*   See also Gellermann: "Moskau ruft Heeresgruppe Mitte," Bernard & Graefe Verlag, Koblenz 1988

Focke-Wulf Ta 154 night fighter (1944) minus its night-fighting antenna (e.g. the so called H*irschgeweih*** for the FuG 220** ***Lichtenstein*** **SN-2). During flight testing of the** ***Mistel*** **combination, it was discovered that the weight difference between the Ta 154 bomb and the Focke-Wulf Fw 190 control plane was too little to ensure a clean separation of the two aircraft.**

this unorthodox weapon arrive on the scene too late, but the technology employed limited the maneuverability of formations, and the efforts which went into planning were in no way justified by the results achieved. It was a failed investment similar to that of the He 177.

It is amazing that, in spite of the precarious situation, other companies focused seriously on the *Mistel* concept. One of these was Focke-Wulf, as seen in a brief description of the Bremen factory form 14 July 1944. In this case, an Fw 190A-4 or A-8 would serve as the control plane for a Focke-Wulf Ta 154 wooden twin-engined night fighter "bomb."

The T*echnisches Amt* stopped this program in November 1944 after the construction of seven night fighter prototypes and the start of a production run of A-0 and A-1 variants at three separate plants; the *Technisches Amt* had based its decision on problems with the wooden construction and shortages of material. Weighing 15.13 metric tons in their *Mistel* configuration, six of the completed Ta 154As were to also have served in another capacity as "aerial torpedoes," or as "formation busters" to break up heavy bomber formations. Focke-Wulf reckoned with the total loss of the aircraft in such a capacity, but felt that the Ta 154 was particularly suited to the role thanks to its cheap and Dural-saving construction. Many of these were test flown by the end of 1944, but as a result of unfavorable flight test results, these aircraft never achieved their operational purpose.

The Messerschmitt company also devoted much time and resources to developing a *Mistel* project. The project suffered technological problems surrounding transfer of throttle and control inputs from the pilot in the control plane to the unmanned *Mistel* aircraft, and the problems associated with using different fuels in engines which, for the most part, varied considerably between control plane and *Mistel*. In conclusion, the *Mistel* concept was a failed attempt at taking existing aircraft and creating a new weapons system under the motto of "new from old." It was a weapons system that, although of great interest from a technological standpoint, in no way could be justified in light of the actual requirements needed in the air at the time—a hopeless undertaking involving great expense, high risks, and many victims!

18. What's Left of Bomber Planning?

The developments just described show how far removed the realistically attainable availability of modern air power assets were from the bomber planning "wish list" drawn up by the *General der Kampfflieger*. That this was directly attributable to the military situation which, despite the greatest efforts of the armed forces, rapidly developed to Germany's disadvantage from 1943 onward is obvious. Nor is it any wonder that Hitler's political style and the pleasure he derived from making decisions ("*Entscheidungsfreudigkeit*") down to the detail of specific weapons obviously was a major contributing factor in the lack of goal-oriented, coordinated military planning. Ultimately, however, it was the Allies with their ever increasing superiority—especially their growing air power—who were the ones dictating the course of events. What course had the war taken in the interim, and what effect was it having on Germany's potential air power?

Situation Winter 1944/1945

In the East

After the failure of Germany's offensive along the central sector of the Eastern Front in July 1943, it was the Soviets' turn to begin their general offensive. Hitler's order of 12 August 1943 calling for "the immediate establishment of an eastern wall" did little to delay the onslaught. The Kuban bridgehead had to be vacated in October 1943, the Crimea in May 1944. After struggling for almost three years, the Red Army was able to push a large percentage of German forces back to the western border of the Soviet Union after their March 1944 spring offensive. German counterattacks could only throw the advancing Soviets back in isolated areas and, even then, could only temporarily stop their superior numbers and massed air support.

The Red Army's offensive against Romania in August 1944 led to a coup d' etat within Romania on 23 August. On 25 August Romania switched sides and declared war against Germany. At the end of August the Soviets occupied the Romanian oil fields and marched into Bucharest. This shift in power led in turn to an accelerated withdrawal of German forces from Greece and Albania. After a change in government, Bulgaria also declared war on Germany on 8 September 1944. On 19 September Finland signed a cease-fire agreement with the Soviet Union, and on 23 December a Hungarian counter-government agreed to a cease-fire with the Soviets and declared war on Germany. As a result, not only did Germany lose the bulk of her allies in late 1944/early 1945, it also lost the most important sources of raw material in southeastern Europe, as well as strategically important positions in the north. Germany's retreat to the "inner line" was unavoidable and in full swing!

In the South

Once Army Group Africa capitulated in May 1943, the Allies landed in Sicily on 10 July 1943, from whence approximately 40,000 German and 62,000 Italian soldiers were evacuated with their supplies and equipment without loss in August. On 3 September 1943, while British forces under Field Marshall Montgomery were landing in Calabria, the Italians agreed to a cease-fire (which was not announced until 8 September 1943). This had followed Mussolini's removal from office and arrest by King Victor Emanuel III. Under heavy air attack by the Germans, the Italians delivered their fleet of warships to the Allies in Malta, where the ships arrived on 10 September and surrendered, thus ending the war in the Mediterranean for all practical purposes. The battleship *Roma* (46,215 tons) had been sunk the previous day by two Fritz X remotely guided bombs dropped from high altitude by two Do 217K-2 bombers; its sister ship, the *Italia*, and other ships were damaged, but nevertheless made it to Malta's harbor. On the same day, the 9th of September, the Americans landed in the Gulf of Tarento. On 12 September German paratroopers freed Mussolini from his prison at a mountain hotel in the Abruzzis and flew him out in a Fieseler Fi 156 St*orch*. On 15 September 1943 he formed a "Republican-Fascist" Italian counter-government with its capital at Salo, on Lake Garda. On 20 September Sardinia was evacuated, with Naples following on 1 October. Corsica followed suit on 5 October, and on 13 October the Italian Badoglio government also declared war against Germany.

History of German Aviation: Bombers and Reconnaissance Aircraft

The *Luftwaffe* fought incessantly, almost always in support of army troops, but on 2/3 December German bombers succeeded in attacking the harbor facilities at the important supply port of Bari. Direct hits and subsequent explosions on an ammunition ship destroyed 19 freighters with a total tonnage of 73,343 BRT and severely damaged seven more.

Meanwhile, the Americans and British pushed up from the Volturno to the Sangro. American forces landed on 22 January at Anzio and Nettuno, south of Rome, behind the Germany's southern front and, after unsuccessful German counterattacks, were able to break out of the bridgehead in May. Their initial attempts at breaking out around Monte Cassino failed, but on 18 May 1944 they were able to take the town and its Benedictine abbey after heavy fighting. On 4 June 1944 they occupied Rome. On 18 July the British launched a full-scale attack along the Adriatic seaboard, and Ancona had to be surrendered. With the evacuation of Florence on 10 August 1944, the German retreat (which had begun from Rome on 4 June) was, for the time being, finally able to stop at the "Green Line" (southeast of La Spezia - Appenines - north of Rimini)

In the West and Over the Reich

Here the air war had increased in its intensity and ruthlessness. From 24 to 30 July 1943 Hamburg was bombed as part of the graduated Allied air offensive agreed upon at Casablanca on 10 June 1943 (Operation Gomorrha). The Allies suffered minimal losses during the raids as a result of dropping millions of strips of aluminum foil, rendering German radar systems inoperative.

Improvements to the German air defense systems continued at a feverish pace. Allied bomber losses soon increased, particularly on 14 October 1943 during the attack on Schweinfurt, and led to the "autumn crisis of Allied daylight bombing." Nevertheless, the Allies' raids of destruction continued without interruption, the Americans by day and the British by night. The American air forces in Europe, now under a central command for coordinating the strategic air war against Germany, established as its goal the systematic destruction of Germany's aviation industry, and on 11 January 1944 attacked aircraft production facilities in Halberstadt, Braunschweig, Magdeburg, and Oschersleben. Of 663 bombers involved, 64 (approximately 10%) were shot down by fighters armed for the first time with rockets. 40 German fighters were lost.

On the German side, the "A*ngriffsführer England*" gathered together all bomb platforms and, beginning on 21 January 1944, launched a series of large-scale attacks on London, as well as targets in southern and southeastern England. The British soon adapted themselves to these raids and called them the "Baby Blitz," recalling the period in September 1940 when London had earlier been the target.

On 20 February 1944 the Americans began their so-called "Big Week," during which they concentrated their attacks on aircraft industrial centers in Leipzig, Braunschweig, Gotha, Wiener Neustadt, and other cities.

As a long overdue step Göring now agreed to a proposal by Milch for the creation of an inter-ministerial "Fighter Staff" which, beginning on 1 March 1944, was tasked with the responsibility of using all means necessary to increase fighter production. Of particular importance was the dispersal of production facilities, to include locations underground. Karl Otto Saur became its de facto director. As a result, fighter production was given absolute priority over all other aircraft production programs, with the emphasis focusing on construction of jet aircraft, especially the Me 262. In April 1944 the Allies broke their previous daily record of 4,000 tons of bombs dropped on German cities and industrial centers (Messerschmitt in Wiener Neustadt on 12 April, ball bearing factory in Schweinfurt on 13 April). Beginning on 12 May 1944 the Allies concentrated almost exclusively on synthetic oil processing plants in order to strangle German fuel supplies. In April the Americans had launched raids from Italy against oil refineries near Vienna, Budapest, and in Upper Silesia, as well as the Romanian oil fields around Ploesti, and in May this continued to be the focus of heavy bomber raids. By now fuel supplies were the number one bottleneck! Allied strategic planners paid particular attention to Germany's synthetic fuel processing facilities. Bombers succeeded in destroying Leuna by 60%, Tröglitz by 100%, Böhlau by 50%, and the synthetic oil processing plants near Prague by 100%. Much effort went into repairing the damage, only to see that effort wasted under renewed attacks. The catastrophic consequences for Germany's ability to continue the war as a result of the successful Allied raids on the synthetic oil processing plants in May and June 1944 were outlined in a memorandum Albert Speer sent to Hitler in June 1944, in which he explained that more would follow, since fuel supplies became more precarious month by month.

Stalin had been pushing for a second Anglo-American front against the Axis forces to relieve the hard-pressed Soviet Union ever since Germany's 1942 summer offensive. Initially, the Anglo-Americans viewed their intensified aerial offensive as being something of a "second front" already, and the landing operations on the Southern Front seemed to be a way to meet Stalin halfway. But on 6 June 1944 the "second front" truly began with the long awaited invasion on the northern French coast under the most massive aerial umbrella ever seen up to that point, with 14,674 sorties flown by the Allies on the day of the invasion. Compare this with a total of just 319 sorties by the air fleet of the western German commanding authority! This absolute air supremacy played out in favor of Operation Overlord and its landing operation, Neptune. Over 6,400 ships of all kinds were used, landing in the first week 326,000 men, 104,000 tons of materials, and 54,000 vehicles between the mouth of the Orne northeast of Caen and Cherbourg. Despite the invasion and the shift of the Allied air war focus to a tactical-operative role, the strategic air raids against German synthetic oil processing plants continued in their intensity almost without pause. On 21 June 1944 144 B-17 Flying Fortresses hit the Ruhland refinery hard, then flew with some of their escort fighters to the Soviet airbase at Poltava. On the following night, about 200 German bombers which had been gathered together for a strike on Soviet industrial centers succeeded in making a surprise raid on Poltava. Under the light of flares they destroyed 42 B-17s, 15 P-51 Mustangs, and a number of Soviet aircraft on the ground.

The *"Schnellstbomber"* That Never Was

Hitler's direct involvement and amateurish decision to build the Messerschmitt Me 262 jet fighter as a high-speed bomber (*Schnellbomber*, also called the *Kleinstbomber, Blitzbomber,* or *Schnellstbomber*) was made after a review and flight demonstration of modern weaponry at Insterburg in East Prussia in late November 1943. The Me 262, still under development, was part of the display, and in response to a query by Hitler, Prof. Messerschmitt assured him that it would be no problem for the Me 262 to carry two bombs weighing 250 kg each. Messerschmitt had already addressed that very issue with Göring on 2 November 1943 when Göring had paid a visit to the Regensburg plant, even going so far as to mention specific bomb caliber and giving an estimated conversion period of about 14 days.* With all signs pointing toward an imminent invasion, Hitler's idea involved having a *Blitzbomber* drop bombs—or at least a bomb—onto the staging areas of the invasion troops with the goal of causing confusion and chaos. Accuracy was not of primary concern. Although Hitler again categorically ordered the Me 262 to be produced exclusively as an ultra high-speed bomber in May 1944, by the end of May it was obvious that the aircraft had only been developed as a high-speed bomber, with nine examples being built up to that point. It was not until then that concrete thought was given to what measures were needed to make the Me 262 into an operationally capable bomb platform

In fact, by the time of the invasion (six months after Hitler's November decision) there had been no development of the structural reinforcements and fittings for this relatively lightweight fighter; it was not yet deployable as a bomber. The Jäger*stab* and the *General der Jagdflieger* fought against Hitler's decision, lobbying instead for other production priorities dictated by the troubled situation in the skies over Germany. As they saw it, the Me 262 should primarily serve in the role of fighter.

In July an aircraft was hurriedly configured as a bomber and tested. Results from the evaluation showed that the jet had a top speed of 850 km/h at 7,000 m in clean configuration, but with a 500 kg bomb at this altitude could fly at just 750 km/h; with two 250 kg bombs it reached a speed of 713 km/h at 5,000 m. As a fighter-bomber, it had a penetration radius of around 170-200 km with two 250 kg bombs.

The first fighter-bomber Me 262s became operational in August with the so-called E*insatzkommando Schenck* in northern France, going into action with varying degrees of success. In October 1944 the type began flying from Hopsten near Rheine in larger numbers, primarily conducting individual sorties against the Allies advancing through Holland and Belgium. They focused particularly on the British-held road bridge over the Waal in Nijmegen. Although lack of a suitable bombsight meant that results were not altogether satisfactory, these missions were flown without a single loss due to enemy aircraft.

The tactics used on these missions varied considerably. During the day, pilots would attack by putting the planes into a shallow dive from an altitude of about 7,500 m, dropping the bombs at about 5,500 m as a rule. At dusk, or in poor weather, however, the approach was made at low level (around 300 m), descending to about 150 m to release the bombs. At night, the approach took place at 3,500 to 4,000 m, with a shallow dive down to about 2,500 m when the bomb was released. To be sure, bombing accuracy suffered considerably at these speeds and altitude differentials, but these also gave enemy air defenses practically no chance to bring an Me 262

---

* These discussions in Regensburg, mentioned in detail in other sources, and the breadth of necessary work involved lead one to the conclusion that the 14 days referred only to designing the configuration of the bomb shackles (which would be supplied from subcontractors) and their aerodynamic skinning, to include the jigs necessary for assembling the skin. In no way was this a reference to the assembly or fitting of these parts (according to information provided by Wolfgang Degel on 19 October 1988).

down. We will take a look at the Me 262's continuing role as fighter-bomber and reconnaissance aircraft shortly.

In its fighter-bomber configuration the pilot suffered from poor visibility and, without an actual bombsight, the CEP could be up to 3 kilometers in diameter. In light of this, the General der Kampfflieger—in line with Hitler's incessant push for an Me 262 bomber—called for having a proper bombsight so that the bombs would fall where they were supposed to. The plane needed no armament in his view, but its range needed increasing. Messerschmitt estimated a development time of six months for a requisite all-around view canopy and the availability of a suitable bombsight, as well as to find a way of controlling the centerpoint and ballast in order to turn the auxiliary fighter-bomber into an ultra high-speed bomber. This meant that Hitler's *Blitzbomber* would not be available before 1945. On the other hand, of the 1,384 high-speed fighter and fighter-bomber versions promised by the end of 1944, 564 were actually built and 385 (including the two-seat conversion trainer variants) were delivered to field units despite technical and delivery problems with the still-untried Junkers engines. The difference between the number built and the number delivered were generally used as test aircraft which, like the trainers, suffered a high loss rate; this aerodynamic design was a demanding one to fly. By April 1945 the number of Me 262s delivered by the company had risen to a total of 1,433 examples, this despite constant air raids necessitating the dispersal of the production facilities.

Messerschmitt Me 262A-2a

Unofficially known as the "S*turmvogel*" (Stormbird), this fighter-bomber version differed from the Me 262A-1a ("*Schwalbe*" - Swallow) solely by the installation of the fuse activation circuitry and two external racks for two 250 kg bombs or one 500 kg bomb. The textbook attack in such a configuration involved, as a rule, a 30% descent from higher altitudes until reaching a speed of between 850 and 900 km/h, with bomb release and pullout at about 1,000 m. Accuracy was about the same as with the Fw 190 fighter-bomber, but acquiring the target was much more difficult due to the high approach speeds. Once sighted, a target was normally first approached in level flight until it had passed under the right or left engine intake. The pilot then nosed the plane over into the 30% dive, building up the speed mentioned above until bomb release and pullout. To be sure, when approaching the target in level flight the bomb-laden Me 262 fighter-bombers could be intercepted by

Messerschmitt Me 262A-2a fighter bomber with two 250 kg bombs (1944).

faster enemy fighters, but this was almost impossible once they began their high-speed descent. Several aircraft were fitted with a TSA system (TSA = *Tief- und Sturzfluganlage*, low-level/dive system) in the nose in an attempt to improve the bombing accuracy. Such planes carried the designation of Me 262A-2a/U1, and had their armament reduced from four to two 30 mm MK 108s. One version, designated Me 262A-2a/U2, was fitted with a new wooden nose to accommodate the gyroscopic stabilized Lotfe 7H bombsight and a prone bombardier. It carried the same payload as the A-2a fighter-bomber, but had all its guns removed.

The Me 262A proved to be ill-suited for strafing; the muzzle velocity of its four or two 30 mm MK 108 onboard cannons—at 520 m/s—was too slow, and the rate of fire of 600 rounds per minute too low, to be effective at altitudes of any higher than 500 m in attacks on ground targets. Furthermore, the plane carried just 360 rounds (2x 100 and 2x 80 per barrel), far too little to effectively cover a target area from an extremely fast moving aircraft. Plus, there was absolutely no extra armor provided to protect the pilot and vital areas of the airplane. Accordingly, an Me 262 was designed specifically for low-level attacks, the Me 262A-3a, with armor for the fuel tanks, as well as the underside and sides of the cockpit. It retained the same weaponry, but was never put into production.

Another type for possible ground attack was the "bomber-destroyer" concept, a cannon version of the Me 262 developed from the Me 262A-1 with a 50 mm Rheinmetall BK 5 cannon whose barrel jutted out over 2 meters from the nose of the plane.

---

*   Not a company or RLM designation.

Messerschmitt Me 262A-1a (V-083) with BK 5 cannon (1944). Even with the improved 50 mm Rheinmetall-Mauser MK 214A the variant never went into production—despite good results with the BK 5 during test firings at Lechfeld.

Thus configured, there was surprisingly little loss of performance despite a counterbalance installed in the tail to compensate for the shift in the center of gravity with such a weapons installation. The nose wheel also rotated through 90° when retracting to lie flat beneath the cannon. Air-to-ground gunnery trials from altitudes of 1,200-1,500 m, conducted against rectangular target approximately 32 m long (roughly the wingspan of a Liberator bomber), resulted in 25-27 hits out of 30 fired. Due to the smoke buildup, however, only one round was fired per firing pass. This quite stable gun platform was followed by an improved version with a 50 mm Rheinmetall-Mauser MK 214A on 23 March 1945, but the variant had not been completed when the war ended just a few weeks later.

A reconnaissance fighter was also developed from an Me 262A-1a. This type, the Me 262A-1a/U3, was fitted with two aerial cameras in the nose in place of its armament and delivered to the field units in small numbers. A somewhat more advanced development was the Me 262A-5a using the same two Rb 50/30 or Rb 20/30 and Rb 75/30 cameras side-by-side as the A-1a/U3, but included an observation window in the cockpit floor, as well as two 30 mm MK 108 cannons. Retaining the external pylons (called "*Wikingerschiffe*"- Viking Longboats) of the A-2a, the A-5a could carry either two 300 liter or one 600 liter external fuel tank to improve range.

Summary

Hotly contested over its roles as fighter and bomber, in late 1944/early 1945 the fighter-bomber Me 262 was a design which, thanks to its high speed, was generally immune to interception. Nevertheless, it left much to be desired with regard to its bomb load and accuracy; the Me 262 modified as a bomb platform was a far cry from the ideal concept of a "*Schnellstbomber*"—despite Hitler's naive illusion and his explicit orders. Ultimately, there remains considerable doubt whether the aircraft, conceived from the outset for the fighter role, could ever have actually been redesigned as an effective bomber.

Making Due with What's On Hand

The above situational overview to late 1944/early 1945 addressed the jet aircraft in its role as fighter-bomber and reconnaissance platform as part of overall offensive air power. But Hitler was not yet willing to elevate the "jet bomber" to the level where he would specifically advocate its availability for widespread use with the main bomber units (those organized roughly along the lines of the centrally controlled British Bomber Command). Too many uncertainties about this direction of development existed at the time. Bear in mind that the *General der Kampfflieger* had only briefly mentioned the jet bomber in his bomber planning paper of 5 October 1943; consideration would be given to the type once available, and only then would the number of bombers with standard engines be reduced accordingly. The Gen. d. K. was realistic enough to promote the previously mentioned—with varying degrees of effectiveness—He 177 and Ju 290 as long range bombers and the Ju 88/188 as the backbones of the operative air war. The Do 217, He 111, Me 410, and He 219, all of which he discounted as being uneconomical aircraft types, remained in the bomber inventory for an astoundingly long period of time, since the heavy wear and tear on aviation in a multi-front war necessitated more than ever the uninterrupted output of ongoing production to plug any gaps cropping up. With one exception (which will be examined in depth later and which paved the way for future development of bomber aircraft after 1945), in 1944 all production capacity was directed towards fighter aircraft and missiles, particularly the so-called "V-weapons."

Let us then, over the next few sections, take a look at those aircraft labeled "less desirable" by the Gen. d. K. and examine their wartime development and use.

Dornier Do 217
Development of the Do 217 as a high altitude bomber had already begun in September 1941, when the 21st Do 217E production aircraft was diverted for redesign. Powered by turbocharged Daimler-Benz DB 603 engines, the Do 217H (as it was now designated) carried out high altitude flight trials at the Echterdingen test center. Before Do 217 production ceased in 1943 the Do 217H had evolved into the Do 217P combat reconnaissance aircraft, a variant based on the Do 217E-2 airframe further developed for high altitude flights.

History of German Aviation: Bombers and Reconnaissance Aircraft

Two DB 603B engines served as powerplants, these being boosted by a DB 605T in place of the turbochargers. Known as the HZ system (HZ = Höhenlad*erzentrale*, or centralized high altitude boost), this arrangement provided the first prototype, the Do 217P V1, with a takeoff rating of 2,575 kW/3,500 hp and a maximum output of 2,735 kW/3,720 hp at 2,000 m and 2,380 kW/3,240 hp at 5,700 m. For climbing and combat boost this engine configuration gave 2,120 kW/2,880 hp at 13,700 m; the maximum authorized military setting at this altitude was an impressive 1,940 kW/2,640 hp. The DB 605T was installed in the center of the fuselage and drove a two-stage charger, fed by air from large boost air intermediate radiators beneath the wings between the fuselage and engine nacelles, as well as from intakes for the engines and boost air located in the lower fuselage just behind the wing trailing edge. The four-man crew was housed beneath an extensively glazed canopy enclosing a pressurized cockpit designed to be separable at a point ahead of the wings. The first flight of this interesting development took place on 6 June 1942, but without the boost, and subsequent flights could only attain a maximum altitude of 15,200 m using the initially quite troublesome HZ system. By 1 April 1943 the P V1 had made a total of 23 flights, chalking up 34 hours, most of these being conducted at Daimler-Benz for testing the HZ system.

Two other prototypes, the Do 217P V2 and P V3, began their flight testing operations in the summer and autumn of 1942, respectively. They differed externally from the V1 by having longer wings (a wingspan of 24.50 m and an area of 67 m^2). The P V2 had to be written off on 27 March 1943, when it was almost totally destroyed during weapons release trials at Cazaux. Three additional prototypes, the P V4 through V6, never made it to flight testing and were chopped up for scrap in accordance with an RLM directive of 17 March 1944. There were, however, three Do 217P-0 combat reconnaissance aircraft, construction of which had begun in early 1943. The first entered flight testing in the early summer of 1943, and the other two followed in the autumn of that year. In late 1943 they went to the Rechlin test center for official evaluation, with a follow-on reassignment to Daimler-Benz at Echterdingen. It was there that two of these high altitude testbeds, along with a Henschel Hs 130E-0 (also a high altitude reconnaissance platform equipped with the HZ system), fell victim to a strafing attack on 5 September 1944. All three aircraft were destroyed.

The P-0 was armed with two 7.9 mm MG 81s in the nose (*A-Stand*) and two aft-firing MG 81s in both the upper rear cockpit and under the fuselage. These, however, could not be fired at high altitude due to the problem with sealing off the pressurized cockpit. With regard to ordnance, the Do 217P-0 could carry two 500 kg bombs externally under the two outer wing sections, or two 900 liter drop tanks instead. Primarily conceived as a high altitude reconnaissance aircraft, it had an Rb 20/30 aerial camera installed in the fuselage just behind the pressurized cockpit section, together with two automatic Rb 75/30 cameras in the aft part of the center fuselage section. The Do 217P-0 attained a maximum speed of 585 km/h at 14,000 m with a weight of 13,270 kg; maximum altitude with the same weight was somewhere above 16,000 m.

The Do 217P-1 and P-2, with a wing area of 100 m^2 and thus possessing somewhat better high altitude performance, were in an advanced stage of construction when the RLM rejected them because of the enormous amount of effort involved in building the new wing. In spite of its quite impressive performance potential, the Do 217P had just as little chance of going into production as did the previously mentioned Henschel Hs 130E or the "Bomber-B" conceived Do 317.

Although these futuristic high altitude planes from Dornier never went into production (mainly because of the problems with

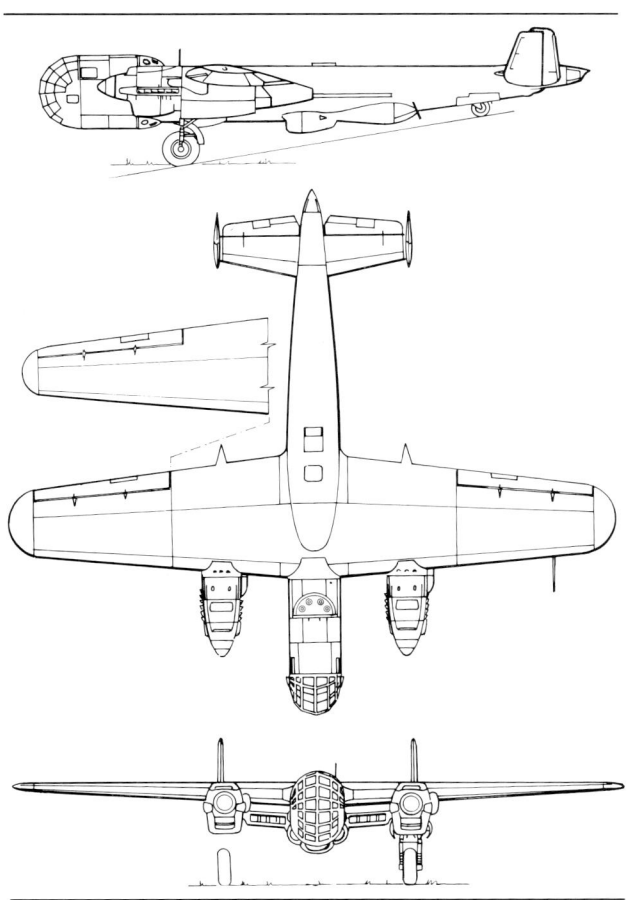

Dornier Do 217PV1 (1943) with Daimler Benz DB 605T engines having a central high altitude turbocharger. The Do 217P-0 was fitted with the longer wings shown.

the engines, especially the HZ system), the Do 217's developmental potential had in no way been exhausted by this point.

The disadvantages of the Do 217E series, already discussed earlier, led to the development of the K-series. Initially planned as a night bomber, the Do 217K V1 started its flight testing on 31 March 1942 at the company's airfield in Löwenthal. Ten pre-production Do 217K-0 aircraft began rolling off the assembly lines in the late summer of 1942, and these also served as additional prototypes. The Do 217K V1 was initially fitted with the standard single rudder, but later acquired twin rudders and served as a torpedo carrier testbed for four L5 torpedoes, although it never saw operational use as such. The Do 217K-01 served as a testbed for conventional and dive flight trials, as did the K-02 and K-03. Beginning in 1943, however, these latter aircraft carried out other types of testing, including trials with a single-seat DFS 228 high altitude long range reconnaissance aircraft mounted piggy-back. The DFS 228, weighing 5,500 kg itself, would be released at an altitude of 10,000 m, at which point its Walter HWK 109-509 liquid-fuel rocket engine would propel it to an altitude of over 23,000 m. The DFS 228 could maintain this service ceiling for about 45 minutes. Its prone pilot would take his infra-red images during a 300 km long shallow descent (glide ratio of 1:20), during which the plane cruised along at about 610 km/h.

Other than the pressurized cockpit in the forward fuselage, this experimental plane was built almost entirely of wood. Two prototypes were constructed and initially tested purely for aerodynamics without the powerplant. Despite the fact that trials had not yet been completed and that preliminary results showed unsatisfactory performance during the climbing phase, the Schmetz company in Griesheim, near Darmstadt, initiated production of a batch of ten DFS 228A-0s. These pre-production aircraft were never completed, however.

But back to the Do 217K, whose ten pre-production aircraft served as testbeds for ordnance dropping, load trials, crew conversion, and field unit evaluation. The subsequent Do 217K-1s differed markedly from their predecessors. The two-step arrangement of canopy, with its limited view, had already given way on the K V1 to a canopy with clean lines with optimal perspectives for both the pilot and the bombardier. The BMW 801ML engines were soon replaced on the Do 217K-1s being built at Wismar with the more powerful (by a good 75 kW/100 hp) BMW 801D. This engine burned 96 octane fuel and provided a takeoff output of 1,250 kW/1,700 hp and 1,060 kW/1,440 hp at 5,700 m altitude. Armament was also brought up to date over that of the E-series, with ammunition in some cases doubling; it consisted of:

Dornier Do 217K-1 with blackened undersides for night operations. The clean lines of the canopy and some of the plane's defensive armament can be seen to good effect in this photo.

one 7.9 mm MG 81Z (twin gun) with 1,000 rounds in the *A-Stand* nose gun position
one 13 mm MG 131 in an electrically driven WL 131/1 turret with 500 rounds in the *B-Stand* upper fuselage
one MG 131 in the *C-Stand* lower fuselage with 1,000 rounds
two MG 81s on either side of the cockpit, each with 750 rounds (later increased to four MG 81s), and
two MG 81s fixed in the tail (on later production numbers).

Thus armed, the Do 217K-1 was primarily used as a night bomber over England.

Beginning in December 1942 the Do 217K-2 was produced as a bomber specially equipped to drop the remotely guided FX 1400 free-fall bomb. The K-2 had a wingspan increased from 19.0 m to 24.5 m and an area of 67 m^2 vice 57 m^2, and as a rule was fitted with field conversion kits 15, 17, and 19 (for specifics, see portion of "Basic Data for Do 217E-2 and Sub-Variants" section dealing with field conversion kits). In addition, there were plans to install fittings for a pair of fixed 250 round MG 81s firing to the rear, housed in the aft section of each engine nacelle. The entire six gun battery of fixed 7.9 mm MG 81s firing aft from the base of the tail and the back of the engine nacelles was sighted and controlled by the pilot via an RF 1A or 2C periscope sight combined with a PV1B sighting unit, or by the gunner from the *B-Stand*.

The Do 217K-3, following shortly thereafter in July 1943, was equipped with the newer FuG 203c or 203d Kehl IV for operations both with the FX 1400, as well as the Hs 293 (see volume 7 of this series).

The Do 217M, first flown on 16 July 1942, was built parallel to the Do 217K from 1942 onward. This was almost identical to the K-series, but because BMW engines were in short supply for bombers, was powered by the somewhat more powerful 12-cylinder DB 603 inline engine (1,287 kW/1,750 hp on takeoff). This is considered the most advanced of the Dornier bombers, of which 230 were built before all Do 217 production ceased in June 1944. By this time a total of about 1,730 Do 217 bombers/reconnaissance aircraft and 364 night fighters had been produced. But back to the M-series.

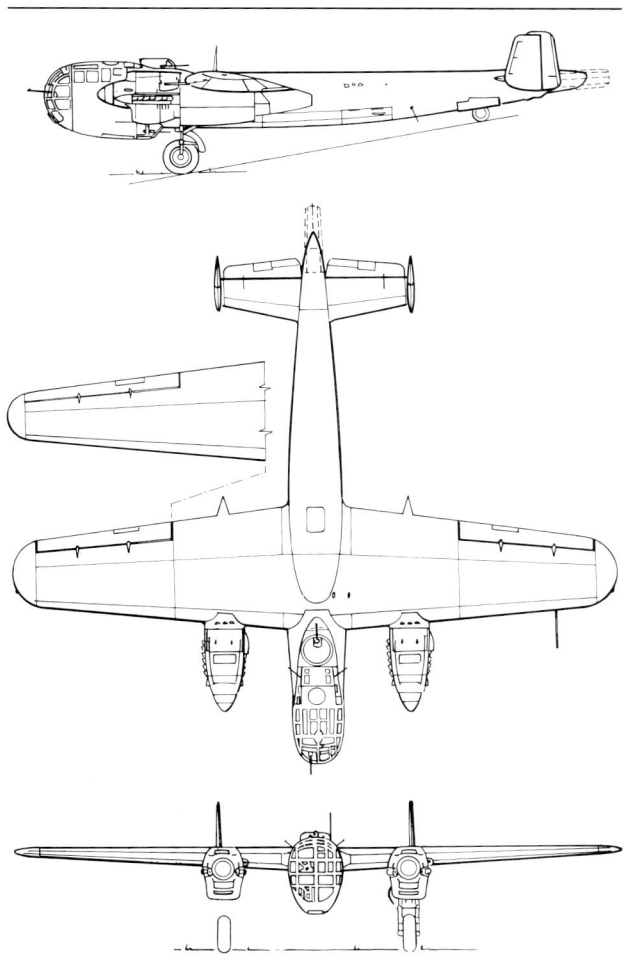

Dornier Do 217M-1/M-11 (1942-1944). The M-1 was a heavy night bomber with the shorter wings of the K-1, while the M-11 was the control aircraft for the FX and Hs 293 with larger wings (19.0 m span and 57 m2 area increased to 24.5 m span and 67 m² area).

This version, too, was built with different wing sizes depending on intended role. The standard bomber version, produced in rather large numbers, was the Do 217M-1 with the short wingspan of the Do 217K-1, while the other series, the Do 217M-11, had the same longer wings of the Do 217K-2.

The M-1 was conceived as a heavy night bomber with a payload capacity of up to 4,000 kg of bombs, of which a maximum of 2,500 kg could be carried internally, and at maximum overloaded capacity had a takeoff weight of 16,700 kg. The M-11, on the other hand, was a long range missile platform, capable of launching the FX 1400 or Hs 293A. Both types of ordnance were carried semi-internally, since there was not enough clearance for them to be carried entirely externally. Flight characteristics of the M-11 were considered quite good; maneuverability, ease of control, pulling potential of its fully automatic four-bladed VDM propellers with their 3.8 m diameter, and well balanced rudder forces all added up to an excellent weapons platform at any speed. Unfortunately, by early 1943 Do 217 production had in the main shifted over to the more urgently needed Do 217N night fighter variant.

For the sake of completeness, few interim sub-variants of the M-series are listed below. All of these, however, were one-off single variants and mostly served as testbeds.

- Do 217M-1/U1 with the wing center section of the Do 217R, was a design utilizing the larger wingspan and area (67 m2) and the DB 603 and T9 exhaust turbine in the aft engine nacelles; development broken off in 1943.
- Do 217M-2, tested as a torpedo bomber, never went into production due to the RLM's decision to make the Ju 88 the standard torpedo bomber.
- Do 217M-3 was a heavy dive-capable bomber with a new 59 m² wing, parabrake in the tail, and triangular twin rudders. Never entered production.
- Do 217M-4, planned from September 1944 as a high altitude bomber, also never went into production.
- Do 217M-5, envisioned with an Hs 293 half-buried in the fuselage, was another version which was never produced—its performance failed to live up to all the RLM's requirements.
- Do 217M-8 also had the larger wing area, the DB 603A with exhaust turbocharger, the new triangular rudders of the M-3, and improved pylons for the FX and Hs 293; the RLM postponed its production on 20 May 1943 and eventually canceled it altogether.
- Do 217M-9 had the large wing area, DB 603A, triangular rudders, but had a slightly modified forward fuselage and aerodynamic improvements to the B- and *C-Stand* gun positions. Designed for operations with FX and Hs 293.

History of German Aviation: Bombers and Reconnaissance Aircraft

Summary

With the M-11 series, the Do 217 attained the level of a highly effective all-weather platform for the most advanced—if not always reliable—ordnance of the period. Today, we would undoubtedly call the Do 217M-11 "a pilot's aircraft." In spite of this, as early as late 1943 the Gen. d. K. had effectively written off the Do 217 in favor of the Ju 88/188 follow-on development as part of a necessary commonality program.

Heinkel He 111

Probably the most heavily tasked workhorse of the Luft*waffe* was undoubtedly the He 111, a type which should have long since been pulled from front-line operations. Even the Gen. d. K. wanted little more to do with the He 111. Yet it would remain in service almost up until the end of the war, going far beyond its—relatively modest—performance level as a bomber. Thanks to its highly reliable flying qualities, it was particularly useful for maritime operations as a torpedo bomber and even a minesweeper. It also flew as a combat transport for resupplying military bases, a tow plane for the DFS 230 transport glider (offering excellent flight stability in such a configuration), and finally as a flying launch platform for V-weapons.

From late 1942 onward there was also the curious five-engined He 111Z, made by combining two He 111H-6 airframes to create a two plane for heavy transport gliders like the large-capacity Messerschmitt Me 321 Giga*nt*.

The variants following the He 111H-6 included:

He 111H-7: a variant of the H-6 lacking the MG 17 or the fittings for it in the tail.

He 111H-8: modified He 111H-3 and H-5 with the Jumo 211D and balloon deflector frame ahead of the engines and cockpit, to include cable cutters. Approximately 30 such aircraft were built. Due to the added 250 kg weight incurred by the kit and the associated counterweight in the tail, this aircraft was relatively clumsy and was pulled out of service following a few costly missions over England. The remainder were reconfigured as the He 111H-8/R2 and served as tow aircraft.

He 111H-9: with Jumo 211F-1s (like the H-6) and various equipment suites for special missions.

He 111H-10: with Jumo 211F-2s and a *Kuto-Nase* for cable cutting operations.

He 111H-11: as H-10, but with a fully enclosed *B-Stand* including bullet-proof glass and the WL 131AL turret for a single 13 mm MG 131. Improved armor and defensive armament in the aft ventral gondola, where the MG 15 was replaced by two 7.9 mm MG 81s. In the field, the side MG 15s could be replaced with the MG 81Z twin gun; thus modified, the H-11 was designated the He 111H-11/R1. If in addition it was fitted with the couplings for glider tow, the type was then designated the He 111H-11/R2.

He 111H-12: without ventral gondola, in early 1943 this was the first He 111 built especially for operations with the Hs 293A glide bomb. The transmitter (FuG 203b Kehl III) was installed in the *B-Stand*, while the control input operated by the observer was located on the starboard side of the cockpit. The Hs 293s were dropped individually when the mother ship was flying at a speed of about 340 km/h. The He 111 then assumed a position to the left of the missile and reduced its speed to about 260 km/h during the control stage. Only a small number was built exclusively for trials and crew training—the H-12 never saw front-line service.

He 111H-14: like the H-10, but kitted out as a pathfinder aircraft with specialized navigation/communications systems (FuMB 4 Samos radio receiver operating in the 80-480 MHZ range with the APZ 5 direction finder, plus the FuG 351 Korfu, which functioned as the FuMB 11 radio listening receiver in the 2,500-3,750 MHz range and, from 1944 onward, as the FuMB 15 in the 7,500-11,000 MHz range). 30 He 111H-14s were built with this equipment and assigned for radio reconnaissance duties over the Atlantic, operating from southern France as part of a so-called "*Sonderkommando Rastedter.*" A further 20 examples were built without this specialized radio equipment, instead being fitted with tow couplings and designated He 111H-14/R2 for operations in the East.

He 111H-15: also a pathfinder version, but trial configured for dropping BV glide bombs (similar to the Hs 293).

He 111H-16: this is the third actual large-scale production variant, after the H-3 and H-6. It stemmed from the H-6 with Jumo 211F-2s and incorporated as standard all modifications dictated by operational requirements. Defensive armament was the same as for the H-11. Fuselage bomb bays were retained. When carrying fuel tanks in the bomb bays it was fitted with one or two ETC 200 externally.

Heinkel He 111H-16 (in production from 1942) being loaded with 250 kg bombs. Indispensable workhorse during the 1944 Allied operations to strangle *Festung Europa*.

History of German Aviation: Bombers and Reconnaissance Aircraft

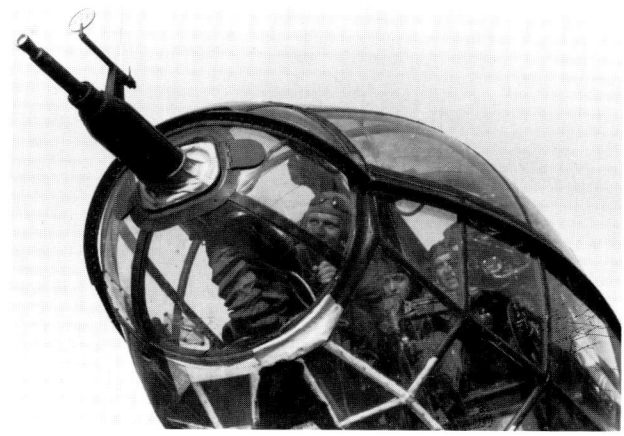

Heinkel He 111H-16 (1943) with the ideal all-round canopy for pilot and observer; the third crewman in the photo is the flight engineer, who also manned the gun in the ventral gondola. This variant came standard with the balloon cutting *Kuto-Nase* frame and the 20 mm MB FF cannon with attached sight.

With different configurations, such as the electrically driven dorsal fuselage turret (*B-Stand*) with the 13 mm MG 131, it was delivered as the He 111H-16/R1. The He 111H-16/R2 had a tow shaft in the tail as a rigid-tow coupling for transport gliders, and the He 111H-16/R3 carried additional armor with reduced payload capacity for its pathfinder role.

He 111H-18: specially configured for night missions, it had exhaust flame dampers. This variant was equipped like the He 111H-16/R3 and incorporated the radio systems of the H-14. Some were kitted out as torpedo bombers with the FuG 200 Hohentwiel surface search radar, serving mainly with the *Sonderkommando Rastedter*, as well.

He 111H-20: designed for a broad spectrum of operational roles for which the He 111 V46, V47, and V48 had served as testbeds. On the basis of the H-16 airframe, it was supplied in four different operational configurations:

- **He 111H-20/R1** with a three-man crew, an assault transport for 16 paratroopers with drop slides in the floor and two external racks for carrying 800 kg supply canisters
- **He 111H-20/R2** with a five-man crew, a transport (to include fuels) and transport glider tow aircraft
- **He 111H-20/R3** night bomber with heavy defensive firepower, consisting of one 13 mm MG 131 in the A-, B-, and C-Stand gun stations; the D-Stand (forward ventral gondola) and both side fuselage gun stations each had a 7.9 mm MG 81. External racks for a total payload capacity of 2,000 kg of bombs
- **He 111H-20/R4:** served as a nuisance bomber for night missions with twenty 50 kg bombs carried externally.

Up to and including the He 111H-20 series all aircraft were powered by the Jumo 211 engine. In early 1944 the Jumo 211 gradually became replaced by the more powerful Jumo 213, and subsequent He 111 variants were powered by this engine. These included the following:

He 111H-21: based on the H-20/R3, but with structural reinforcements to accommodate the new engine; however, because of shipping delays with the Jumo 213 the first 22 examples of this new series were still powered by the Jumo 211F with exhaust turbocharger. The first H-21 machines with the Jumo 213 became available in the late fall of 1944. Fitted with the Jumo 213E engine, which offered a takeoff rating of 1,287 kW/1,750 hp, takeoff weight could be increased to 16,000 kg and the bomb load to a maximum of 3,000 kg. Without bombs the H-21 reached a maximum speed of 480 km/h.

He 111H-22: several He 111H-21s were built as launch platforms for the Fieseler Fi 103 V-weapon (generally known as the V-1) and could carry one of these on each of the two pylons located between the fuselage and engine nacelle. Thus configured, following a brief period of crew training the H-22 was deployed from Holland against London from late July 1944 onward. Operations took place at night using radio beam guidance against a large area target. The V-weapon was pointed and released at an altitude of about 500 m, flying to the target under rocket power and lacking any kind of course correction. As such, accuracy rate was low, but not the psychological and destructive effects. This type of operation had been experimented with during the winter of 1943/1944 at Peenemünde using the He 111, and had been shown to be technologically and methodologically sound. It was designed to be a fall-back plan for Operation Rumpelkammer, the bombing of British targets suing V-1s, in the event that the Fi-103's launch ramps along the English Channel were lost as a result of the anticipated Allied landings. By the time the Normandy invasion took place on 6 June 1944 the operational potential of launching V-1s from He 111s had been fully evaluated. As He 111 production gradually drew to a close, a further batch of He 111H-16s and H-20s were modified at Oschatz as V-weapons platforms and supplied to the field. From late July 1944 until 14 January 1945, the day of the last He 111/V-1 mission, over 1,700 Fi 103s had been air-launched against England. 77 He 111s were lost on these missions, some when the V-1 exploded just after the mother ship had taken off.

He 111H-23: the last He 111s coming off the assembly line in the fall of 1944 were designed as specialized aircraft for inserting agents by parachute. These had the same drop slides as the H-20/R1. Powerplant was the Jumo 213A-1, rated at 1,308 kW/1,776 hp on takeoff and 1,177 kW/1,600 hp at an altitude of 5,500 m. When these were delivered to the field units, the H-23s were converted in the field back to bombers and served as such.

Photo of a license-built Spanish He 111H-16 painted in early German markings for the British film "Battle of Britain." In 1972, this aircraft could be found in the Southend Historical Museum near London.

Summary and the He 111R

After nine years of production, the He 111H-23 was the last variant to be built in Germany. Spain was the only country which license-built the type, producing the He 111H-16 until the 1950s. The Spanish Air Force was the sole operator of the type, and several had a starring role in the 1968 British film "Battle of Britain." Shot in Spain and England, the film was quite realistic, although the He 111s were powered by the British Rolls-Royce Merlin engines with which the He 111s built in Spain were fitted. Designated as C.2111, they were produced until 1956.

The problems with the "Bomber-B" program prompted the design of an interesting variant of the He 111 in 1943: the He 111 as an interim high altitude bomber until the perfect high altitude bomber became available. The first project aircraft, the He 111R-1, was originally designed with Jumo 211F engines, but estimated performance was expected to be too low, and in 1944 the He 111R-2 was proposed with DB 603U engines and turbochargers. The DB 603U provided 1,330 kW/1,810 on takeoff and 1,177 kW/1,600 hp at 12,800 m, offering the He 111R-2 an estimated maximum speed of 500 km/h. Maximum airborne weight was 15,000 kg. An He 111H-6 served as the prototype for the planned design, and underwent trials as the He 111V32 beginning in early 1944 with the DB 601U engines and TK 9AC turbochargers. Further development came to a standstill shortly afterward, though, and no He 111Rs were ever built.

The five-engined He 111Z mentioned in the beginning was purely a tow plane for heavy-lift transport gliders, and as such falls outside the scope of this volume. However, as a result of its proven reliability in service, the *Technisches Amt* authorized two variants which are worth looking at (although neither of these were ever built). These were as follows:

He 111Z-2 as a long range bomber with either four 1,800 kg bombs or two LMA III mines plus two 1,800 kg bombs, or six 1,000 kg or four Hs 293 glide bombs on external racks. With the four glide bombs the aircraft was expected to have a range of 1,100 km at a cruising speed of 310 km/h—somewhat meager for long range operations.

He 111Z-3 as a long range reconnaissance aircraft with four 900 liter drop tanks. With full tanks, it would have had a maximum overloaded weight of 33,000 kg compared to the maximum weight of the Z-1 tow plane at 30,000 kg. Maximum range was expected to have been 4,300 km, with a maximum speed of 476 km/h at 4,900 m without external tanks.

Undoubtedly interesting figures, but despite this impressive data it would have been a not altogether maneuverable "barn door" for enemy fighter defenses. It was a good thing that this curiosity never became operational as a bomber or reconnaissance aircraft and restricted itself to its role as tow plane, where it unquestionably performed well.

All in all, it can be said that the Heinkel He 111 was a reliable and multi-faceted bomber despite its mediocre performance in comparison with the Ju 88/188 and the Do 217. It remained in service until the last days of the war. More than 7,300 He 111s were built over a period of nearly ten years, including the following numbers just prior to and during the war:

1939 = 1,399 acft	1940 = 827 acft
1941 = 930 acft	1942 = 1,337 acft
1943 = 1,408 acft	1944 = 714 acft

The figures for 1942 and 1943 reveal the extent to which wear and tear on aircraft in the East forced continued construction of an airplane that had been obsolete since 1940.

Messerschmitt Me 410 and Heinkel He 219

The *General der Kampfflieger*, in his bomber planning paper, had rejected the Me 410 high-speed bomber and even the He 219 as being uneconomical. Unlike the Do 217 and He 111, which contin-

Messerschmitt Me 410A-1 (1943) is being armed and loaded with two 250 kg bombs. The double bomb racks are fitted to the bombs, which are carried horizontally inside the fuselage.

ued to serve in the bombing war, the Me 410 and He 219 in fact only served in a subordinate role or were never considered at all.

How the Me 410 became a high-speed bomber in the first place has already been discussed. The first prototype, the Me 410V1, was a modified Me 210A, and began its flight testing in the fall of 1942. From a flying standpoint it exhibited none of the problems associated with its Me 210 predecessor.

This spawned the Me 410A-1 high-speed bomber, which went into production at Augsburg form January 1943 onward. The A-1 was a two-seater (back-to-back), with Kut*o-Nase*, dive brakes, and the Stuvi 5B dive bombing sight. The two DB 603A-1s each provided 1,290 kW/1,750 hp on takeoff and, with the GM-1 system engaged, gave the A-1 a top speed of 615 km/h at maximum pressure altitude. The variant's maximum weight was 11,244 kg, which included one 1,000 kg or two 500 kg, or up to eight 50 kg bombs in the fuselage bomb bay.

The Messerschmitt Me 410A-3 reconnaissance version (1944). The blanked over openings for the absent MG 17 guns, removed to make way for the camera installation, can be clearly seen here.

The port side gun (a 13 mm MG 131) of an Me 410 is being fed an ammunition belt.

However, payloads over 500 kg were classed as being overloaded, so that any additional load would need to be compensated for by reducing the fuel carried. The practicable bomb payload thus remained relatively moderate for a bomber. In late April 1943 48 Me 410As were delivered to the *Luftwaffe* and initially began replacing the Do 217Es, followed in the late summer by replacement of the Ju 88A-4. By the end of 1943 the Augsburg factory had built 457 Me 410As. Using different conversion kits, the Me 410A-1/U1 was supplied as a reconnaissance aircraft with a vertical camera in the fuselage (either an Rb 20/30, 50/30, or 75/30), whereas the Me 410A-2/U2 was a heavy fighter with a WB 151 gun pod. This was a drum-shaped pod with two 20 mm MG 151/20s, each with 250 rounds, to supplement the normal forward-firing armament. This consisted of two 20 mm MG 151/20s and two 7.9 mm MG 17s fixed in the nose. In addition, the Me 410A-1 was fitted with two 13 mm MG 131s to cover the rear in remotely operated barbettes on either side of the center fuselage. The A-1 could also, if needed, carry external ETC racks beneath the wing roots for a total of four 50 kg bombs in tandem.

The Me 410A-1/U1 proved to be somewhat inadequate in the reconnaissance role with its single camera, and this led to the development of a specialized reconnaissance version. The Me 410A-3 carried two Rb 75/30 aerial cameras for taking pictures from higher altitudes, but the two MG 17s had to be removed to accommodate this arrangement. Beginning in early 1944 the A-3 served with several long range reconnaissance squadrons, primarily in the South, where it enjoyed considerable success.

Also in early 1944 Dornier began producing the Me 410A-1, as well. This variant, mainly flown with success over southern England, was replaced on both assembly lines in April 1944 by a new variant, the Me 410B-1 high-speed bomber (and the B-2 heavy fighter). It differed from the A-1 by its more powerful DB 603G engines (rated at 1,400 kW/1,900 hp on takeoff) which, with the GM-1 switched on, provided a maximum speed of 630 km/h—somewhat greater than that of the A-1. Like with the A-1 series, the B-series had its own reconnaissance version, the Me 410B-3, which differed from the A-3 only in the type of engine used.

History of German Aviation: Bombers and Reconnaissance Aircraft

In the meantime, the situation in the air had become precarious. This, plus the advances made in the direction of jet bombers, prompted the OKL on 8 May 1944 to order that all A-1 and B-1 high-speed bombers be reequipped for the heavy fighter role. In reality, this directive affected just two bomber groups, both of which were operating from France, and in any event had been slated for conversion to the Me 262 ultra high-speed bomber in the late fall of 1944. From this point on the Me 410 flew exclusively as a heavy fighter and reconnaissance platform—as a high-speed bomber it had only served for a brief period.

With regard to the heavy fighters, it should be mentioned that in the spring of 1944 a special unit was formed at Lorient for the anti-shipping role. The unit was equipped with several Me 410B-6s, which had been fitted with the FuG 200 Hohentwiel surface search radar. In place of the two 7.9 mm MG 17s they were armed with two 13 mm MG 131 and carried the WB 103 weapons pod with two 30 mm MK 103 cannons as standard armament. But in the summer of 1944 these machines, specially configured for the maritime strike role, were assigned to the role of air defense. The same fate befell the Me 410B-5, specially designed as a torpedo bomber for engaging Allied shipping. This latter variant never even made it beyond initial flight testing and a few weapons trials. In addition to the familiar LT 5 aerial torpedo, these trials included the use of new bomb torpedoes (BT 200 and BT 400), an SB 800RS Kurt rolling bomb, an SB 1000/410 with stabilizer 'chute, and the LT 950/L10 Fried*ensengel* glide torpedoes—interesting weapons developments which never made it to the field.

In September 1944 Me 410 production ceased altogether in favor of the emergency fighter program.

The Heinkel He 219 was designed as a twin-engined night fighter and optimally tailored for this role. Production, which took place at a feverish pace, began in early 1944 at Vienna-Schwechat. With regard to the He 219, we only need to mention that the proposed He 219A-3 bomber, a three-seat fighter-bomber with two DB 603G engines, and the He 219A-4 high altitude reconnaissance aircraft, with two Jumo 222 engines and enlarged wings, never made it beyond the design stages. Although a planned He 219C-2 fighter-bomber variant was built as a prototype, the lack of Jumo 222s precluded any type of flight trials for the type. An He 319 high-speed bomber project also never left the drawing board.

Summary

The Me 410 can be considered the best performing twin-engined universal aircraft in its weight class serving with the Luftwaf*fe*. As a high-speed bomber it was pulled out of its operative role from May 1944 onward in favor of air defense. Employed in its original capacity as a heavy fighter against Allied bomber formations (a difficult task in any event), it suffered heavy losses at the hands of enemy escort fighters, particularly America's P-51 Mustang, which had been flying cover for U.S. bombers as far as Berlin since March 1944. Since no further replacement Me 410s came off the production lines from September 1944 onward, the process of attrition became so grave that by late 1944/early 1945 there was just one heavy fighter group left operating the type. Based in Norway, it had converted to Me 410s from Me 110s and Ju 88C-6s in November 1944. Otherwise, the Me 410 flew in an operative capacity only with a handful of long range reconnaissance squadrons.

All told, 1,160 Me 410s were produced, of which 702 of these were built in the nine months of 1944 until production ceased— 444 at Messerschmitt's Augsburg and Regensburg plants, and 258 at Dornier in Oberpfaffenhofen. The Me 410 served just long enough in its role as high-speed bomber to demonstrate that it was extremely difficult for enemy fighter defenses to catch, but that its limited front-line strength and small payload reduced its effectiveness to little more than harassment, without any decisive impact.

The He 219, also mentioned briefly, only really served in the role of night fighter and never achieved status in its planned roles of bomber or reconnaissance platform.

This, then, concludes our treatment of "available types" and their development. Those aircraft types identified by the Gen. d. K. as part of the commonality program were, in part, eliminated by the progress of the war, while others—such as the Do 217 and He 111— were forced to remain in service an astoundingly long time due to a lack of better bombers being produced in quantity. These better bombers were undoubtedly on par with the Ju 88/188 from a performance and technological standpoint, but the available production capacity of the aviation industry was severely hampered by the enemy's bombing campaign. Furthermore, clinging to the He 177 program—considered a failure—for an overly long period of time effectively stymied further progress in the industry, which increasingly switched over to pure fighter production from March 1944 onward.

A new era of high-performance aircraft based around the jet engine got off to a rough start with the Messerschmitt Me 262 and the altercations between the bomber and fighter pilots over how the new type should be used. Ultimately, Hitler's ignorant decision to make a high-performance fighter into an ultra high-speed bomber not only delayed utilizing the type to its fullest potential, it also had far reaching consequences for Germany's air defenses.

19. Last Hope - Gaining the Upper Hand through Superior Technology?

The *General der Kampfflieger's* chief demand in his bomber planning called for increasing monthly production from the roughly 380 bombers of varying types in late 1943 to 600 modern bombers. Yet neither the numbers of Ju 290s built for the so-called long range wings, nor the Ju 88/188 and He 177 aircraft planned for the normal bomber wings even came close to fulfilling his wishes. The same applied to the He 177s he had foreseen for special wings, torpedo bombers in wing strength for carrying out an effective multi-ocean maritime war. This was, in effect, the death sentence for the operative *Luftwaffe* envisioned by the Gen. d. K., which he had hoped to model along the lines of the centrally led and tightly controlled British Bomber Command. Despite unrelenting support, the chief of the command staff, Gen*eralleutnant* Karl Koller, was the one who presided over Germany's abandonment of its air attack capability as it related to an operative air war. Koller, who soon became chief of the *Luftwaffe's* general staff, was opposed to weakening the bomber arm, and on 19 May 1944 issued a study on the requisite minimum strengths the *Luftwaffe* needed to maintain control of Central Europe. This study went beyond the support role for the *Heer* and *Marine*, instead spelling out the need for an operative *Luftwaffe*, i.e. one capable of waging an offensive air war, which would determine the need to act. Koller maintained that the minimum strength should not be allowed to drop below 40 bomber groups (= a minimum of 1,200 operational aircraft). However, even this figure was almost half of what the Gen. d. K. had determined necessary for his requirements (24 wings = a minimum of 2,300 operational aircraft).* With the creation of the previously mentioned "*Jägerstab*" (Fighter Staff) in early March 1944, it was apparent that fighter production was to be given top priority, and this soon led to a noticeable increase in fighter output. On 12 June 1944 Hitler had a meeting with Koller. During the meeting Hitler clearly revealed he was at the end of his patience when it came to the unsatisfactory development of the He 177 situation. When he learned that the He 177B-5 with its four single engines (which was expected to alleviate the problems of the double-engined He 177) could not be produced in numbers before 1946, Hitler stated that "it mattered not in the least bit" to him whether or not this bomber was ever produced. Two days later he clarified his position in the daily situational report: "Our situation depends on building fighters and more fighters. This includes high-speed bombers. The air umbrella over our homeland and our infantry must ultimately be made secure. If this entails doing without an operative *Luftwaffe* for several years, then that is something we will just have to accept."

On 18 June 1944 Göring ordered that "all production of bombers, torpedo bombers, and like aircraft be stopped, with the same applying to all training in these areas." This initially resulted in the largest, material- and fuel-consuming aircraft types being stricken from the production program. Those affected included the latest He 177A-5, the four-engined He 177B-5, and the Ju 290. He 111 production was expected to expire in December 1944. Then, in early September 1944, Hitler ordered "the immediate disbandment of all bomber units," which included abandoning operations with those modern high-speed bombers the Gen. d. K. had called for: the high-speed and outstanding Ju 88 S-series, the Ju 188 (also possessing above-average speeds) and its large payload capacity for bombs and torpedoes, and the Ju 388 high altitude bomber and reconnaissance plane. The last of these modern bombers rolled off the assembly lines as late as the beginning of 1945 but, other than the previously mentioned Ju 88 M*istel* units and a few high-speed jet bombers, an operative bomber arm had virtually ceased to exist.

The True High-Speed Bomber

In addition to the Me 262 (covered in a previous section), Hitler's idea of a "*Schnellbomber*" was the Arado Ar 234 project, a pure jet aircraft originally conceived as a reconnaissance platform, heavy fighter, and also as a night fighter. What follows is a chronology of events leading to the type becoming the first true frontline jet bomber in the world.

---

* As a rule, a Kampf*geschwader* (bomber wing) of the Luftwaffe at the time consisted of three *Gruppen* (groups) plus a wing commander's unit (six aircraft). Each *Gruppe* comprised three *Staffeln* (squadrons) of 9-12 aircraft each and a group commander's unit (three aircraft). A *Gruppe* therefore included 30-39 aircraft, while a *Geschwader* had 96-123. It was also common for a fourth *Gruppe*, an operational training unit, to be available to a *Geschwader* for operational missions.

Arado Ar 234

Foundations and General Project Description

As early as the late autumn of 1940 the Arado company was already working on a faster high altitude reconnaissance platform as an eventual replacement for the Ju 86P. Arado's design, part of a study commissioned by the RLM for a long range reconnaissance aircraft, was to make use of the new jet engines being developed by BMW and Junkers. Under the direction of Arado's technical director, Walter Blume, designers *Dipl.-Ing.* Kosin, *Dipl.-Ing.* Reberki, Wenzel, and Eckstein (who went to the Henschel company in January 1941) all collaborated on a series of preliminary designs which were submitted to the RLM. In early 1941 the design labeled E 370 was determined to show the greatest promise; its layout was frozen shortly thereafter, and the project was officially designated 8-234. In addition to being jet powered, another requirement was a range of 2,200 km (to be able to reconnoiter the Scapa Flow area), and the team of designers came up with the following, relatively orthodox project layout:

A single-seat mid-wing design with extensive glazing for the canopy, no wing sweep, engines suspended beneath the wings in pods, simple empennage; unlike the original concept of a retractable nose gear, for landing it was decided to make use of a hydraulically extendable main skid under the fuselage and support skid beneath the slender nacelles. This was due to the wings being too thin to accommodate a landing gear and for weight saving reasons. Takeoff would be made with the aid of RATO packs (*R-Geräte*) in conjunction with a trolley dropped by parachute once airborne. Takeoff weight with trolley was estimated at 9,000 kg, once released it would be about 8,500 kg.

As original as the trolley idea might seem at first glance, the principle had already been employed by Wilbur and Orville Wright during the first takeoffs of their Wright Flyer on 17 December 1903 at Kitty Hawk in the U.S. The only difference was that the Wright brothers used a takeoff rail instead of a trolley.

But let us return to the Arado project: on 24 October 1941 the *Technisches Amt* decided in favor of the design, and after consultation with *Oberst* Rowehl the go-ahead was given for further development based on the E 370/8-234 concept with an eye toward the *Luftwaffe's* reconnaissance needs. After the death of Udet on 17 November 1941, however, the future of Project 8-234 seemed to be in question for a period. Udet's successor, Erhard Milch, visited the Arado company near Berlin on 4 February 1942. Shown the project, he was impressed enough to issue a contract to Walter Blume for building a wooden mockup and continue with the structural development of the Ar 234. A good two months later, Arado received the preliminary decision calling for construction of six prototypes, with work expected to begin on these in the spring and summer of 1942.

Twin-Engined Prototypes

Work on the first two prototypes, the Ar 234V1 and V2, was conducted at such a pace in 1942 so that the planned newly developed Jumo 004A jet engines could be installed immediately upon delivery—something the company would be forced to wait almost a year for!

Arado dismissed the idea of installing piston engines as an interim solution, since the ground clearance of the propellor blades would be too little using the intended skids. On 28 December 1942 Arado's contract was raised from six to 20 prototypes (V7 through V20), but with the stipulation that these additional prototypes would be fitted with a retractable undercarriage in place of the skid arrangement with a view to a bomber version. To this end, the RLM awarded Arado a contract on 9 February calling for two prototypes to be built as high-speed bombers with external racks. Finally, that same month saw the shipment of the sixth and sixteenth Jumo 004A-0 pre-production jet engines, which were immediately fitted to the Ar 234V1. However, these two engines were not cleared for flight operations, since the first of the flight authorized jet engines were initially to go to the Me 262 program. The jet engines assigned to the Ar 234 program served solely for static run-ups and taxi trials. With the takeoff trolley weighing almost 600 kg, taxi trials were somewhat of an awkward affair; nose steering and braking of the tricycle trolley alone proved to be no simple undertaking. The trolley was designed to be electrically jettisoned on takeoff and recovered via parachute, with the necessary hydraulic brake lines being separated by a switch from within the cockpit. Initial plans called for the trolley to be dropped from an altitude of several hundred meters after takeoff until an optimized takeoff procedure could be established. On 18 July 1943 a disassembled Ar 234V1 was transported by road to Rheine, where it was reassembled preparatory to beginning its flight testing. Here the initial engines were replaced by two flight authorized Jumo 004As. There followed static run-ups for calibration, as well as more taxi trials, during which the trolley's nose wheel was discovered to flutter as speeds approached takeoff velocity. The port engine also had to be removed, repaired, and reinstalled because of faulty insulation. Finally, on the evening of 30 July 1943 pilot Selle took the Ar 234V1 up on its 13-minute maiden flight. Various sources reporting the Ar 234V1's first flight as 15 June 1943 have missed the mark by a good six weeks! The following are excerpts from Selle's flight report of 12 August 1943 and cover this and his second flight in the V1 on 10 August 1943, which lasted 54 minutes:

History of German Aviation: Bombers and Reconnaissance Aircraft

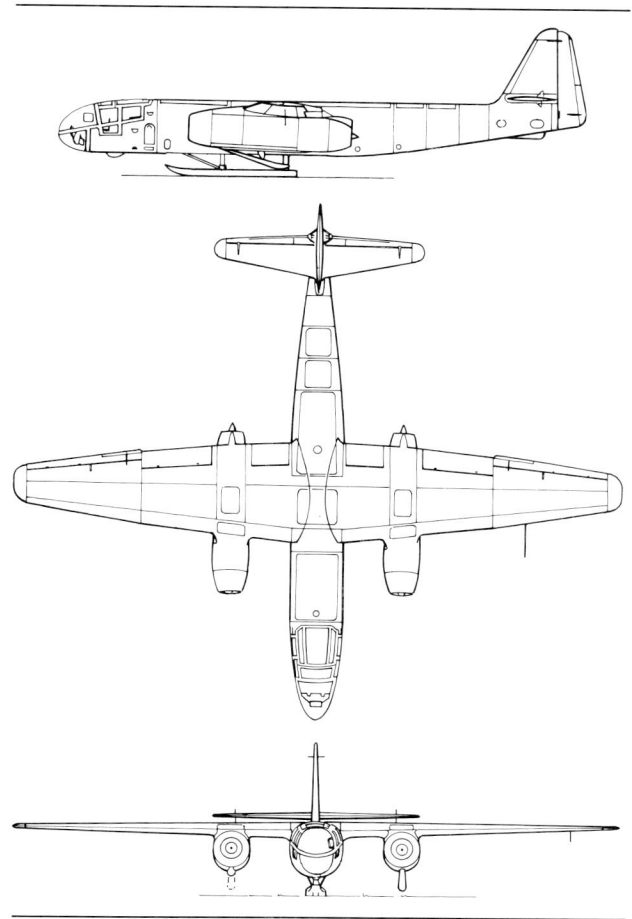

Arado Ar 234V1, first jet bomber in the world with its maiden flight on 30 July 1943.

"...good stability in all three axes. Aileron control is quite good, amazing effect. At speeds of 600 to 650 km/h the control stick forces are sometimes a bit high, but increase steadily.... Even where the control forces have not yet been fully harmonized the aircraft makes a good impression with regard to maneuverability, and in my opinion more than fulfills the requirements of a fighter. The high g-forces are quite noticeable from the pilot's perspective.... The aircraft was flown to speeds of 650 km/h; no complaints were noted.... Taxiing posed no problems, nor did the takeoff using the trolley. The aircraft rests quite peacefully on it and gives a feeling of safety. The better acceleration from a runway compared to from a grass strip is apparent. Takeoffs took place from a concrete runway, sometimes with a light crosswind, under the aircraft's own power without assisted takeoff.

The takeoff run on the first flight (G = 6,330 kg) was approx. 750 m; on the second flight (G = 7,400 kg) it was a good 1,000 m, with V_a-rotation = 160 to 170 km/h.

Climb angle after takeoff was initially quite shallow. This will significantly improve once the trolley can be dropped immediately upon rotation as planned in the future. The takeoff trolley was dropped from an altitude of 500 and 650 m, respectively, at a V_a = 230 km/h. Separation functioned well and with no complaints. Both trolleys were destroyed, however, as the parachute did not deploy as planned.

Both landings were made using the skid with flaps fully extended (45%). The approach angle on landing, even at idle, is relatively shallow, similar to that of a sailplane. Both approaches were flown at a Va = 200 km/h. Touchdown on the skid occurred at a V_a = 150 and 165 km/h, respectively.

Landings are not problematic. The pilot is required to do some rethinking, as the aircraft must be flown quite close to the ground before touchdown, with the aircraft being set down as level as possible on the skid. Pitch tendency is quite low. Rollout required a length of 300-400 m (on dry grass).

Landing approach could be improved by increasing the glide angle through greater flap angle or by extending a spoiler.

Engines. Both engines operated without problem. Their quiet running compared to standard engines is quite amazing. The pilot is not affected by engine noise, and there is hardly any airframe vibration...."

So much for excerpts from the first two flights. After the takeoff trolleys were destroyed when dropped from 500 and 650 meters, for his third flight in the V1 on 29 August 1943 Selle tested a takeoff procedure proposed earlier, and reported:

"At 1656 hrs on 29 August 1943 I took off on a factory test flight for the purpose of evaluating the new takeoff procedures. Prior to takeoff I taxied along the runway to check the smooth handling of the new trolley. I then took off and released the trolley just after rotation, at a Va = 160 km/h. As on earlier tests, the trolley separated without problems, and after the parachute deployed it rolled about 150 m to the right before coming to a stop. In the aircraft itself, no changes in trim were noted. Much improved over the old takeoff procedure was the rapid buildup of speed and improved climb angle following the trolley's release.

I feel that this takeoff method is the only one. Takeoff poses no problems for an average pilot. Release itself occurs mechanically via a cable winch. The strength required is within the capabilities of any pilot...."

Although the takeoff method seems to have been brought under control at this point, on landing Selle moved the throttle back to idle, resulting in a simultaneous flame separation in both engines. Selle was unable to make the runway and had to land outside the airfield perimeter, resulting in considerable damage to the V1.

In the meantime, work had progressed on the Ar 234V2 such that Selle was able to make initial flights on 13, 14, and 15 September 1943 of 48, 38, and 23 minutes' duration, respectively, with the first flight made using RATO packs. Single engine flight was tested without complications, but on the third flight the right engine gave out for real at an altitude of 4,200 m and could not be restarted. The subsequent single engine landing occurred without problems. During another test flight on 1 October 1943 the V2 was destroyed when the left engine flamed out, then caught fire at an altitude of 1,500 m. Selle was killed in the crash.

His place as director of flight testing at Alt-Lönnewitz (Saxony) was taken by F*lugbaumeister* Walter Kröger. Prior to the crash Selle had ferried the Ar 234V3 to Alt-Lönnewitz on 29 September 1943 and began testing it the next day, 30 September. This was the model for the planned Ar 234A series, with ejection seat and pressurized cockpit. On its second flight on 30 September the trolley was released at a height of just 2 meters. It rolled briefly, flipped over, and was damaged. While the takeoff trolley might be an acceptable solution for the Ar 234's flight test program, it was certainly out of the question for field unit operations (the intended recipient of the A-series), where landed aircraft would block runways and provide beautiful targets for low-flying enemy strafers.

The Ar 234V3 continued its flight trials in November 1943, primarily at the hands of company pilots Walter Kröger and Ubbo Janssen, who filed this report on takeoffs using the R-*Gerät* rocket assisted takeoff method:

"The R-*Gerät* is engaged at the moment the pilot has effective rudder control. Acceleration is good, and takeoff with the system is pleasant. From a functional standpoint there are no complaints. Turns with empty packs, which are large drag-inducing bodies, can be made without difficulty. Control inputs are acceptable in all axes up to V_a = 280 km/h (with flaps retracted). Speeds in this configuration were not flown below V_a = 280 km/h. The R-*Geräte* were jettisoned at V_a = 300 km/h and posed no problems."

Janssen, too, carried out problem-free single-engine landings (as a result of engine failures), as well as landings using a toothed ice skid on iced over runways with snow to the depth of 3-6 cm without mishap, but warned of too much airframe stress if the skid's teeth were any larger. Later, on 22-25 February 1944, Janssen tested parachute brakes when landing the V3, which provided almost 40% reduction of the rollout. However, too much stress was placed on

Arado Ar 234V4 (first flight on 18 December 1943), with which the takeoff had been so refined and so mastered that the takeoff trolley could be released and its chute deployed immediately after rotation.

the skid undercarriage, leading to the support skids breaking when a crosswind caught the parachute and damaging the wingtips. Ubbo Janssen also took the Ar 234V4 up on its maiden flight on 18 December 1943.

This was immediately followed by the Ar 234V5, flown by Janssen for the first time on 22 December 1943. The V5 was the first prototype to be powered by pre-production Jumo 004B-0 engines which, although having the same static thrust of 8.9 kN/910 kp at 8,700 min-1 as the first of these TL engines,* was approximately 90 kg lighter. The V5 was built without a pressurized cockpit and designed primarily for compressibility tests; on 7 April 1944 it attained a speed of 725 km/h at an altitude of 2,100 m, but by this time it was already apparent that the Ar 234A would never go into production—its skid undercarriage was impractical for use in a wartime environment.

Subsequent prototypes were initially also built with skids, with the Ar 234V7 being the first to be fitted with production Jumo 004B-1 turbojet engines planned for the Ar 234B series. There is conflicting data derived from the apparently brief life of the V7, which we will not go into detail with here. One thing is certain, though; it was turned over to the military as a reconnaissance platform at quite an early stage in its development.

The Ar 234V6 and V8 will be discussed later, as they fall under the four-engined prototypes for an Ar 234C version. The C-series was to have been a high-speed reconnaissance and bomber aircraft and resulted from the discovery that the Ar 234 airframe required much more thrust than that provided by the two Jumo 004B engines in order to reach the intended Mach boundary.

The Ar 234V9 was also powered by two Jumo 004B-1s and first flew on 10 March 1944, with flights two and three on 15 March—all with Janssen at the controls. The V9 had an ejection

---

* TL = T*urboluftstrahl (triebwerk)*, or turbojet (engine)

seat and pressurized cockpit. It was the first Ar 234 with retractable landing gear, and Janssen gives an account of its qualities:

"Aircraft has become more sensitive in the vertical axis on landing approach, aircraft makes small movements around the vertical axis during extension and retraction (struts and doors follow one another). Travel time good. Takeoff and landing characteristics good."

This, then, reveals that the big step in integrating a functioning retractable tricycle undercarriage in the Ar 234 was a success. Just one month later the V29 had completed 23 flights and 18 flying hours without a single problem with the undercarriage, and six months later, on 5 September, it made its 110th flight—an excellent testimony for the quality of the landing gear design. The V9 became the first prototype for the Ar 234 B-series, whose assembly line had already been spooled up in late 1943 at Alt-Lönnewitz (near Falkenburg on the Elster). In fact, the first pre-production Ar 234B-0 took to the air on 8 June 1944, just ten weeks after the V9's first flight—and two days after the Allied landings at Normandy.

In the meantime, the Ar 234V10 was able to begin its flight trials on 2 April 1944, followed by the Ar 234V11 on 5 May. By this time the primary roles of the prototypes, now fixed at 25 examples, had also been established as follows:

V9 airframe testing and cockpit configuration
V10 stability and flight handling
V11 commonality of parts
V12 bombing trials at Rechlin
V13 precursor for the C-series (paired twin BMW turbojets)
V14 radio communications testing at Rechlin
V15 for BMW (two turbojets mounted singly)
V16 swept wooden wing trials
V17 as V15
V18 testbed for modified bomb racks and parachute brake trials
V19 pressurized cockpit testing
V20 as V19
V21 general testing
V22 evaluation of the armed pressurized cockpit for operations as a bomber and high altitude fighter
V23 through V25 as V22.

Four-Engined Prototypes

The Ar 234C series was to have been fitted with four turbojet engines, these being the somewhat lighter BMW 003A models. The Ar 234V6 and Ar 234V8 (and later the V13 and V19) were built with the intent of testing this engine configuration, with the V8 making its first flight on 1 February 1944 with Janssen again at the controls. Problems with its landing flaps delayed a second flight until 4 March 1944. There were again problems with the flaps, but more troubling were those with the fuel transfer pump system for the four engines, which in turn led to undesired shifts in the airplane's center of gravity. By the time of the third flight on 10 March 1944 these problems still had not yet been fully rectified. On Janssen's fourth flight on 30 March 1944 the difficulties with the landing flaps were no better; they failed to retract on the first try and, depending on speed, hung 15 to 100 mm below their retracted position. In addition, this aircraft lost engine no. 2 because of regulator problems, problems which had already cropped up during static run-up tests with the BMW engines. After multiple change-outs of the regulator it was apparent that there were fundamental problems with the design, for either the idle was too high, the engine flamed out when throttled back, or maximum power was too low—there was even trouble with the regulators at altitude. It was quite obvious that the BMW turbojet engines still had many hurdles to overcome! The Ar 234V6, whose four BMW turbojets were housed in four separate nacelles, did not enter flight testing until almost three months later.

A report from 21 April 1944 reads: "Still no flight, constant problems with regulators. Today an engine had to be changed out because of extensive blade damage (sabotage? Same thing happened prior to first flight of V8)," but on 25 April 1944 Janssen was finally able to take the four-engined V6 up on its first flight. He was not, however, entirely enthusiastic about the aircraft, as can be seen in this excerpt from his report:

"On takeoff the engines required much nursing due to knocking, since despite different regulator characteristics the engines can only be operated simultaneously. If the outer engines accelerate rapidly at different rates, there's a strong danger of ground looping during the initial stage of takeoff."

Arado Ar 234V6, second prototype with four separate engines. The first, the V8, had entered flight testing on 1 February 1944.

During his second flight on 27 April 1944, Janssen noted:

"Aircraft drifted sharply to the right upon initiation of takeoff and could only be kept aligned by throttling back on the left outer engine. The entire hydraulic system (switches, etc.) sweats because of too much pressure in the tank; the cockpit is constantly covered in oil. Upon touchdown, the wings bend considerably due to the mass located so far out and thrust also shifted between two engines."

On his third flight on 30 April, Janssen reported:

"Ground loop on takeoff only avoided by throttling back the left outboard engine. This resulted in much too high gas temperature noted shortly after takeoff and engine had to be throttled back sharply. Post-flight it was found that engine parts behind the turbine had been burned away as from a cutting torch. Engine change-out. All other engine observations in brief:
1. Fundamental problems with the regulators; regulators have a mind of their own.
2. Engines cannot be flown to their full performance potential, since at full throttle the gas temperature beyond 9 on the gauge is too high.
3. Airframe cannot be tested to its limit for the present because of BMW engine's state of development.
4. Pilot feels that flight safety is too low.
   Other items: main skid fails to retract; support skids work."

On 12 May, during the fourth flight of the V6, he reports other problems with the engines:

"Engine trials: two fuel transfer pumps shut down due to markedly unequal draining of both fuel tanks and sharp shifting of the center of gravity. One engine flamed out, probably due to insufficient fuel feed. Unequal drainage must be fixed from the ground up; lines from the rear tank to the engine showed large cross cuts."

And on 17 May, the fifth flight, Janssen reported:

"BMW calibration flight. Regulators affect each other, as they are controlled by fuel pressure and hang from a single cord. Again widely varied fuel drainage; fuel lines have now been replaced by ones with larger length on a trial basis. Regulator for engine no. 1 must be replaced. Engine no. 3 lost its aft cover plate in flight; plates on the other three engines have already torn through the screws."

The Ar 234V6's sixth flight on 30 May was fairly uneventful. Not so flight #7 on 1 June. At an altitude of just 500 m engine no. 3 died, followed seconds later by no. 1 and nos. 4 and 2 in short order. Using what velocity he had left, Janssen pulled up to a height of 1,000 m and began making preparations for an emergency landing. Janssen was able to restart the number 2 engine, but by now his speed had tapered off sharply. With this single engine running at full throttle, the pilot managed to clear some woods and set the plane down in a field immediately beyond. It was a smooth landing, and the aircraft was undamaged as it skidded some 500 meters to a halt—a piece of masterful flying by Uddo Janssen! In spite of these altogether quite negative initial results of the four-engined airplane, which the test pilots had deliberately worded in such detailed terms, just three days later—on 3 June 1944—the Arado company met with representatives of the Generalluftzeugmeister at Landeshut. The outcome of the meeting was that the C bomber (four-engined, to be fitted with retractable undercarriage) was to have an overloaded maximum weight of 12,000 kg in order to fully exploit its operational potential with regard to fuel capacity and/or bomb payload, with greater wear and tear on the tires apparently being considered an acceptable tradeoff! This was still a long way off, however, for in addition to correcting all its previously mentioned shortcomings it would be quite a challenge to increase the takeoff weight of a rather petit aircraft weighing all of 8,000 kg by a third to 12,000 kg.

Twin-Engined Follow-On Developments and Operational Testing

But let us return to the not altogether utopian world of the twin-jet Ar 234V9, the prototype which was to spawn the B-series. The military became involved in the V9's test program as early as its 7th, 9th, and 18th flights, and on its 30th flight on 25 April 1944 was flown by *Generalmajor* Dietrich Peltz himself for a good 30 minutes. Peltz's rather brief assessment of the plane was:

"Takeoff and landing quite good, extemely low level flight, sharp turns, steep banks, instrument spirals in a cumulus storm (using voice radio bearings), pushed to 900 km/h. Assessment: no changes needed to flight handling characteristics, instrument flight quite good, easy takeoff and landing, main instrument panel annoying, as is the jettison hatch bulkhead (view forward).

Flights 37 and 38 were carried out by an RLM representative on 30 April 1944, who commented:

"Test flight. Assessment: no changes desired to flight handling characteristics, takeoff and landing quite easing, all in all a 'marked advance in aviation design.' Horizontal view interrupted by bulkhead."

Thus, from the perspective of the end user (*Luftwaffe*) the Ar 234 was considered to be fully acceptable, assuming that further testing proceeded smoothly and it could be developed into an operational weapons system—its flight characteristics met all expectations. It was indeed a "marked advance in aviation design." This acceptance was reinforced by field testing carried out by the

Versuchsverband OKL Oranienburg. On 20 July 1944 prototypes V5 and V7 were deployed to Juvincourt, near Reims, as part of a *Kommando Sperling*, along with two pilots, 18 technicians, and three civilian mechanics to deal with engine and airframe problems. From Juvincourt these two aircraft flew reconnaissance missions over the invasion beachhead, providing in many cases excellent photo intelligence, although evaluation of this data and its translation into military action suffered at the hands of the given situation on the ground and the shortcomings of Germany's air power. Despite considerable organizational and logistical difficulties, both aircraft continued flying (with their questionable skid gear) until 13 September, making a total of 14 reconnaissance sorties without loss in skies where the Allies enjoyed absolute air superiority. The V5, however, was hit by friendly AAA near Chièvres during a transfer flight and was out of action. The pilot did succeed in flying the damaged plane to Oranienburg, where he made a smooth belly landing on the runway; immediately afterward, though, it was rammed by another aircraft taking off and suffered major damage.

The director of the *Kommando Sperling* expressed the sometimes harsh reality of this first use of the Ar 234 in combat in his evaluation report, an excerpt of which follows:

"Operations with the takeoff trolley, particularly from unimproved, bombed airfields, is unsuitable for combat operations. An aircraft is particularly vulnerable to enemy action for too long a period after landing. Constant problems arose with the main and support skids collapsing, which led to frequent damage to the wingtips.... Problems with the cameras occurred because of the mechanism jamming. No frosting or icing was noted.... Enemy action on the ground: in three instances strafing attacks were flown against landed aircraft, and once the V7 narrowly avoided being destroyed by carpet bombing."

But the detachment leader's words were not entirely negative: "The engines were flown by temperature as much as possible. No losses as of yet. Total hours flown without interruption for the V7 is circa 25 hrs, for the V5 approximately 22 hrs. When the thrust nozzle setting is moved to the third position the increase in V_a at 10,000 m is approximately 25 km/h. There were problems shutting down the engines which occurred when we had to land with the wind. Fuel feed pressure was maintained up to altitudes of 11,000 m, the engines never flamed out.... *R-Geräte*. No problems with the RATO packs, parachute failed to deploy only once, could always be jettisoned. Some packs could be used for up to seven takeoffs."

It was quite obvious at this time that, from a developmental standpoint, the Junkers Jumo 004B-0 and B-1 turbojets in the V5 and V7 were well ahead of their BMW counterparts in the V6 and V8.

Arado Ar 234B-1/b reconnaissance version. This particular airplane, along with three other Ar 234B-2 bombers, underwent further testing and evaluation in the U.S. in 1945 after they fell intact into American hands in Austria.

Arado Ar 234B-1

The Arado 234B-1 followed closely on the heels of the B-0. It was specially configured as a reconnaissance platform, but otherwise was identical to the B-0 with the exception of the new PATIN three-dimensional automatic pilot (which provided an enormous relief for the pilot) and the potential of carrying two 300 liter drop tanks to improve range. The conversion kits installed as standard were the following:

"b" for photo-cameras with automatic film advance and 120 m of film

"p" for the PATIN PDS 11 or 12 three-dimensional autopilot with pitch indicator (for descent)

"r" for two 300 liter drop tanks carried on the bomb racks

The B-1 was also only produced in small numbers as an interim series, and was soon replaced by the much more versatile B-2.

Arado Ar 234B-2

The Ar 234B-2 was designed for both photo reconnaissance and as a bomber and pathfinder. It was capable of carrying a maximum of 1,500 kg of bombs on external racks, with one 500 kg bomb under each nacelle and one under the fuselage. Its greatest single payload was the PC 1400 carried on its fuselage ETC. For high altitude level bombing the B-2 was fitted with a Lotfe 7K between the pilot's feet; with the autopilot engaged, the pilot would unlock the joystick and move it forward, then fly the plane to the target using the gauges on the Lotfe for course correction input to the autopilot, which would respond accordingly. Bomb release occurred automatically via the Lotfe, with the pilot only required to keep the crosshairs of the Lotfe's powerful magnifying lens centered on the target, assuming he had entered the necessary course, drift/lead, and groundspeed data into the bombsight's computer. During bombing trials, under average drop conditions from 6,500 m altitude and a speed of 550 km/h, results showed a CEP of 33 m right and 9 m overshot, not a bad average at all! In addition, the B-2 carried the new BZA 1B for shallow dive bombing, which during tests provided CEPs of 13 m left and 27 m short of the target under average

Arado Ar 234B. The pilot's fore- and rear-view periscope above the canopy can clearly be seen in this photo.

drop conditions from an altitude of 1,450 m, a speed of 740 km/h and an 18.5% descent angle, also quite a good result. It was discovered, however, that the dive bombing accuracy was no greater than that with level bombing at altitudes over four times higher. With regard to the BZA 1B it should also be mentioned that the pilot sighted this using a PV 1B periscope, which acted as a fore and aft periscope. When the BZA was turned off the periscope provided a view aft, but when the BZA was on it automatically switched to a forward view.

The Ar 234 had evolved into an impressively performing aircraft, one that could "stay in the fight" despite the enemy's mastery of the air. In its one-man configuration, however, it not only demanded that the pilot had an above average flying ability, but that he also had a solid background in additional weapons system training. The *Luftwaffe* was still able to draw from a pool of experienced pilots who met these prerequisites. Göring had prohibited further training of any new generation of pilots capable of instrument or night flying with parallel training in the skills a bombardier needed to operate the Lotfe sight and the bomb release system. In any event, by the second half of 1944 such training would not have been possible in light of the given situation in the air. What remained of the *Luftwaffe's* bomber elements were recruited from their daily dwindling numbers. Young pilots who completed their brief flying training program of 50 to 70 hours went to the fighter units for the defense of the *Reich*, where fighter wings were filled or created anew as fast as possible from new personnel or crews from dissolved bomber units. There remained only a modest core of flyers for the actual bomber role, and these had already been earmarked for Me 262 fighter-bomber operations and the *Mistel* teams. Could the Ar 234, with its performance far superior to anything the Allies then possessed, still turn the tide in the air war?

Beginning on 24 August 1944, an increasing number of experienced pilots from Do 217, He 111, and Ju 88 units were brought together at Alt-Lönnewitz and schooled on handling the Ar 234B-0 jet. These days were not without losses; for many this new high performance airplane was too revolutionary. In the spring of 1945 a few Ar 234B-1a two-seat trainers became available and led to a reduced number of training accidents.

By 30 November 1944 all 20 Ar 234B-0s and 95 Ar 234B-1s and B-2s had been delivered, out of a total contract number of 190. Nevertheless, the RLM's plans did not include increasing the production of the twin-engined Ar 234 beyond a total of 210. Its ideas were aimed more optimistically with an eye toward the future, perhaps even along the lines of the original bomber planning, but in a modified and more modern form?

We have already looked at the problems facing the four-engined Ar 234 prototypes in detail. The individual configuration of the Ar 234V6's four BMW engines was fraught with difficulties, while the paired configuration of the V8's turbojets seemed to be a much more sensible approach from a purely flying perspective.

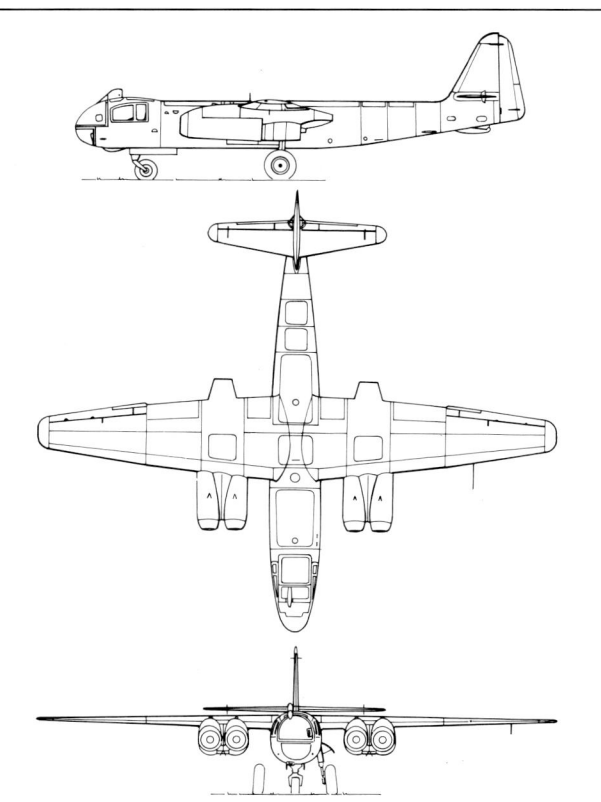

Three-view of the Arado Ar 234C showing the paired arrangement of the four jet engines.

Four-Engined C-Series

Other prototypes had been built and flight tested in the meantime using the paired layout, but unlike the V6 and V8, these all had retractable landing gear. These included the Ar 234V13 in late August 1944, which reached an altitude of 12,800 m, the Ar 234V19 in late September 1944, and the Ar 234V20 and V21 in late October 1944, the latter being assigned to Küpper-Sagan in Lower Silesia where they became the prototypes for the four-engined C-series.

Multiple dispersals of the final assembly facilities led to delays in the construction and development of the four-engined type,

Static photo of the Arado Ar 234C-1 with the paired BMW 003 jet engines (1944)

but the vision senior leaders had for the Ar 234C's use in place of the twin-engined B-series was quite clear. Plans called for the four-engined type to be produced as:

Ar 234C-1 as a pure reconnaissance platform with the same camera suite as the B-1, but with defensive armament consisting of two 20 mm MG 151/20s firing aft, each with 250 rounds.

Ar 234C-2 as a pure bomber with no defensive armament, payload was one 1,000 kg bomb plus two 500 kg bombs. Takeoff run loaded was barely 900 m, or approx. 600 m with two RATO packs.

Ar 234C-3 as a multi-role aircraft for bombing, night fighting, or strike. Armament same as C-1, but with two additional 20 mm fixed forward-firing MG 151/20s.

Ar 234C-4, again a pure reconnaissance platform, but with more powerful BMW 003C-TL engines (a subsequent company brochure from 26 September 1944 lists four BMW 109003-A$_1$)

Ar 234C-5, pure bomber, but the first operational version of the Ar 234 with a two-man crew, since the load placed on a single pilot in a bombing role for a single-seat, four-engined all-weather bomber had become evident to the planners by this time.

Ar 234C-6, reconnaissance version, similar to the C-5 in that it had two seats.

Ar 234C-7, a two-seat night fighter with more powerful Heinkel-Hirth TL He S 011 or, if unavailable, Jumo 004C-TL. Anticipated armament included one 20 mm MG 151/20 with 300 rounds in the nose and a flat profile pod beneath the center fuselage housing two 30 mm MK 108s with 100 rounds each. The C-7's takeoff weight with RATO packs was 10,700 kg, or 10,400 kg without the packs.

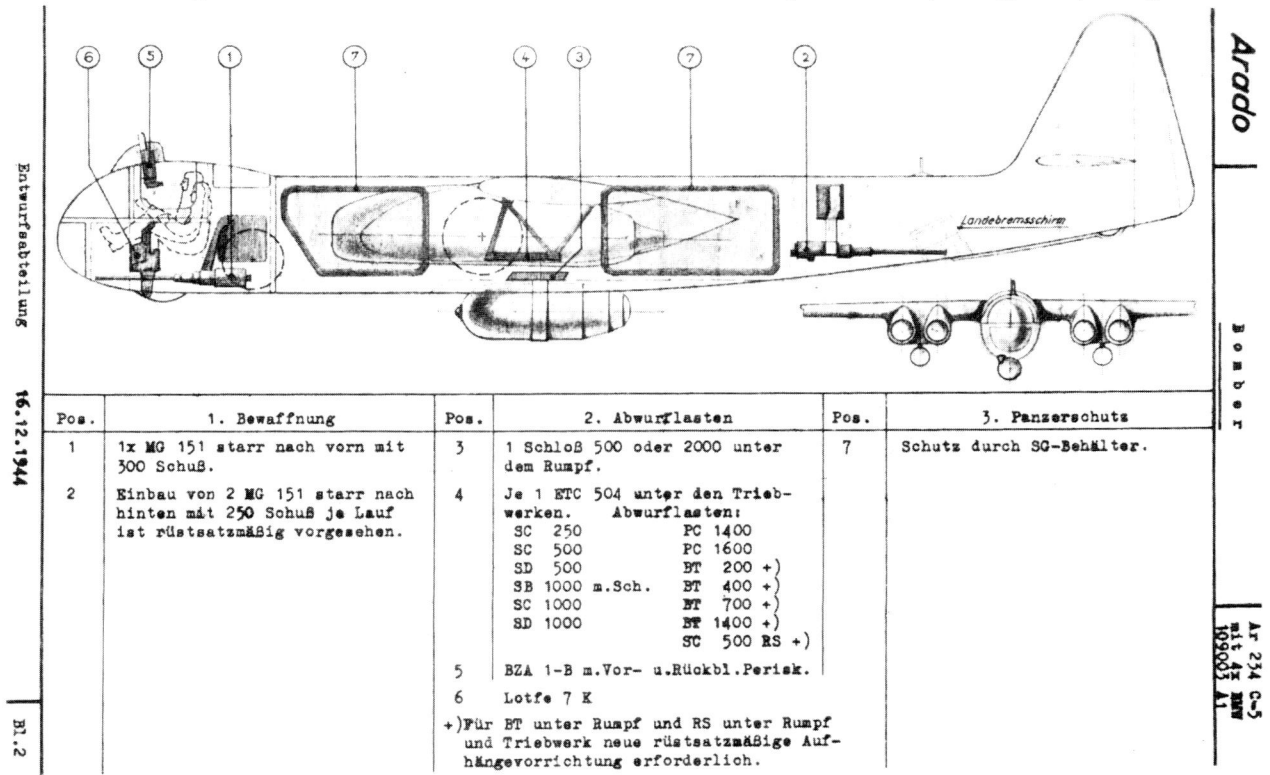

Arado Ar 234C-5 bomber (1944) taken from a company prospectus dated 16 December 1944 showing armament, ordnance, and armor protection.

Arado	Bomber	Ar 234 C-5 mit 4x BMW 109003 A₁

Gewichtsaufstellung

		Bel.Fall I 1xSC 500 Kraftst. 3600 kg	Bel.Fall II 1xSC 1000 Kraftst. 3100 kg	Bel.Fall III 1xSC 1000+ 2xSC 500 Kraftst. 2100 kg
Flugwerk	kg	2750	2750	2750
Triebwerk	kg	2935	2935	2935
Baugewicht	kg	5685	5685	5685
Ausrüstung (mit Bewaffnung nach vorn und hinten)	kg	885	885	885
Rüstgewicht	kg	6570	6570	6570
Führer	kg	200	200	200
Kraftstoff $\gamma = 0{,}84$	kg	3100	3100	2100
Zusatzkraftstoff	kg	500	-	-
Abwurfbehälter	kg	50	-	-
Anlaßkraftstoff	kg	30	30	30
Schmierstoff	kg	90	90	90
Munition	kg	160	160	160
Bomben	kg	500	1000	2000
Zuladung	kg	4630	4580	4580
Startgewicht ohne Starthilfe	kg	11200	11150	11150
Startgewicht mit Starthilfe	kg	11500	11450	11450

Triebwerk:	4x BMW 109003 A₁
Besatzung:	2 Mann (Druckkabine mit Vollsicht)
Steuerung:	Dreiachsensteuerung
Landehilfe:	Einfache Spaltklappe
Fahrwerk:	Bugradfahrwerk
Kraftstoffanlage:	3700 l in 2 geschützten Behältern. 2 abwerfbare Zusatzbehälter können unter den Triebwerken an den Bombenschlössern aufgehängt werden.
FT:	FuG 15, FuG 25a, FuG 102 A, FuG 136, FuG 217, FuG 142
Bewaffnung:	1 MG 151 starr nach vorn mit 300 Schuß. Einbau von 2 MG 151 starr nach hinten mit 250 Schuß je Lauf ist rüstsatzmäßig vorgesehen.
Bomben:	Normal 1000 kg bei vollen Kraftstoffbehältern. Höhere Bombenlasten und Variationen s. Bl. 2.
Zielgerät:	Lotfe 7K und BZA 1-B mit Vor- und Rückblickperiskop.

Arado Ar 234C-5 bomber (1944) taken from a company prospectus dated 16 December 1944 showing general description and weight breakdown.

Ar 234C-8, again a single-seat bomber armed with two 20 mm fixed forward-firing MG 151/20s, each with 250 rounds, and the even more powerful Jumo 004D turbojet. Payload was estimated at 1,000 kg of bombs.

The RLM also apparently had ideas with regard to specific numbers, although these did not include any particular time frame and there are no independent sources to back these figures up (see Bateson, p. 95):
C-3 with 1,795 examples of this single-seat multi-role variant
C-4 with 330 examples of this single-seat reconnaissance variant
C-5 with 1,395 examples of this two-seat bomber variant
C-7 with 290 examples of this two-seat night fighter

for a total of 3,810 Ar 234Cs, of which 3,190 were high-performance bomb platforms. By applying a bit of logic, we can assume that the C-1, C-2, C-6, and C-8 were considered as small series or interim variants, while the C-3 and C-5 would form the bulk of the modern four-engined jet bomber fleet—an impressive armada! Two production lines for the C-series were set up, one in Alt-Lönnewitz and one in Brandenburg-Neuendorf. By June 1945 these two sites were to have achieved a monthly tempo of 350 aircraft.* Several prototypes associated with these variants were built at these two facilities, as follows:

Ar 234V21 through V25, for the Ar 234C-3 series
Ar 234V28 for the Ar 234C-5 series
Ar 324V29 for the Ar 234C-6 series
Ar 234V27 for testing various types of air brakes, and
Ar 234V26 and V30 for evaluating new wing designs.

In addition to these ten prototypes, built shortly before the end of the war, a further 14 Ar 234C-3 pre-production and Ar 234C-1 production aircraft were built. Many of these lacked engines, however, in order to begin their flight test program. None of these last few aircraft ever made it to the field.

---

* Based on information provided by *Dipl.-Ing.* Kosin on 22 August 1988.

History of German Aviation: Bombers and Reconnaissance Aircraft

Summary and the Dornier Do 435, Do 635, and Do 335

The Ar 234B-1 performed admirably in its reconnaissance role over England, the invasion areas in the West and South, and taking pictures as far as Scapa Flow. The Ar 234B-2, too, flew numerous bomber missions after being assigned to a bomber unit reactivated in October 1944, operating from Achmer and Rheine against Allied point targets on the few clear days during Germany's Ardennes Offensive begun on 16 December 1944. Yet despite all these efforts this jet bomber, technologically far superior to anything else in the skies, was unable to turn the tide of the war in the air. In the end, the actual numbers of aircraft were too small, the enemy's air superiority (even at the local level) too great, and Germany's fuel shortage so grave, that its operational potential—both in the air attack as well as the air defense role—could only be brought to bear sporadically.

In March 1945 the last of the fuel reserves were mustered and the Ar 234 unit was able to fly up to fifty sorties a day, many of these between 7 and 17 March against the Ludendorff Bridge at Remagen. But heavy losses on the ground at the hands of strafing aircraft, the carpet bombing of airfields (such as the one on 21 March against Achmer, which destroyed no less than ten Ar 234s), and the Ar 234's vulnerability during landing to enemy fighters besieging their staging airfields all took their toll on Arado units. By 10 April 1945 there were only 38 serviceable aircraft left, 26 reconnaissance planes and 12 bombers.

The Allies succeeded in notching up only a relatively few Ar 234s as true air kills—the Ar 234 Blitz (Lightning) was simply too fast for most of them! With a maximum speed of 1,000 km/h at its recovery altitude of 1,000 m, which could just barely be attained at 100% thrust and a dive angle of 40°, if a pilot recognized the threat in time he could effortlessly leave any enemy plane "in the dust."

In April and May 1945 several flyable Ar 234s fell into British, American, and Soviet hands. These bomber and reconnaissance variants were evaluated in these countries, where they were held in awe for their advanced technological and flight characteristics. Characteristics which had been achieved under the most trying of times by German designers, technicians, test pilots, and—last but not least—the front-line pilots themselves.

Probably one of the best qualified foreign test pilots to evaluate the Ar 234 was the British pilot Eric "Winkle" Brown, director of Royal Aircraft Establishment's Aerodynamics Flight at Farnborough. He flew no less than 55 different German types, some quite extensively, including Ar 234s he personally ferried to Great Britain from Norway, Denmark, and Germany. He assessed the Ar 234 Bli*tz* as follows:

---

* Internally, the Ar 234 was known as the "H*echt*," or pike.

"Here, once again, the German aircraft industry had produced a very superior aircraft too late and in too small numbers seriously to affect the course of the war. It was a magnificent aeroplane of which no real equivalent existed in the Allied order of battle, so it may be said without fear of contradiction that the *Blitz* was truly in a class of its own."

This assessment might also represent the opinions of many other pilots who were similarly impressed with the Ar 234 and its combat potential.

In conclusion, we should perhaps mention briefly the Dornier Do 435 and the Do 635. The Do 435 was derived from the Do 335 fighter and was a bomber project with a fuselage slightly longer forward and a bit wider than the Do 335, allowing the two-man crew to sit side-by-side and slightly offset from one another. Planned payload was 1,000 kg. The Do 435 never progressed beyond the design and mockup stages, however, and the project was canceled in the fall of 1944.

The Do 635 was a twin-plane design for long range missions, made by combining two Do 335 fuselages joined by a rectangular center wing section and powered by four DB 603E engines, each rated at 1,325 kW/1,800 hp. The Do 635 remained on the drawing board.

The Do 335 actually became a reality, with the V1 flying for the first time on 26 October 1943 with pilot Hans Dieterle at the controls. A total of approximately 40 aircraft were built: 14 prototypes starting in 1943, followed by 10 Do 335A-0 pre-production aircraft and 11 Do 335A-1 fighter-bombers, plus a handful of two-seaters with dual controls and tandem seating for pilot conversion training. 14 examples of the A-0 and A-1 are known to have been produced before the production facilities were overrun by American forces.

The Do 335A-1 fighter-bomber, the first example of which rolled off Dornier's Oberpfaffenhoffen assembly line in the late fall of 1944, was capable of carrying a 500 kg bomb load in the fuselage and two 250 kg bombs on wing hard points. It was powered by two DB 603E-1 engines, each with an output of 1,325 kW/1,800 hp and offering a maximum of 1,400 kW/1,900 hp at 1,800 m. Armament was one 30 mm MK 103 cannon firing through the propeller hub with 70 rounds, plus two 15 mm MG 151s mounted above the engine, each with 200 rounds. None of the Do 335s, however, ever saw service. The most prominent features of the type were:

tricycle undercarriage, ejection seat, tandem engines—the forward engine driving a tractor propeller and the aft engine, buried in the central fuselage, driving a pusher propeller in the four-part tail via an extended shaft. Thanks to its aerodynamically clean lines and low frontal drag resulting from the tandem arrangement of the engines the Do 335 was the fastest propeller-driven airplane of its day, clocking a maximum speed of 780 km/h when tested at Rechlin.

History of German Aviation: Bombers and Reconnaissance Aircraft

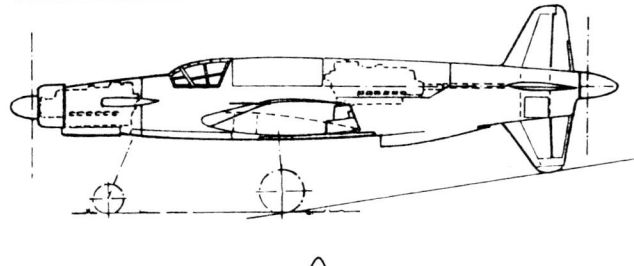

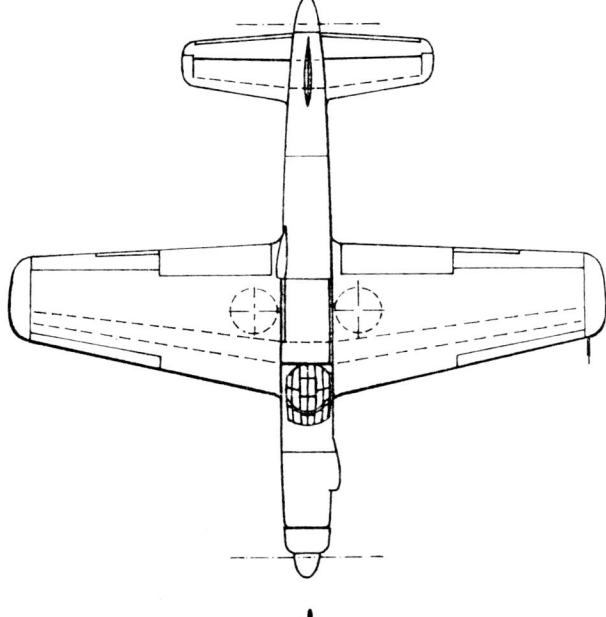

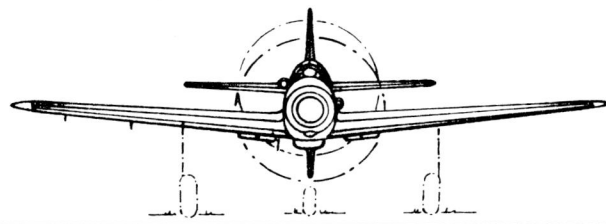

Dornier Do 335V1, first flight 26 October 1943.

Right: The Dornier Do 335 restored in 1974 at Dornier Oberpfaffenhofen is currently on display at the Deutsches Museum in Munich.

The cross-like empennage and aft pusher propeller gave the Dornier Do 335 its characteristic appearance.

It had a takeoff weight of 11,510 kg (maximum), a maximum range of 2,060 km (internal fuel only), and a service ceiling of 11,500 m.

An unarmed long range reconnaissance version, the Do 335A-4, was begun in the fall of 1944, but was never flight tested.

The Do 335 and the Focke-Wulf Ta 152 high altitude fighter were the last high-performance piston-engined propeller-driven aircraft being produced as the jet age was dawning. And while the Do 335's unique layout may have been the final development for piston aircraft, it most certainly offered much in the way of individual development potential.

A Do 335 captured by the Americans in 1945 and flight tested in the States was loaned to the Federal Republic of Germany in 1974. After restoration by the Dornier company, it found a new home in the Deutsches Museum in Munich.

20. Death of the German Bomber and the End of an Era

The decision to shift the focus of production to fighters and Göring's order in the summer of 1944 to disband all but a few numerically insignificant specialized bomber units had been made in the face of the Allied stranglehold in the air, and it was these decisions which also spelled the end for the once impressive offensive power of the *Luftwaffe*. On the other hand, the destructive potential of the American and British heavy bomber units was growing by leaps and bounds, with four-engined heavies operating virtually without interruption both day and night.

On 5 April 1944 the Allied air offensive from bases in Italy against the Romanian oilfields at Ploesti began in earnest. Oil refineries and synthetic oil processing plants near Vienna, Budapest, and in Upper Silesia were subjected to systematic bombardment. Tonnage dropped on German cities reached its high-water mark on 18 April 1944, with a record 4,000 metric tons of bombs falling that day for a loss of only 40 Allied aircraft. These numbers could lead to the conclusion that one four-engined heavy bomber was being lost for every 100 tons of bombs dropped, but this ratio in fact varied from mission to mission.

An example of this is the nighttime British raid on Nuremberg on 30/31 March 1944. 106 bombers from the 570 Lancaster and 212 Halifax bombers used on the raid were lost, 79 of these to night fighters alone. This equated to a loss rate of 13.6% for these 782 heavies, which had a combined payload that night of 2,565 metric tons of high explosive bombs and 318 metric tons of incendiary bombs. The ratio of bomb tonnage (about 2,900 metric tons) to aircraft lost (106), therefore, was 3.66 aircraft for every 100 tons—or, on what was probably the RAF's most disastrous night operation, one bomber written off for every 27 tons dropped!

One thing, however, was certain in this year of the unimaginable escalation of the bomber war, and that was the fact that the Americans dropped almost 600,000 tons of bombs and the British about 525,000 tons on German and German occupied territory, for a loss of 7,749 and 2,770 aircraft, respectively. During these twelve fateful months of 1944 the Americans, operating almost exclusively by day, lost an average of one plane for every 77 tons of bombs dropped, while the predominantly night-flying British wrote off one aircraft for every 106 tons dropped. Gruesome data which only hints at the number of civilian and military victims directly affected by these raids.

The instrument of war which had a major bearing in this increasingly one-sided struggle, and ultimately the deciding factor, was the offensive air force. Over the course of the war the Allies consequently placed the highest priority on building heavy four-engined bombers and, despite high losses, employing these in ever increasing and concentrated numbers against German and German occupied territories. Such strategy was made possible by factors which had exactly the opposite effect on Germany:

- a constantly expanding aviation industry generally free of problems with raw material supplies and almost entirely free of enemy influence
- a growing supply of trained bomber and reconnaissance crews who could be trained world-wide, free of attacks by enemy aircraft and part of a gigantic training organization capable of providing the necessary schooling under almost peacetime conditions

Compared to this, in 1944 Germany's aviation industry might have been able to produce its greatest number of aircraft under unspeakable stress, shortages of raw materials, and constant harassment by the enemy, but such statistics easily belie a combat strength that was actually non-existent. With very few exceptions, Germany's aviation industry had almost completely switched over to fighter production, and by late 1944 its industrial capacity was shrinking markedly. The power source of Germany's offensive fleet, the construction of bombers, was increasingly drying up throughout 1944. Not even the fast and technologically advanced Arado Ar 234 could bring about a change; compared to the heavy four-engined bombers of the Allies, it was produced in too few numbers and had a relatively low payload. The same could be said for the other extreme, the Ju 88 *Mistel* bomb which, in its piggy-back configuration was too unmaneuverable, too slow, and therefore too vulnerable. Germany's offensive potential had unilaterally switched to the use of surface-to-surface rockets, the V-1 and V-2 V-weapons, but neither their range, their concentration of force potential, nor their accuracy could even remotely match those of the Allied heavy bombers.

Operating under an almost seamless umbrella of air cover, Allied forces continued their relentless advance on all fronts against tough resistance by German troops, troops who could no longer rely on friendly air support in their bitter struggle on the ground. The Allied air umbrella was, however, occasionally breached by German air defenses, the last time being on 1 January 1945 during Operation Bodenplatte where over 1,000 aircraft—mostly fighters—surprised the Allied airfields in low-level attacks behind the Western Front. 479 enemy aircraft were destroyed that day, for a loss of 277 German aircraft.

With constant support from Soviet strike planes and bombers, on 12 January 1945 the entire Eastern Front between Memel and the Carpathians erupted with the great Russian offensive, and in just two weeks had pushed back Germany's defensive lines to Upper Silesia and the Oder. Between 4 and 12 February Stalin, Roosevelt, and Churchill met at the Yalta Conference to coordinate final military operations and map out the details of Germany's surrender.

In the West, the temporary military balance achieved by the Ardennes Offensive was again lost when a *Panzer-Armee* was pulled out on 12 January 1945 and transferred to Hungary for the purpose of retaking Budapest and protecting the Hungarian oil fields. The Western Allies were then able to resume the offensive with uninterrupted air support. Up until the end of January 1945 they intensified their air raids on fuel production facilities such as those at Brüx, Zeitz, Wanne-Eickel, and Leuna.

On 3/4 March 1945 German harassment bombers, after a long break, again dropped bombs on targets in Great Britain—one of the last notable missions flown against the island nation—while on 14 March the British dropped super-heavy 10 metric ton bombs for the first time on a railroad viaduct at Bielefeld. A last futile operation took place on 7 April when 183 Rammj*äger*, protected by jet fighters, intercepted a large formation of American bombers. For a loss of 133 German aircraft and 77 pilots killed, only 23 U.S. bombers were successfully rammed and destroyed in the skies over the Steinhuder Meer; the jet escorts were able to shoot down a further 28 bombers. The philosophy of using what amounted to suicidal *Rammjäger* is clear evidence that the air war had degenerated to one of desperation. The essence of Germany's flying units—with the exception of a few jet units—had been drained away. Some personnel were thrown into the ground war, poorly trained and even more poorly armed. The Luft*waffe* was no longer capable of mounting any kind of large-scale operation.

On 16 April 1945 the Soviets launched their drive to Berlin from along the Neiße and the Oder bridgehead.

On 23 April Hitler released R*eichsmarschall* Göring from all offices because of his "betrayal," and on 29 April kicked him out of the Nazi Party.

On 25 April Soviet and American soldiers met at Torgau on the Elbe, and on 30 April Hitler learned that Soviet troops had reached Potsdamer Platz and American forces had occupied Munich. That same day, around 1530 hrs, he committed suicide in his Berlin command bunker after he had appointed *Großadmiral* Karl Dönitz the previous day as the "*Reichspräsident und Oberster Befehlshaber der Wehrmacht*" (Reich President and Commander-in-Chief of the Armed Forces).

The RAF Bomber Command carried out its last bombing mission against Germany on 2/3 May 1945, targeting the harbor facilities at Kiel, and on 7 May 1945 at 0241 hrs G*eneraloberst* Alfred Jodl, on behalf of Dönitz, signed the unconditional surrender of Germany's armed forces in the presence of General Dwight D. Eisenhower in the latter's headquarters at Reims. This was followed by another signing ceremony on 9 May at the Soviet headquarters in Berlin-Karlshorst by *Generalfeldmarschall* Wilhelm Keitel, *Generaladmiral* Hans-Georg von Friedeburg, and *Generaloberst* Stumpff. At 0016 hrs all three signed their names to the following text:

"We, the undersigned, acting on behalf of the High Command of the German Armed Forces, hereby submit to the commander-in-chief of the Allied Expeditionary Forces and, at the same time, the Red Army High Command all armed forces on land, sea, and air presently under German command...."

This document effectively closed the book on ten years of military aviation under the auspices of the German *Luftwaffe*.

On 23 May 1945 the Dönitz government and the remainder of the OKW were arrested in Flensburg and made prisoners of war. With the elimination of the central government, which still existed *de jure* at the time and was treated by the Allies as such, the era of Hitler's Germany had finally come to an end.

21. The Great Leap to the Present

Road to Alliance

The ten postwar years—from the end of the Second World War and that of the Bismarckian German Reich to the Federal Republic of Germany's admittance into NATO—are outlined in the following timeline insofar as these dates relate to West German military aviation and its associated aviation industry:

7 May 1945	Germany's surrender - End of the Second World War in Europe
15 August 1945	Japan's surrender following America's dropping atomic bombs on Hiroshima and Nagasaki on 6 and 9 August 1945, respectively - End of the Second World War
17 March 1948	Defense agreement between Belgium, France, Luxembourg, the Netherlands, and Great Britain (Brussels Pact)
24 June 1948	Start of the 323-day blockade of West Berlin by the Soviets - Western Powers air bridge for resupplying the city
September 1948	Creation of a "Western Union Defense Organization" based on the Brussels Pact
4 April 1949	Signing of the North Atlantic Treaty
23 October 1954	Signing of the Paris Accords - End of the occupying regime in the Federal Republic of Germany - the Federal Republic and Italy join the Brussels Pact, which then becomes the "Western European Union" (WEU). Within this Union, the Federal Republic obligates itself to provide a defense contribution as defined by agreement.*
5 May 1955	The Federal Republic of Germany becomes a member of the North Atlantic Treaty Organization (NATO) - The Federal Republic's future armed forces become a part of NATO forces
14 May 1955	Warsaw Pact comes into being
1 November 1955	Initial establishment of the Bundeswehr with a *Bundesluftwaffe*, *Generalleutnant* Josef Kammhuber becomes first Inspector of the *Luftwaffe*
24 September 1956	Handover of the first American aircraft to the *Bundesluftwaffe* at Fürstenfeldbruck

Allied Army, Initial Reequipment of the Air Arm, and Joint Aviation Programs

On 1 November 1955 the *Bundeswehr* came into being as part of the Federal Republic of Germany's military contribution to NATO. Initially, the *BundesLuftwaffe*'s flying units were equipped in the main from American inventory as part of the Mutual Defense Assistance Program (MDAP), with the aircraft being provided free of charge—Germany did not yet have its own ability to build aircraft. Other sources adequately discuss the toilsome and complicated buildup of West Germany's aviation industry. With the exception of Dornier and the Vereinigte Flugzeugwerke (VFW)-Fokker, all other German aircraft manufacturers merged with this company in the years 1968 and 1969. In practice, at least with regard to bombers and reconnaissance aircraft, Germany's aviation industry was initially limited to maintenance and overhaul work on these types of aircraft of American and British make. It was, however, able to build and/or reactivate a core of experts and, step by step, become involved in joint ventures and license manufacturing programs, thus acquiring valuable technical "know how."

Initial Reequipment of the Air Arm
Under the auspices of the MDAP the Allies initially provided the Bundesluftwaffe with the North American F-84F Thunderstreak fighter-bomber and the RF-84F Thunderflash single-seat medium-range reconnaissance aircraft, and the air arm was assigned a purely tactical role. The *Marineflieger*, the naval air arm of the *Bundesmarine*, received as its initial stock British aircraft—the Hawker Siddeley Sea Hawk (a single-seat naval bomber, of which

---

* According to the protocol of the WEU treaty on arms control (Protocol A, No. 5) "Armament Restrictions for the Federal Republic of Germany," the German government obligated itself in Article 2 to cease further production of certain types of weapons (listed in Appendix III) in its territory, unless the WEU council, upon the recommendation of the Federal Republic of Germany, overrules the NATO commander-in-chief with a two-thirds majority vote. As far as the theme of this book is concerned, one of these weapons types was "strategic bomber aircraft."

History of German Aviation: Bombers and Reconnaissance Aircraft

Republic RF-84F, single-seat tactical reconnaissance plane of American origin. Here follows some performance parameters based on operational service as a basis for the operational performance profile of this aircraft generation, which assumes a maximum speed for the RF-84F of 1,093 km/h.

Actual operational speed at low altitude was 420 kts (= 780 km/h), with a tactical radius of 590 NM (= 1,100 km) and an endurance of 1 hr 24 min with a fuel reserve of 1,500 lbs (= 680 kg; this was dependent on air temperature, e.g. between +10 and +15% = 870 liters) upon landing. In order to attain the above performance the RF-84F was normally flown with two drop tanks, each holding 450 U.S. gallons (= 1,700 l). At optimum altitude (service ceiling was 14,000 m) the RF-84F had a range of 3,500 km with engine throttled back to minimum cruise, i.e. three times the range at low altitude.

This data shows how much the performance values for jet aircraft fluctuated, and the care which must be taken when dealing with purely tabular data in order to get a realistic idea of the true performance of a high performance aircraft.

Oberpfaffenhofen. Other companies involved in license building were:
- Messerschmitt, for the construction of the entire fuselage forward section, including cockpit and armament, as well as the aft fuselage
- Heinkel, for the construction of the wings and empennage.

Furthermore, 27 two-seat G.91/T3s were produced in the Federal Republic between 1969 and 1972 in order to provide conversion training, with the same breakdown of responsibility as for the single seater.

Fiat G.91/T3 (1969), the two-seat trainer version of the single-seat Fiat G.91/R3 light strike bomber (first flight on 9 August 1958). This type was numbered among the second generation of the new *Bundesluftwaffe*. Beginning in 1978, the G.91 was phased out in favor of the Alpha Jet.

34 Mk 100s were provided to the navy), as well as 16 examples of the Fairey (later Westland) Gannet (a three-seat anti-submarine aircraft with two turboprop engines). Both types were carrier-capable aircraft, but in this case flew from land bases. Seaplanes and flying boats no longer played a part in the new German naval aviation.

Second Generation of Bombers and Reconnaissance Aircraft

It was not until the Bun*deswehr's* second generation of bombers and reconnaissance aircraft that the German aviation industry became involved in international joint venture programs, as follows:

Fiat G.91 Light Bomber

Developed by the Italian company Fiat in response to a 1956 NATO requirement, this short range ground attack and reconnaissance aircraft made its first flight on 9 August 1958. As part of a "G.91 Work Group" between 1959 and 1966 the Dornier company built the fuselage center section under license and, as the main contractor, took over final assembly and flight testing for 294 G.91/R3s at

Bréguet Br 1150 Atlantic Maritime Reconnaissance and Anti-Submarine Aircraft

This French airplane was developed in answer to a NATO requirement from 1957 as part of an international joint venture program. It made its first flight on 21 October 1961. The Dornier company was involved in designing and assembling the aft fuselage and lower section. A total of 87 of these aircraft were built, going into service with the forces of France, Italy, the Netherlands, and the Federal Republic. 20 of these went to the B*undesmarine*, where they replaced the obsolescent Gannet. By the end of the 1970s it was becoming apparent that the Br 1150's submarine and maritime surveillance capabilities could only be maintained until a suitable replacement came along if the critical operational components were modernized. Accordingly, 1978 saw the introduction of a combat improvement program, with German companies involved in replacing the radar and EW systems, as well as expanding the navigational and underwater location systems. The upgrade program was completed in 1987 and gave the Atlantic the capability to operate effectively well into the 1990s, at which time a follow-on platform was expected to replace it in the inventory of Germany's naval forces. The Military Tactical Objective (*Militärisch- Taktische*

History of German Aviation: Bombers and Reconnaissance Aircraft

Zielsetzung - MTZ) was approved in December 1986 to determine a potential successor under the project designation MPA-90. The two contenders were the four-engined Lockheed P-3C Orion turboprop and the Dassault Bréguet Atlantique 2, with the latter having more powerful turboprop engines and a takeoff weight increased to about 50,000 kg. Once the winner was selected, it was expected that 12-18 aircraft would be produced beginning in 1997.

Lockheed F-104G Starfighter Medium-Range Multi-Role Strike Fighter and Tactical Reconnaissance Aircraft

First conceived in 1951 as a supersonic interceptor, the U.S. designed XF-104A completed its maiden flight on 7 February 1954. The Federal Republic was among the ten European and four non-European countries to purchase the type in 1958, but instead of the interceptor, opted for the second generation tactical bomber (F-104G) and reconnaissance (RF-104G) versions. The G model, tailored to European requirements, made its first flight in 1960. The *Luftwaffe* began taking deliveries of the G series in April 1961, with the *Marineflieger* following suit in the spring of 1963. Although not an original German design, the problems associated with this advanced weapons platform deserve a closer look, and what follows are some of the more prominent performance characteristics of what was to become a highly controversial aircraft.

F-104G Performance Characteristics

The G series was a follow-on development of the original F-104 day interceptor, designed primarily in response to German requirements. It had a structurally reinforced airframe to accommodate loads of up to 2,200 kg on five wing and fuselage hardpoints. For its optimized role of engaging point targets using both conventional and nuclear weapons it was fitted with a Litton LN-3 inertial navigation system and a new NASARR F fire control and multi-function radar. Because of the European low-level requirements, the American interceptor's downward-firing ejection seat was replaced by a rocket-propelled C-2 ejection seat firing upward through the cockpit canopy. The engine, too, was strongly influenced by German technology. Built by MTU in Munich, the J1K is an improved version of the General Electric J79-GE-11A providing a boost in performance to 7,235 kp/70.28 kN with afterburner. With four jettisonable drop tanks on wing hardpoints (= 2,764 liters) plus 6,615 liters of internal fuel, the F-104G has a ferry range of up to 3,500 km. Its operational radius at a maximum takeoff weight of 13,054 kg is over 800 km using a Hi-Lo-Hi operational profile, i.e. high altitude ingress to the target area, low-altitude in the target area itself, and a high altitude return leg. The F-104G reaches an altitude of 12,000 m in less than 2 1/2 minutes, flies at 1.2 Mach at low level and 2.2 Mach as its maximum speed at 11,000 m; its service ceiling is 18,300 m.

Lockheed RF-104G, the reconnaissance version of the F-104G Starfighter, shown here with two wingtip tanks in the standard configuration for a recce flight at low level using three Dutch Oude Delft TA7M 70 mm vertical cameras. Normal operating speed at low level was around 450 kts (= 835 km/h), although the aircraft could quite easily be flown at much higher speeds; this limit was set based on the ability of the human eye to acquire a target.

In 1978 the Marineflieger (Naval Aviation) received an improved camera suite for naval reconnaissance which included the following:
1. Low altitude camera with five 57 mm lenses in tandem.
2. Oblique angle camera with a focal length of 457 mm, obviating the need to fly directly over naval targets.
3. Infra-red camera, which responded exclusively to thermal radiation and provided a relative temperature entropy diagram using standard film.

Pilot training and weapons familiarization of German crews on what is a demanding high-performance interceptor takes place almost exclusively in the United States. Training in Germany is limited to a "Europeanization" program to acclimate the crews to local weather and flight safety conditions. Such data highlights the leap in performance that the *Luftwaffe* and *Marine's* flying and ground crews were compelled to make when cross training to an aircraft of this performance class. 186 F-104Gs were supplied by Lockheed, with a further 2,000-plus examples produced by Canadair (110) and European license manufacturers. Four work groups, two in the Federal Republic, were formed to deal with European license production and were run under the coordination of a NATO Starfighter Management Office (NASMO). The German (or predominantly German) work groups include:

- The "*Arge Nord*," consisting of the companies Hamburger Flugzeugbau, Focke-Wulf, Weser-Flugzeugbau, Fokker Amsterdam, and Aviolanda Belgium.
- The "*Arge Süd*," comprising the companies Heinkel, Messerschmitt, Siebel/ATG, and Dornier. From 1 December 1960 to 1972 alone, this work group built a total of 210 F-104Gs and, beginning in 1961, reassembled a further 66 airframes that Lockheed had shipped disassembled.

The Bundeswehr acquired approximately 780 F-104Gs, which went to replace the Luftwaffe's F-84F and RF-84F aircraft and the Bundesmarine's Sea Hawks. Born from a series of mergers, the new company of Messerschmitt-Bölkow-Blohm (MBB) assumed responsibility for overseeing technological development and the industrial systems management of the F-104G weapons system, the above mentioned license production, as well as the industrial maintenance and overhaul programs. The last F-104Gs were phased out of the Luftwaffe's inventory in 1987 after 29 years of service. In summary, we can safely say that the above mentioned Starfighter license production work groups, with their Belgian-German-Italian-Dutch sponsored European acquisition program, played a key role in an extensive transfer of technology from across the Atlantic.

Germany's Aviation Industry Catches Up With the West

The license production and/or component assembly of the Bundeswehr's three foreign second-generation bomber and reconnaissance aircraft (G.91, Bréguet Atlantic, F-104G) was the final chapter in the first phase of the new Germany's aviation industry development, a chapter which can best be categorized as a learning period. This industry was now fully capable of standing on an equal technological footing with other joint venture development programs on the international stage and infusing them with products of its own design.

This capability of creating its own products had already borne fruit in other areas, such as the Ministry of Defense's design competition for an interceptor aircraft. In response to this competition, the Entwicklungsring Süd (EWR) in Munich built the VTOL VJ 101C-X1 and C-X2 and successfully tested them between 1963 and 1971. A fundamental shift in NATO's tactical concepts, however, led to the program being officially dropped in 1963, and no further development took place.

A similar fate befell the VAK 191B, a VTOL single seat bomber designed by the Vereinigte Flugtechnische Werke (VFW) Bremen as a successor to the Fiat G.91. The company, formed in 1964 with the merger of the Focke-Wulf-Werken and the Weser-Flugzeugbau, built three prototypes with separate propulsion systems—one engine for forward flight and one for lift. First flight of the V-1 took place on 10 September 1971 in Bremen; the V-2 and V-3 followed that same year, but in 1972 further development ceased, and the project was finally dropped in 1974.

Dornier's VTOL combat transport, the Do 31, which demonstrated its VTOL capabilities for the first time on 14 July 1967, was also a non-starter; the military requirements to which this weapons system had been tailored had been revised and the type was no longer needed.

The national developmental potential had not yet reached the stage where it could push airborne weapons platforms meeting the alliance's requirements through to production maturity. Nevertheless, these independent designs created by German designers and aircraft manufacturers were part of an emancipation process necessary for achieving technological parity, and therefore equal status as partner and full participant on international projects.

NATO's Operational Concept and Doctrine

The NATO Treaty as Foundation

Before delving into the tactical operational principles of the Alliance and the aircraft requirements being driven by these principles, we should perhaps take a look at some of the general tenets of NATO's strategy which dictate the organization's overall armament needs. NATO's strategy is based on the North Atlantic Treaty, whose Article 5 states:

"The Parties agree that an armed attack against one or more of them in Europe or North America shall be considered an attack against them all, and consequently they agree that, if such an armed attack occurs, each of them, in exercise of the right of individual or collective self-defense recognized by Article 51 of the United Nations Charter, will assist the Party or Parties so attacked by taking forthwith, individually and in concert with the other Parties, such action as it deems necessary, including the use of armed force, to restore and maintain the security of the North Atlantic area...."

and Article 6 follows up with the definition of "armed attack" as:

"For the purpose of Article 5, an armed attack on one or more of the Parties is deemed to include an armed attack on the territory of any of the Parties in Europe or North America,... on the forces, vessels, or aircraft of any of the Parties, when in or over these territories or any other area in Europe in which occupation forces of any of the Parties were stationed on the date when the Treaty entered into force or the Mediterranean Sea or the North Atlantic area north of the Tropic of Cancer."

Military Strategic Concept

The NATO strategy derived from this clearly defined wording on the issue of defense is based on the ability to repel an attack at

the border in a cohesive manner—the so-called forward defense—as well as the capability to maintain or reestablish security in the North Atlantic region. The goal of the military strategic concept is to prevent war by a demonstrative ability and willingness to defend, thus deterring a potential enemy from making an attack and convincing him that such an attack would involve an incalculable risk. In a nutshell, NATO's defense concept can be defined as the strategy of deterrence or flexible response, and consists of three fundamental defense components:
- conventional forces
- short- and medium-range nuclear weapons systems, and
- intercontinental strategic nuclear weapons.

These three components make up the NATO triad, and form the backbone for the implementation of the Alliance's strategy. In order to flexibly react to any kind of attack in a calculated manner, the operational concept of NATO involves the following types of response:
- direct defense
- premeditated escalation
- general nuclear response

These are not necessarily graduated steps, but are independent responses structured to correspond to the level and persistence of a given act of aggression.

The type and extent of the weapons systems needed to give credibility to this military strategic concept are solely dependent on the overall combat potential of any perceived threat, which during the Cold War was seen as coming from Warsaw Pact member states and, at the nuclear level, from the Soviet Union.

Within the scope of a desirable détente policy are ongoing disarmament negotiations, focusing on eliminating unilateral arms policy changes and encouraging a balanced, simultaneous reduction of forces and weapons with unhindered mutual inspection access—the so-called "verification" process. If these negotiations prove successful—and there are good signs that they will—in reducing arsenals and troops down to a level where neither side had the capability to attack the other, then a large percentage of the weapons systems tailored to the deterrence concept could almost certainly be dismantled. Until such time of a mutually guaranteed non-aggression capability, however, a full defense capacity must be maintained at any given stage of disarmament. A particularly difficult goal, but one that good intentions on both sides can easily ensure—intentions that will become readily apparent as time goes on.

The Federal Republic's bomber and reconnaissance aircraft play an important role in this deterrence, both now and in the future. In what tactical operational concept of NATO should they then be classed?

General Operational Concept of NATO's Air Arm

The former Luftwaffe's L. Dv. 16 "Luftkriegführung" guidelines compelled the Luftwaffe of the thirties and forties to be universally operable (although not *expressis verbis* in the strategic arena, but most certainly operative in the strategic sense), and at the same time function as a cooperative and support weapon for the two other branches of service—an operational spectrum of such breadth that the Luftwaffe found itself hopelessly overtaxed. The NATO guidelines, drawn up collectively by all the Alliance's partners, take the opposite approach: these spell out the focus of responsibility for the air forces of the individual partners. Only the U.S. and, to a lesser extent, the British have strategically deployable air forces in any appreciable quantity. Within the scope of operative and tactical forces, though, practically all of the Alliance's partners come under Allied operational command. Above all, there are no longer any purely national air operations in the European theater, since in the event of a defense situation all operations would fall under the exclusive control of the operational command authority of the responsible NATO commander. In this context the shift from operative air operations in its former sense to tactical operations is fluid and, with regard to the penetration range, naturally depends on the given geostrategic situation, as well as on the types of aircraft available.

NATO's regulations for prosecuting a tactical air war, the so-called ATPs (Allied Tactical Publications), are roughly defined as:
- engaging enemy air forces
and
- supporting friendly ground and naval operations.

The *Bundeswehr's* bombers and reconnaissance aircraft must be able to function in both capacities.

Engaging enemy air forces (offensive counterair operations, or OCA) normally takes the highest priority in the event of an armed attack as defined in Article 6 of the NATO Treaty, and would certainly be conducted with massive air support. Destroying an enemy's staging bases would be the primary focus of such operations prior to any new aerial attack waves being launched.

Supporting ground and naval operations is divided into three areas:
- Close Air Support (CAS, or Lu*ftnahunterstützung*/LNU)
- Battlefield Air Interdiction (BAI)
- Air Interdiction (AI), also called Deep Interdiction.

These tactical operations concepts/types are only applicable to naval warfare operations in a limited sense, since the use of aircraft is considered naval warfare from the air. In the case of the *Bundeswehr*, naval aviation has been fully integrated into the *Bundesmarine* to ensure efficient prosecution of such warfare. All these operational roles are accompanied by the ongoing battle in the electromagnetic spectrum, which each opponent strives to exploit for his own purposes while denying it to the enemy; "electronic warfare" (EW, or *Eloka - Elektronische Kampfführung*) and all its variants is probably the most well-known term spawned by this genre of combat, and its effects are worth considering when equipping an air force's aircraft.

So much for the general framework under which the Allied forces in Europe would function in the event of a conflict, as well as the operational roles of the offensive air forces (which obviously includes tactical reconnaissance aircraft)

Roles of Offensive Air Forces

These operational responsibilities within the framework of the NATO Alliance's strategy, which also falls to the Bu*ndeswehr*, result in a logical breakdown of the air force's tasking. This also includes the role served by the *Heer's* anti-tank helicopters, as this role seamlessly merges into combat against enemy forces on the battlefield (close air support - CAS).

In turn, it is these roles that drive the requirements for the airborne weapons systems and also the division of tasking between the *Luftwaffe* and the *Heer*. However, these should not be seen as fixed and unchanging, but are in fact ultimately determined by the offensive potential of an aggressor, as well as the technological potential for optimization of weapons and command systems needed for striking at the enemy.

Third Generation of Bomber and Reconnaissance Aircraft

These governing rules for NATO requirements fall under the auspices of defense, but go beyond to include the physical wear and tear of weapons systems by more capable replacements (insofar as the political security situation deems this necessary) and, do not forget, the retention of high value future-oriented jobs in research, development, and production. By meeting the demand for third-generation bomber and reconnaissance aircraft (with one exception), Germany's aviation industry—which owes much of its technical know-how to the knowledge and experience gained from the license manufacture of the F-104G—has evolved into an equal and fully functioning NATO partner.

McDonnell Douglas F-4 Phantom

The one foreign originated exception among the third-generation aircraft is the McDonnell Douglas F-4 Phantom II. The Phantom is indisputably the most successful American multi-role combat aircraft. It flew for the first time on 27 May 1958, and approximately ten years later the Bu*ndesluftwaffe* became interested in the reconnaissance version of the Phantom, the RF-4C, as a replacement for the RF-104G which no longer met the requirements for a modern-day reconnaissance platform. On 24 October 1968 the *Bundestag's* Defense Committee made the decision to acquire 88 RF-4Es, costing a total of DM2.2 billion, built to German specifications for the *Luftwaffe's* heavy reconnaissance wings.

The twin-engined, two-seat RF-4E flew for the first time on 1 August 1970; with a total thrust of 16,240 kp, it can fly at twice the speed of sound. German units began taking their first deliveries of the type on 20 January 1971, and by the fall of 1972 had completely converted over to the RF-4E. Unarmed and equipped with modern reconnaissance sensors to include side-looking radar and infra-red sensors, this "thoroughbred recce bird" is expected to continue flying well into the Nineties. The German aviation industry

Roles of the air forces (offensive).

McDonnell Douglas RF-4E Phantom (from January 1971) with its nose lengthened by 0.84 m in comparison with the fighter-bomber version of the F-4. In addition to the search radar, it also housed the reconnaissance sensor packages.

has only marginally been involved in the development of this exceptional reconnaissance system, although the industry did work on several "upgrade packages" for the type over the last eight to ten years which will enable the RF-4E to fly as a potent reconnaissance system with some measure of security well past the year 2000. However, because of the minimal contribution by Germany's aeronautical industry, this front-line aircraft will not be discussed in further detail from a technological standpoint.

The same generally applies for the F-4F, which replaced the *Luftwaffe's* F-104G as an interceptor. As a fighter-bomber, it also replaced the Starfighter and, to a limited extent, the G.91. 175 Phantom aircraft were acquired from the U.S. between May 1973 and Jun 1976. The only thing of note here is the fact that the F-4F version incorporated the German requirement for automatically activated leading edge slats, giving this Phantom series a markedly improved turn capability over its American predecessors. As a result of the German requirement and the *Luftwaffe's* good experience, part of the American Phantom inventory was subsequently also converted over to automatic slats. In November 1986 the MBB company was given a developmental contract for carrying out upgrades on the F-4F. Beginning in 1989 the F-4Fs would undergo a four-year period of phased upgrades designed to extend the variant's life by a good ten years, thus ensuring that it, too, would see the year 2000 as a front-line aircraft. For the 110 interceptors these upgrades included a new internal navigation system and new aeronautical computer, as well as the installation of a mission computer (LITEF), a new pulse-Doppler radar (APG-65), and the integration of the associated medium-range AMRAAM air-to-air missile system. Even before this developmental program, though, Germany's aviation industry was involved in improving the F-4F's performance both in the air-to-air role, as well as its air-to-ground (fighter-bomber) capability, programs which were completed in 1984. In all, MBB was not only involved in assembling airframe components up until 1977, but then went on to take over a large part of the responsibility for overseeing both the F-4F's, as well as the RF-4E's technological development, management of the industrial weapons systems, and industrial maintenance and overhaul.

Yet this participation in performance and upgrade programs was focused primarily on what were, in effect, originally foreign aircraft types. With the next third-generation successor, however, German aircraft technology and its inherent design potential was brought fully to bear on what became the replacement for the Fiat G.91 ground attack/close support and light reconnaissance aircraft.

Alpha Jet

Foundations, Selection and Role Establishment

This Franco-German joint development, originally designed as a jet trainer, was initially conceived in about 1968. After reviewing a project proposal by the Dornier company with options for subsonic and supersonic capabilities, in October of that year the BMVg decided upon a pure subsonic aircraft of standard layout as a trainer, but with a backup role as a ground attack platform. At the same time in France there was a need for a modern jet trainer, and the Dornier and Dassault-Bréguet companies agreed to continue their project work together. Both France and Germany issued a requirement in May 1969 as the result of an agreement between their defense ministers signed on 1 May, prompting a response by the following three business consortiums:

- VFW/Fokker with their T 291 design
- MBB/Aérospatiale with the E 650 Eurotrainer
- Dornier/Bréguet with their TA 501 (from Trainer **A**ttack Bréguet 126 plus Dornier P 375 - 126 + 375 = TA 501).

After the T 291 and E 650 projects were eliminated, on 23 July 1970 Dornier and Dassault-Bréguet received contracts for working out a joint definition phase for their TA 501 proposal. As a result, France decided upon a pure trainer suitable for weapons training, while the Federal Republic pushed for a replacement for its Fiat G.91/R3 light bomber when it became obvious that the NKF—now conceptually altered into the MRCA—would no longer fit this bill (more on this in the section on the Panavia Tornado).

The definition phase concluded in February 1971 with a joint basic conception calling for both companies to pursue their own role-specific designs. On 16 February 1972 a contract was issued for four prototypes, as well as a fuselage for fatigue testing and an airframe for static trials. The general distribution of component assembly agreed upon by the two companies was broken down as follows:

- Dornier would build the wings and empennage, as well as the fuselage aft section and install the engines
- Dassault-Bréguet was responsible for the forward and center fuselage sections and final fuselage assembly
- SABCA in Belgium, as a subcontractor for Dornier and Dassault-Bréguet, was to build the nose and landing flaps.

Overall final assembly would take place in France, Germany, and in the event of a Belgian delivery contract, in Belgium, as well. This agreement established an approximate development and assembly ratio between the French and Germans at 50/50. Because of the 90% involvement in the developmental costs of the French GRTS (a consortium formed by Turboméca and SNECMA) LAZARC 04 jet engine, the French companies and the German companies MTU (Motoren- und Turbinen-Union) and KHD (Klöckner-Humboldt-Deutz), Cologne, agreed on an approximate 56.6% participation by the French companies and 43.4% for the German industry. From this point on the joint aircraft project was to be known as the Alpha Jet.

Prototypes

Prototype 01 made its maiden flight on 26 October 1973 at the French flight test center at Istres, near Marseille, with Dassault-Bréguet company pilot Jean-Marie Saget at the controls. Prototype 02, assembled by Dornier, followed on 9 January 1974, with its first flight at Oberpfaffenhofen piloted by Dornier's Dieter Thomas. Thomas flew this plane to Istres on 17 January for further testing alongside the 01. On 6 May 1974 Thomas also took prototype 03 up on its first flight; this was to be the prototype for the German ground attack/close support version. Dornier assembled prototype 04 and flew this aircraft for the first time, as well, on 11 October 1974 at Oberpfaffenhofen.

The 04 was to have been the prototype for the French trainer version and initially served as a testbed for structural and systems trials. However, on 23 June 1976 it crashed at Mont-de-Marsan after striking the ground during a low-level flight. 03 later served as an experimental platform for testing the Direct Side Force Control (DSFC) system, which began on 22 March 1982 and enabled the aircraft to shift laterally in flight without changing direction. The DSFC concept was never implemented on the production line, however.

Alpha Jet Production

On 30 September 1975 a Franco-German governmental agreement cleared the way for production which, following initial preparations in early 1976, got underway in the fall of that year with the assembly of production components. In the summer of 1977 Dornier delivered its first production components for final assembly to the Dassault-Bréguet company, which flew the first production aircraft, an Alpha Jet E (E stands for *école* = trainer) on 4 November 1977 at Istres. The German LNU version (*Luftnahunterstützung*, or close-air support) was designated the Alpha Jet A (A for *appui* = support aircraft), and flew for the first time on 12 April 1978 at Oberpfaffenhofen. The Alpha Jet 1B was the designator of Belgium's replacement for its Lockheed T-33 jet trainers; this version completed its maiden flight on 20 June 1978. 200 production aircraft went to France, 175 to the Federal Republic, and 33 to Belgium. Other countries ordering the type included Egypt, the Ivory Coast, Cameroon, Morocco, Nigeria, Qatar, and Togo. In the meantime, France had initiated testing and assembly of the export Alpha Jet MS2 in 1982, which was fitted with a modern navigation and fire control system and came to be viewed as a fully capable front-line aircraft. Germany's units had completed their conversion to the type by the end of 1982, and on 26 January 1983 production officially ended at Oberpfaffenhofen of the 175 German Alpha Jets.

By 1983 over 500 examples of the Alpha Jet were flying in ten countries.

Alpha Jet Technical Data and Details
Cantilever mid-wing two-seat design with simple aft empennage and two jet engines, appearing in two versions: the Alpha Jet A (German Air Force) light fighter-bomber, and the Alpha Jet E (French Air Force) advanced jet trainer.

> **Airframe**
> *Fuselage:* comprising three main sub-assemblies: forward section with pressurized cockpit and nose gear; center section with fuel tanks, main undercarriage, and side inlets; aft section with engines, air brakes, and empennage.
> *Undercarriage:* low-pressure tires fitted to nose and main gear, designed as rocker arm retraction/extension retracting forward via hydraulic power. In the event of hydraulic system outage, all three units extend and lock by gravitational force.
> *Control surfaces:* simple vertical stabilizer consisting of relatively large fin and narrow rudder. Two VOR antennas located in upper section, VHF antennas housed inside tip. Single-unit horizontal stabilizers with negative dihedral, set lower than the wings to provide good stability around the pitch axis.

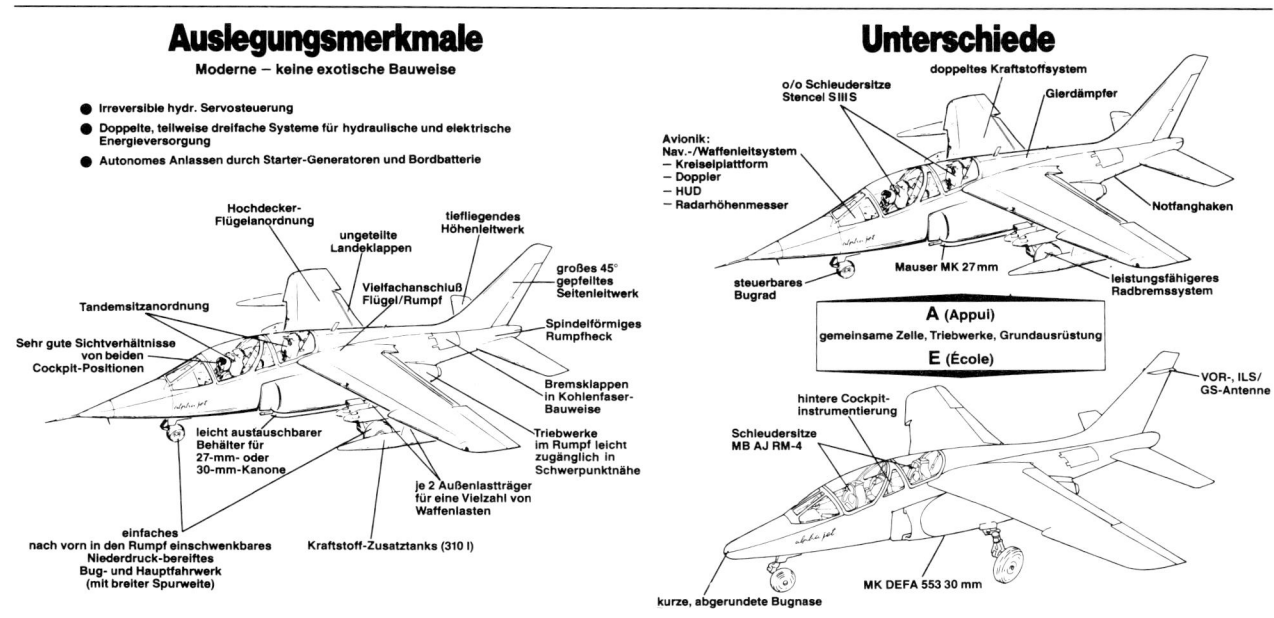

Features of the Alpha Jet showing the differences between the German A version and the French E version.

Control systems: ailerons and horizontal stabilizer controlled by joystick, vertical stabilizer via adjustable rudder pedals with toe brakes. Electronically adjustable trim tabs; hydraulically activated flaps and air brakes.
Wings: twin-spar, separate wings having an area of 17.5 m² and an average sweep of 30% with a 6% negative dihedral. 4.8 wing aspect ratio. Each of the two wing halves is fitted with pylons for external loads, two on the French and four on the Belgian and German versions. Torsion box comprising milled ribs and panels, forming an integral fuel tank.

Powerplants
Two GRTS (Groupement Réacteurs Turboméca/SNECMA) LAZARC 04, later C1, dual cycle non-afterburning engines each providing 13.24 kN/1,350 kp static thrust.
Weight: 290 kg each
time required for engine change
1 hour
bypass ratio
1:1.13
specific fuel consumption
21.2 g/kNs (0.74 kg/kph)
Fuel system: 2,050 liters (Alpha Jet A) in integral tanks in the fuselage center section and torsion box. For improved range, potential of carrying two 310 liter wing mounted drop tanks = total of 2,703 liters maximum, of which 2,670 liters can be used.
Performance:

strength: design load factor	+12 g/-6.4 g
operational load factor	up to + 7.33 g
sustained turn load factor depending on aircraft weight	up to +6 g
aerobatic potential:	fully aerobatic (with no external loads)

Performance in the air
maximum speed

without external loads	1,019 km/h
with cannon pods also	1,019 km/h
with external loads	927 km/h
on single engine	650 km/h
maximum cruising speed	850 km/h

range and endurance

action radius, depending on configuration	430 to 1,100 km
ferry range	3,000 km
endurance	approx. 3 hrs
rate of climb	57 m/s max.
service ceiling	14,630 m
on one engine	8,300 m

takeoff roll

empty	400 m
with 7,200 kg takeoff weight	approx. 1,150 m
with 7,500 kg (max. takeoff weight)	approx. 1,300 m

Dimensions and Weights
Main dimensions

wingspan	9.11 m	wing area	17.5 m²
length	12.29 m	with pitot tube	12.47 m
height (with static compressed struts)	4.19 m		
wing sweep	30%	aspect ratio	4.8
wheel track	2.71 m	wheel separation	4.72 m

History of German Aviation: Bombers and Reconnaissance Aircraft

Weights
standard empty — 3,500 kg
takeoff weight (no external load, only internal fuel) — 5,200 kg
max. takeoff weight with external load — 7,500 kg
max. external load — 2,300 kg

Armament (fighter-bomber version)
weapons pod on centerline station with 27 mm Mauser BK 27 drum cannon with 150 rounds (rate of fire selectable between 1,700 and 1,000 rounds per minute)
4x external pylons

Features of the Alpha Jet

Head-up Display (HUD)
A glass-plate head-up display (HUD) is installed in the front cockpit above the instrument panel at the eye level of the pilot and provides all relevant data needed in the various stages of flight in the form of symbols, letters, and numbers. The HUD enables the pilot to devote his full attention to the outside and not become distracted by having to check the instruments on the instrument panel. He can select five different display modes, as follows:

Navigation: the HUD provides information for terrestrial navigation, i.e. direction, distance, and flying time from the current position to any given waypoint/target previously entered into the navigation system's computer. The computer receives some of the information needed for navigating independent of ground points via a Doppler radar. These parameters include relative ground speed, calibrated airspeed (CAS), radar altitude up to approx. 150 m, barometric altitude, as well as wind direction and strength with drift index.
Bombs: the HUD provides information for delivering all types of bombs.

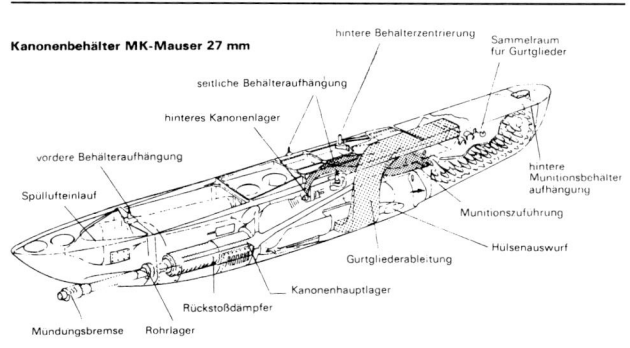

Cannon bay for the MK Mauser 27 mm of the Alpha Jet.

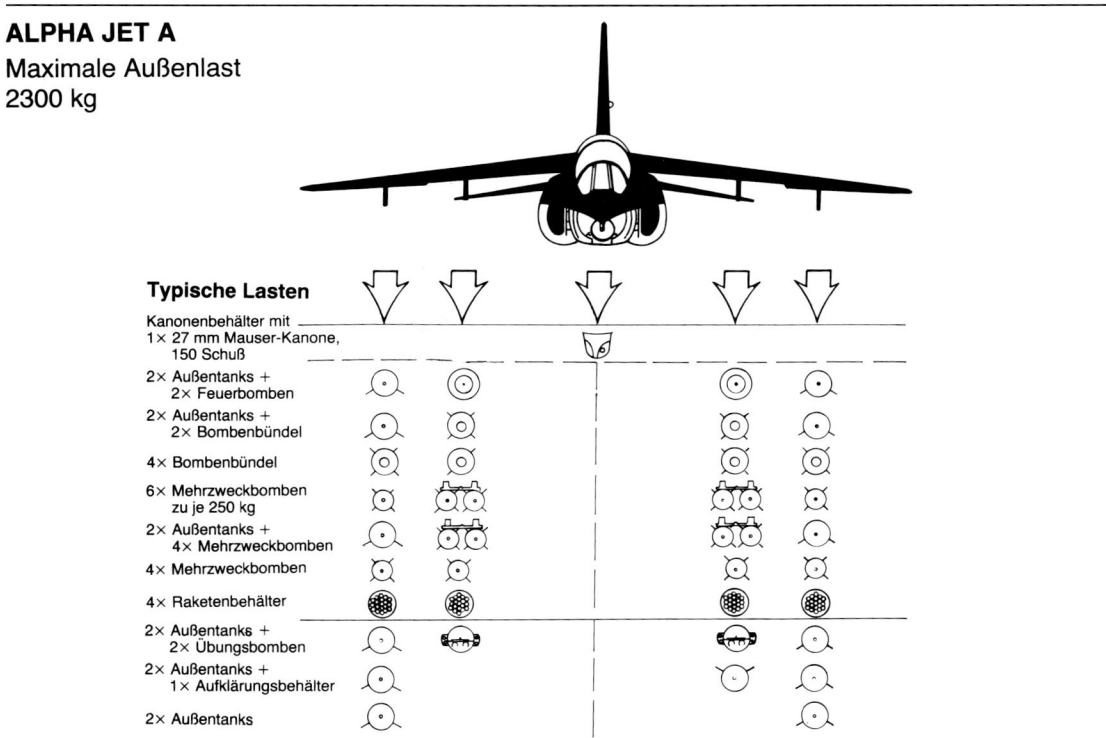

Air-to-Ground Cannon: the HUD provides information on using the BK 27 cannon against ground targets.

Rockets: the HUD provides information on using unguided rockets against ground targets.

Clarification on the three previous modes (Bombs, Air-to-Ground Cannon, Rockets) for the use of air-to-ground operations: when one of these modes is selected, the weapons computer maintains a running calculation of where the weapon will actually hit (the so-called CCIP, Continuously computed impact point) and reflects this in the HUD. In addition, the HUD provides information on the maximum/minimum possible release range of the weapon and the time at which pullout must occur in order to avoid damaging the aircraft.

Air-to-Air Cannon: the HUD provides information for using the BK 27 cannon against air targets in air-to-air combat.

Supplemental information: in addition to navigation and target information the HUD also provides information on the status of the onboard gun, a warning cross when the aircraft drops below a predetermined altitude, and the outage of any onboard components critical to operations, such as the navigational computer, the Doppler radar, etc. A 16 mm gun camera is also integrated into the HUD which films the entire glass plate, as well as the background at the moment the weapons system is activated; it thus provides an outstanding means of evaluating weapons training without requiring the expenditure of ammunition.

Navigational System

The Alpha Jet is fitted with the MITAC (miniature TACAN) navigation system, which offers the pilot constant direction and distance to a ground station. A new gyroscopic platform and improved autopilot ride are examples of additional improvements to navigation provided by the MITAC.

Automatic Flight Data Processing

Even the most modern of navigational equipment can only provide reliable navigational information if the pre-flight data is correspondingly precise. A computer controlled system for navigational planning of missions, called AFA (*Automatisierte Flugdatenaufbereitung*, or automatic flight data processing), has been developed for the Alpha Jet in order to save time and increase the accuracy of mission planning. The pilot supplements these values for his particular mission with wind data, flight altitudes, required time over target (TOT), etc. and then plans his flight path by digital inputs for the desired waypoints/target(s) on the AFA's plotting board. The printer or the AFA's screen then provides him with a flight log for his mission. In addition, via the AFA's MDA (*Missiondatenausgabe*) he receives the MDT (*Missiondatenträger*) with the stored coordinates of the waypoints and/or target(s). With the aid of the MDT the flight path data is immediately fed into the Alpha Jet's navigational computer, which then flies the planned flight program.

Improving the Breed

The fatigue life of the Alpha Jet has been set at 10,000 flying hours, which corresponds to a front-line service of about 30 years, i.e. theoretically well beyond the year 2000.

Naturally, upgrades are indispensable over such a long period of time. The general validity of this requirement led to the then-I*nspekteur der Luftwaffe*, *General* Eberhard Eimler, making the following points in 1986:

"The life expectancy of the *Luftwaffe's* weapons systems today has been set at over 30 years—calculated from the start of development onward. Such a long period of time understandably precludes specific predictions regarding a shift in threat, as well as tactical conceptions and technological innovations in the threat's defenses. The consequence of this is to carry out performance enhancing improvements during the use period of the weapons system. For economic reasons and in order to avoid limiting the availability of the system as much as possible, individual upgrade measures are combined. Upgrade to or maintenance of a system's combat potential assumes, however, that the system has a corresponding growth potential built in. Such measures require that pre-planning takes place in the initial design of a system. New technology demands a careful assessment of the risk. Only those tested components which satisfy the strict criteria of front-line operations are to be considered. This effectively avoids cost-driven changes during the development or introduction of a weapons system; nevertheless, the lifetime costs have already been established for the most part during development. The goal here is also to limit the time required for development and implementation. Combat potential upgrades during the use phase are then possible, and form the basis for effectively responding to the changing threat and exploiting the available potential. The decision to upgrade must be justified by comparing the expense to use—referring to the remaining useful life of the system. This often means that desired ideas are red-lined. In the future, when new weapons systems are laid out more emphasis will be placed on concentrating on what is feasible. The field must have systems which meet their mission objectives; an efficient means to this end are pre-planned follow-on performance upgrades."

The Alpha Jet also embodies a corresponding improvement upgrade program, but one which involves more than just techno-

Even with a full weapons load the Alpha Jet's performance suffered little. This air-to-air photo shows the strike fighter carrying a BK 27 under the fuselage, two 310 liter fuel tanks on the outboard wing pylons, and two general purpose bombs on a double ejector rack beneath each wing.

logical improvements, some of which have already been realized in the area of electronics and electronic countermeasures. As part of technological improvements, the Alpha Jet's undercarriage and brakes have been strengthened and a voice warning system has been installed to alert the pilot when flying below certain altitudes. A new gyroscopic system and better three-dimensional stabilization of the flight control system contribute to improved navigation. The nav-attack system has been improved with a software upgrade, and the planned re-engining with the C-20 turbojet is expected to bring a 10% improvement in performance combined with a drop in operating costs. A more effective electronic self-defense will increase the survivability and life expectancy of the Alpha Jet so that it can remain fully operational well into the new millennium.

Conclusions

The German variant of the Alpha Jet is an advanced technology weapons system primarily optimized for close air support of ground forces. Its good low-level flight characteristics, impressive maneuverability, and high subsonic speeds make the Alpha Jet most suitable for engaging attacking combat helicopters and for airfield defense. Its quick reaction time is of particular value during the initial phase of a surprise attack if NATO ground forces are not yet fully capable of functioning in their forward defense areas.

The easy access to all components within the aircraft, their simple maintenance, and the quick turnaround time following a sortie (e.g. by using pressurized refueling the Alpha Jet can be fully refueled in ten minutes), combined with a low outage rate due to technical problems offer a high sortie rate per operating period.

The flexibility of the weapons system enables it to operate from forward airfields and highway strips without modification in a number of configurations with a wide variety of loads optimized for the type of intended target. Whether additional upgrades to the Alpha Jet will be cost effective remains to be seen, for attacks against moving targets ultimately are directed by the human eye—and the Alpha Jet provides an excellent platform for this!

There is a certain rivalry between the Heer and the *Luftwaffe* in the matter of responsibility for close air support over the battlefield out to a distance of approximately 100 km, but in the long term—at least until the new millennium—priority will be given to the more cost-effective solution. This may mean that a weapons system such as the Alpha Jet might be forced to compete with the new Franco-German PAH-2/HAC anti-tank helicopter and its eight HOT or PARS-3 anti-tank missiles, which could potentially enter service with the Heer*esflieger* sometime in 1997. Perhaps there is a critical need for both aircraft types to operate on or above the battlefield; a union of the two weapons systems would be able to provide

A model Alpha Jets flying in close formation. This four-ship formation is flying in clean configuration, meaning without any external stores.

ground forces maximum firepower in both indirect and direct air support with minimal response time and the fastest transit speeds to the target area.

Numerically far superior, the threat posed by Soviet helicopter regiments and their offensive helicopters plays a major role in influencing any decision on the matter discussed above and cannot be underestimated.

In conclusion, it should be noted that international cooperative programs such as the Alpha Jet effect a multitude of economic ties extending beyond national borders, ties which undoubtedly will become the wave of the future for developments in the marketplace—for joint venture programs, as well. No less than 74 European companies have contributed to the manufacture of the Alpha Jet, as evidenced by the following list from 1980:

> **The following European companies are involved in the Alpha Jet program:**
>
> | | |
> |---|---|
> | Abex GmbH | Messier-Hispano |
> | ABG-Semca | Metallschlauchfabrik Pforzheim |
> | AEG-Telefunken | ML Aviation |
> | Air Precision | MTU München GmbH |
> | Alfred Teves GmbH | Nordmicro GmbH |
> | Apparatebau Gautin GmbH | OMERA-SEGID |
> | ATEI | Pierburg Luftfahrtgeräte Union GmbH |
> | Autoflug GmbH | Precision Mécanique |
> | Becker Flugfunk GmbH | Rhein-Flugzeugbau GmbH |
> | Behr | Rheinstahl Gießerei AG |
> | Bodenseewerk Gerättchnik GmbH | Rohde & Schwarz |
> | Brion Leroux | SABCA |
> | Bronzavia | SAFT |
> | DEFA | SEB |
> | Deugra GmbH | Secan |
> | Diehl Luftfahrgeräte | SEMMB |
> | Dornier System GmbH | SFENA |
> | Drägerwerk AG | SFIM |
> | E.A.S. | Siemens AG |
> | Kurt Eichweber Präzisionsgerätewerk | H. Sinn |
> | Eros | SOCRAT |
> | Elektronik System Gesellschaft mbH | SNECMA |
> | Fabrique Nationale Herstal S. A. | Standard Elektrik Lorenz AG |
> | Feinmechanische Werke Mainz GmbH | Starec |
> | Otto Fuchs Metalwerke | Steinheil-Lear-Siegler |
> | Internationale Fluggeräte- und Motorengesellschaft mb | Teldix GmbH |
> | Intertechnique | Tital |
> | Kablewerke Reinshagen GmbH | Thomson-CSF |
> | Klöckner-Humboldt-Deutz AG | TRT |
> | La Jonchére | Turboméca S. A. |
> | Le Bozec & Gatuier | VARTA AG |
> | Liebherr Aerotechnik GmbH | VAW Leichtmetall GmbH |
> | LITEF | VDO-Luftfahrtgerätewerk GmbH |
> | LMT | Westfälische Metallindustrie KG |
> | Louis L'Hotellier S. A. | Zenith Aviation |
> | Ludolph | Zollern Stahl und Metall GmbH |
> | Muser-Werke Oberndorf GmbH | |

Panavia Tornado

Foundations

Up until December 1967 NATO's strategy focused on massive retaliation (MC 14/2 from 21 March 1957), and alongside other member nations the Federal Republic of Germany was also expected to drop nuclear weapons in the event of an aggressive act. Among other things, this had led to the acquisition of the F-104G Starfighter from a selection of eight American, three British, and one Swedish aircraft type. But a shift in NATO strategic doctrine to that of flexible response led to a focus on strengthening its conventional capabilities (MC 14/3 from 13 December 1967) and, when coupled with the natural aging process, necessitated finding a replacement for the Starfighter which would be optimized for such a doctrine. A part of the F-104's former operational spectrum had in the interim been picked up by the American RF-4E Phantom heavy reconnaissance aircraft beginning in January 1971. The F-4F, too, assumed the role of fighter-bomber and interceptor in May 1973, but for the bulk of F-104s there was no real successor for close air support, air superiority, and counterair operations.

National considerations initially concentrated on a single seat, single-engine aircraft, one which would be small, light, and maneuverable, have a combat radius of about 300 km, and cost no more than DM10 million. A thousand of these so-called "N*eues Kampfflugzeug*" (NKF, or New Combat Aircraft) were expected to be built. An attempt to draw in partner countries involved in the F-104 consortium for a joint successor aircraft led to Germany broadening its tactical requirements for the 104's successor with the goal of arriving at a common denominator among the various ideas from those countries showing initial interest.

V/STOL Options

At the time there was still ongoing development of the previously mentioned VJ 101, FAK 191, and Do 31 VTOL aircraft, as well as a joint German/American project begun in February 1965, the so-called AVS project (AVS = Advanced V/STOL vertical/short take-off and landing). Regarding the AVS, the *Luftwaffe* envisioned a Mach 2 aircraft with a payload of six metric tons with a total weight of 21 to 23 metric tons and a combat radius of 800 km, while the USAF was more interested in a subsonic medium range bomber with variable geometry wings (VG), i.e. swivel wings and V/STOL capabilities. From among the various company proposals one design entered the definition phase in March 1967. It was a two-seat low-level bomber with a combat radius of 600 to 850 km and a maximum takeoff weight of 20.6 tons, capable of flying Mach 2 at higher altitudes. The definition phase came to an end in late 1967 with the sensible realization that even with this V/STOL project the technical problems would not be satisfactorily resolved, since the high payload requirement was not compatible with vertical take-off performance. An alternative solution, the "rolling vertical take-off and landing" (RTOL) concept, was given the cold shoulder since this approach would require hard surface runways at least 150 m in length instead of the 15x15 m surfaces envisioned for V/STOL operations. Work on the AVS project was halted on 19 March 1968; German industrial development potential involved in the project was now free to concentrate on the NKF program.

From NKF to MRCA

For several hundred German technical experts in the aviation industry, the civilian arms community, and the *Luftwaffe*, the abandoned AVS project was nevertheless an intensive exercise in project management.

The AVS project would have been an expensive one, with the costly equipment involved in testing and evaluating new aviation technology leading to a per unit price of about DM30 million. This reinforced the *Luftwaffe* senior leaders' support of the NKF as a simple and relatively cheap aircraft to replace the F-104 sometime in the mid-Seventies.

Franco/British efforts to entice the Federal Republic to join in two previous bilateral combat aircraft projects had met with failure. These were the AFVG project (Anglo-French VG aircraft), born in 1965, and the Jaguar ground attack plane. France abandoned the AFVG project on 29 June 1967 due to budgetary constraints, and in December 1968 the Federal Republic let it be known that the NKF concept took priority over other programs. On the recommendation of *General* Johannes Steinhoff, since 12 September 1966 the *Inspekteur der Luftwaffe*, a "Successor Model F-104" working group was formed from within the F-104 consortium, to which belonged Belgium, the Federal Republic, Italy, and the Netherlands. Probing the British, French, and Canadians for interest in a cooperative project along the lines of the NKF led to Britain and Germany drawing closer together, as the United Kingdom now viewed its security as coming from Europe. This followed on the heels of the failed German/American AVS project and the British government's decision to pull back from the area "east of Suez" by no later than 1971. Another victim of the change in policy was the TSR 2 (TSR = tactical strike reconnaissance) tactical and strategic low level bomber and reconnaissance platform replacement for the British fleet of V-bombers. The TSR 2 had been designed for conventional and nuclear missions and flew for the first time on 27 September 1964. The Anglo-French AFVG project was to have replaced the TSR 2, but the French—as mentioned earlier—withdrew from the program in mid 1967.

In this situation Great Britain was faced with the question of whether it would be possible to continue with a VG project working in cooperation with other European countries, since—by going it alone—such a technologically advanced project would undoubtedly tax Great Britain's economic resources to the breaking point. The British shift toward non-French Europe was accelerated by the government's cost-cutting programs, which blocked the acquisition of 50 American F-111 bombers in early 1968. For the time being, Great Britain's efforts for procuring an advanced medium range bomber with good all-weather capability for the RAF seem to have been unsuccessful.

On 5 March 1968, following a conference by the air force chiefs of Belgium, the Federal Republic, Italy, Canada, and the Netherlands, a working group was formed out of these five countries with the intent of formulating common goals for an F-104 successor, with the general basis of Germany's NKF concept being used as a springboard for working out the basic data for an MRA-75 (multi-role aircraft for 1975). This "Hornet Working Group" attempted to harmonize the ideas of the Canadians and Dutch, who were intent on a high speed interceptor, with those of the German NKF concept tailored to close air support of the Heer, and thus work out common "operational equipment objectives" (OEO). By broadly expanding the original NKF concept the working group submitted its proposals for common military objectives in April 1968. These OEOs, submitted to the aviation industry's "Arb*eitsgemeinschaft Neues Kampfflugzeug*" on 22 April, led to a concept proposal in early May 1968 which called for the MRA-75 to be twice as heavy—and accordingly more expensive—as specified in the OEO. The industry was unwilling to reduce the takeoff weight (limited to 15 metric tons) and the maximum price (set at DM10 million in 1968 marks) by picking and choosing from the OEO's specifications, especially since the military of the five countries was unable to come to a consensus on the matter in its joint working group. This seemed to have worsened Germany's prospects for maintaining its original NKF requirements in an acceptable manner even further, especially since Great Britain's accession to the joint working group on 1 June 1968 shifted the focus markedly. In the eyes of the British, the joint aircraft should be a complex, medium bomber capable of operating at extremely low altitudes with optimum penetration for conventional and nuclear weapons delivery. Furthermore, it should be a two-seater, equipped with state-of-the-art avionics enabling it to successfully carry out high-risk operations deep in enemy territory. This expanded working group now faced the difficult task of developing common "operational equipment requirements" (OER); also, the difference in introduction dates had to be ironed out, since the Federal Republic required the new aircraft by the middle of the 1970s and the British did not need the type until after 1980. In spite of these problems, a memorandum of understanding (MoU) was signed in Rome on 17 June 1968 calling for the joint definition, development, and production of the MRCA. This did not meet the original German intentions of a fundamental agreement for the overall program, but it did initiate a phase 1, the so-called concept phase for the MRCA program (here identified as "multi-role combat aircraft"/MRCA for the first time) during the period from 1 July through 31 December 1968.

Turbulence

The Belgians and the Canadians did not want to ally themselves with the program in July 1968 and, following a two-month "wait and see" period, left the project altogether. With the differences of the four remaining parties still quite varied, the MRCA's conceptual development distanced itself even further from the Luftwaffe's NKF vision. During a visit by the *Inspekteur der Luftwaffe* on 16 August 1968, industrial representatives (led by the German MRCA project manager *Dipl.-Ing.* Helmut Langfelder of MBB) announced that the MRCA would be heavier and more complex and the NKF's price proposal would be exceeded by at least 75%.* As a result, *General* Steinhoff and his command staff searched for various alternatives. He stressed the need for the MRCA's good combat capability, evaluated the Swedish SAAB 37 Viggen with its canard wing principle, and looked at the American F-14 and F-15 air superiority fighters. He specified the exact minimum values needed for the MRCA to maintain air superiority based on the data available for the Soviet MiG-23 FLOGGER and, in the end, joined with the British in accepting a French government offer in February 1969 for trinational production of the Mirage G. This followed a related proposal a few months earlier made by the Avion Marcel Dassault company to Messerschmitt-Bölkow. A demonstration of the Mirage G in late February 1969 in France for members of the *Bundestag's* defense committee went so well, and after seeing this advanced swing-wing aircraft fly many officials felt that the old NKF was "dead." Despite agreeing in principle, the British joined in with an international feeling among the three largest national development and manufacturing houses in Europe that this French proposal was nothing more than an attempt to undermine the Anglo-German MRCA project. Instead, Britain decided to torpedo the Mirage G project, seen as being under French control and without any binding internal French need for the type. Germany, too, was somewhat skeptical of France's intentions and, following a meeting between the two defense ministers Pierre Messmer and Gerhard Schröder on 1 May 1969 (General de Gaulle had resigned two days earlier as president of the French Republic following a people's referendum), decided to take up the joint development of a jet trainer with France. With regard to the MRCA, Germany would renew contacts with France on the matter if the latter country decided to become involved in the program. This in effect swept both the Dassault initiative as well as France's involvement in the MRCA project from the table, but it did point the way for the joint development of the future Alpha Jet close air support airplane.

Nor did the Americans remain inactive on the matter of developing a purely European combat aircraft, and may have assumed, with some measure of confidence, that they too would be involved in any successor to the American F-104 generation. After the German-American AVS project was abandoned in March 1968 they had no real successor to throw into the ring, instead contracting with McDonnell Douglas in January 1969 for working out in short order a project study for a simplified, "unrefined," and cheaper version of the F-4E Phantom with an NKF look, but one which was hardly able to match the NKF's weight and price requirements. Even the American aviation industry brought pressure to bear in the form of the Northrop P-530 Cobra, a design tailored to the ground attack role and a simple but better performing fighter successor to its F-5 Freedom Fighter. It was also the forerunner of the F/A-18. However, neither this nor ministerial involvement was able to sway the Federal Republic's generally positive opinion of the MRCA. Some of the American pressure had nevertheless been neutralized earlier by the purchase of 88 RF-4E recce Phantoms in late October 1968 and the delivery of an additional 175 F-4Fs beginning in May 1973. These purchases were made, in part, to offset the cost of stationing American troops in the Bundesrepub*lik* and were the result of a currency compensation agreement.

There were also problems on the part of those companies involved in the MRCA program itself. On 23 August 1968 these companies (British Aircraft Corporation [BAC], Canadair, Fokker, Messerschmitt-Bölkow, and SABCA Belgium) met with Fokker in Amsterdam-Schiphol. Items on the agenda included questions about work distribution and the creation of an industrial management structure, with British and German objections causing particular concern and requiring over six months of difficult negotiations before being resolved. The Chief of the L*uftwaffe* was also pushing for the concept phase to end as planned by the end of 1968, and for the BAC and Messerschmitt-Bölkow companies to submit a joint execution study by that time. In order to achieve this, in October 1968 at Fürstenfeldbruck the *Inspekteur der Luftwaffe* was compelled to

---

* Mechtersheimer MRCA, p. 60-62.

make concessions to his project partners with regard to weight limits and several key data. In practice, the difficulties the military working group had in coming up with common operational equipment requirements (OER) were foisted off onto the aircraft designers. Even so, by the end of 1968 there were still no industrial proposals forthcoming.

General Agreement - Who Picks Up the Tab?

During February 1969 several weeks of cooperative work between German, Dutch, and Italian design engineers in conjunction with their British counterparts at BAC's Warton plant finally resulted in an agreement being hammered out. On 14 March the concept phase drew to a close with the issuance of a so-called "baseline configuration brochure," which outlined the configuration of the MRCA in very general terms. Here, too, there were several options differing with regard to the powerplant and the potential operational roles.

The two engines planned for the MRCA were to be the newly developed Rolls Royce RB 199, but at Germany's insistence the option was kept open for the installation of an available American powerplant. As a result, Rolls Royce presented an even newer version, which at the time only existed on paper. This was the RB 199-34R with improved thrust, documented in an Annex B dated 28 March 1969. Known from this point on as the "Annex B Aircraft," the MRCA was a swing-wing aircraft optimized for the long range fighter-bomber role (interdiction/strike = IDS) with a two-man crew and designated the PANAVIA 200. A simplified lighter single seat version for the German Luftwa*ffe* went by the designation of PANAVIA 100. There was approximately 85% commonality between the PANAVIA 100 and the 200, and the performance of both was almost identical.

Also in March 1969—before the Franco-German decision for the Alpha Jet as a ground attack aircraft had been made—the Inspe*kteur der Luftwaffe* specifically pleaded with the defense committee, asking that the F-104's successor mainly have the capability of establishing air superiority. Along with this, he wanted the type to be able to provide ground support to include interdiction, as well as being able to meet the MiG-23 FLOGGER on its own terms. But could such a role spectrum ever be covered by a single aircraft type?

Steinhoff's efforts for improving the MRCA's air combat capabilities found an ally in his British partner, Air Chief Marshall Sir John Grandy, who had hoped to limit the type's maximum speed to Mach 1.5 for his RAF. It was based on this that *General* Steinhoff gave his approval for continuation of the MRCA program in mid-April 1969. On 22 April defense ministers Schröder and Denis Healy therefore authorized the work on the joint combat aircraft program to continue, with both parties estimating the type to enter service by the mid-Seventies. Apparently, though, BAC chose to place little weight on the RAF leadership's efforts to limit speed in favor of improved agility, instead envisioning changes to the concept to be made at the expense of its maneuverability. To be sure, Steinhoff protested energetically against these changes, but ultimately the success or failure of the cooperative effort with the British was at the forefront of politics, and in Bonn on 14 May 1969 the defense ministers of the Federal Republic, Great Britain, and Italy signed a governmental agreement for the entire program, as well as an MoU for a definition phase retroactive to 1 May 1969, limited to a year (i.e. 30 April 1970). The Dutch were not signatories to this agreement since the defense minister of the Netherlands had bowed out from the program on 29 April 1969 due to insurmountable disagreements with the conceptual design of the MRCA.

A Large-Scale Project Is Organized

The government-level agreements initially stipulated the multirole combat aircraft project to be divided as follows:
- 600 examples for the Federal Republic (480 for the Luft*waffe* and 120 for the *Marineflieger*)
- 385 for the Royal Air Force
- 200 for the Italian Air Force

Several institutions/organizations were established in order to achieve what was becoming the largest international aircraft project to date, which ultimately would involve over 500 British, German, Italian, and even American companies. These organizations, both governmental and military, were to become responsible for the necessary requests for tender, contract issuances, program control and monitoring, as well as the overall coordination of the program. They included the following:

- NAMMO = NATO MRCA Management Organization, an international management organization consisting of a board of directors from the three governments (those with overall responsibility for the MRCA program) and its executive organization, the
- NAMMA = NATO MRCA Management Agency (successor to the former tri-national governmental management organization IMO) headquartered in Munich. NAMMA is the contractor and, among other things, reviews development and production and is responsible for bookkeeping and accounting. The first general manager was *Generalmajor* Horst Krüger (*Bundesluftwaffe*), fol-

lowed by *Generalmajors Dipl.-Ing.* Birkenbeil and Obleser, a later *Inspekteur der Luftwaffe*, and minister director Hans Ambos; *Generalmajor* Gülzow became general manager in 1984, and from 1989 until 1993 was, as the manager of this NATO agency, the replacement for minister director Hans Rühle, the director of the planning staff within the BMVg.

- PANAVIA Aircraft G.m.b.H., founded on 26 March 1969 with its seat in Munich. This organization is the system director, industrial coordinator, and prime contractor for the airframe, and consists of three companies whose percentage of involvement is broken down as follows:
  - 42.5% BAC = British Aircraft Corporation, later British Aerospace (BAe)
  - 42.5% Messerschmitt-Bölkow, later Messerschmitt-Bölkow-Blohm (MBB) G.m.b.H., Aircraft Division (formerly, *Entwicklungsring Süd*, which transferred to MBB in 1969)
  - 15.0% Aeritalia S.p.A.

  First managing director was Prof. *Dipl.-Ing.* Gero Madelung from *Entwicklungsring Süd*, followed in 1978 by *Dr.- Ing.* Carl Peter Fichtmüller, then by Hans-Joachim Klapperich in 1982, who for many years had been the vice managing director and director for finances and contracting.

- TURBO-UNION Limited, founded in early October 1969 in London with a branch in Munich, as the main contractor for powerplants, consisting of the following companies and contractors, broken down as follows:
  - 40% RR = Rolls Royce in Filton/Bristol, which assumed responsibility for the system management
  - 40% MTU = Motoren- und Turbinen-Union in Munich
  - 20% FIAT in Turin

- AVIONICA Systems Engineering G.m.b.H., established on 28 August 1969 in Munich, became responsible for the quite costly on-board electronics systems of the MRCA and consists of the following electronics systems companies:
  - ESG = Elektronik-System-Gesellschaft m.b.H. (= Siemens, AEG-Telefunken, Standard Elektrik Lorenz [SEL], Rhode & Schwarz)
  - EASAMS = Elliot Automotive Space and Advance Military Systems Limited.
  - SIA = Società Italiana Avionica, not founded until 14 November 1969 by about ten Italian electronics firms; when AVIONICA was founded only two of SIA's startup companies were involved.

However, AVIONICA was dissolved in June 1970 without ever having fully assumed its responsibilities. This means that, in place of a single multinational subcontractor responsible to PANAVIA for the overall avionics, the three individual national avionics con-

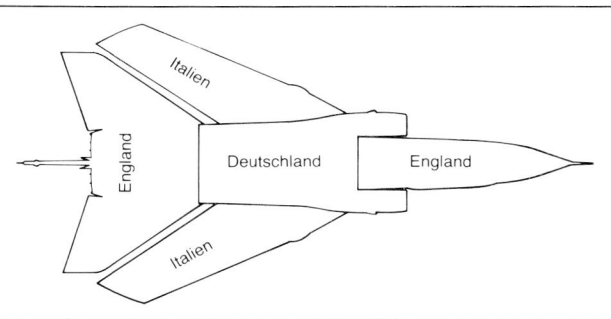

Panavia Tornado - breakdown of development and manufacture among the countries involved in the program.

sortiums became involved within PANAVIA immediately and are each represented by their national aircraft construction companies.

- IWKA = Industriewerke Karlsruhe Augsburg is the contractor for the Mauser-Werke in Oberndorf subcontractor, responsible for the BK 27 on-board cannon.

Eventually, about 500 companies were contracted with to supply parts for the program—almost seven times more than for the Alpha Jet project, giving an idea of the difference in scale between the two programs. NAMMA and PANAVIA are responsible for their selection and supervision of contracts.

The Definition Report and More Stumbling Blocks

In early February 1970 PANAVIA submitted the definition phase report, comprising over 18,000 pages, to the three governments for review. The report confirmed the similarities between the single seat PANAVIA 100 with the two-seat PANAVIA 200 with regard to performance, weight, and cost, with differences fluctuating at about the 5% margin. The previous intent to develop both versions was dropped, which for the Germans meant that they would generally be without the close air support (CAS) and air superiority (AS) roles against an enemy's tactical fighters. This shift to a clear accent on the interdiction role virtually invalidated the earlier NKF concepts and requirements, but it did alleviate the conceptual differences between the Luftw*affe* and *Marineflieger* needs.

On 24 March 1970, therefore, the newly appointed defense minister Helmut Schmidt officially notified Denis Healey that Germany would not be purchasing the single seat version. This consequently led to a reduction in the number of aircraft Germany required, dropping from an original 600 to

- 322 examples for the *Luftwaffe* (reduced from an original 480), and
- 122 examples for the *Marineflieger* (increased from an original 120).

These numbers meant that the *Luftwaffe* would be unable to use this expensive, high-tech plane to fly the CAS role for the *Heer*—except perhaps in emergency situations. It was around this time—on 23 July 1970, to be precise—that the close air support role was assigned to another joint venture program with the award of a contract to Dornier and Dassault-Bréguet for the joint Franco-German TA 501 project, later to become the Alpha Jet.

On 30 April 1970 the Italians also agreed to the two-seat version of the MRCA, albeit somewhat reluctantly. After a review of the definition report, according to plan the definition phase was to have begun on 1 May 1970, but the decision of 24 March 1970 necessitated a reworking of the report to include limiting the MRCA exclusively to the two-seat version and incorporating the swing-wing principle. Detailed calculations and research by the design team showed that a fixed wing aircraft with the same performance would have been larger, heavier, more difficult to fly, and would have had a much greater fuel consumption rate.

In the meantime, there were political hurdles to overcome, such as the June 18th 1970 change of government in the British lower house in the midst of Germany's parliamentary approval process. This led to a renewed proposal on 3/4 July 1970 by France's defense minister Michel Debré for a joint Franco-German program on the basis of the Mirage G8, this time with a more concrete foundation than the earlier attempt to make the Mirage G palatable. The Germans and Brits, however, stood by their cooperative venture, which finally led to a signing of an MoU on 22 July 1970—almost three months after the targeted 1 May—for the first developmental stage up to and including maiden flight. This MoU obligated the Federal Republic to pay its portion, amounting to over one billion Deut*schmark* (cost in 1970), retroactively from 1 May 1970 to 30 September 1974. Not only did this in effect mean that the aircraft's development was a done deal, it also set the mold for its production, even when the magnitude of such a decision and a governmental crisis following concession to a determination deadline precluded Italy from joining the MoU until 29 September 1970. At this point, the trilateral development of the MRCA was secured.

Development, Maiden Flight, and Production

The developmental phase was to have terminated on 30 September 1974 with the first flight of the MRCA prototype. In Germany this time span was punctuated by and heated, often emotional and controversially publicized, parliamentary and political discussions. The entire MRCA program was seriously questioned as a result of its almost immediate cost increases, which were not inconsiderable in their rise. Without delving further into the pros and cons of these discussions, it should be mentioned that a good portion of these arguments suffered from a questionable knowledge of the subject matter.

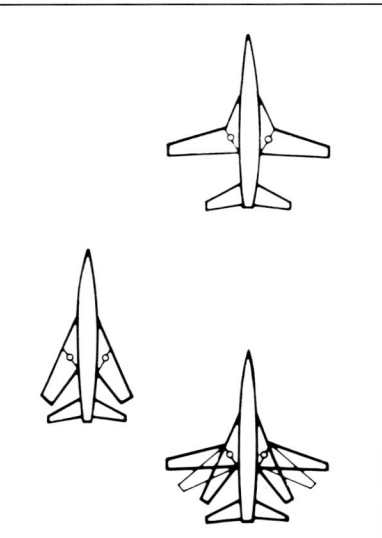

- Höchstauftrieb für Kurzstart und -landung
- Niedrige Landegeschwindigkeit zur Erhöhung der Flugsicherheit
- Große Überführungsreichweiten
- Lange Warteflugzeiten
- Gute Wendigkeit im Unterschallflugbereich

- Kleinster Widerstand im schallnahen und Überschallflugbereich (große Reichweiten, Höchstgeschwindigkeit, Beschleunigung)
- Gute Manövrierfähigkeit bei Überschallflug in Bodennähe
- Höchstmöglicher Flugkomfort bei Schnellflug in böigem Wetter

- Optimale Anpassung an unterschiedliche Einsatzprofile
- Größte Flexibilität in der Benutzung

Panavia Tornado - explanation of the variable geometry features required by the MRCA and sketches showing the optimal sweep angle utilizing the swing-wing principle, which the pilot could set at any sweep angle.

History of German Aviation: Bombers and Reconnaissance Aircraft

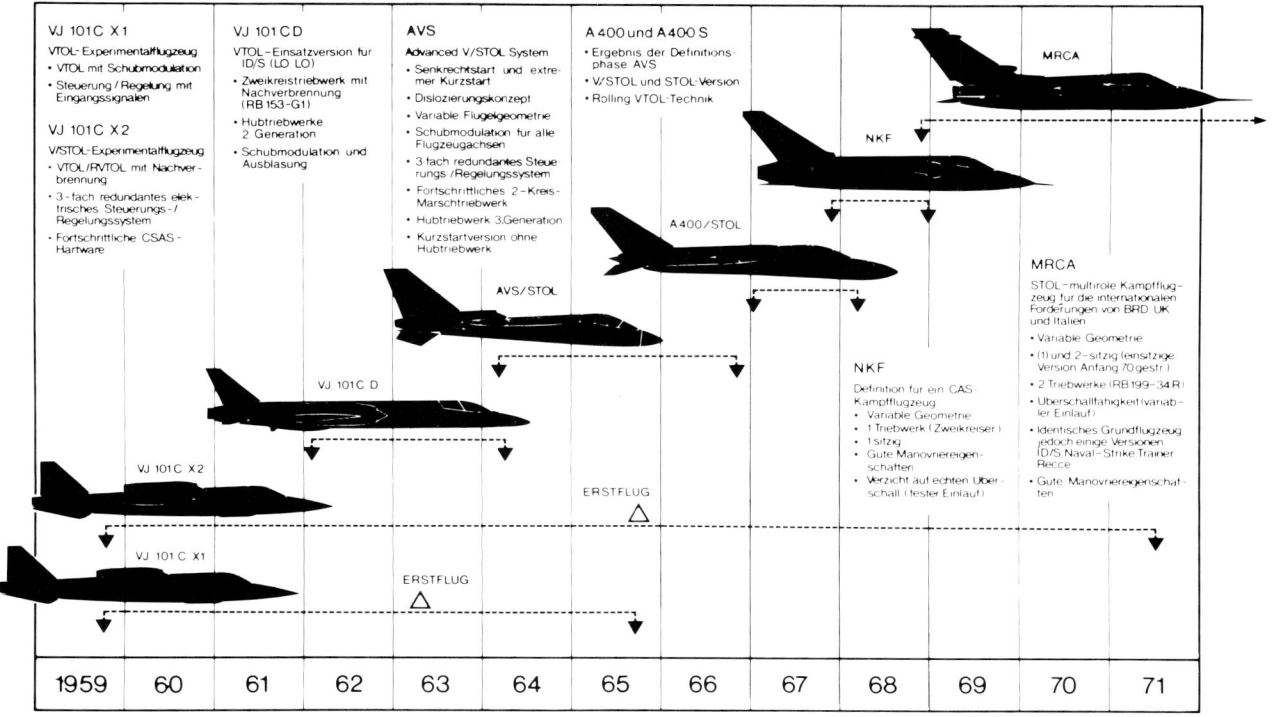

Genesis of the development leading to the MRCA.

Development and Layout

Development of aircraft began without much fanfare, and the designers at AERITALIA, BAC, and MBB took into account the following operational requirements for the weapons system, requirements which corresponded to the latest developments in aircraft construction. MRCA called for the airplane to:

- have the ability to deliver on target a sufficient quantity of conventional ammunition in all weather, including night and under foggy conditions, with the accuracy required of conventional weapons. This necessitated a high emphasis on avionics.
- have the ability to get through to the target in the face of enemy defenses. This meant a high transonic speed and automatic terrain following.
- be highly survivable on the ground, making use of Europe's defensive terrain. This referred to the aircraft's takeoff and landing capabilities and the ability to operate from short segments of runway, while at the same time having a short landing capability.
- also have these characteristics apply equally to the maritime air war, which additionally dictated that the aircraft be highly maneuverable, be equipped with stand off weapons for avoiding shipborne defenses, and be able to carry reconnaissance sensor packages.

In all areas, particular emphasis was placed on logistics, meaning:

- high functional and operational safety and short turn-around times with high operations tempo
- modular concept for accelerated switching of defective equipment and components, with a work height for all critical systems and refueling points no higher than 1.70 m
- ease of service, visual access to fitted monitoring devices, simplified and standardized plug-ins
- markings for particularly critical maintenance and repair points, e.g. special checking of the seals on the integral fuel tanks, and internal engine monitoring using bore scope technology (endos copy)
- dispensing with periodic maintenance, with a shift to "on condition maintenance" for planned maintenance in the area of not less than 300 flight hours. In the area of unplanned correction of problems, the requirements are: correction of 50% of the problems within 45 minutes on the aircraft, with the overall ratio of

this type of maintenance corresponding to no more than seven direct man hours per flight hour.
- in order to achieve maximum flexibility with the weapons system, planned maintenance/checking on the flight line is specified and defined as follows:
  - no more than two hours man hours required for postflight inspection
  - sustaining the postflight inspection through 24 hours by preflight inspections, wherein the time for an external preflight inspection not including fueling/refueling is not to exceed more than five minutes
- particular stress is placed on a ratio of 35 direct man hours per flight hour, with 25 flight hours being standardized per month at an average duration of one hour per sortie.
- in order to comply with a high readiness state for the weapons system, vertical removal and installation of the engines were specified, capable of being carried out by three crewmen in three hours

All these requirements and the definitive configuration are reflected in the design and technological layout of the MRCA, a relatively compact design despite these many requirements.

The most important features distinguishing the MRCA are as follows:

> **Airframe**
> The **airframe** is of integral construction with a "safe-life" layout. This means high structural redundancy despite large openings in the structure for easy access to components, with numerous access panels mostly fitted with speed screws. Design life is 16,000 flight hours, with standard design life for fighters normally set at 12,000 hours. Having a quadruple safety factor, the MRCA's users estimate 4,000 flight hours. The wing pivot box for the swivel wings, necessitated by the requirement for transonic low-level speeds as well as short takeoffs and landings, has been manufactured of titanium using electrobeam welding techniques.

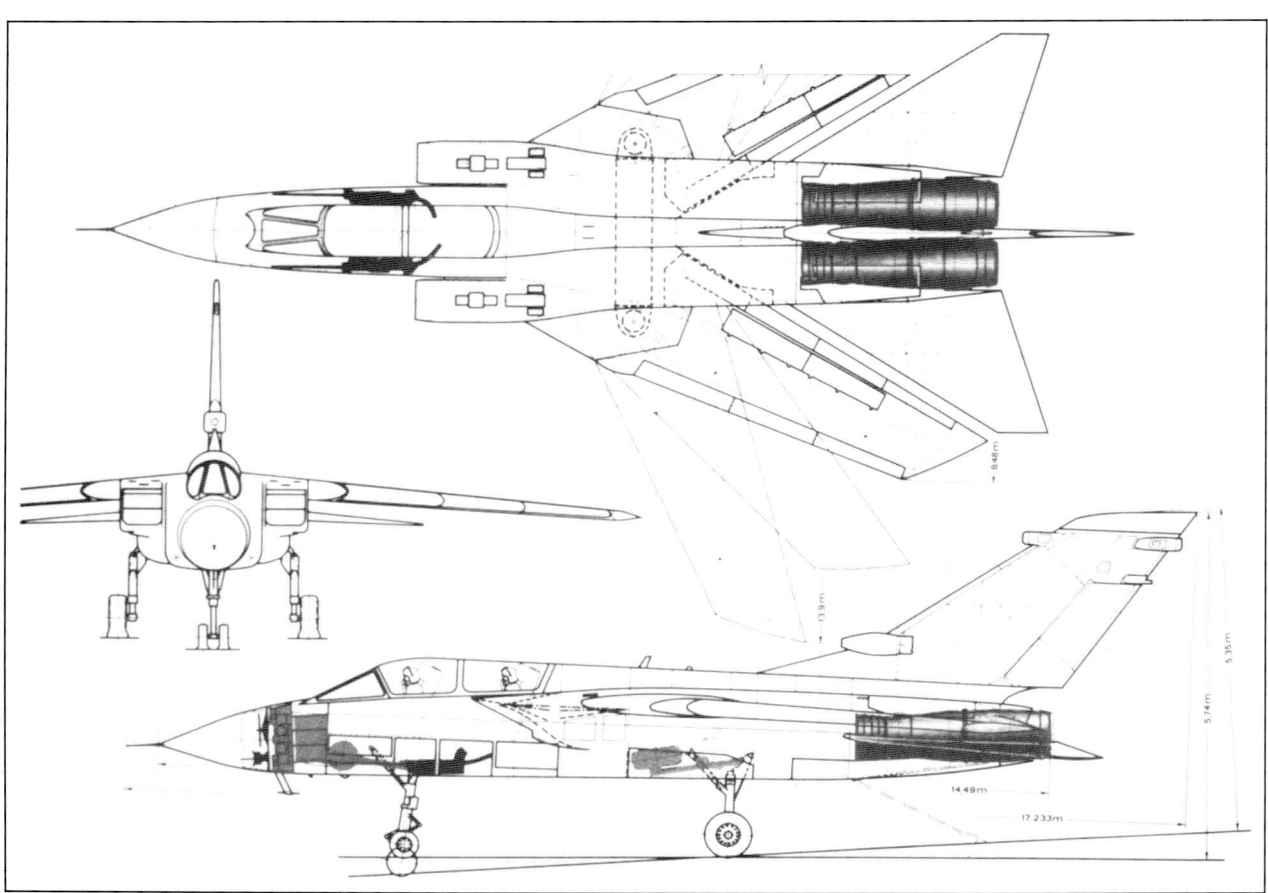

Three-view of the MRCA.

The **flight systems**, with swing wings, tailerons (all-moving tailplanes) for control in the pitch and roll axes (= lateral and longitudinal axes), and conventional vertical stabilizer are enhanced by spoilers for improved roll control at minimum wing sweep. The flight control system is computer controlled for automatic control and stabilization of the rate of movement in all three axes. All control commands are transmitted electronically via a self-monitoring triple-redundant fly-by-wire system. In the event of an emergency, the aircraft can also be controlled mechanically.

Secondary control systems include the four flaps, three slats, as well as Krueger flaps on the non-moving part of the wing. The flaps are computer activated depending on the given wing setting. The flight control system offers the MRCA a high degree of stability and insensitivity to ground gusts up to Mach 0.9 and at altitudes below 60 m/ 200 ft. Thus, the aircraft had become an outstanding weapons platform for low-level attack. The high-lift features not only contribute to short takeoff and landing runs, but are also a critical factor for meeting the requirement for high maneuverability.

The **auxiliary power unit** with integral motor ensures that the requirement for general independence of ground equipment on the flight line has been met. It consists of a small gas engine attached to the auxiliary power unit of the starboard engine and two motors for starting the engines on the ground, as well as for driving the electrical generators and the hydraulic and fuel pumps during maintenance. Electric fans have been installed in the cooling circuit to keep the avionics at cool temperatures when the engines are not running. This makes the aircraft not dependent on external cooling sources.

Electrical supply is accomplished by means of two brushless constant frequency synchronous generators, each with an output of 40 kVa. This offers a sufficient growth potential for additional electronic components.

Weapons and armament of the aircraft suit the multi-role operational spectrum of the MRCA. A so-called stores management system has been designed on-board, which covers both current and newly developing weapons systems, such as the MW 1 multi-purpose weapon (discussed later). Comprehensive studies and trials led to the selection of a 27 mm caliber for both on-board cannons. The Mauser company was the winner in the selection process, beating out British and Swiss competitors with its BK 27 (also installed in the Alpha Jet).

Powerplant

The demands made on the weapons platform are also reflected in the specific requirements for engines, which are met by the Rolls Royce RB 199 engine as follows:

Specific requirement	Design feature
- high thrust/weight ratio	- three-spool design
- small dimensions	- compact, integral afterburner
- high supersonic thrust realm	- new materials/production methods
- brief max. thrust in transonic	- high compression ratio and high turbine intake temperature
- low fuel consumption at cruise	- design as a bypass engine with electronic performance control

Minimal down time for maintenance and repair is achieved by using a modular design for the basic engine, sensibly arranging the fittings so that they can be swapped out with the engine still installed, and applying preventive checks using endoscopy, oil analysis, and electronic monitoring. The requirement for rapid removal and installation of the engines has been met in practice. The technologically advanced use of thrust reversing not only shortens the landing roll-out significantly, it also has a positive effect on the turnaround time by dispensing with the need to repack and install a parabrake (such as on the F-104, for example).

Avionics

The MRCA avionics system reveals a number of innovations as a result of its requirements both from a technological standpoint, as well as in its system design. Of especial interest are the use of digital data processing and transfer technology, the high degree of integration of the individual systems, the widespread use of self-checking systems, and a miniaturization of components corresponding to the latest state of technology. These systems fulfill the following tasks:

- precision navigation, independent of ground stations, by day and night and in any type of weather
- accurate target acquisition, identification, and engagement through the use of optimized attack methods and in poor weather conditions
- option of manual or automatic terrain following control at extremely low altitudes
- protection against enemy defense systems through the use of electronic protection and combat systems
- optimal communications
- self-checking avionics system

The heart of the entire avionics package is the central computer. It consists of a freely programmable, digital on-board calculator and associated analog and digital converters, the so-called interface units. By entering operational flight data, course headings and strike data can be pre-programmed into the system. As a rule, flight data is entered using the voice recorder; data can be accessed via a keypad by the weapons systems officer during flight. The cassettes are programmed by means of a "cassette preparation ground station" (CPGS), a ground-based device which significantly eases and speeds up flight preparations, as well as assists in avoiding mistakes in flight preparation and execution. This system corresponds in principle to the "automatic flight data processing" (AFA) system described in the section on the Alpha Jet.

Despite all these irritations the companies involved in the MRCA concentrated their developmental efforts on the program over the following years. Governmentally set milestones were established as interim checks on cost, aircraft performance, technical development, and work distribution; if these were not met any country was free to leave the program without having to compensate the other partners. Confronted by difficulties on all sides, these were more or less met, albeit with much trouble and beset with delays, such as the bankruptcy of Rolls Royce and the company's reorganization under the former head of the Royal Air Force, Air Chief Marshal Sir Denis Spotswood. Engine development initially lagged behind the intended goals; there were problems during turbine test runs, but these were eventually ironed out. The anticipated first flight of the MRCA by no later than September 1974 was postponed. An optimistic MRCA report by the director of the BMVg's armament division, minister director *Dipl.-Ing.* Albert Wahl, and the new inspector of the air force, *Generalleutnant* Gerhard Limberg, renewed hopes in late June that the project would indeed be completed on time. Limberg had replaced *Generalleutnant* Günther Rall (inspector from 1 January 1971 to 31 March 1974) on 1 April 1974 and had been closely involved with the MRCA project since its birth.

The Tornado Flies

By 14 August 1974 the time had come. MRCA prototype P 01 took to the air for the first time at Manching, near Ingolstadt. To be sure, the engines were not yet capable of operating at full thrust, but the much criticized bird flew nonetheless! Pilot was BAE chief test pilot Paul Millet, with MBB test pilot Nils Meister acting as his navigator. Five weeks later MBB chief test pilot Hans Friedrich "Fred" Rammensee flew the first official display and demonstration of the P 01 at Manching, conducted before a full house of prominent figures and press officials. Defense minister Georg Leber expressed himself optimistically, "There is certainly no other aircraft which fills the role as well as the MRCA and does it as cheaply."

Yet the MRCA was still not a weapons system, but only a well designed high performance aircraft whose swing wing technology gave it impressive flight handling characteristics, remarkable even with engines incapable of running at full power.

The skeptical and often mocking critics in the press over the last few years had melted away, and for the opponents of the project, the MRCA's successful first flight took the wind out of their sails.

On 11 October 1974 the three governments obligated themselves in three MoUs to support developmental phase 3B, in which flight testing would be conducted and the manufacture of components for series production would begin. A short time later, on 30 October 1974, MRCA P 02 took to the skies for the first time over Warton, with Paul Millet again at the controls. This time, though, his navigator was Italian test pilot Pietro Paulo Trevisan, who later flew the P 05 on its maiden flight on 5 December 1975 at Caselle, near Turin. In the interim, P 03 had flown on 5 August 1975 at Warton and P 04 on 2 September 1975 at Manching. MBB pilot Fred Rammensee was at the controls, with Nils Meister again as navigator; later Meister would be the test pilot for MRCA P 07's first flight on 30 March 1976 at Manching. The total of nine prototypes rolled off the assembly lines with regularity and were assigned to flight testing, which was planned at over 5,000 flying hours. The tenth machine was pulled off for static tests, and with aircraft number 11 began the first pre-production batch of six aircraft. Number 11 first flew on 5 February 1977, again with Rammensee at the controls at Manching, but with Kurt Schreiber in the second cockpit of this dual-controlled conversion trainer. There followed number 13 with Friedrich Soos, test pilot of *Erprobungsstelle* 61 Manching, and Rainer Henke on 10 January 1978, and as the last "German" aircraft of the pre-production run, number 16 with Armin Krauthann and Fritz Eckert on 26 March 1979. This latter machine was fitted with a production forward fuselage section. In the meantime, Warton and Caselle also saw the remaining prototypes and pre-production models complete their initial flights and enter the testing program as planned.

Production Gets Underway

A change in government in Great Britain in 1974, cost cutting measures by Harold Wilson's government, technical problems with the engines, and strikes in England all led to major delays, and by the milestone of 1 May 1975 only 50% of the intended flight testing program had been completed. Two prototypes were out of commission for awhile—a bird strike at Warton damaged the P 02's engine, and P 05 made a hard landing at Caselle, causing considerable damage and leading to toilsome repair work. Nevertheless, flight testing continued at a relatively smooth pace, and the British stuck to their special air defense version of the MRCA, the ADV (= Air Defence Variant). Britain had determined that 165 out of a total of 385 MRCAs for the RAF would be these ADVs, tailored for the long range interception role and built at Britain's own expense. On 9 August 1979 the first ADV prototype rolled off the assembly line as the Tornado F-2.

On 7 April 1976 the German cabinet also approved further production of the Tornado, as the MRCA became officially known following the acceptance of series production. Flight testing at this time was not even close to completion, however. The aircraft's empty weight had increased by about 10% and fuel consumption rate by 7%, mainly caused by additional components, such as the automatic terrain following system. In spite of this, on 19 May 1976 the defense committee approved the procurement proposal, followed by the unanimous approval of the B*undestag's* budget committee on 2 June. On 29 July the agreement for series production was signed. That same month the *Inspekteur der Luftwaffe*, *Generallleutnant* Gerhard Limberg, flew the Tornado for the first time and found that his expectations he had harbored for the success of the MRCA program had been exceeded: "The MRCA program is a complete success!"

The first batch of 40 aircraft, agreed upon on 29 July 1976, was followed by the signing for a second batch on 31 May 1977 in Munich, calling for another 110 Tornados, and a third batch in February 1979 for 164 aircraft. The fourth batch, also comprising 164 machines, was agreed upon in August 1980. Series production was truly underway. By mid 1979, with the first flights of production aircraft for the RAF and the *Bundeswehr* (the latter receiving Tornado GT 001 on 6 June 1979), the 15 prototypes and pre-production versions had completed a good half of their 5,000 hour test program. The first flight of Italy's production aircraft took place in the latter half of 1981.

On 2 July 1982 the *Marineflieger* became the first operational unit within the *Bundeswehr* to begin flying the Tornado, with the *Luftwaffe's* first wing becoming operational in the summer of 1983. Flying units began converting about 12 years after the Tornado's developmental phase had begun in May 1970, and from 1982 onward the type gradually began replacing the F-104G. The Starfighter's last operational mission with the *Luftwaffe* took place on 23 October 1987. Around the time the first units took initial delivery of the type—on 31 March 1982, to be more precise—the prototypes, pre-production aircraft, and those production versions in flight testing had together logged 14,000 flying hours and garnered a considerable amount of flight experience. This experience found a home in the Trinational Tornado Training Establishment (TTTE), a joint British/German/Italian training center located in Cottesmore, Great Britain, and founded on 31 March 1982. At the TTTE Tornado crews are put through a 13-week course where they are schooled in the theoretical fundamentals of the aircraft (over the first four weeks), followed by a good 35 flying hours for the pilot converting over to the new type (during the next nine weeks). After about 15 flights in the front seat there follows the test flight, with issuance of the type certification to the pilot. Subsequent to this is combined training with the navigator/observer, who logs a total of 27 flying hours. After this standardized training the graduates of the TTTE return to their respective national weapons training facilities, where follow-on tactical training with weapons under operational conditions takes place. In the Federal Republic this occurs at the so-called Tornado WaKo (from *Waffenausbildungskomponente*, weapons training component), which was first activated at Erding on 16 February 1982 and received the initial 15 Tornados. It was the facility which trained the first unit crews for the *Marineflieger* after their completion of the TTTE course. In all, an outstanding incremental conversion training course which also contributes to commonality of flight performance and interoperability of the four Tornado users in Europe.

System Leaders

In the execution of the mission of NAMMO/NAMMA, it was the system leaders who played a critical role in the technical management of their respective companies and thus wielded the greatest influence on the development, startup, and follow-through of production during the decisive years. This was the triple team of B. O. Heath at British Aerospace, R. A. Mautino at AERITALIA, and Dipl.-*Ing.* Helmut Langfelder and *Dr.-Ing.* Otto-Ernst Pabst at MBB. Prof. Dipl.-Ing. Gero Madelung, PANAVIA's first general director from 1969 to 1978, played a central role in not only ensuring the type's technical beginnings were coordinated and economically handled, but also brought his influence to bear to see it through. The same applied to his successor *Dr.-Ing.* Carl Peter Fichtmüller. Both were supported in all their economical and financial negotiations by Hans-Joachim Klapperich, who in 1982 succeeded Fichtmüller as PANAVIA's general manager. These are the individuals, too, who made sure that the Tornado was technologically capable of meeting what was required of it, as evidenced by the graphical chart on the following page. This chart shows an assessment of problems with the airframe, engine, general equipment, and avionics in the time frame from January 1981 to June 1982, during which 15,000 flying hours were logged. The Tornado has indeed become "manageable" for the front-line units operating it!

The Tornado and Operational Doctrine

Much has been discussed—and dissected—regarding just how well the Tornado fits into Germany's operational vision. The heated protagonists of the position that the Tornado's introduction was apparently a mistake seem to take the viewpoint that there will be a scenario involving a threat to national interests, which would re-

Fault overview for 15000 Tornado flying hours in 1981/1982.

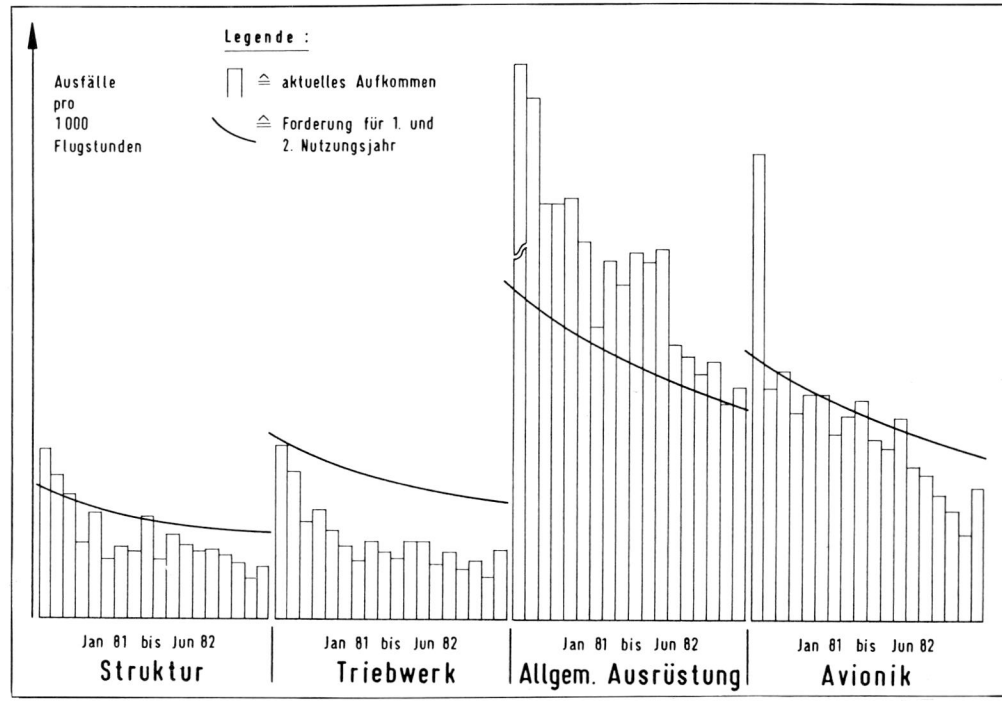

quire a national response. They utterly fail to comprehend the alliance's interrelationship and the allied command function in which the Germans play a major role in the event of a military conflict. When on the defense, an allied air force commander would have at his disposal the entire complement of every country's air assets without regard to national origin in order to be able to meet his defensive role as tasked by the NATO governments. Naturally, a British Harrier, an American A-10, or a German Alpha Jet is more suited to the close air support role on the battlefield. And it is more feasible to utilize an American, Belgian, Danish, or Dutch F-16 or German Phantom F-4F for air interdiction than a British or German Tornado, which only then would have been required to meet the NKF standard—assuming, of course, that there was no other allied aircraft available for these roles. As long as the WP retains the capability of attacking/invading and continues to pose a latent threat with its superior numbers of conventional forces, NATO must be prepared for the eventuality of a conventional attack by these WP forces. Its defenses not only include the immediate support of NATO ground troops, but also the suppression of the enemy's offensive air forces, as well as the prevention of follow-on forces from joining in the attack. These operational roles, which were briefly addressed at the outset, can be summed up as

- engaging enemy air forces (offensive counterair = OCA)
- interdiction operations (air interdiction = AIA) behind the forward battlefield area

and require an airplane capable of surviving well behind enemy lines in what has become the most concentrated air defense network in central Europe. Not only that, but this aircraft must also be able to deliver its weapons with precision accuracy in any weather, day or night. By availing itself of improved conventional weaponry, the current NATO strategy contributes to raising the nuclear threshold, which may one day lose its significance altogether if negotiations between the U.S. and the Soviet Union continue to enjoy the success of recent years. Until the day when such negotiations result in a mutual state of non-aggression between the WP and NATO, it therefore becomes all the more critical to optimize conventional forces capable of efficiently carrying out the follow-on forces attack (the FOFA concept). And it is precisely in this arena that the Tornado acts as the most suitable weapons system. It is the legacy of visionary politicians, soldiers, and designers to have seen the type through to operational maturity in the face of all the inner-German resistance to the idea. The Tornado is one of the most reliable guarantees for continued assurances of peace, for it is the only conventional weapons system that has the same quality of deterrence as the former nuclear deterrence. What, then, are the Tornado weapons system's particular qualities which make it so well suited to its role?

Unique Features of the Tornado and Its Weaponry

For an aircraft penetrating enemy airspace, the chances of surviving the enemy's air defense weapons multiply with increased speed and lower operating altitudes, combined with the use of electronic deception and/or countermeasures preventing acquisition by radar systems. The Tornado meets these needs by having
- excellent high speed handling qualities achieved by a smooth sweep of its wings from 25° to 67° (measured on the leading edge)
- reliable, high performance RB 199-34R dual-flow three-stage jet engines, for combat in the supersonic realm with high thrust (71.0 kN = 7,245 kp with afterburner) and low fuel consumption on long range low-level flights not requiring afterburner
- terrain following radar for automated flights hugging the ground profile at extremely low levels, independent of weather and light conditions
- electronic deception and combat sensors, leading to the neutralization of enemy radar acquisition

Not only that, but such an aircraft must be able to fly with pinpoint navigation independent of ground stations in order to accurately approach point targets over long distances. For this, the Tornado makes use of an avionics system based on its central computer. For navigational purposes the system employs four main sensors, as follows:
- Inertial navigation system
- Doppler radar
- Air data computer (ADC)
- Secondary attitude and heading reference (SAHR)

What, then, makes the high-performance TORNADO into such a highly efficient weapons platform?

The Weapons Systems

With its swing-wing features, its suitability for automatic flight at extremely low altitudes in any weather condition, and its precision navigation, the Tornado's combat potential has already been well illustrated. All these capabilities, however, are simply the foundation for its versatile weapons systems, of which but a few are highlighted below.

The **MW-1 dispenser** is an airborne, conventional multi-purpose weapons system for engaging ground targets at extremely low altitudes, and is the weapon of choice for German and Italian IDS Tornados. With its various types of sub-munitions the MW-1 is a highly effective piece of ordnance for neutralizing airfields, armored, and mechanized targets. Launched from the sides of the MW-1 canister, the sub-munitions are optimized for two main target groups (*Hauptzielgruppen*, or HZG) based on their effectiveness. These groups are

HZG I = armored and mechanized targets, against which are used the following types of sub-munitions:
- KB 44, a hollow-charge bomblet for attacking armored targets, with supplemental fragmentation effect
- MIFF, an armor mine with a highly refined sensor system
- MUSA, a fragmentation charge for attacking semi-hardened ground targets.

HZG II = airfields, attacked with
- STABO, a runway denial bomb for destroying paved surfaces. By using a stabilized drogue parachute it impacts the surface of the runway at an angle, penetrates, and explodes.
- MUSPA, a fragmentation mine for engaging aircraft taxiing, taking off, or landing.
- MIFF, as under HZG I
- MUSA, as under HZG I

The **BL 755** is a cluster bomb dispenser with small caliber fragmentation bomblets or other types of sub-munitions.

Kormoran and **Sea Eagle** are self-propelled standoff missiles with electro-optical and radar seeker heads, optimized for the anti-shipping role from extremely low levels (sea skimming). Sea Eagle is the British weapons system, whereas the Kormoran is the main weapon of the German *Marineflieger* (although the Italian Tornados also began carrying the type in 1983). The Kormoran cruises at about 1,100 km/h and has a range of about 30 km. The *Marineflieger* aircraft carry four Kormorans operationally, and became fully armed with the weapon in 1982.

HARM and **ALARM** (British) are anti-radiation missiles which are fired without the need for target detection by the Tornado's on-board sensors or its crew; instead, once launched they home in on the signals "radiating" from the target (e.g. anti-aircraft radar equipment). A new digital MIL BUS 1553 data bus effectively manages the interaction between the Tornado's computer, the weapons computer of the HARM (called the MCU = missile control unit), the launch decoder unit, and the missile itself. HARM is the primary weapon of the ECR Tornado, which will be covered in greater detail later.

Cerberus II (for German Tornados) and **Sky Shadow** (for British) are electronic warfare pods for active electronic jamming.

Chaff/Flare dispensers are containers for ejecting foil strips and jamming devices for self-protection against enemy radar acquisition and infra-red sensors.

History of German Aviation: Bombers and Reconnaissance Aircraft

1. Air data probe
2. Radome
3. Lightning deflector
4. Terrain-following radar antenna
5. Ground mapping radar equipment, antenna
6. Radar equipment, hinged module (opened)
7. Radome (opened position)
8. IFF antenna
9. Radar drive mechanism
10. Radar equipment bay
11. UHF/TACAN aerial
12. LASER rangefinder and designator (Ferranti), starboard
13. Cannon muzzle
14. Ventral nose Doppler antenna
15. Pitot tube
16. Canopy emergency release
17. Avionics equipment bay
18. Forward pressure bulkhead
19. Windscreen rain dispersal air duct
20. Front windscreen (Lucas-Rotax)
21. Retractable telescoping aerial refueling probe
22. Refueling probe retraction strut
23. Hinged windscreen panel for access to instruments
24. Head-up display (Smiths)
25. Instrument panel
26. Head-down display
27. Instrument panel shroud
28. Control column
29. Rudder pedals
30. Battery
31. Cannon barrel
32. Nose gear door
33. Landing/taxi light
34. Nose gear strut (Dowty-Rotol)
35. Pivoting torque link
36. Forward retracting dual tires (Dunlop)
37. Nose gear steering arm
38. Nose gear door
39. Electric equipment box
40. Ejection seat rocket propulsion unit
41. Throttles
42. Wing sweep setting lever
43. Manual radar control lever
44. Side console
45. Martin-Baker Mk 10 pilot ejection seat
46. Safety belt
47. Head support unit
48. Canopy (Kopperschmidt)
49. Brace for one-piece canopy
50. Navigator's radar display
51. Navigator's instrument panel and weapons operations equipment
52. Foot rests
53. External canopy control panel
54. Total pressure head
55. 27 mm caliber Mauser cannon
56. Ammunition feed chute
57. Heat exchanger air intake
58. Ammunition magazine
59. Liquid oxygen converter
60. Cockpit air conditioning system
61. Weapons selector system
62. Engine air intake
63. Intake lip
64. Cockpit bulkhead structure
65. Navigator's Martin-Baker Mk 10 ejection seat
66. Engine air intake, starboard
67. Air intake bypass
68. Canopy actuator
69. Canopy hinge
70. Aft pressure bulkhead
71. Intake ramp actuation arm
72. Navigation light
73. Two-dimensional adjustable intake ramp
74. Intake suction relief doors
75. Wing glove Krueger flap
76. Bypass ducts
77. Intake ramp hydraulic jack
78. Forward fuselage fuel tank
79. Wing sweep actuating screw jack (Microtechnica)
80. Flaps/slats actuation drive shaft
81. Central control and drive unit for wing sweep and high lift devices
82. Wing pivot box integral fuel tank
83. Air system ducting
84. Anti-collision light
85. UHF aerials
86. All-through wing pivot box made of electro-beam welded steel
87. Wing pivot bearing (starboard)
88. Flex-section of flap/slat actuator screw jack
89. Drive actuator/screw jack for wing setting (starboard)
90. Wing leading edge sealing sleeve
91. Wing root glove fairing
92. External fuel tank, 1500 liter capacity
93. AIM-9L air-to-air missile for self defense
94. Canopy in opened position
95. Canopy jettison unit
96. Pilot rear view mirror
97. Three-section leading edge slats, extended (starboard)
98. Leading edge slat screw jack
99. Leading edge slat drive shaft
100. Wing pylon pivot shaft
101. Inboard pylon pivot bearing
102. Integral wing fuel tank
103. Access panel to wing fuel tank
104. Outboard pylon pivot bearing
105. Marconi Sky Shadow ECM pod
106. Outboard pivoting wing pylon
107. Navigation and formation lights
108. Wing tip antenna fairing
109. Double slotted Fowler flaps, extended
110. Flap guide rails
111. Spoilers, extended (starboard)
112. Landing flaps drive shaft
113. External fuel tank stabilizer fins
114. Wing trailing edge housing (at full sweep)
115. Dorsal fuselage fairing
116. Aft fuselage fuel tank
117. Vertical stabilizer root fairing
118. HF antenna
119. Heat exchanger air intake
120. Maximum sweep position of wing (starboard)
121. Air brake, extended
122. All-moving tailplane (taileron, starboard)
123. Air brake actuator
124. Main heat exchanger
125. Heat exchanger exhaust
126. Engine bleed air piping
127. Vertical stabilizer attachment point
128. Air brake ribbing
129. Vertical stabilizer heat shield
130. Vortex generators (turbulence plates)
131. Vertical stabilizer integral fuel tank
132. Fuel vent system tubing
133. Fin structure
134. ILS antenna
135. Fin leading edge
136. Forward housing for passive ECM system
137. Fuel jettison system
138. Fin tip fairing, antenna housing
139. VHF antenna
140. Tail navigation light
141. Rear radar warning receiver
142. Obstruction warning light
143. Fuel jettison
144. Rudder
145. Honeycomb internal structure of rudder
146. Rudder hydraulic actuator cylinder
147. Spine end fairing
148. Thrust reverser bucket door (in position)

Panavia Tornado cutaway showing breakdown of its components (with British external ordnance).

History of German Aviation: Bombers and Reconnaissance Aircraft

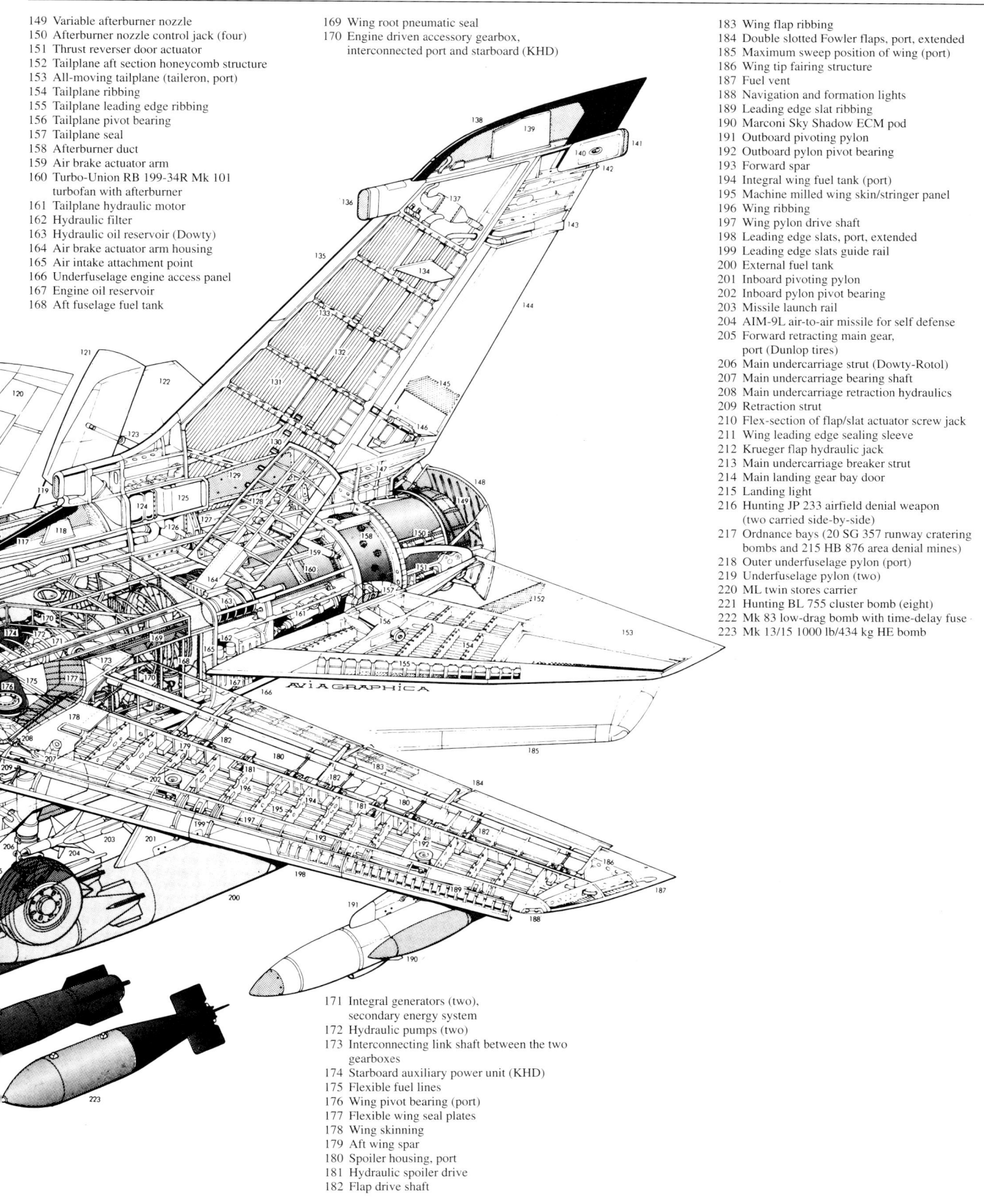

149 Variable afterburner nozzle
150 Afterburner nozzle control jack (four)
151 Thrust reverser door actuator
152 Tailplane aft section honeycomb structure
153 All-moving tailplane (taileron, port)
154 Tailplane ribbing
155 Tailplane leading edge ribbing
156 Tailplane pivot bearing
157 Tailplane seal
158 Afterburner duct
159 Air brake actuator arm
160 Turbo-Union RB 199-34R Mk 101 turbofan with afterburner
161 Tailplane hydraulic motor
162 Hydraulic filter
163 Hydraulic oil reservoir (Dowty)
164 Air brake actuator arm housing
165 Air intake attachment point
166 Underfuselage engine access panel
167 Engine oil reservoir
168 Aft fuselage fuel tank
169 Wing root pneumatic seal
170 Engine driven accessory gearbox, interconnected port and starboard (KHD)
171 Integral generators (two), secondary energy system
172 Hydraulic pumps (two)
173 Interconnecting link shaft between the two gearboxes
174 Starboard auxiliary power unit (KHD)
175 Flexible fuel lines
176 Wing pivot bearing (port)
177 Flexible wing seal plates
178 Wing skinning
179 Aft wing spar
180 Spoiler housing, port
181 Hydraulic spoiler drive
182 Flap drive shaft
183 Wing flap ribbing
184 Double slotted Fowler flaps, port, extended
185 Maximum sweep position of wing (port)
186 Wing tip fairing structure
187 Fuel vent
188 Navigation and formation lights
189 Leading edge slat ribbing
190 Marconi Sky Shadow ECM pod
191 Outboard pivoting pylon
192 Outboard pylon pivot bearing
193 Forward spar
194 Integral wing fuel tank (port)
195 Machine milled wing skin/stringer panel
196 Wing ribbing
197 Wing pylon drive shaft
198 Leading edge slats, port, extended
199 Leading edge slats guide rail
200 External fuel tank
201 Inboard pivoting pylon
202 Inboard pylon pivot bearing
203 Missile launch rail
204 AIM-9L air-to-air missile for self defense
205 Forward retracting main gear, port (Dunlop tires)
206 Main undercarriage strut (Dowty-Rotol)
207 Main undercarriage bearing shaft
208 Main undercarriage retraction hydraulics
209 Retraction strut
210 Flex-section of flap/slat actuator screw jack
211 Wing leading edge sealing sleeve
212 Krueger flap hydraulic jack
213 Main undercarriage breaker strut
214 Main landing gear bay door
215 Landing light
216 Hunting JP 233 airfield denial weapon (two carried side-by-side)
217 Ordnance bays (20 SG 357 runway cratering bombs and 215 HB 876 area denial mines)
218 Outer underfuselage pylon (port)
219 Underfuselage pylon (two)
220 ML twin stores carrier
221 Hunting BL 755 cluster bomb (eight)
222 Mk 83 low-drag bomb with time-delay fuse
223 Mk 13/15 1000 lb/434 kg HE bomb

History of German Aviation: Bombers and Reconnaissance Aircraft

Air-to-air photo of a Panavia Tornado armed with four Kormoran anti-ship missiles and two *Eloka* pods on the outboard wing pylons.

Panavia Tornado with seven general purpose bombs, wings in full sweepback for maximum speed. As the wings pivot, the external stores pylons remain parallel to the longitudinal axis of the aircraft.

Recce pod, of modular design, contains within its central module the following reconnaissance sensors
  - LHOV (Trb 60/24) Zeiss camera system, which can be smoothly focused in flight from 300 m to infinity. Depending on the type of photo required it can be aimed vertically, at an angle, or horizontally.
  - LLDC (Krb 6/24) provides a distortion-free picture of the terrain from horizon to horizon, even at high-speed passes both day and night.
  - Texas Instruments RS-710 infra-red line scanner (IRLS), which identifies heat sources on the camera film by day and night. Its coverage is similar to that of the LLDC, so that comparative information can be derived.
  - the LITEF RIS (reconnaissance interface system) data system, which automatically controls the entire equipment suite and acts as the interface between the sensor data and the aircraft data. The RIS "stamps" a data block on the film in the form of digital information showing such things as the geographic coordinates of the targets, aircraft altitude and ground speed, heading, and time.

In the recce pod's nose module is the climate control system, which maintains the temperature for the reconnaissance package at a constant 35° Celsius. Air is supplied from the engine's bleed air. The reconnaissance suite is supplemented by the instrument panel of the WSO (= weapons systems officer, the current designation for the former observer/navigator), as well as two camera sights for the pilot.

For standard unguided bombs, so-called "iron bombs" (of which the Tornado can carry eight on external pylons—weighing a maximum of 454 kg/1,000 lb each), there have been three primary types of bombing methods employed over several thousand hours of practice runs. These are:
- Lay-down, or blind bombing from horizontal high-speed low-level flight
- Shallow dive, where the target is visually acquired
- Toss bombing, or releasing the bomb on pull-out. Also stand-off bombing from a distance of 5-6 km to the target.

The average results have been impressive: for blind bombing the average impact radius is less than 30 m. To avoid damage to the aircraft, this method makes use of retarded bombs with delayed action fuses. When the target is visually acquired from a shallow dive the average impact radius falls under 25 m, and for stand-off bombing it is around 60 m. These two latter methods employ ballistic bombs having a ballistic trajectory.

With such accuracy, it is little wonder that Tornado crews have repeatedly emerged winners in various American-sponsored competitions. These multi-week contests took place with stiff competition from B-52 and F-111 crews under all types of weather and visibility conditions.

In general, to provide the flexibility needed for countering the threat of aggression regardless of its scale, it should be noted that 75% of the *Luftwaffe's* operational Tornado units are trained in the dual role of conventional and nuclear operations.

Technical Data and Details of the Panavia Tornado
First flight on 14 August 1974

Manufacturer: Panavia Aircraft GmbH with shares of British Aerospace (Great Britain), Messerschmitt-Bölkow-Blohm (Federal Republic of Germany), Aeritalia (Italy)

Crew: 1 pilot, 1 weapons system officer

Engines: 2 Turbo-Union RB 199-34 R, dual cycle regenerating three spool bypass turbofan engines, by Rolls Royce, Motoren Turbinen-Union, and Fiat-Aviazione.

	Multi-role IDS	ADV fighter
Static thrust:		
no afterburner	kN (kp) 40.0 (4,080)	
with afterburner	kN (kp) 71.0 (7,240)	
Dimensions:		
Length m	16.70	18.06
Height m	5.7	
Wingspan m	13.90/8.60	
Wing area (approx.) m^2	30	
Aspect ratio	6.4/2.5	
Leading edge wing sweep deg.	25 to 67	
Weights and Loads:		
Empty kg	13,600	14,100
Takeoff kg	18,600	
W/external payload kg	28,500	
Load factor at 25 deg. sweep	approx 7.5	
Wing loading kg/m^2	620	
Flight performance:		
Max. speed at altitude	Mach 2.2+	
Max. speed at sea level	Mach 1.2 (1450 km/h)	
Landing speed km/h	220	
Action radius km	1,390	
incl. 2 hr loiter time		
Ferry range km	5,000	
Takeoff run		
depending on weight m	366 to 900	
Landing roll-out m	400	

Batches 5 through 7 and Maintaining Combat Effectiveness

On 19 August 1982 a contract was issued for Batch 5 with 171 Tornados, of which 119 were IDS and 52 ADV. Yearly production had stabilized at 110 aircraft: 44 in Great Britain, 42 in the Federal Republic, and 24 in Italy. With Batch 6 of 155 aircraft, of which 63 were IDS and 92 ADV, production neared the original goal of 805 Tornados (with 165 being British ADV variants), which was increased to a total of 809 by refitting four pre-production aircraft to bring them up to the latest standards.

In the meantime, logistical methods had come into play, temporary replacement part bottlenecks had been eliminated, and by 1984/85 the first post-1,200 hr depot inspections (DIs) were being routinely carried out at the MBB Works in Manching.

An aerial tow target system, the DATS-3 (for Dornier Aerial Target System), was developed in 1984 and consisted of a winch and tow cable together with a tow body carried beneath the Tornado's fuselage. Following operational live-fire testing of the target, towed on a length of cable approximately 500 m long, this aerial tow target system was cleared for use within Tornado units for air-to-air gunnery training with the BK 27 in 1986.

Although the Tornado's avionics, developed some 15 years earlier, was the *non plus ultra* in aircraft avionics at the time, if one also considers that over the preceding 15 years computer and storage capacity had quadrupled, then it is understandable to learn that the first upgrade measures for the Tornado took place primarily in its avionics suite. The introduction of the HARM anti-radiation missile in the *Luftwaffe* and *Marineflieger* and the similar ALARM in British service, plus the MAVERICK used by the Italians prompted the three countries to agree on a "first upgrade" modification program. This first upgrade focused on doubling of the main computer's capacity, improved software, and a standardized interface with the HARM. In addition, the terrain following radar and the flight control system were improved, as was the electronic warfare potential.

NAMMA issued a contract for batch 7 in May 1986, which added a further 124 Tornados to the previously contracted 809 aircraft. These included:

- 72 aircraft for Saudi Arabia, with BAE as primary contractor (48 IDS + 24 ADV), of which the first had previously been delivered from British stock on 27 March 1986
- 8 aircraft (ADV) for the Sultanate of Oman, had also been initially supplied from the RAF contingent "before the fact."
- 9 aircraft as backup reserves for the RAF
- 35 Tornado ECR for the *Bundesluftwaffe*

Batch 7 was to be delivered between 1989 and 1992. This effectively secured, for the first time in the Tornado's history, a spot on the international market for exports to countries outside the original partnership.

Tornado ECR

ECR stands for "electronic combat and reconnaissance," and this latest version of the Tornado is based on the operational attributes of the Tornado IDS heavy fighter-bomber. It has the capabilities of both passive and active electronic warfare and complements the RF-4E Phantom in the latter's role of penetration tactical reconnaissance. The ECR Tornados began joining existing wings in northern and southern Germany in 1989.

This ECR version is tailored to the following roles: stand-off reconnaissance and border patrol, and armed reconnaissance for acquiring photo and electronic images combined with the use of anti-radiation missiles. To this end, the Tornado ECR makes use of the following additional components:

- ELS (emitter locator system) for determining position of ground-based radar systems
- permanently installed infra-red sensors (IIS and FLIR = infra-red imaging system and forward looking infra-red)
- on-board system for processing, storing, and transmitting reconnaissance data (ODIN = operational data interface)
- advanced tactical displays for both pilot and weapons systems officer

Thus equipped, the Tornado ECR is not only capable of supplementing/replacing current tactical reconnaissance aircraft, it can also supplement or replace NATO's older electronic warfare systems (such as the F-4G) serving in the suppression of enemy air defense (SEAD) role.

Determining Location and Contract

At the time of writing, follow-on development of the Tornado weapons system has already gotten underway with the first upgrade of Batches 1 through 6 and the Tornado ECR of Batch 7. An eighth batch is planned to include eight Tornado IDS aircraft for Jordan, 35 IDS for the *Bundesluftwaffe*, and 16 ECR Tornados for Italy, plus a further 15 Tornados for British reserves, meaning that it is entirely possible more than 1,000 Tornados will be built altogether.

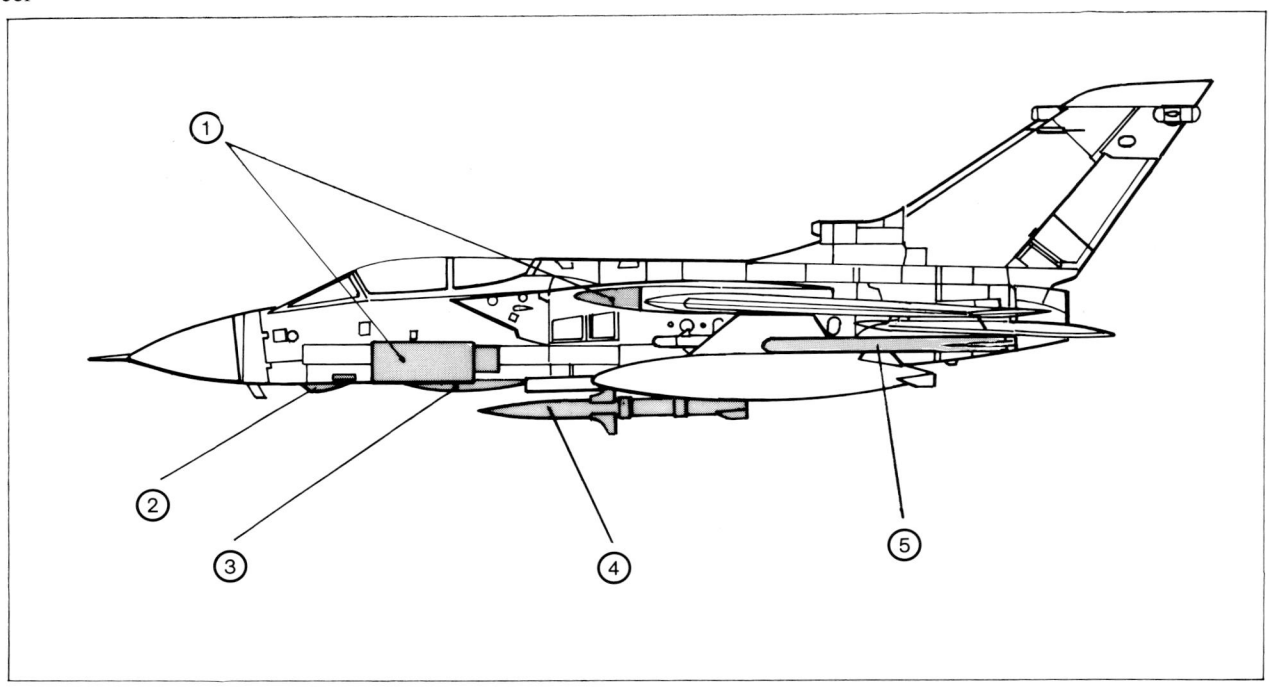

Sensors and weapons of the Tornado ECR for specialized missions, particularly for attacking enemy ground based air defense elements.
1. ELS = Emitter locator system
(for searching out ground based radar sites)
2. FLIR = Forward looking infra-red
(navigational aid system)
3. IIS = Infra-red imaging system
(180 degree field of view in relation to the aircraft's longitudinal axis)
4. HARM missile = High velocity anti-radiation missile
(for engaging "radiating" targets)
5. AECM pod = Active electronic countermeasures pod

Future upgrades for the Tornado weapons system will undoubtedly orient themselves toward the existing threat potential. The *Luftwaffe's* planning will be an influencing factor in these upgrades, which will concentrate on avionics, the powerplants, and ongoing further development of so-called smart weapons using integrated targeting sensors. Combined, all these give the Tornado excellent growth potential.

In the meantime, the air forces of the Warsaw Pact have developed the capability to conduct independent air operations well behind NATO's front lines. This means that, from the outset of a conflict, they threaten not only the NATO troops along the border, but also the key elements for a successful forward defense: the reinforcements and reserves, NATO's air forces, and the remaining escalation potential left to NATO following the INF treaty.

Retaining the option of flexible response means maintaining and strengthening the conventional forces within the alliance, thus minimizing the need for dependence on nuclear weapons at an early stage of any conflict.

A key role for NATO is the "follow-on forces attack" (FOFA), officially defined as:

- attacking enemy follow-on units with the intent of delaying, exploiting, and defeating them before they can reinforce the attack on the battlefield
- destroying the lines of communication necessary for controlling combat forces

These definitions, in turn, have spawned the following tactical/operative objectives:

- blunting the pressure of the enemy's attack with the forward defense by weakening and exploiting follow-on forces
- denying Warsaw Pact forces the freedom to operate behind enemy lines
- thus gaining time for negotiations and for making decisions on whether to escalate the conflict

However, a constant and effective engagement of follow-on forces, both behind the lines as well as in key areas of an enemy's attack, is only possible assuming there is a sufficient quantity of suitable weapons platforms. Given the above-mentioned conceptual doctrine of NATO, it goes without saying that the Tornado weapons system plays a critical role in combination with other weapons systems extending into the target area of FOFA operations. The Tornado is an instrument of peace of the first order; its operational availability and flexibility, its upgrade capability, and the skilled mastery of future-oriented technology of this weapons system by its flying crew all combine to make any type of aggression a high risk proposition. Even over the course of dismantling the Warsaw Pact's invasion potential, the Tornado is still an effective instrument of monitoring international armament agreements—for trust is good, but monitoring is better!

Nevertheless, in order for bombers and recce platforms to operate effectively they must have their backs free, i.e. their staging airfields must be protected along with other sensitive areas in the overall defense structure. One component of air defense is the fixed, ground-based systems, although these can be swept aside by enemy air power, paving the way for follow-on bomber units. The other component of air defense is a fully flexible fighter arm, superior to the enemy's attacking forces and capable of intercepting inbound aggressors and defending friendly staging airfields.

This vastly improved capability of Soviet bombers to penetrate deep into enemy airspace has led to the European Fighter Aircraft project (EFA) to counter such a threat. Following the example of the Tornado program, this time carried by four nations instead of three, this NATO project has been approached with the same goal of ensuring peace as does the Tornado and the men who fly her.

22. Bombers and Reconnaissance Aircraft from Political and Military Perspectives

Results of an Era

Given all the evidence detailed in this volume, it goes without saying that this category of offensive military weaponry is a critical, if not the most critical, component of modern air power. These aircraft have come a long way on a journey marked by differing political influences and the resulting military strategic considerations; in today's doctrine they comprise an integral element, bound by treaty, of allied air forces maintaining a defensive posture on the world's stage.

Douhet's philosophy was only able to play a role in the initial phases of the German L*uftwaffe's* development in the 1930s, whereas the British and Americans took Douhet's lessons of air warfare to heart by employing the heavy bomber in their offensive bombing campaigns, and thus ultimately contributed immensely to the surrender of Germany.

Why did the concepts originally embraced by Wever take a course which, from 1942 onward, led to disaster? It was not entirely due to Wever's premature departure from responsibility or the suicides of Udet and Jeschonnek, as is often thought to be the case, but stems from the fact that too few of the men making the decisions had Wever's intuition and vision. The majority of those responsible for planning the buildup of air power came from the *Heer*, or to be more precise, from the *Heer's* general staff. Certainly, high profile fighter pilots from the First World War were placed in key positions, but is army and fighter pilot experience alone enough to sensibly build an air force? The scientists, engineers, technicians, and even those officers who had cut their teeth on this technology were only invited to share their opinions on a few rare occasions.

To be sure, at the time there was the appearance that, in spite of everything, all had been planned and organized in a sensible fashion, such as the application of the dive bombing concept by several proponents of this tactic. The ideas held by those advocates of the cooperative role, i.e. using the L*uftwaffe* as primarily a tactical tool directly serving other *Wehrmacht* forces, seem to have been borne out by initial successes. These included the unanticipated support that Germany's untried flying forces provided in the Spanish Civil War from late 1936 onward and the virtually uninterrupted *Luftwaffe* involvement in Poland, Norway, and the West. The subsequent course of the war, however, soon revealed the narrow limitations of this doctrine. The Battle of Britain in 1940/41 and the expansion of the air war into the Balkans, the Mediterranean, and North Africa in 1941/42 fragmented Germany's air power on a series of multi-front campaigns and made demands of the German *Luftwaffe* far beyond its tactically oriented mission.

With complete disregard for military reality, the campaign on the Eastern Front burst its chains in the summer of 1941, and the W*ehrmacht* in general, and the *Luftwaffe* in particular, now found itself fighting a war on all sides. Unable to come to grips with armament industry issues, personnel, and training, the situation was exacerbated by the Allied invasions in the south and west and the intensification of the bombing campaign. There was no way out, not even by using the so-called "wonder weapons."

The political goals of those in power at the time—measured by Germany's resources and given military potential—were considered excessive in every respect. The enormous effort put forth by the aviation industry and its outstanding designers, technicians, test pilots, and workers, as well as the courage of the flying crews was, in the end, unable to compensate for this shortsightedness. A shift in the military situation compelled a change in the development of Germany's air resources, as they found themselves in direct confrontation with the Allies' concept of air warfare and their superior bomber fleets operating almost exclusively in the operative-strategic role. The fiasco with the Heinkel He 177 heavy bomber and the entire Bomber B program—where success would have been a prerequisite for any operative-strategic prosecution of the war in the air—soured the opinions of many who had been straddling the fence on the strategic bomber issue. With the exception of the Arado Ar 234, a futuristic jet bomber arriving on the scene too late and in too few numbers, Germany's bomber forces ultimately found themselves reduced to what was feasible at the time: the underdeveloped concept of the fighter-bomber.

The Fighter-Bomber

Josef Kammhuber, former General *der Flieger*, was a strong advocate of the fighter-bomber. A series of studies on the air war during WWII was commissioned in the late 1940s, and he concluded his particular study on the use of the fighter-bombers with the following comments:

"The purpose of this study was to show that the fighter-bomber is the way of the future, despite its hermaphroditic position between fighter and bomber, and indeed, that it appears on the verge of becoming the most flexible type of aircraft for which new roles are constantly being devised. The fighter-bomber had become the absolute best close combat aircraft as early as the Second World War; thanks to its speed, maneuverability, and climb capability, and the massive triple-effect firepower it embodies, it is capable of mastering all tasks in the tactical and operative arena. Furthermore, the hermaphrodite disadvantage immediately disappears when you avoid the mistakes Germany all too often made in the Second World War. These mistakes were the result of missing the mark when it came to setting objectives and tasking and the inability to settle on a distinct solution to the problem. On the other hand, if you formulate unambiguous ideas, if you know what you can and should do, and if you unerringly pursue objectives which have been clearly thought out, then all the hermaphrodite disadvantages the Germans imbued the fighter-bomber with during World War II fall by the wayside.

Thus, we see that the fighter-bomber has become a new type of weapon that, although closely related to its older brother the fighter, now falls under a fully separate class within the air force. It is one of the *Luftwaffe's* pillars of support which can—and must—take on greater significance in the future."

In his unique factual-sensible way, Kammhuber thus recognized the importance of the fighter-bomber at an early stage.

It is no wonder, then, that the Bunde*stag's* personal review committee appointed General Kammhuber the first inspector for the *Bundesluftwaffe* when it was resurrected on 1 November 1955. This air force had wisely imposed upon itself a considerable amount of self-moderation with a clear understanding of what was politically and economically practical—and cognizant of the WEU's armament control guidelines monitored by NATO and its accepted fighter-bomber role based on its initial aircraft inventory. Kammhuber's early and correct recognition of the fighter-bomber's developmental potential is evidenced by the broad spectrum of roles played by today's fighter-bomber, the Tornado, and particularly the ECR Tornado.

Application of Reconnaissance Systems

A final word on the role and the importance of today's reconnaissance in Germany: in a defensive posture, it is based on the RF-4E Phantom, the "thoroughbred recce bird," working alongside the ECR Tornado to provide all-weather penetration reconnaissance using its recce system with side-looking radar and infra-red sensors. Over the battlefield, aerial reconnaissance is supplemented by unmanned reconnaissance drones and visual reconnaissance by all aircraft operating in the area. Other allied partners have at their disposal strategic high altitude reconnaissance platforms, reconnaissance satellites, specialized aircraft for electronic reconnaissance (ELINT), for communications reconnaissance (COMINT), as well as airborne early warning aircraft (AWACS). During peacetime, NATO places a particularly high value on border area surveillance. It is an instrument for:

- monitoring applicable armament control agreements
- early warning in the event of enemy attack preparations

both of which are extremely time-sensitive factors for maintaining peace. Consequently, there should be a supernationally controlled and therefore economical pooling of all reconnaissance resources, as sometimes currently occurs with NATO's AWACS aircraft. The timely transmittal of information and centralized analysis would offer the same type of security as is already the case with NATO's integrated air defense network. In the event of an act of aggression, this network is capable of immediate response due to the data interlinking of all sensors, even in peacetime, and by placing airborne and ground-based air defense assets under the control of NATO's air defense commander. The timely, centralized evaluation of all reconnaissance data offers the best guarantee for keeping the peace, since today's reconnaissance sensors are fully capable of picking up any incongruity on the other side of NATO's borders. This precautionary monitoring, which could easily be mutual, plays a part in efforts to build trust—for vigilance is the price of peace in a free world!

We continue to shoulder our share of the burden in maintaining western-style freedom—to us as important as life itself—and ensuring that this freedom, if necessary, is defended with the best means at our disposal. Modern reconnaissance aircraft and bombers will continue to be an indispensable part of this defense so long as the latent potential of the Warsaw Pact's forces to attack and invade the territories of NATO member states remains in place.

# Appendix

Appendix 1
Table of German Combat Aircraft

```
LC II,1                Geheim              den 8. Juni 1936.
LC II Nr. 3459/36 1 zbV. geh.

                    V e r z e i c h n i s

              der deutschen Militärflugzeuge nach dem
                 Entwicklungsstande vom 1. 6. 1936,
        (entsprechend den vom A-Amt gestellten Entwicklungsrichtlinien.)
```

Lfd. Nr.	Baumuster	Verwendungszweck	Hersteller
1	Ar 64	Jagdeinsitzer (nur bei Schulen verwendet)	Arado
2	Ar 65	Jagdeinsitzer	"
3	Ar 68	dto.	"
4	Ar 76	Übungseinsitzer	"
5	Ar 80	Verfolgung Jagdeinsitzer	"
6	Ar 81	Schweres Stuka	"
7	Ar 95	Küsten- u. Bordaufklärer	"
8	Ar Proj.	Trägermehrzwecke	"
9	Ar Proj.	Kleiner Borderkunder	"
10	Bf 109	V.J.	Bayrische Flugzeugwerke
11	Bf 110	Flugzeugzerstörer	"
12	Bf 161	Höhenfernerkunder	"
13	Bf 162	Schnellbomber	"
14	Bf 163	Verbindungsflugzeug	"
15	Do 11	Mittlerer Schulbomber	Dornier
16	Do 17 E	Mittlerer Bomber	"
17	Do 17 F	2-motoriger Fernaufklärer	"
18	Do 18 B	dto. See	"
19	Do 19	Grossbomber	"
20	Do 23	Mittlerer Bomber	"
21	Do 24	3-motoriger Fernaufklärer	"
22	Fi 98	Leichtes Stuka (nicht weiter verfolgt)	Fieseler
23	Fi 156	Verbindungsflugzeug	"
24	Fi Proj.	Trägermehrzweckeflugzeug	"

Lfd. Nr.	Baumuster	Verwendungszweck	Hersteller
25	FW 56	Übungseinsitzer	Focke-Wulf
26	FW 57	Flugzeugzerstörer	"
27	FW 58 B	Übungsbomber	"
28	FW 159	V.J. jetzt Frontjagdflugzeug	"
29	FW 187	V.J. 2-motorig	"
30	FW Proj.	Trägerflugzeug	"
31	" "	Kleiner Borderkunder	"
32	FWH Proj.	Verbindungsflugzeug	Flugzeugwerk Halle
33	Ha 137	Leichtes Stuka	Hamburger Flugzeugbau
34	Ha 138	3-mot. Fernaufklärer	" "
35	Ha 140	See-Mehrzwecke m. Schwimmern 2-Mot.	" "
36	He 45	Fernaufklärer	Heinkel
37	He 46	Nahaufklärer	"
38	He 50	Land-Sturzbomber	"
39	He 51 Land u. See	Jagdeinsitzer	"
40	He 59	See-Mehrzwecke	"
41	He 60	Küsten- u. Bordaufklärer	"
42	He 111 B	Mittlerer Bomber	"
43	He 112	V.J.	"
44	He 114	Küsten- u. Bordaufklärer m. Schwimmern	"
45	He 115	2-motor. See-Mehrzwecke	"
46	He 118	Schweres Stuka	"
47	He 119	Schnellbomber	"
48	He Proj.	Trägerflugzeug	"
49	He Proj.	Kleiner Borderkunder	"
50	He 45/DB 600 (He 45 G)	Nahaufklärer	"
51	Hs 121	Übungseinsitzer (nicht weiter verfolgt)	Henschel
52	Hs 122	Nahaufklärer	"
53	Hs 123	Leichtes Stuka	"

Appendix 2
Consolidated Aircraft Type Program

```
                              - 3 -
--------------------------------------------------------------
Lfd.
Nr.    Baumuster    Verwendungszweck              Hersteller
--------------------------------------------------------------
54     Hs  124      Flugzeugzerstörer             Henschel
55     Hs  125      Übungseinsitzer (nicht weiter
                                    verfolgt)         "
56     Hs  126      Nahaufklärer (Weiterentwicklg.
                                 Hs. 122).            "
57     Hs  127      Schnellbomber                     "

58     Ju  86 A     Mittlerer Bomber              Junkers
59     Ju  87       Schweres Stuka                    "
60     Ju  88       Schnellbomber                     "
61     Ju  89       Grossbomber                       "

    In diesem Verzeichnis nicht aufgeführt sind Behelfsflug-
zeuge wie Ju 52, He 70 usw.

Verteiler:
Kdo. d.E-Stellen Rechlin
    (2 Ausf.z.Weiterltg. an
       Travemünde)            4 x
PfL                           2 x
DVL                           2 x
L A                           5 x
L C                           1 x
L C  Chef.Ing.                3 x
L C  I                        2 x
L C  III                      5 x
L C  IV                       5 x
L C  II Ch.                   1 x
L C  II J.                    1 x
L C  II,2                    10 x
     II,3                    10 x
     II,5                    10 x
     II,6                    10 x
     II,7                     3 x
     II,1 L                   1 x
     II,1 a                   1 x
     II,1 b                   1 x
     II,1 c                   1 x
     II,1 d                   1 x
     II,1 e                   1 x
     II,1 f                   1 x
     II,1 g                   1 x
     II,1 h                   1 x
     II,1 zbV                 1 x
     II,1 z.d.A.              1 x
Reserve                      20 x
```

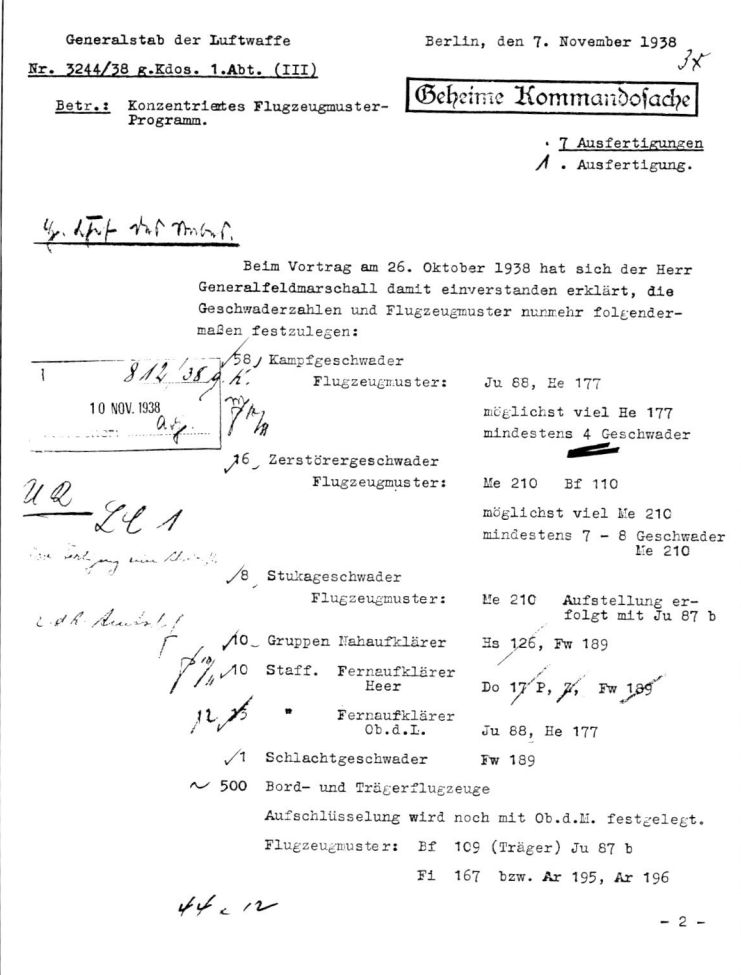

Generalstab der Luftwaffe   Berlin, den 7. November 1938
Nr. 3244/38 g.Kdos. 1.Abt. (III)

Betr.: Konzentriertes Flugzeugmuster-
       Programm.

Geheime Kommandosache

· 7 Ausfertigungen
1 . Ausfertigung.

Beim Vortrag am 26. Oktober 1938 hat sich der Herr
Generalfeldmarschall damit einverstanden erklärt, die
Geschwaderzahlen und Flugzeugmuster nunmehr folgender-
maßen festzulegen:

58 Kampfgeschwader
    Flugzeugmuster: Ju 88, He 177
                       möglichst viel He 177
                       mindestens 4 Geschwader

16 Zerstörergeschwader
    Flugzeugmuster: Me 210   Bf 110
                       möglichst viel Me 210
                       mindestens 7 - 8 Geschwader
                                            Me 210

8 Stukageschwader
    Flugzeugmuster: Me 210   Aufstellung er-
                                 folgt mit Ju 87 b

10 Gruppen Nahaufklärer   Hs 126, Fw 189
10 Staff. Fernaufklärer
        Heer   Do 17 P, Fw 189
   "   Fernaufklärer
        Ob.d.L.   Ju 88, He 177

1 Schlachtgeschwader   Fw 189

~ 500 Bord- und Trägerflugzeuge
      Aufschlüsselung wird noch mit Ob.d.M. festgelegt.
    Flugzeugmuster: Bf 109 (Träger) Ju 87 b
                       Fi 167 bzw. Ar 195, Ar 196

- 2 -

Appendix 3

Ordnance in 1939

The following ordnance was cleared for purchase and produced between 1933 and 1939:

Year of introduction	Designation	Type	Weight in kg
1933	B1E	electron incendiary bomb	1
	SD 10	fragmentation bomb	10
	SC 50	aerial mine(bomb)	50
	SC 250	aerial mine(bomb)	250
1935	SC 500	aerial mine(bomb)	500
1936	LMA	air-dropped land mine	500
	LMB	air-dropped land mine	1,000
1938	LT 5	aerial torpedo	800
	SD 50	general purpose bomb	50
1939	B13E	electron incendiary bomb, steel nose	1.2
	SD 500	general purpose bomb	500
	PC 500	armor piercing bomb	500

The size of the most common bombs and dimensions of their craters are shown in the following drawing. "o.V." stands for *"ohne Verzögerung"* (no delay), and m.V. for *"mit Verzögerung"* (with delay).

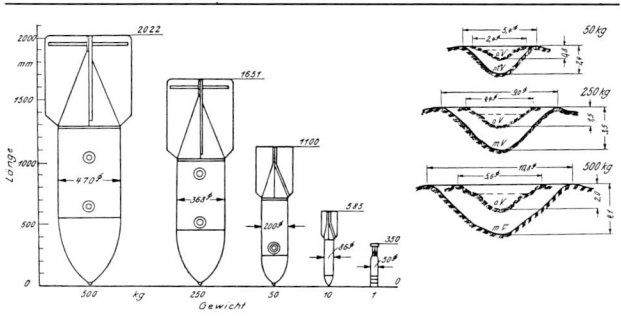

Appendix 4
Aircraft Armament from 1936 through Today

A. Available in 1939

The following aircraft armament was available in 1939 and was used in bomber and reconnaissance aircraft entering front line service at the time:
1. The **MG 15** began trials in 1933 as a flexible mounted machine gun; caliber 7.92 mm; 1,000 to 1,050 rounds per minute rate of fire; 755 m/s muzzle velocity; weight of the weapon was 8.2 kg; ammunition feed via dual drums containing 75 rounds; was used in practically all bombers and reconnaissance aircraft until replaced by the MG 81.
2. The **MG 17**, developed from the MG 15 and cleared for use in late 1934, generally fixed mounted parallel to the aircraft's longitudinal axis; 7.92 mm caliber; rate of fire (uninhibited) was about 1,200 rounds per minute; 755 m/s muzzle velocity; weapon weighed 10.2 kg; ammunition feed via variable length *Gelenkgurte* 17 ammunition belt, which,

History of German Aviation: Bombers and Reconnaissance Aircraft

for example was 4,800 mm long for 250 rounds, later also via disintegrating belt. Used in the Hs 123, Hs 126, and Ju 87 (in some cases also as a fixed tail gun in He 111s)

3. The **MG-FF**, introduced in 1936; a heavy machine gun which could be flex-mounted, but was generally fixed mounted parallel to the aircraft's longitudinal axis; 20 mm caliber, 520 rounds per minute rate of fire; 600 m/s muzzle velocity; weapon weighed 26.3 kg; ammunition feed via 15 round clip magazine or drum magazine with 30, 45, 60, and 100 round capacity. Fitted to the Hs 123 underneath the wings, later also in the He 111 and Ju 88 as a flex-mounted nose gun (*A-Stand*).
4. The **MG 81**, *Technisches Amt* approved design of gun in 1938, mass production began in 1939; fixed or flex-mounted machine gun, known as "I*lling*" when fitted singly. Also came in a twin barrel "*Zwilling*" version, the MG 81Z; 7.92 mm caliber; Illing rate of fire was 1,600 rounds per minute, while Zwilling's was 3,200 per minute; 705 to 825 m/s muzzle velocity; weight of Illing was 6.5 kg, Zwilling 12.9 kg; initially planned mainly for installation in the Ju 88
5. The **MG 131**, envisioned as an improved performing on-board weapon compared with the MG 15 and MG 17, undergoing testing in 1938 but had not yet been cleared for use by 1939. This heavy machine gun would be used in a series of different bomber types.

B. Introduced between 1939 and 1945

1. Entering mass production in 1939, the **MG 81** (7.9 mm) was fitted in practically all bomber, dive bomber, and reconnaissance aircraft between 1939 and 1944.
2. Not yet cleared for use at the outbreak of the war, the **MG 131** was introduced in late 1940. In addition to being fix-mounted in fighters and in barbettes on the Me 210 and Me 410 heavy fighters, it was also used as a flex-mounted gun in the He 111 (*A-Stand*), Do 17Z (*B-Stand*), Ju 88 (*B1-* and *B2-Stand*), and in the He 177 quad-mounted in the tail (trial installation).
13 mm caliber; 900 rounds per minute rate of fire; 710-750 m/s muzzle velocity; weapon weighed 16.6 kg; ammunition feed via disintegrating or non-disintegrating belts of varying capacity, e.g. 300 and 450 rounds.
3. The **MG 151** (15 mm) and the **MG 151/20** (20 mm) were ultra-heavy machine guns primarily fix-mounted in fighter and heavy fighter aircraft, but also in the Ju 88 night fighter, either as an engine gun, synchronized to fire through the propeller arc, or in wing troughs or gun pods (WT 151). For bombers, it was used in the Fw 200C-3's *A-Stand* (hand-operated, flex-mounted MG 151/20) and in the *B1-Stand* (flex-mounted MG 151 in the hydraulically operated HD 151/1 turret); the 20 mm weapon (MG 151/20) also trial fitted in the FHL 151 remote controlled tail barbette in the Ju 288C-1, as well as a twin barrel version in the tail of the He 177.

	MG 151	MG 151/20
caliber	15 mm	20 mm
rate of fire	600-700 rounds/min	630-720 rounds/min
muzzle velocity	850-1,020 m/s	695-785 m/s
weight of gun	42 kg	42 kg

ammunition feed via disintegrating belt of varying length, e.g. 100, 130, 150, 200, or 250 rounds.
4. The **MK 101**, a machine cannon developed in 1936 and tested in a gun pod beneath the fuselage of a Bf 110B. During an inspection of the Rechlin Test Center on 3 July 1939, such a configuration made a most favorable impression on Hitler and Göring. Conceived especially for engaging tanks and ships, it was used on bombers and reconnaissance aircraft only in limited numbers in the Do 24 flying boat (long range maritime reconnaissance platform) and the Hs 129 strike fighter. Never fully lived up to its anti-tank role expectations.
30 mm caliber; 230-260 rounds per minute rate of fire; 725-960 m/s muzzle velocity; weapon weight 139 kg; ammunition feed via 6 round magazine or 30 round drum.
5. The **MK 103** was derived from the MK 101 in 1942 and was fitted in the Hs 129's ventral bay. Other aircraft armed with this cannon were the Me 410B-6 anti-shipping aircraft, which carried two guns (each with 100 rounds) in a WB 103 pod, and the Do 335 as an engine cannon with 70 rounds. Operational performance was markedly better than that of the MK 101.
30 mm caliber; 380-420 rounds per minute rate of fire; 860-940 m/s muzzle velocity; weapon weight 145 kg; ammunition feed via belt with 100 or 70 rounds.
6. The **BK 3.7**, a modified Flak 18 anti-aircraft cannon, was first tested on a Ju 87 in the summer of 1942 and put into production with the Ju 87G-1 for anti-tank duties. Also in 1943, the Hs 129B-2/Wa strike fighter was equipped with this cannon when the 30 mm gun proved inadequate for effectively attacking more modern Soviet armor (T-34). In the late fall of 1943 the Ju 88P-2 and P-3 were fitted with two BK 3.7 guns mounted side-by-side in the belly.
3.7 mm caliber; rate of fire theoretically 140 rounds/min, in practice 80 rounds/min; 795-860 m/s muzzle velocity; weapon weight 272 kg; ammunition feed via 6 round clip magazine.
7. The **MK 108**, primarily designed for installation in fighter and heavy fighter aircraft against bombers, became available in 1943, and in 1944 was fitted in the fighter-bomber version of the Me 262, the Me 262A-2a, which had four fixed-mounted weapons in the nose with 2x 100 (upper guns) and 2x 80 rounds (lower guns).
30 mm caliber; 600 rounds per minute rate of fire; 520 m/s muzzle velocity; weapon weight 58 kg; ammunition feed via belt with 60 rounds.
8. The **BK 5** began weapons testing in 1943, followed by trial installation that same year in a handful of He 177s in the anti-armor role. Beginning in March 1944 the cannon was production fitted to versions of the Me 410 as a specialized anti-bomber weapon. In the late fall of 1944 the BK 5 also armed the Ju 88P-4 anti-tank version, and in March 1945 was test fitted to the Me 262A-1 (which did not see operational service).
5 cm caliber; 45 rounds per minute rate of fire; 920 m/s muzzle velocity; weapon weight 540 kg; ammunition feed via ring magazine with 22 rounds.
9. The heaviest of on-board armament, the **BK 7.5** was derived from the proven 7.5 cm PaK 40/L. It was evaluated on a Ju 88A-4 during the summer of 1942 and fitted to a small batch of Ju 88P-1s (approx. 30 in number) that same year. In 1943 five He 177A-3/R5 (*Stalingradtyp*) aircraft were also trial-fitted with the same gun. Due to considerable structural stress on the He 177s and tactical considerations no further BK 7.5s were installed. In 1944 about 25 Hs 129B-3/Wa strike fighters were fitted with the BK 7.5.
7.5 cm caliber; 30-40 rounds per minute rate of fire; 704 m/s muzzle velocity; weapon weight 705 kg; ammunition feed via 9 round clip or 12 round drum.
Note: Data for rate of fire and muzzle velocity is dependent on type of round and fitting point (e.g. synchronized MG 151s and MG 151/20s rates of fire were approximately 200 rounds per minute less)

C. German Guns on *Bundeswehr* Combat Aircraft

While the *Bundeswehr's* foreign procured bombers and reconnaissance aircraft are armed with foreign guns (not reviewed in this book), German developed guns have been installed in the joint projects

- Alpha Jet A (German fighter-bomber version), which is armed with a single **Mauser MK 27 mm** cannon
- Panavia Tornado, fitted with two **IWKA-Mauser BK 27 mm** cannons.

The Alpha Jet carries its cannon in a pod beneath the fuselage, which also holds 150 rounds of ammunition, and uses it in both the air-to-ground and air-to-air roles. The Tornado's two MK 27 mm guns, however, are integrated within the fuselage beneath the cockpit, and carry 180 rounds of ammunition. These **IWKA Mauser BK 27** mm single-barrel drum cannons, specially developed for the Tornado, completed prototype testing in early 1976. At 1,025 m/s they have an extremely high muzzle velocity; the pilot can select two rates of fire: 1,700 rounds per minute for air-to-air combat, or 1,000 rounds per minute for air-to-ground operations. Additional data for both guns: 27 mm caliber; weight of gun is approximately 100 kg; and ammunition feed via disintegrating belt.

Notes on this appendix: these weapons and the ammunition they use are covered in greater detail in Hanfried Schliephake's work *"Flugzeugbewaffnung - Die Bordwaffen der Luftwaffe von den Anfängen bis zur Gegenwart"* (see bibliography).

Appendix 5

Reconnaissance Systems from 1935 Onward

Aerial Photography Equipment

All aerial photography equipment contracted for by the RLM was built from the early 1930s onward by the Zeiss company.

The **Handkamera 13 x 18** (picture format), a handheld camera, contained 60 photos in a daylight cartridge, designed to eject the exposed film. This was followed by the **HK 12.7** with a picture format of 7 x 9 cm, which became the preferred camera for tactical reconnaissance due to its ease of operation. Later the **HK 12.5** and **HK 19** were used in the Fw 189 and the Me 109.

For automatic aerial cameras, also called serial photogrammetry cameras (R*eihenmeßkammern*, or RMK), which became available in the field from 1935 onward as the **Rb 20/30, Rb 50/30,** and the **Rb 75/30**, the focal length determined the size of the camera housing and thus the type capable of being installed in a given aircraft. The focal length is indicated by the first two numbers in the camera designations listed above. The "30" following the slash shows that the picture format is 30 x 30 cm. The camera shutters were converted from the standard slit type to central shutters with selectable exposure times of 1/25 to 1/350 seconds, which in favorable light conditions enabled sharp photos to be produced even when taken from low flying altitudes. Shutter tension and release, as well as film advance for these Rb cameras were automatic via a motor drive powered by the aircraft's electrical system. Compressed air held the film snugly against a glass plate, so that it lay just as flat as with a plate camera. The normal advance speed was one photo every two seconds, enabling the photogrammetry cameras to provide outstanding coverage of target areas and terrain or photographic plan. With the picture interval regulator, though, other intervals between shots could be set, as well. The Rb cameras were, as a rule, loaded with a film cassette containing 60 m of perforated 32 cm wide film.

The **Rb 21/18** and **Rb 15/18**, both having smaller formats and shorter focal lengths, were also carried in the Fw 189 and Me 109.

Specially optimized for nighttime photography in 1943 were the N**Rb 35/25, 40/25,** and **50/25** cameras, which were fitted in the Ju 188D and F versions, among other aircraft types.

Photo Processing Equipment

It goes without saying that photo processing equipment is needed along with photo cameras, automatic aerial cameras, and photogrammetry cameras. Such processing equipment is designed to convert, correct, and analyze the perspective distortion caused by the disparity between the photographic axis and the vertical line in order to provide an accurate scale rendition of the terrain.

Distortion correction equipment, such as the **Zeiss SEG I** (weighing 435 kg) and the **SEG IV** (210 kg) were available in 1939, as were more complicated systems for fine analysis such as the **Zeiss-Stereoplanigraph**. This was a dual projector used for precision analysis of stereoscopic photographs taken from any angle, as well as vertical and oblique exposures.

This device, weighing in at 1,750 kg, was suited to producing highly accurate planimetric and elevation maps. Location accuracy was potentially quite good, and corresponded to a theoretical graphical accuracy of A1/10 mm. Elevation precision varied depending on film rating and base ratio between 1/2,000 and 1/5,000 of the flying altitude above ground.

In addition, other stereoscopic mapping equipment at the time also included **radial triangulators, anaglyph analyzers,** and the **Zeiss multiplex aeroprojector,** as well as easily transportable observation and analysis equipment characterized by their

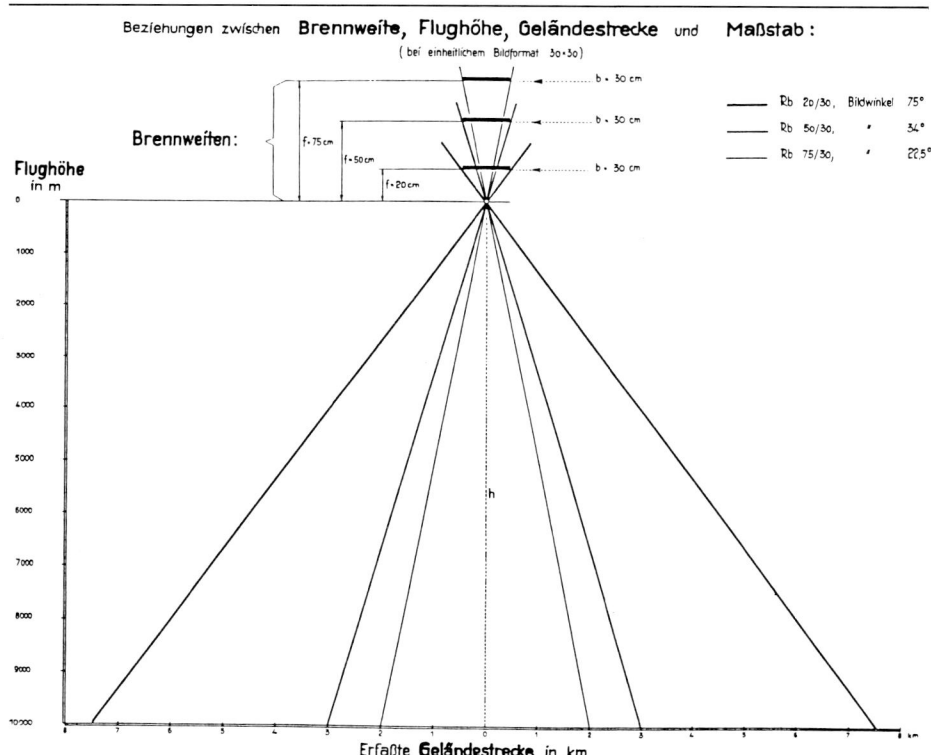

simplicity of design and ease of use. One example of this were the stereoscopes, ranging from simple pocket types to folding mirror stereoscopes with marked stereometers, parallel construction, and field glass magnifiers attached.

A more detailed treatment of these photo processing devices would be outside the scope of this brief synopsis of reconnaissance equipment and is best left to more specialized publications dealing with this specific subject.

Relations and Basic Concepts

As the diagram on page 294 shows, for the family of photogrammetric cameras with focal lengths of f=20 cm, f=50, and f=75 cm there is a relationship between focal length, flying altitude, area covered, and scale when using the standard picture format of 30 x 30 cm.

Some basic concepts are needed to understand the different types of aerial photographs:
- with regard to the direction of the photograph, there are the following differences
  a) vertical exposures, where the camera axis deviates little or none from the vertical plane. These show the photographed area similar to a map, with standard albeit so-called undefined scale (e.g. 1:11,325)
  b) oblique exposures, where the camera axis deviates considerably from the vertical plane. These provide a perspective shot of the photographed area. Oblique exposures are divided into low-oblique photographs, which deviate up to 30° from the vertical, medium-oblique photographs from 30° to 60°, and high-oblique photographs of greater than 60° from the vertical plane.
- with regard to the arrangement, there are the following differences
  a) single frame photography
  b) strip photography, designed to seamlessly cover the area along the flight path. A lateral overlap of individual frames of 20 to 30% is sufficient for correcting distortions when photographing level areas. Of more value, however, is an overlap of at least 60%, as this not only ensures good spacial coverage of level areas, but also offers more accuracy for determining position and height in mountainous terrain.
  c) Areal photography, comprised of individual photo passes by a camera-equipped aircraft. Each pass should have an overlap of about 30%.

Appendix 6

Aircraft Utilization, Stress, and Class Groupings

For determining structural load, aircraft were categorized based on utilization, stress, and function.

1. Main Utilization Groups

H High performance and experimental aircraft (***H****ochleistungsflugzeuge* = high performance aircraft)
G Aircraft used for the commercial transportation of cargo (***G****üter* = goods)
P Aircraft used for the commercial transportation of people (***P****ersonenflugzeuge* = airliner)
R Aircraft for personal use, not to include commercial applications (*Private **R**eiseflugzeuge* = private touring aircraft)
S Trainer aircraft for training Class A pilots (***S****chulflugzeuge* = trainer aircraft)
K Aircraft primarily tailored to aerobatics (***K****unstflugzeuge* = aerobatic aircraft)

2. Stress Groups

	Stress
1	very low
2	low
3	average
4	high
5	very high

examples:
standard commercial airliner: P 3
single seat fighter: H 5
airplane with limited aerobatic capability: S 4, K 4

Seaplanes were classified into the following two groups based on the stress of the float structure:

Group I Seaplanes designed primarily for takeoff and landing in calm seas. (Formerly inland bodies of water up to Beaufort scale 3: WB)

Group II Seaplanes designed for landing and taking off on medium/choppy seas. (Formerly high seas up to Beaufort scale 5: WH)

Classification

In addition to distinguishing aircraft based on stress and utilization, the various aircraft were also classified according to their size and type.

The aircraft class divisions logically correspond to the different types of pilot's licenses. According to air traffic regulations, these classes are:

1. Land Planes

Class A1: Aircraft for one to two persons with a gross weight of up to 500 kg, whose landing roll is no greater than 300 m after clearing a 20 m obstacle.

Class A2: Aircraft for one to two persons with a gross weight of 500 to 1,000 kg, with a landing roll no greater than 450 m, or aircraft for one to two persons with a gross weight less than 500 kg and a landing roll greater than 300 m but less than 450 m, or aircraft for three persons with a gross weight of up to 1,000 kg and a landing roll no greater than 450 m.

Class B1: Aircraft for one to three persons with a gross weight of 1,000 to 2,500 kg, or aircraft for one to three persons with a gross weight less than 1,000 kg and a landing roll greater than 450 m.

Class B2: Aircraft for four to six persons with a gross weight up to 2,500 kg

Class C1: Single-engined aircraft for more than six persons or single-engined aircraft for less than six persons but weighing more than 2,500 kg.

Class C2: Multi-engined aircraft for more than six persons or multi-engined aircraft for less than six persons but weighing more than 2,500 kg.

2. Seaplanes

Seaplanes are divided into the same categories as in section one. The corresponding weight limits are:

for class A1	600 kg gross weight
for class A2	2,200 kg gross weight
for the B classes	3,500 kg gross weight
for the C classes	3,500 kg gross weight

Regarding the landing and takeoff capabilities of land planes of all classes, it should be noted that according to "Construction Specifications for Aircraft" certain conditions must be met before they can be certified.

Takeoff: maximum run of 600 m from full stop to 20 m altitude
Landing: maximum roll of 600 m from 20 m altitude to full stop

No such specifications exist for seaplanes; the only requirement is that takeoff cannot be longer than 75 seconds. The class of aircraft is entered on the certification document.

Appendix 7
Ju 88 Chronology

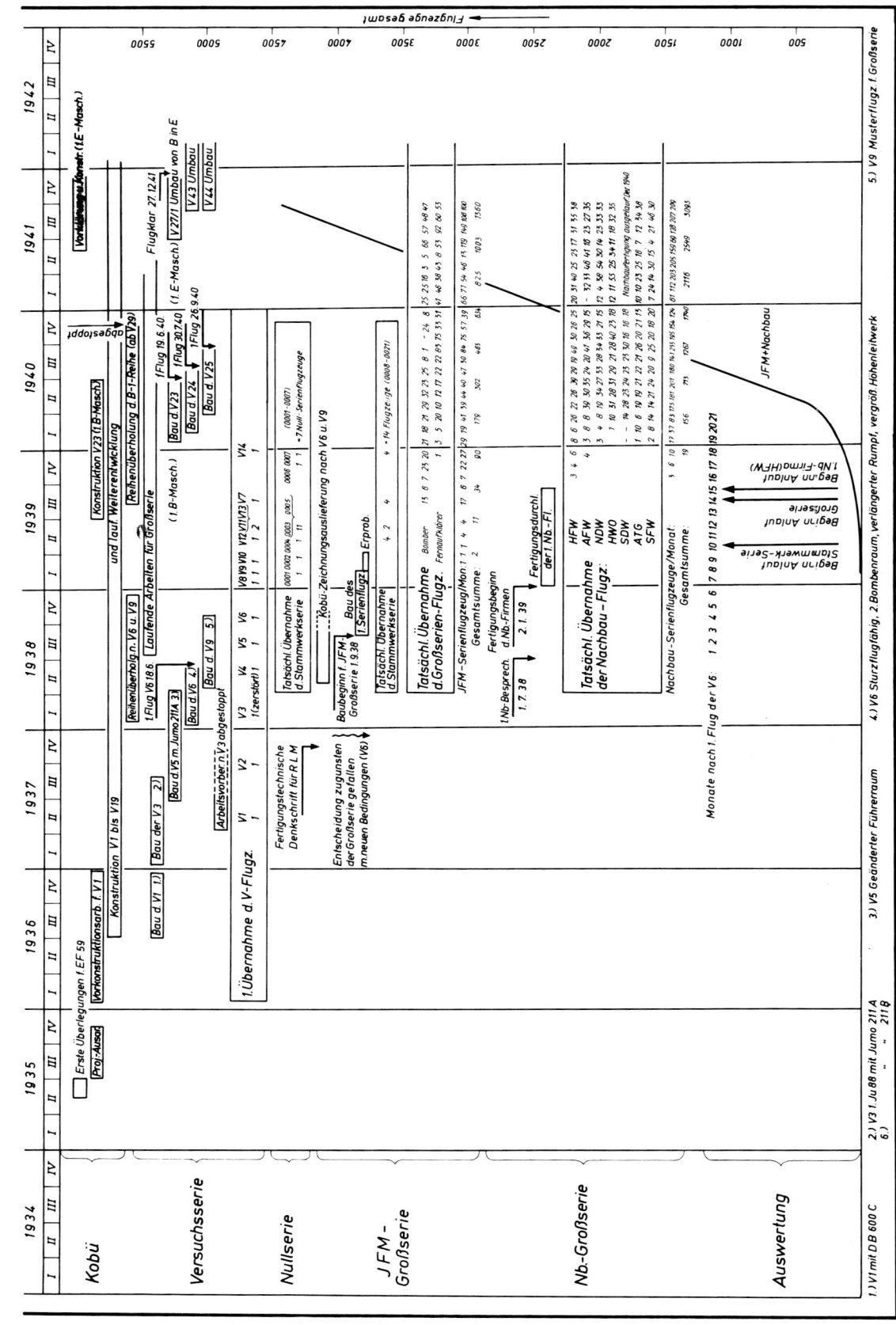

History of German Aviation: Bombers and Reconnaissance Aircraft

Appendix 8

Performance Table for German Bombers, Dive Bombers, and Reconnaissance Aircraft (as of 1 June 1937)

Author's note: During the thirties and forties, spectacular record-setting flights by specially built aircraft, state-sponsored politically motivated propaganda methods, and sales-boosting company announcements and publications were often used to disseminate aircraft performance values among the public. In many instances, however, these performance values failed to stand up under critical observation, nor did they reflect the performance achieved in day to day operations.

Additionally, front-line training often prevented the best performance to be wrung from military aircraft. Also, there were certain values attained by test aircraft that frequently could not be approached by production and license-built planes. This was especially true of aircraft that were flown in a fully equipped state.

For these reasons the Luf*twaffe's* general staff issued revised performance figures, some classified "top secret." These values had been tried and tested with production aircraft in operational training wings and at the Rechlin test center, and gave a realistic approximation of the performance potential for warplanes in service at the time.

The following excerpt was taken from the publication "Der Reichsminister der Luftfahrt und Oberbefehlshaber der Luftwaffe General Stab der Luftwaffe, Nr. 1440/37 g. Kdos. 1. Abt. II, 1, Az 40 a 2, Betr.: Leistungstabelle der deutschen Kriegsflugzeuge" dated 6 September 1937. This performance table, effective as of 1 June 1937, was prefaced by the following clarifications:

"**Clarification:**   **Best values** are the best performance figures outlined in the type sheet for a given aircraft as provided by the LC.*

Normal values are those data determined by usage in the field, which are to serve as the basis for operations. These do not take tactical safety buffers into account....

Maximum speed is covered by best and normal values on the basis of field trials.

With regard to r**ange**, it was discovered that the **normal values** were 10% less than the **best values**. However, based on experience, the values could be achieved both in solo flight and while flying three-ship *vics*. The values for range are based on a given cruising speed.... In this case, **normal values** refers to maximum authorized cruise setting and **best values** applies to the most favorable throttle setting.

The bomber, dive bomber, and reconnaissance aircraft discussed in this book are reflected by the performance data in the following table; these figures provide a picture of the data foundation used for planning purposes at the time.

*LC = Te*chnisches Amt*/Technical Office

Type	Crew	Weight (kg)	Armament	Engine # & Type	Max. Speed, Best Value (solo) 0 m	4000 m	5000 m	7000 m	Payload	Range (solo)	Fuel Capacity (liters)	Climb Rate, Best Value		Best Value (m) Normal Value 4,000	5,000	6,000	Service Ceiling (m)
Medium Bombers																	
Do 17E	3	7,000	1-2 mg	2x BMW VI 7.3	354	315	300	-	400 750	1,275 750	1,300 750	1,150 620	1,300 720	1,400 850	18'	29'	-	5100[1]
Ju 86	4	8,000	3 mg	2x Jumo 205C	285	270	240	@5,500	750 1,000	1,250 825	1,125 725	1,000 650	950 825	900 600	25'	45' @5,500		5,500
He 111	4	9,400	3 mg	2x DB 600C	360	365	315	@8,000	1,000 1,400	900 825	1,000 875	850 575	850 575	1,400 1,000	14.2' 18'	18.4' 23'	45' 28.5'	@8,000 8,250[1]
Bombers with Supplemental Fuel Tanks																		
Ju 86	4	9,000	3 mg	2x Jumo 205C	-	-	230	-	500[2]	-	3,700 @5,000 m	-	-	1950	-	-	-	5,000[3]
He 111	4	10,600	3 mg	2x Jumo 211A	-	-	370	-	500[2]	-	3,200 @6,000 m	-	-	244-	-	-	-	6,000[3]
Dive Bombers																		
Hs 123	1	2,160	2 mg	1x BMW 132	288	278	265	240	100 200	825 530	890 570	500 325	750 480	420 270	12.6'	19.8'	40'	6,000
Ju 87	1-2	3,420	1-2 mg	1x Jumo 210D	280	290	275	-	500/250	610	710	525	600	440	18'	21'	31'	7,000
Reconnaissance Aircraft																		
Do 17F	3	7,000	2 mg	2x BMW VI 7.3	354	315	300	-	-	2,000	2,000	1,650	2,000	2,150	18'	29'	-	5,100[1]
Hs 126	2	2,930	1 mg	1x Bramo 323A	288	345	340	-	100	-	950	-	855	550	7.3'	-	11.6'	8,900[3]

1) With variable pitch propellers
2) Greater payloads possible using accelerated takeoffs
3) Calculated value not using accelerated takeoff

Appendix 9

New Ordnance Types from 1940 Onward

The campaigns in Poland and Norway revealed the disadvantages in the lack of heavier bombs; for example, at the outbreak of the war the only bomb available against armored ships was the PC 500 (a 500 kg armor piercing bomb). This prompted an accelerated development of new bomb types, many of which had passed their evaluation stage by 1939 and were cleared for production in quantity. These bombs included the "heavy caliber":
- PC 500 RS (same bomb as the PC 500, but with rocket propulsion to enhance the penetration force when making diving attacks on battleships up to 35,000 metric tons)
- SC 1000 (1,000 kg aerial mine against naval bases, port facilities, and for retaliatory raids)
- SC 1800 (1,800 kg aerial mine against fortified facilities, naval bases, and for retaliatory raids)
- SD 1700 (1,700 kg general purpose bomb)
- PC 1400 (1,400 kg armor piercing bomb for high altitude and dive bombing against older-type battleships with armor thickness up to 130 mm, as well as fortified facilities, underground air-raid facilities, subways, and tunnels).

With regard to lighter or special purpose bombs, the following were cleared for production in 1940:
- SD 2 (2 kg strike fighter bomblet)
- Sbe 50 (50 kg concrete/fragmentation bomb)
- Flam C 250 (250 kg incendiary bomb)
- Flam C 500 (500 kg incendiary bomb)

The two latter bombs were primarily used against metalworks, steel and rolling mills, blast furnaces, oil refineries, warehouses, sugar factories, etc., as well as against other soft targets. With regard to initial operations with heavy caliber bombs, it should be noted that bombers had been pulled back to the West well before the end of the Norwegian campaign insofar as these aircraft were fitted with external racks capable of carrying the heaviest payloads, such as the SC 1800, SD 1700, and PC 1400. Thus configured, these bombers and their crews were made available on short notice for operations against well fortified, hardened facilities—especially those along the French Maginot Line—and therefore were no longer available for the final phase of the Norwegian campaign.

From 1941 onward   kg

B2E(Z)	electron incendiary bomb	2
Sprbrd C 50	combination incendiary-demolition bomb	50
Strbd C 500	scatter incendiary bomb	500
SD 250	general purpose bomb	250
PC 1000 RS	rocket-propelled armor piercing bomb	1,000
BM 1000	bomb mine	1,000
SC 2500	aerial mine	2,500
SB 2500	area denial bomb	2,500

Notes on the PC 100 RS: It was designed for high penetration. Its use was limited to battleships, aircraft carriers, and heavy cruisers. It was dropped by specially trained Ju 88 crews from a height of no less than 1,400 m at a dive angle of at least 60 degrees, providing the greatest potential penetration force while at the same time avoiding skipping/ricocheting and damaging the bomber.

Notes on the SB 2500: This was the *Luftwaffe's* largest bomb, designed for maximum generation of shock waves against industrial sites and cities. There were two types:

a) SB 2500 (AI). One-piece body made of cast aluminum or of a casing of welded aluminum plating with an aluminum copper magnesium alloy (Bondur) nosepiece welded on and heavy-gauge metal ring-type guide fins welded to the casing. 3,895 mm overall length, 825 mm diameter. Could only be hung horizontally from mounts. Removable lifting eyes and guide pilots were screwed directly into the bomb casing. Production ceased, and the limited numbers on hand were used up.

B) SB 2500. Body consisted of a casing made of welded steel plating with cast steel nosepiece welded on and heavy-gauge ring-type guide fins. On both types of bombs, two fuse receptacles were welded in to accept the two fuses. The SB 2500's dimensions were reduced so that it could be loaded into newer aircraft types, such as the He 177 and Do 217. 3,693 mm length, 785 mm diameter.

	SB 2500 (AI)	SB 2500
c) weight	2,500	2,370
explosive	Fp. 60/40	Fp. 60/40
amount	2,000	1,570 kg

Beginning in December 1942 the SB 250 was sometimes filled with Trialen 105. These bombs were reserved exclusively for attacking merchant ships. With regard to weight, they were approximately 140 kg heaver than when filled with Fp. 60/40.

The standard SB 2500 had the following effect:
50-70 m radius: total loss, or such extensive structural damage that buildings would have to be torn down. 100-200 m radius: destruction of window frames and doors.
Up to 1,000 m radius: glass shattered.
The following illustration shows its shape and dimensions:

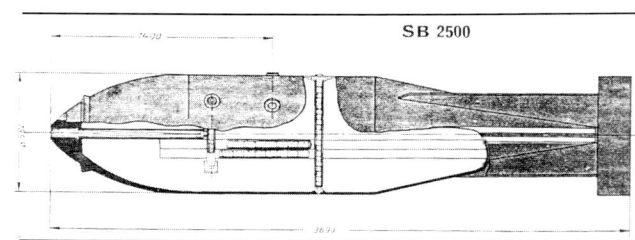

From 1942 onward   kg

SD 1	fragmentation bomb	1
SD 4 Hl	hollow charge bomb	4
SD 70	general purpose bomb	70
SD 1000	general purpose bomb	1,000
PC 1400 X	remote controlled armor piercing bomb	1,400

Note on PC 1400 X: also known as FX 1400 or Fritz X; covered in detail in the text.

From 1943 onward   kg

Brand C 50	high intensity incendiary bomb	50
Brand C 250	high intensity incendiary bomb	250
SD 9/SD 15	fragmentation bomb containing 8.8 and 10.5 cm HE bomblets	
SD 65	general purpose bomb	65
Brand 10	liquid-filled incendiary bomb	10
SB 1000	area denial bomb	1,000
SC 2000	aerial mine	2,000
PD 500	armor piercing bomb	500
PD 1000	armor piercing bomb	1,000
PC 1800 RS	rocket-propelled armor piercing bomb	1,800

General Characteristics
of the bombs in use
1. **Shape and construction** of the cylindrical types of bombs differentiate themselves as follows:

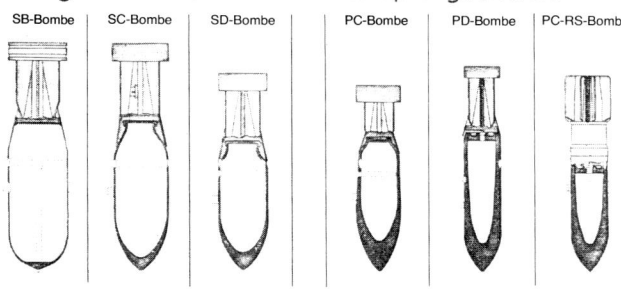

Der grundsätzliche Aufbau der Sprengbomben.

2. **Different Grades** of bombs resulted from the differences in manufacture, as shown in the following example for the SC 250:

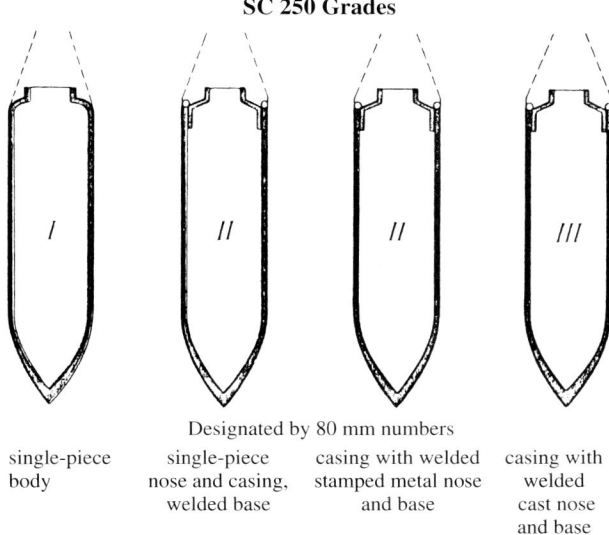

3. **Marking and painting** of cylindrical bombs were based on the following marking example:

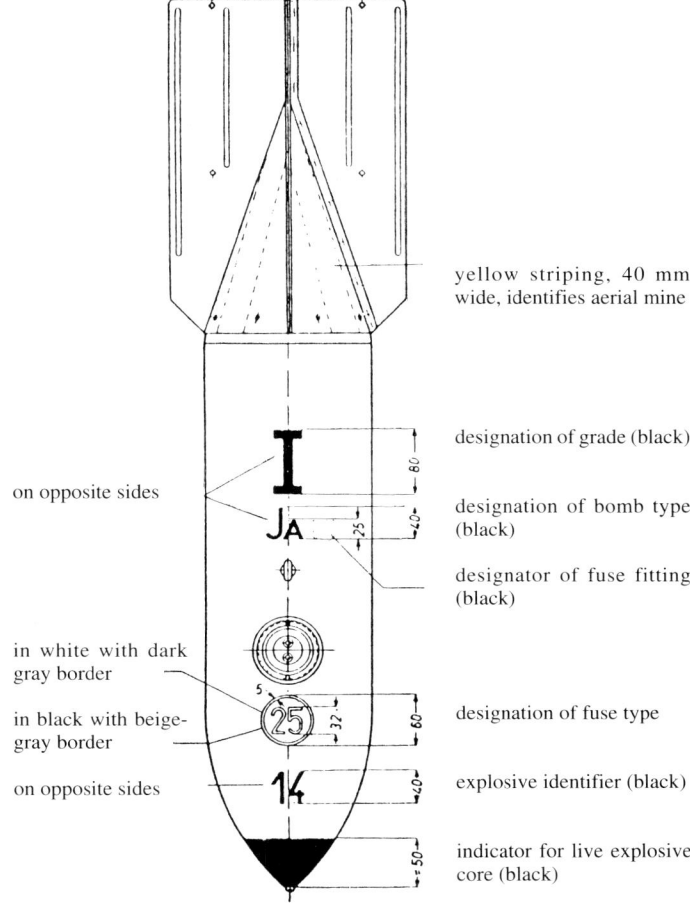

4. Armor piercing bombs had the following characteristics:
   a) **Rocket-propelled PC bombs** (e.g. PC 1000 RS)
   Thick-walled short bomb body with less than 15% explosive charge and rocket propulsion unit in extended base section. Could only be used effectively in steep dives providing additional impact force against armored targets.
   b) **PD bombs** (e.g. PD 500)
   Thick-walled slender bomb body with less than 10% explosive charge. Low percentage was offset by using high-quality explosives. When dropped from level flight from altitudes of at least 4,000 m the stream lined shape provided the same penetration force as PC and PC RS bombs of the same caliber.
   Due to having too low of an impact velocity, it was useless when dropped from a dive.
   Identification: 1 blue stripe between each guide fin.

5. **Incendiary bombs:** The body of the incendiary bomb was mainly filled with highly flammable liquid or a solid which ignited as a result of the high thermal development on impact. It was used primarily against buildings and, usually dropped *en masse*, was intended to produce long lasting conflagrations. When introduced, the **electron incendiary bombs** were cylindrical bodies stabilized by guide fins and having a flammable electron skin and a thermite ignition element.
Designation:
   B   for identification as *Brandbombe* (incendiary bomb), followed by number indicating weight.
   E   for identification as electron (replaced the previous El to avoid confusion with El - electric)
   Z   when necessary, identified that the bomb was fitted with a supplemental demolition charge, e.g. B1E, B1EZ, B1.3E, B1.3EZ, B2EZ.

6. **Drop canisters** were cluster bomb units fitted with an opening fuse and filled with small-caliber fragmentation bomblets. These were used mainly by strike fighters. An example of their designation for an AB 70 is found on the next page:

   AB =   *Abwurfbehälter* (drop canister)
   70 =   70 caliber
   -1 =   development number one
   50SD1 =   canister filled with 50 SD 1 bomblets

The AB 250 and AB 500 were also quite commonly used.

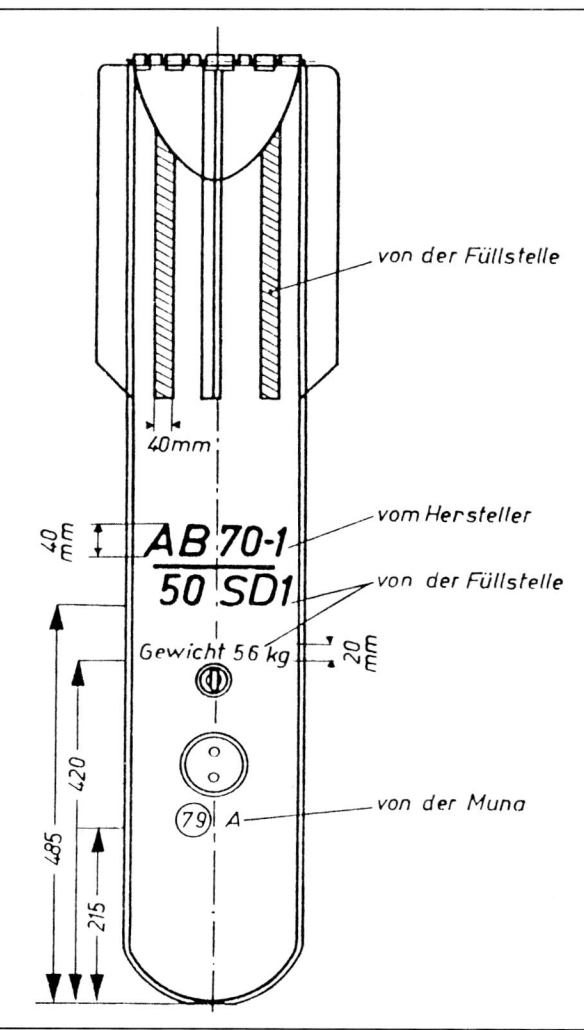

- Zünder 26 (Land) for the Flam C 250 and C 500
- Zünder 55 (land) for SD 50, Sbe 50, SD 250, SD 500
- Zünder 41 (Land) for low-level attacks using the SD 2
- Zünder 59A for LC 50 flares
- Zünder 79 for AB drop canisters

b) **Land targets from shallow dives up to 30 degrees and steep dives from 50 to 70 degrees**
- Zünder 15 (Land) as under a)
- Zünder 25 (Land) ditto
- Zünder 25A (Land) ditto
- Zünder 55 (Land) ditto

c) **Naval targets from level flight**
- Zünder 28A (See) for SC 250 with nose ring, SC 500 with nose ring, and SC 2500
- Zünder 28B (See) for SC 250, SC 500, SC 1000, SC 1800; neither Zünder 28 was suitable for low-level attacks due to the danger of damage to the drop aircraft
- Zünder 38 (See) for SC 250 with nose ring, SD 250, SC 500 with nose ring, SD 500, suitable for merchant shipping

d) **Naval targets from shallow dives up to 30 degrees and steep dives**
- Zünder 28A (See) as above
- Zünder 28B (See) ditto
Neither fuse was suitable for extreme low-level drops, since the bomb detonated immediately upon impact
- Zünder 38 (See) for SC 250 with nose ring, SD 250, SC 500 with nose ring, SD 500, suitable for merchant shipping

e) **Land and naval targets (combined) from level flight**
- Zünder 25B (Land/See) for SC 50, SD 50, SC 250, SD 250, SC 500, also suitable for low-level drop
- Zünder 35 (Land/See) for PC 1000, PC 1400, not suited for low-level due to danger of damage to drop aircraft

f) **Land and naval targets (combined) from shallow dives up to 30 degrees and steep dives**
- Zünder 25B (Land/See) as under e)
- Zünder 35 (Land/See) ditto
- Zünder 49B, C for PC 1000 RS, PC 1800 RS

Note: These fuses were used based on type of bomb, type of attack, and drop altitude, and further differentiated themselves by their time delay:
- impact fuse (*Aufschlagzünder* - A.Z.)
- time delay fuse (*Zeitzünder* - Zt.Z.)
- long delay fuse (*Langzeitzünder* - L.Z.Z.) for delays of 1-100 hours to detonation

without delay (*ohne Verzögerung* - o. V., detonation upon impact) or delayed action (*mit Verzögerung* - m. V., 0.2- 2.0 seconds), or with longer delayed action (*mit längerer Verzögerung* - M. Vz., 5-14 seconds). For optimum effect of the bomb at the target, as well as to preclude damaging or destroying the bomber aircraft, it goes without saying that much coordination was required between mission control, bomb technicians, and the bomber crews.
This naturally also applies to modern ordnance of the latest generation.

7. **Henschel Hs 293 glide bomb**, virtually the first standoff weapon, was developed especially for use against shipping targets. Due to its operational criteria, a more detailed explanation can be found in the text.

8. **Fuses:** Each bomb is equipped with one or more electronic fuses, activated via a fuse arming panel (*Zünderschaltkasten* - ZSK). 150 V was provided for level and shallow dives, while 240 V was used for dives in excess of 30 degrees. The choice of fuse was based on mission and type of target, with the following differences noted:

a) **Land targets from level flight:** these were attacked with bombs fitted with the following fuses:
- Zünder 15 (Land) for the SC 50, SD 50, SC 250, SD 250, SC 500; no longer produced after 1941
- Zünder 25 (Land) for the same bombs as the Zünder 15, also not produced after 1941
- Zünder 25A (Land) for the same bombs

9. **Aerial Torpedoes** are discussed in the body of the text. The responsibility for development of aerial torpedoes originally rested with the Reichsmarine.

Bibliography

Bateson, Richard P. *Arado Ar 234 Blitz*. Aircraft Profile Pulbications, Ltd., Windsor, Berks

Baumbach, Werner. *Zu spät?* Motorbuch-Verlag, Stuttgart 1977

Bekker, Cajus. *Angriffshöhe 4000*. Stalling Verlag, Oldenburg 1964

Below, Hicolaus v. *Als Hitlers Adjutant 1937-1945*. v. Hase & Koehler Verlag, Mainz 1980

Benecke, Theodor, ed. *Flugkörper und Lenkraketen*. Bernard & Graefe Verlag, Coblenz 1987

Boehm-Tettelbach, Karl. *Als Flieger in der Hexenküche*. v. Hase & Koehler Verlag, Mainz 1981

Bolz, Rüdiger. *Synchronopse des Zweiten Weltkriegs*. ECON Taschenbuch Verlag, Düsseldorf 1983

Bongartz, Heinz. *Luftmacht Deutschland*. Vol. 1: Werden und Aufstieg. Essener Verlagsanstalt, Essen 1941

Boog, Horst. *Die deutsche Luftwaffenführung 1935-1945*. Deutsche Verlags-Anstalt, Stuttgart 1982

Brown, Eric. *Wings of the Luftwaffe*. Jane's Publishing Company, London 1979

Brütting, Georg. *Das Buch der deutschen Fluggeschichte* (3 vols.). Drei Brunnen Verlag, Stuttgart 1979

Buckley, Christopher. *Greece and Crete 1941*. P. Efstathiadis & Sons S. A., Athens 1984

Bülow, Hilmer Frhr. V. *Geschichte der Luftwaffe*. Moritz Diesterweg Verlag, Frankfurt a. M. 1934

Dahms, Hellmuth Günther. *Der Spanische Bürgerkrieg 1936-1939*. Rainer Wunderlich Verlag, Tübingen 1962

Dierich, Wolfgang. *Die Verbände der Luftwaffe 1935-1945*. Motorbuch Verlag, Stuttgart 1976

Dornier GmbH. *Dornier Flugzeuge/Aircraft*. Dornier GmbH, Friedrichshafen 1983

Douhet, Guilio. *Luftherrschaft*. Drei Masken Verlag, Berlin 1935

Duval, General. *Entwicklung und Lehren des Krieges in Spanien*. Paul Neff Verlag, Berlin 1938

Ebert, Hans J. *Messerschmitt Bölkow Blohm*. Motorbuch Verlag, Stuttgart 1973

Estob, Peter. *Legion Condor*. Ballentine Books Inc., USA 1973

Ethell, J. and A. Price. *The German Jets in Combat*. Jane's Publishing Company, London 1979

Faber, Harold. *Luftwaffe - An Analysis by Former Luftwaffe Generals*. Sidgwick & Jackson, London 1979

Feuchter, Georg W. *Geschichte des Luftkriegs*. Athenäum-Verlag, Bonn 1954

Galland, Adlof. *Die Ersten und die Letzten*. Schneekluth Verlag, Darmstadt 1953

Gellermann, Günther W. *Moskau ruft Heeresgruppe Mitte. Was nicht im Wehrmachtsbericht stand: Die Geschichte des geheimen Kampfgeschwaders 200 im Zweiten Weltkrieg*. Bernard & Graefe Verlag, Comblenz 1988

Gersdorff, Kyrill v. and Kurt Grasmann. *Flugmotoren und Strahltriebwerke*. Bernard & Graefe Verlag 1981

Gersdorff, Kyrill v. *Ludwig Bölkow und sein Werk - Ottobrunner Innovationen*. Bernard & Graefe Verlag, Coblenz 1987

Göppel, Gerhard. *Der verzweifelte Todesflieger - Einsatz der deutschen Luftwaffe*. Dr. Gerhard Göppel, Hamburg 1986

Granier, Gerhard. *Das Bundesarchiv-Militärarchiv*. BMVg, Füs I 3, Bonn 1981

Green, William. *The War Planes of the Third Reich*. MacDonald & Jane's, London 1970

Greffrath. *Die deutsche Luftrüstung 1933-1945* (study). Militärgeschichtliches Forschungsamt Freiburg, 1962

Griehl, Manfred. *Dornier Do 217 - 317 - 417*. Motorbuch Verlag, Stuttgart 1987

Groehler, Olaf. *Geschichte des Luftkriegs 1910 bis 1980*. Militärverlag der Deutschen Demokratischen Republik, Berlin 1981

Gundelach, Karl. *Die deutsche Luftwaffe im Mittelmeer 1940-1945* (vols. 1 and 2). Verlag Peter D. Lang, Frankfurt a. M./Bern 1981

Gunston, Bill. *The Encyclopedia of the World's Combat Aircraft*. Hamlyn, London, and New York 1976

Gunston, Bill. *Panavia TORNADO*. Ian Allan Ltd., London 1979

Heeresleitung: Truppenführung (T.F.) H. Dv. 300/1 and /2. Mittler & Sohn, Berlin 1936

Heinkel, Ernst. *Sturmisches Leben*. Mundus-Verlag, Stuttgart 1955

Hermann, Hajo. *Bewegtes Leben - Kampf- und Jagdflieger 1935 - 1945*. Motorbuch verlag, Stuttgart 1984

Hollbach. O. *Deutscher Flugzeugbau, Exporthandbuch der Luftfahrtindustrie - Naturkunde und Technik-Verlag Fritz Knapp*, Frankfurt a. M. 1939

Homze, Edward L. *Arming the Luftwaffe*. University of Nebraska Press, Lincol and London 1976

Hümmelchen, Gerhard. *Die deutschen Seeflieger 1935-1945*. J. F. Lehmanns Verlag, Munich 1976

Irving, david. *Die Tragödie der Deutschen Luftwaffe*. Ullstein Verlag, Perlin 1970

Irving, David. *Göring*. Albrecht Knaus Verlag, Munich/Hamburg 1987

Kens, Karlheinz. *Die Flugzeuge des Zweiten Weltkrieges 1939-1945*. Wilhelm Heyne Verlag, Munich 1968

Kens, Karlheinz, and Heinz J. Nowarra. *Die deutschen Flugzeuge 1933-1945*. J. F. Lehmanns Verlag, Munich 1977

Kober, Franz. *Die ersten Strahlbomber der Welt*. Podzun-Pallas-Verlag, Friedberg 1980

König, Friedrich. *Die geschichte der Luftwaffe von 1910-1945*. Erich Pabel Verlag, Rastatt/Baden 1980

Kosin, Rüdiger. *Die Entwicklung der deutschen Jagdflugzeuge*. Bernard graefe Verlag, Coblenz 1983

Kurowski, Franz. *Der Luftkrieg über Detuschland*. Econ Verlag, Düsseldorf/Vienna 1977

Kurowski, Franz. *Seekrieg aus der Luft*. E.S. Mittler & Sohn, Herford 1979

Krüger, Alfred W. *TORNADO - das Kampfflugzeug der NATO*. Podzun-Pallas-Verlag, Friedberg 1981

Lange, Bruno. *Typenhandbuch der deutschen Luftfahrttechnik*. Bernard & Graefe Verlag, Coblenz 1986

Larrazábal, J. S. *Das Flugzeug im Spanischen Bürgerkrieg*. Motorbuch Verlag, Stuttgart (from Span. Original 1969) 1973

Lusar, Rudolf. *Die deutschen Waffen und Geheimwaffen des 2. Weltkrieges und ihre Weiterentwicklung*. J. F. Lehmanns Verlag, Munich 1962

Mason, Herbert Molloy. *Die Luftwaffe, Aufbau, Aufstieg und Scheitern im Sieg*. Paul Neff Verlag, Vienna/Berlin 1973

Mason, Herbert Molloy. *Die Luftwaffe, Entstehung, Höhepunkt und Niedergang der Deutschen Luftwaffe bis 1945*. Heyne Taschenbuch, Munich 1979

McKee, Alexander. *Entscheidung über England*. Bechtle Verlag, Munich/Eßlingen 1960

Mechtersheimer, Alfred. *MRCA TORNADO*. Osang verlag, Bad Honnef 1977

Michaelis, Herbert, ed. *Der 2. Weltkrieg -Bilder, Daten, Dokumente*. Bertelsmann Lexikon-Verlag, Gütersloh 1968

Michaelis, Herbert, ed. *Ursachen und Folgen vom deutschen Zusammenbruch, vol. 15: Kriegführung gegen die Westmächte 1940, das Norwegenunternemen, der Frankreichfeldzug, der Luftkrieg gegen England*. Dokumenten- Verlag Herbert Wendler & Co., Berlin 1970

Middlebrook, Martin. *Die Nacht in der die Bomber starben*. Ullstein Verlag, Frankfurt/M. and Berlin 1975

Mondey, David. *Pictorial History of Aircraft*. Octopus Books Ltd., London 1977

Moyes, Philip J. R., ed. *Bombers of World War II*. Vintage Aviation Publications Ltd., Kidlington 1981

History of German Aviation: Bombers and Reconnaissance Aircraft

Munson, Kenneth. *Aircraft of World War II*. Ian Allan, London 1972

NATO: Tatsachen und Dokumente. NATO-InformationSservice, Brussels 1976-1988

Nemecek, Václav. *Sowjet Flugzeuge* (from CSSR original). Lufthahrt-Verlag Walter Zuerl, Steinebach 1969

Nowarra, Heinz J. *Die Luftschlacht um England*. Podzun-Pallas Verlag, Friedberg 1978

Nowarra, Heinz J. *Die verbotene Flugzeuge 1921-1935*. Motorbuch Verlag, Stuttgart 1981

Nowarra, Heinz J. *Die deutsche Luftrüstung 1933-1945 (vols. 1-4)*. Bernard & Graefe Verlag, Coblenz 1985-1988

Nowarra, Heinz J. *Focke-Wulf Fw 200 Condor*. Bernard & Graefe Verlag, Coblenz 1988

Pabst, Otto E. *Kurzstarter und Senkrechtstarter*. Bernard & Graefe Verlag, Coblenz 1984

Pawlas, Karl R. *Luftfahrt-Dokumente. Arado Ar 234*. Publ. Archiv Pawlas, Nuremberg 1976

Ponomarjov, Alexander. *Militärflugzeuge - Technische Tendenzen*. Militärverlag der DDR, Berlin 1987

Pörtner, Rudolf. *Sternstunden der Technik*. ECON Verlag, Düsseldorf/Vienna 1986

Price, Alfred. *German Air Force Bombers of World War Two*. Hylton Lacy Publishers, Windsor, Berks. 1969

Price, Alfred. *Bildbuch der deutschen Luftwaffe 1933-1945*. Stalling Verlag, Oldenburg/Hamburg 1975

Price, Alfred. *Handbuch Deutsche Luftwaffe 1939-1945*. Motorbuch Verlag, Stuttgart 1979

Proctor, Raymond L. *Hitler's Luftwaffe in the Spanish Civil War*. Greenwood Press, Westport USA/London UK 1983

Quill, Jeffrey. *Spitfire, A Test Pilot's Story*. John Murray Publishers, Ltd., London 1983

Quill, Jeffrey. *Birth of a Legend - The Spitfire*. Quiller Press Ltd., London 1986

Rieckhoff, H. J. *Trumpf oder Bluff? 12 Jahre Deutsche Luftwaffe*. Interavia-Verlag, Geneva 1945

Ries, Karl, and Ring. *Legion Condor 1936-1939* Verlag Dieter Hoffmann, Mainz 1980

Rose, Arno. *"Mistel" - Die Geschichte der "Huckepack-Flugzeuge"*. Motorbuch Verlag, Stuttgart 1981

Schliephake, Hanfried. *Flugzeugbewaffnung*. Motorbuch Verlag, Stuttgart 1977

Schmidt, Heinrich. *"Tornado" - Militär- und Rüstungspolitik in der BRD*. Verlag Marxistische Blätter, Frankfurt a. M. 1979

Schneider, Helmut. *Flugzeug-Typenbuch 1944*. C. Bange Verlag, Hollfeld (reproduction) 1984

Smith, Peter C. *Stuka (The Stuka at War)*. Motorbuch Verlag, Stuttgart 1980

Speer, Albert. *Erinnerungen*. Ullstein Verlag, Frankfurt-Berlin-Vienna 1969

Stahl, P. W. *"Geheimgeschwader" KG 200*. Motorbuch Verlag, Stuttgart 1977

Stanley, Col. Roy M. *World War II Photo Intelligence*. Sidgwick & Jackson, London 1982

Streit, Kurt W., and J. W. R. Taylor. *Geshichte der Luftfahrt*. Sigloch Service Edition, Künzelsau 1975

Thomsen, O. R. *Fliegen liegt in der Luft*. Verlag Bild und Buch, Berlin 1953

Vanags-Baginskis. *Stuka Ju 87*. Jane's Publishing Company, Ltd., London-Sydney 1982

Völker, Karl-Heinz. *Dokumente und Dokumentarfotos zur Geschichte der deutschen Luftwaffe*. Deutsche Verlags- Anstalt, Stuttgart 1968

Walpulski, Günter. *Verteidigung + Entspannung = Sicherheit*. Verlag Neue Gesellschaft, Bonn 1973

Wood, Tony, and Bill Gunston. *Die Luftwaffe*. Buch und Zeit Verlagsges., Cologne 1979

Zindel, Ernst. *Die Geschichte und Entwicklung des Junkers-Flugzeugbaus von 1910 bis 1945 und bis zum endgültigen Ende 1970* - Study of the Deutsche Gesellschaft für Luft- und Raumfahrt (DGLR), Cologne 1979

Other Sources

A. Luftwaffe regulations and guidelines from 1935 to 1945 (L. Dv. L. Dv. T.), RLM aircraft recognition leaflets, design descriptions, aircraft descriptions, brief descriptions of bomber and reconnaissance aircraft issued by aircraft manufacturers.

B. Works by the *"Studiengruppe der Geschichte des Luftkriegs,"* established by the United States Air Force Historical Division after the Second World War. These studies were published in 12 volumes in the 1950s and 1960s and are available at the *"Militärgeschichtliches Forschungsamt"* (MGFA) in Freiburg i. Br.
The studies reviewed by the author follow in alphabetical order by author with the MGFA study reference number in parentheses.

Deichmann, Paul, Gen. d. Fl. a. D. *Die Unterstützung des Heeres durch die deutsche Luftwaffe im 2. Weltkrieg* (MGFA LW 10)

— *Das von der deutschen Luftwaffe im 2. Weltkrieg angewandte System zur Zielbestimmung* (MGFA LW 32)

— *Das System der deutschen Luftwaffe in der Wahl ihrer Angriffsverfahrn* (MGFA LW 33)

Drum. Gen. d. Fl. a. D. *Die deutsche Luftwaffe im Spanischen Bürgerkrieg (Legion Condor)* (MGFA LW 1)

Felmy, H., Gen. d. Fl. a. D. *Die deutsche Luftwaffe im Einsatz auf dem Mittelmeer-Kriegschauplatz; Die Eroberung der Insel Kreta durch deutsche Fallschirmjäger und Luftlandetruppen* (MGFA LW 8)

— *Die deutsche Luftwaffe bis Kriegsausbruch* (MGFA LW 21)

Hertel, Walter, GenIng. a. D. *Die Flugzeugbeschaffung in der deutschen Luftwaffe* (MGFA LW 16)

Kammhuber, Josef, Gen. d. Fl. a. D. *Das Problem des Jaboeinsatzes* (MGFA LW 28)

Klee, Dr. Karl, Hauptmann. *Der geplante Einsatz der deutschen Luftwaffe im Zusammenhang mit dem Unternehmen "Seelöwe"* (MGFA LW 5)

Marquardt, Ernst, GenIng. a. D. *Die Planung und Entwicklung von Bomben für die deutsche Luftwaffe* (MGFA LW 37)

Schabedissen, Walter, GenLt. A. D. *Mehrfrontenkrieg* (MGFA LW 25)

Speidel, Wilhelm, Gen. d. Fl. a. D. *Die detusche Luftwaffe im Polenfeldzug 1939* (MGFA LW 2)

Suchenwirth, Richard, Prof. Dr. *Göring als Oberbefehlshaber der deutschen Luftwaffe* (MGFA LW 21/3)

— *Staatsekretär Milch* (MGFA LW 21/4)

— *Jeschonnek, Wesen, Wirken und Schicksal* (MGFA LW 21/5)

— *Generalluftzeugmeister Udet* (MGFA LW 21/6)

— *Wever, der erste Generalstabschef der Luftwaffe* (MGFA LW 21/7)

— *Historische Wendepunkte im Kriegseinsatz der deutschen Luftwaffe* (MGFA LW 35)

C. The technical-scientific publication *"Forschung und Entwicklung,"* Messerschmitt-Bölkow-Blohm, annual volumes from 1981

D. Annual theses at the *Führungsakademie der Bundeswehr*, Hamburg

Ahrens, H.W., Hauptmann. "Planung, Entwicklung und Beschaffung von Kampfflugzeugen in der deutschen Luftwaffe 1935-1943" (1980)

Pompe, Wlater, Major. "Der Sturzfluggedanke und seine Durchführung in der deutschen Luftwaffe im 2. Weltkrieg" (1975)

Rammers, Henning, Major. "Die deutsche Entwicklung des Strahlfugzeugs vor dem und im Zweiten Weltkrieg am Beispiel der Messerschmitt Me 262" (1974)

Reisch, Herbert, Major. "Der Sturzkampfgedanke und seine Durchführung in der deutschen Luftwaffe bis zum beginn des 2. Weltkrieges" (1973)

E. Trade publications, regularly reviewed:
Luftwaffe-Forum
Militärgeschichtliche Mitteilungen
Military Technology
Soldat und Technik
Wehrtechnik

History of German Aviation: Bombers and Reconnaissance Aircraft

Index of Aircraft Types

(Types in bold are covered in detail in the text)

A-10 280
A 48 34, 88
AAS 01A 208
Alpha Jet 264-269, 274
Ar 65 35, 87
Ar 68 28
Ar 81 18, 35, 88
Ar 198 152
Ar 234 244-253
Ar 340 167
AWACS 289

B-17 Flying Fortress 14, 232
B-52 284
Bf 109 21, 27, 28
Bf 110 32
Bf 161 33, 34
Bf 162 33, 76
Blenheim bomber 109, 119
Bréguet XIX 20
Bréguet 126 263
Bréguet Br 1150 Atlantic 258
BV 141 195
BV 141B 196

C.2111 239
Curtiss Hawk II 34

Dassault Bréguet Atlantique 2 259
DFS 228 236
DFS 230 119, 185, 238
Do 11 52
Do 17 13, 22, 25, 28, 32, 39, 41, **66**, 100, 123
Do 17P 95, **97**
Do 17Z-3 71, 144
Do 19 14, 33
Do 23 52
Do 31 260
Do 215 150, 193
Do 217 144, 201, **234**
Do 217E-2 182
Do 317 167
Do 335 253
Do 435 253
Do 635 253
Do Wal 28
Dornier P 375 263

E 370
E 650 Eurotrainer 263
EF 59 76
EF 61 115
EFA 287

F-4F 262, 271, 280
F-5 Freedom Fighter 271
F-11C-2 Goshawk Helldiver 34
F-13 100
F-14 271

F-15 271
F-16 271, 280
F-84F Thunderstreak 257
F-104G Starfighter 259
F-111 284
Fairey Battle bomber 119
Fairey/Westland Gannet 258
Fi 98 35
Fi 103 (V-1) 239
Fi 156 Storch 230
Fiat CR. 32 21
Fiat G 91 258
Fw 57 31
Fw 189 149, **152**, 195
Fw 189A-1 154
Fw 190 193
Fw 191 167
Fw 200 Condor 39, **112**
Fw 200C-3/U4 **131**, 149, **176**, 193

Gloster Gauntlet 67
Gloster Gladiator 159

Ha 137 18, 35, 88
Ha 141 195
Harrier 280
Hawker Siddely Sea Hawk 258, 260
He 45 25, 101
He 46 24, 41, 101
He 50 35, 87
He 51 20, 26, 28, 87
He 59 26, 28, 41
He 60 26, 41
He 70 16, 25, 41, 95, 157
He 100 46
He 111 18, 22, 28, 32, 39, 41, 46, **59**, 123, **141**, 201, **238**
He 111H-6 169
Hs 115 50, 129
He 118 18, 35, 89
He 119 16, 134
He 170 95
He 177 16, 37, 44, 47, **134**, 169, 179, 201, 243
He 177A-1/R1 180
He 177A-5/R-2 202
He 219 201, 242
He 274 207
He 277 208
He 606 134
Hs 122 102
Hs 123 23, 28, 35, 41, **85**, 148
Hs 124 31, 76
Hs 126 26, 41, 101, 126, 151, 194
Hs 126A 103
Hs 127 33, 76
Hs 128 168
Hs 129 148
HS 129B-1/B-2 188

Hs 130 168, 235
Hurricane fighter 123, 159, 162

I-15 Chato 21
I-16 Rata 21

Ju 49 115
Ju 52 20, 28, 52
Ju 85 75
Ju 86 18, 22, 32, 39, 41, 45, **52**
Ju 86G-1 57
Ju 86P 116, 150, 193
Ju 87 18, 23, 35, 39, 41, **87**, 94, 120, 187
Ju 87B-1 90, 145
Ju 87D-1/D-7 146, **185**
Ju 88 18, 33, 40, 46, **74**, 120, 125, **137**, 150, 193
Ju 88A-4 170, 212
Ju 88A-5 138
Ju 88 Mistel 226
Ju 88S 218
Ju 89 14, **15**, 33
Ju 90 16, 208, 209
Ju 186 194
Ju 187 188
Ju 188 201, 219, 222
Ju 188 201, 219, 222
Ju 188A/E 219
Ju 288 163, **164, 221**, 223
Ju 290 201, 208
Ju 388 202, 222, 224
Ju 488 225
Ju 390 202, 210

K 47 34, 87

Martin B-26 Marauder 75
MC 202 162
Me 108 *Taifun* 119
Me 109 46, 120, 124, 148, 150, 155, 194
Me 110 46, 122, 147, 151, 194
Me 210 47, 163, 174
Me 210Ca-1/C-1 175
Me 262 (*Jabo*) 199, **232**
Me 264 210
Me 321 *Gigant* 238
Me 410 174, 201, 240
MiG 23 Flogger 271
Mirage 271
Model 299 14
MRA-75 270
MRCA 271, **274**

Ni-51 20

P-38 Lightning 207
P-47 Thunderbolt 207
P-51 Mustang 207, 232
P 76 149

P-530 Cobra 271
PAH-2/HAC 268
Panavia 100 272, 273
Panavia 200 272, 273
Panavia Tornado 269-279, 282, 283
Projekt 11 89
Projekt 1041 16, 134

RF-4C 262
RF-4E 262, 263, 271, 289
RF-84 Thunderflash 257, 258

Saab 37 Viggen 271
SB-2 Katyusha 21
S.M. 81 21
Spitfire fighter 123, 193

T-33 264
T 291 263
Ta 152 254
Ta 154 228, 229
Ta 400 211
TA 501 263, 274
Tornado ECR 285, 286, 289
Tornado F-2 278, 284

Tornado IDS 280, 284
TSR 2 270

U-2 200

VJ 101 260
VAK 191 260

W 33 100
Wellington bomber 109
Wright Flyer 244

XF-104 259

General Index

Adlertag 122
Afrikakorps 198
aerial resupply 197, 203
aerial torpedo 229, 292, 300
aerial torpedo bomber 203
aerial torpedo wings 201, 243
afterburning 259
air attack components 41, 163
air attack potential 169
air war 27, 36, 200, 261, 288
air warfare, thoughts on 36
aircraft acquisition program of 1938 42
aircraft acquisition programs 32, 162
aircraft carrier 157, 160
aircraft losses 124
aircraft production in 1941/1942 164
aircraft weaponry 48, Appendix 4
Akademie der Luftwaffe 12
all-moving tailplane 277
aluminum, requirements for 163
Amerika Bomber 208
angle velocity 129
Angriffsführer England 199, 200, 205, 231
Annex B Aircraft 272
anti-radiation missile 281, 285
anti-tank (aircraft) 12, 187, 189, 216
anti-tank cannon 203, 204
Arge Nord 259
Arge Süd 259
armament configuration 142, 193, 238, 266
armament control 257, 289
armed attack, NATO response to 260, 261
armor piercing bomb 298
armor piercing shell 217, 218
army(Heer), support of 37
Army Air Corps 14
atomic bombs, dropping of 257
attack, concept of 31
 on maritime targets 127
Aufklärer F-Land 42
Aufklärer H-Land 42, 96
Aufklärungsgruppe des Ob. d. L. 101, 145
Austria 56
automatic aerial camera 48, Appendix 5, 96
automatic dive recovery system 78, 82, 84, 92, 94
automatic flight data processing 267, 277

auxiliary bomber 52, 112, 114, 131
auxiliary fighter-bomber 233
auxiliary power unit 277
auxiliary reconnaissance aircraft 95
aviation industry 38, 41
avionics 270, 277
AVS project 269, 275

Baby Blitz 205, 231
Balkans 105, 158
balloon deflector 212
balloon cable deflector 140
Barbarossa 161, 197
Battle of Britain 122-125, 288
Battle of the Atlantic 132
battlefield interdiction 261
battlefield reconnaissance 289
bauxite 163
Beethoven-Gerät 227
Beethoven program 227, 228
Belgium 105, 119
Big Week 231
biplane dive bomber 87
biplane fighter 28, 35, 50, 67, 87
Blitz 253
Blitz-Bomber 199
Blitzstrecken 52
Bläserstart 139
Bodenplatte 255
bomb drop 48, 130
bomb platforms 108, 109, 234
bomb release point 130
bomb torpedoes 242
bomb trapeze 91
Bomber A project 16
Bomber B 163, 169
Bomber B concept 168
Bomber B program 222, 288
Bomber B project 168, 212, 219
Bomber B prototype 164
Bomber Command 198, 243, 256
bomber gap 46
bomber pilots 12, 18
Bomber Planning 1943 201, 202, 208, 212, 243
bombsight 113, 139
Bomberversuchsabteilung 88

borescope technology 275
British forces 110, 120
Brückenbevollmächtigter 228
Brussels Pact 257
Bulgaria 101
Bundesluftwaffe 257, 288
Bundeswehr 257, 261

C-Amt 102
cable cutter 140
cannon pod 187, 266
Casablanca 198
catapult-launched takeoff 94
Central Committee of the Communist Party of the Soviet Union 161
central computer 277, 281
chaff 199, 284
Chile 55
chin gun turret 206, 222
China 34, 65
close air support 261, 264, 268, 274
close combat tactics 36
combat transport 238
Combined Bomber Offensive 199
command aircraft 69, 70
comparison criteria 106
comparison flights 88
concept phase 271, 272
Condor Legion 22-25, 35
Consolidated Aircraft Type Program 43, 44, 46, Appendix 2
Constant Luftwaffe 1942 43
control plane 228
Crete 158, 193

Denmark 105, 110
definition phases 263, 273
delivery plan No. 1 31
delivery plan No. 10 81
delivery plan No. 11 81
delivery program 38
depot inspection 285
Deutsches Museum 177, 254
developmental work 131, 162
developmental periods 39
developmental phases 274, 278

History of German Aviation: Bombers and Reconnaissance Aircraft

diesel aircraft engine 23, 53, 115
Directive No. 17 122
Directive No. 18 157
Directive No. 20 157
Directive No. 21 157
Directive No. 22 158
Directive No. 25 159
Directive No. 33 161
dive, shallow 139
dive bomber pilots 12
dive bombers 22, 34, 85, 145, 185
dive bombing capability 35, 82, 86, 116
dive bombing methods, Ju 88 125-127
dive bombing operations 18
dive bombing principle, embracing 220
dive bombing principle, thoughts on 34
dive bombing principle, concept of 288
dive bombing tactics 82
dive brakes 78, 81, 88, 91, 92
Doppelreiter 227
Doppler radar 281
double engine 134, 135, 181
double engine problems 208
drift 48
dual cycle engine 281

East, campaign in the 161, 288
Echolot 176
Egypt 159, 193
Einsatzkommando Schenck 232
Eisenhammer 206, 227, 228
electronic countermeasures 268, 285, 286
emergency fighter program 208, 242
emergency program 35, 85
endoscopy 275
Erprobungskommando Ju 88 81
escalation 261
escalation potential 287
exhaust turbine 115, 116
exhaust turbocharger 194, 235
expeditionary corps 110, 111

F-104 consortium
Fall Gelb 110, 121
Fall Grün 44
Fall Rot 120, 121
Fall Weiß 45
Father and Son 226
Fernaufklärungsgruppe Ob. d. L. 115
Feuerzauber 20
field conversion kits 184, 189, 236, 249
field testing 89
fighter pilots 26, 288, 289
fighter program 168, 195, 204
Fighter Staff 231, 243
fighter-bomber 12, 147, 289
fighter-bomber, as an arm of the Luftwaffe 288
fighter-bomber, role of 151, 289
Finland 110
fire control radar 259
fires, problems with 206
firestorm 199
fixed-tow (coupling) 238
flares 96

flash bomb 48
Fliegerstaffel z. b. V. 100
flight control system 277
flight performance 99, 106, 140, 172
Flugzeugträger A 94
Flugzeugzerstörer 32
fly-by-wire control 277
Flying Aquarium 152
Flying Asymmetry 195
Flying Pencil 13
Flying Powerplant 168
FOFA concept 280, 287
follow-on units 287
formation busters 229
Formidable 160
foreign air forces 105
foreign interest 54, 65, 69, 104
Förstersonde 190
four-engined jet bomber 251
Four-Year Plan 40
Fowler flaps 203, 204
France 95, 105, 108, 110 , 112, 121
free-fall bomb 180, 204
French campaign 113
French forces 110
Friedensengel 242
Fritz X bomb 230, 298
Führerweisung No. 1 37
Führerweisung No. 17 122
emergency fuel jettison system 209
Fueréas Aereas Españolas 20
fuses, types of 300

General der Jagdflieger 199
General der Kampfflieger 200, 201, 210, 233, 240, 243
Generalluftzeugmeister 162, 164
general staff 12, 18, 32, 288
Green Line 231
glide bomb, Hs 293 176, 180, 204, 300
glide bomber 80
glide torpedo 242
GM 1 (boost system) 175, 194, 240
Gomorrha 231
Göring Programm 162
Graf Zeppelin 94
Great Britain 105, 108, 112
Greece 104, 159, 193
Greek government 104

handheld camera 48, Appendix 5
harassment bombing 118
head-up display 266
heavy bomber 169, 176, 185, 202
heavy dive bomber 77
Hecht 253
Heeresgruppe Afrika 198, 230
Herkules 198
high altitude bomber 115, 210, 225, 234
high altitude reconnaissance aircraft 115, 150, 168, 235, 289
high altitude research aircraft 168
high altitude strategic reconnaissance aircraft 118, 236
high performance fighter 242
high speed bomber 31, 32, 33, 69, 127, 174, 241, 242, 243

high speed commercial airliner 66
high speed reconnaissance aircraft 70, 100, 174, 246
Hohentwiel radar system 176, 179, 209, 212, 221, 241
Holland 119
Home Fleet 111
Hubertus Program 222
Hungary 55, 95, 101, 175
Hornet Working Group 270
HZ high altitude turbocharger system 168, 169, 235
HZ testing 235

Illustrious 158
industrial council 163
industrial maintenance 260, 263
industrial systems management 260, 263
INF Treaty 287
infra-red imaging 236
infra-red sensors 259, 284, 286, 289
instrument flying schools 50, 118
interim balance 1940 112
internal navigation system 259, 263
International Flying Meet 1937 67, 69, 78
iron bombs 284
Italy 121, 146

jamming device (extrusion system) 143, 144, 170, 284
Japan(ese) 34, 162
Japanese Navy 112
Jericho trumpets 146, 147
jet bomber 202, 234, 241, 245, 253, 288
jet trainer 264
joint developments 260, 263
joint program 257
Joint Working Group 270
Ju 88 Chronology 81, Appendix 7

K-types 59
Kampfgruppe K/88 22, 28
Kaperflugzeug 113
Kleinstbomber 232
Kommando Sperling 248, 249
Krueger flaps 277
Kuto-Nase 141, 212

L.Dv. 16 31, 36, 200, 261
landing operation Neptune 232
large scale production 38, 170, 238
Lehrgruppe Ju 88
level bombing 34, 36, 176
license company 81
license groups 80
Lieferprogramm 12 46
Liesendahl method 129
load factor on pullout 191
London, attack on 122
long range bomber 131, 133, 202, 204, 206, 210, 234, 240
long range bomber-destroyer 211
long range bomber requirement 32
long range heavy bomber 15
long range reconnaissance aircraft 96, 133, 143, 208, 211, 240
long range wings 201, 210, 243
Lotfernrohr 176, 249
low-level/dive system 233

Luftkommandoamt 12, 13
Luftschutzamt 11
Luftwaffe 12, 41, 43, 121, 261, 288
Luftwaffe, limits of 120, 288
Luftwaffen-Dienstvorschrift 200, 261

Mach limit 247, 310
Maginot Line 109
mailplane 66
main target groups 281
Malta 157, 159, 162
Marineflieger 257, 273, 279
maritime reconnaissance 19
MC 14/2 NATO Strategy 269
MC 14/3 NATO Strategy 269
Mediterranean Fleet 160
medium bomber 35, 52, 137
medium multi-seat bomber 42
Merkur 159
milestone 278
military strategic conception 260
military tactical objectives 259
Mini-Blitz 205
mission computer 263
Mistel (team) 226
modification program 285
modular construction method 277
modular concept 275
Multhopp flaps 168
Multi-role weapon 281
Munich Accords 44
Mutual Defense Aid Program 257
MW 50 system 179, 220, 224

NACA cowl 86, 103
NATO commander 261
NATO air forces 261, 287
NATO air defense 289
NATO strategy 260, 269, 280
NATO triad 261
NATO treaty 260
naval war from the air 157
Neptune 232
Netherlands 105, 119
Neues Kampfflugzeug 269
Neuhammer training grounds 95
New York Bomber 208, 210
night reconnaissance 224
night bomber 222
night strike aircraft 200
night strike pilots 12
NKF standard 280
Non-Aggression Pact 45
North Atlantic Treaty 257, 260
North Atlantic Treaty Organization 257
Northern France 119
Nordmark 109
Norway 110, 288
numbers comparison 1938/1939 49

offensive aircraft 31, 32
offensive potential 255
Operation Bodenplatte 255
Operation Dynamo 120

Operation Eisenhammer 206, 227, 228
Operation Gomorrha 231
Operation Herkules 198
Operation Marita 157
Operation Merkur 159
Operation Overlord 232
Operation Rumpelkammer 239
Operation Steinbock 205
Operation Sonnenblume 158
Operation Taifun 161
Operation Zitadelle 198
operational concept 260
operational doctrine 260, 279
operational flight performance 106
operational profile 259
operative 30, 36, 200, 261, 288
operative air war, prosecution of 36, 37, 130, 234, 243
operative Luftwaffe 34, 66
ore, supply of 110
ordnance 47, appendix 3
outside curve attack 128

Pact of Steel 45
parabrake 183
parachute retarded flare 142, 143, 205
Paris Accords 1926 11
Paris Accords 1954 257
pathfinder aircraft 69, 70, 238
pathfinder role 249
payload requirements 165
Pearl Harbor 162
penetration range 106, 261
periscope sight 164, 165, 166, 224, 236, 250
piggyback configuration 226, 236, 255
picture interval regulator 249
pirate units 44
photo processing equipment 294
photogrammetric camera 294
photoreconnaissance 48, Appendix 5
pilot training schools
 A and B class 95
 C class 50, 118
plan study 1939 37
planning criteria 43
planning parameters 1938 42
Poland 105, 107, 288
Portugal 55
post-flight inspection 276
pre-flight inspection 276
pre-production series 39, 180
pressure chamber 117, 118
pressurized cabin/cockpit 115, 116, 204, 235
primary contractor 258
priorities 40, 202, 261
 shift in 162
prioritization class 131, 219
prioritization list 131
production memorandum 77
production program 1937 41
pulse doppler radar 263

qualitative situation 106
quality bombing 18
quantitative situation 105

R-Gerät 244, 246, 249
Rammjäger 256
radar equipment 199
radio reconnaissance 238
range 158
range, flying 172
range, maximum flying 172
reactions, types of 261
rear warning radar 224
Rechlin method 129
reconnaissance aircraft 24
 HQ staff assigned 73, 100
reconnaissance bomber 235
reconnaissance fighter 234
reconnaissance pilots 19
reconnaissance systems 48, appendix 5 289
reconnaissance version 151
record setting flights 78
Reich, defense of the 250
Reichsluftwaffe 11
remote-controlled free-fall bomb 204
Rhineland program 11, 38
Risikoluftwaffe 11
rocket-assisted takeoff 139
Roma 230
Rostock system 176
Royal Air Force
Rumpelkammer 239
Rüttelfalke 101

safety control 93
Scapa Flow 227, 253
search radius 179, 193
second front 158, 160, 231, 232
Seeaufklärungsstaffel AS/88 26
Seelöwe 122, 124
Seeschlange 176
selection competition, dive bomber 35
series maturation 39
series production 81
sewing machines 200
sickle cut plan 110, 119
single engine flight 170, 171
single engine flight capability 142, 169
Sino-Japanese War 65
Sitzkrieg 109
six-engined aircraft 212
slats 263, 277
skid-type undercarriage 246
smoke pods 154
smoke generating system 153
Sonderkommando Rastedter 238
Sonderstab W 20
Sonnenblume 158
South Africa 56
Soviet Union
Spanish Civil War 20-23, 31, 35, 101, 288
special authority 80, 163
special units 201
special wings 201, 208, 243
speed records 77
Sperling 248, 249
spoilers 277
spray pods 153

Stalingrad-Type 203
standard bomber 137
standard empty weight 117
Steinbock 205
Störtebeker 225
strategic air war 36, 130
strategic reconnaissance 19, 100
strategic reconnaissance aircraft 95, 149, 193
strategic reconnaissance version 25, 150
stress groupings 78, Appendix 6
strike aircraft 31, 148, 185
strike pilots 12
structural soundness tests 113
Studiengruppe der Geschichte des Luftkriegs 108
Stuka 12, 34, 35, 85, 87, 188
submunitions 281
Suez Canal 158, 159
surface radiators/coolers 61
surface cooling 134
Sweden 54, 73, 105
swing-wing (principle) 271, 277, 281
system director 279
systems management 260

tactical 36, 130, 261, 288
tactical reconnaissance aircraft 102, 151, 194
tactical reconnaissance 19
tactical reconnaissance pilots 101
tactical requirement 75
Taifun 161
tail dive brakes 183
tail gunner's position 202, 203, 206, 222
tail turret 221
taileron 277
takeoff trolley 245
tandem configuration 241

tank buster 187
tanker aircraft 210
technical-tactical requirement 75
Technisches Amt 38, 43, 75, 102, 166
technological development, oversight of 260, 263
terminal velocity dives 86
terrain-following flight 275
terrain-following radar 281
Thousand Bomber Raid 198
three-dimensional autopilot 249
three-front air war 163
three-front war 161
thrust reversing 277
Tornado WaKO 279
torpedo bomber 213, 221, 237, 238, 239, 242
transport glider 142, 185, 238
transport glider tow plane 142
TSA system 233
turbocharger 240
turbojet engine 246, 247, 251
Turkey 65
turnaround times 275
two-front war 124
two-seat dive bomber 42

Uerdinger annular springs
ultra-high speed bomber 232, 233, 234, 241, 242
ultra-high speed fighter 232
universal airplane
"unmasking" of the Luftwaffe 12, 101
upgrading 267, 287
upgrade program 258
Ural Bomber 14, 16, 32
U.S. Army Air Forces 199
U.S. Army Bomber Command 198
USA 105, 162
Usaramo 20

V-weapons 234
V-weapons platform 239
variable geometry 274
ventral dustbin turret 22, 45, 58, 63, 64
ventral radiator 61
verification 261
Versailles, Treaty of 12
Versuchsbomberstaffel VB/88 22, 23, 28, 61
Versuchsstaffel 210 174
Versuchsstelle für Höhenflüge 193
Versuchsverband Ob. d. L./des OKL 193, 194, 224, 248
VTOL (aircraft) 260

warhead 255
Warsaw Pact 257
water-methanol fuel injection 179, 224
water-methanol fuel injection system 220
weapons pod 221, 241
Weserübung 100
West, battle plans for the 110
West, campaign in the 119-121, 288
West Berlin, blockade of 257
Western European Union 257
Wikingerschiff 234
Wilde Sau 199
Wild Weasel role 286
wonder weapons 288
world record 78, 134

X-Uhr 130
X-Verfahren 130

Y-Verfahren 130
Yalta 256
Yugoslavia 69, 159

Zerstörer 32, 216
Zitadelle 198

Index of Personnel

Abos, Hans 273
Armengaud, André 108
Armin, von 198

Birkenbeil, Heinz 273
Blomber, Werner von 11, 20
Blume, Walter 244
Brown, Eric 253

Chamberlain, Arthur Neville 44
Chiang Kai-shek 65
Churchill, Sir Winston 110, 122, 198, 256

Daladier, Edouard 44
Debré, Michel 274
Dieterle, Hans 253
Douhet, Giulio 12, 21, 36, 77, 288
Duval, General 28, 30

Eckert, Fritz 278
Eckstein 244
Eimler, Eberhard 267
Eisenhower, Dwight D. 198, 256
Etzdorf, Marga von 75
Evers, Heinrich 75

Fath 66, 70
Felmy, Hellmuth 36, 44, 130
Fichtmüller, Carl Peter 273, 279
Francke, Carl 135
Franco, Francisco y Bahmonde 20, 22, 28, 45
Friedeburg, Hans-Georg von 256

Galland, Adolf 199
Gassner, Alfred 75
Gaulle, Charles de 198, 271
Göring, Hermann 11, 20, 36, 40, 46, 47, 51, 80, 163, 174, 181, 197, 206, 231, 243, 250, 255, 256
Grandy, Sir John 272
Gülzow, Hartmut 273
Günter, Walter 59, 134

Haber, Fritz 226
Halifax, Lord Edward Wood 122
Harris, Sir Arthur T. 198
Healey, Denis 272
Heath, B. O. 279
Heinkel, Ernst 16, 65, 134, 206
Heintz 78
Henke, Rainer 278
Herrmann, Hajo 199
Hertel, Heinrich 134, 164, 180, 203, 222
Hesselbach 89
Hindenburg, Paul von 11
Hinrichs 89
Hitler, Adolf 11, 40, 44, 80, 110, 122, 131, 157, 161, 180, 197, 205, 230, 242, 256
Holzbauer, Siegfried 165, 226

Janssen, Ubbo 246-248
Jeschonnek, Hans 34, 37, 44, 180, 199, 288
Jodl, Alfred 256
Joop 165
Junkers, Hugo 38

Kammhuber, Josef 44, 121, 288, 289
Keitel, Wilhelm 256
Kesselring, Albert 16, 20
Kindermann, Karlheinz 16, 76
Klapperich, Hans-Joachim 273, 279
Koller, Karl 243
Konrad 211
Koppenberg, Heinrich 40, 77, 79, 163
Korten, Günter 199
Kosel, Erwin 152, 168
Kosin, Rüdiger E. 244
Kraft 210
Krauthann, Arnim 278
Kreft 88
Kröger, Walter 246
Krüger, Horst 273

Langfelder, Helmut 271, 279
Leber, Georg 278
Limberg, Gerhard 278, 279
Limberger 77

Madelung, Gero 273, 279
Mautino, R. A. 279
Mehlhorn 168
Meister, Nils 278
Messerschmitt, Willy 77, 163, 232
Messmer, Pierre 271
Milch, Erhard 11, 36, 44, 46, 163, 169, 180, 199, 231, 244
Millet, Paul 278
Montgomery, Viscount Bernard L. 198, 230
Mussolini, Benito 44, 230

Neuenhofen, Willi 88
Nicolaus, Friedrich 102, 149, 193
Nischke, Gerhard 60, 134

Obleser, Friedrich 273

Pabst, Otto E. 279
Pancherz 209
Paulus, Friedrich 197
Peltz, Dietrich 201, 205, 248
Plauth, Karl 87
Pohlmann, Hermann 53, 82, 88, 92, 146
Polte 67
Preuschen 165

Quick, August 75

Raeder, Erich 46
Rall, Günter 278

Rammensee, Hans Friedrich 278
Rastedter 238
Rebeski 244
Richthofen, Wolfgang Frhr. Von 28, 29, 36
Rickert 136
Ritter 76
Rommel, Erwin 158, 198
Roosevelt, Franklin D. 198, 256
Rougeron, Camille 13, 77
Rowehl, Theodor 100, 116, 168, 181
Rühle, Hans 273

Saget, Jean-Marie 264
Sander, Hans 168
Saur, Karl Otto 231
Schenck 232
Schiferstein, Karl 226
Schmidt, Helmut 273
Schönebeck, Carl-August von 88
Schreiber, Kurt 278
Schröder, Gerhard 271
Schwärzler, Karl 59
Seibert 28
Selle 244-246
Soos, Friedrich 278
Speer, Albert 164, 231
Sperling 248
Serrle, Hugo 22, 28
Spotswood, Sir Denis 278
Stalin, Josef 161, 231, 256
Stamer, Fritz 226
Steinhoff, Johannes 270
Stumpff, Hans-Jürgen 44, 256
Tank, Kurt 112, 149, 152
Thomas, Dieter 264
Todt, Fritz 162, 164
Trevisan, Pietro Paolo 278

Udet, Ernst 34, 38, 46, 77, 85, 88, 94, 127, 136, 163, 244, 288
Untucht, Robert 66
Ursinus 136

Victor Emanuel III, King 230
Vogt, Richard 88, 195
Volkmann, Hellmuth 28

Wahl, Albert 278
Wendel, Fritz 199
Wenzel 244
Wever, Walter 12, 16, 36, 288
Wilberg, Helmuth 20
Wilson, Harold 278
Wimmer, Wilhelm 16
Wright, Orville 244
Wright, Wilbur 244
Wurster, Hermann 174

Zindel, Ernst 53, 75, 213

Abbreviations

ADC	air data computer
ADV	air defense variant
AFA	*Automatische Flugdatenaufbereitung* (automatic flight data processing)
AFVG	Anglo-French VG (project)
AI	air interdiction
AMRAAM	advanced medium range air-to-air missile
APU	auxiliary power unit
AS	air superiority
ATP	Allied tactical publication
AVS	advanced V/STOL
AWACS	airborne warning and control system
BAI	battlefield air interdiction
BMVg	*Bundesministerium der Verteidigung* (Federal Ministry of Defense)
BSB	*Brandbombenschüttbehälter* (incendiary bomb dispenser)
BT	*Bombentorpedo* (torpedo bomb)
BZA	*Bombenzielanlage* (bombsight)
BZG	*Bombenzielgerät* (bombsight)
CAS	calibrated airspeed
CAS	close air support
CCIP	continuously computed impact point
CEP	circular area of probability
COMINT	communications intelligence
CPGS	cassette preparation ground station
DI	*Depotinspektion* (depot inspection)
DSFC	direct side force control
ECR	electronic combat and reconnaissance
EiV	*Eigenverständigung(sanlage)* (intercom system)
ELINT	electronic intelligence
Eloka	*Elektronische Kampfführung* (electronic warfare)
ELS	emitter locator system
Elvemag	*elektrisches Vertikalmagazin* (electronic vertical magazine)
ESAC	vertical rack for cylindrical bombs
ETC	electromagnetically operated rack for cylindrical loads
EW	electronic warfare
FLIR	forward looking infra-red
FOFA	follow-on forces attack
FuBl	*Funkblindlandegerät* (radio-navigational instrument landing system)
FuG	*Funkgerät* (radio-navigational system)
Gen. D. K.	*General der Kampfflieger* (General of Bomber Forces)
GM1	supplemental SAUERSTOFFTRAEGER fuel injection
GRTS	Groupement Réacteurs Turbomeca/SNECMA
GV	*Goerz-Visier* (Goerz sight)
HARM	high-velocity anti-radiation missile
Hi-Lo-Hi	high altitude approach—low altitude in target area—high altitude egress
HUD	headup display
HZ	*Höhen(lader)zentrale* (centralized high altitude turbocharger)
HZG	*Hauptzielgruppe* (main target group)
IDS	interdiction strike
IIS	infra-red imaging system
INF	intermediate range nuclear forces
Jabo	*Jagdbomber* (fighter-bomber)
KBO	*Kampfbeobachteroffizier* (battlefield observation officer)
L. Dv.	*Luftwaffendienstvorschrift* (Air Force Regulation)
LMA	*Luftmine (Grundmine)* 500 kg (air-dropped land mine)
LMB	*Luftmine (Grundmine)* 1000 kg (air-dropped land mine)
LNU	*Luftnahundterstützung* (close air support)
Lotfe	*Lotfernrohr* (bombsight)
LT	*Lufttorpedo* (air-launched torpedo)
MC	military committee
MCU	missile control unit
MDA	*Missionsdatenausgabe* (mission data readout)
MDAP	Mutual Defense Assistance Program
MDT	*Missionsdatenträger* (mission data carrier)
MGFA	*Militärgeschichtliches Forschungsamt* (Office of Military-Historical Research)
MITAC	miniaturized TACAN
MoU	memorandum of understanding
MRCA	multi-role combat aircraft
MTZ	*Militärisch-taktische Zielsetzung* (Military-Tactical Objectives)
MW 50	supplemental water methanol (50:50 ratio) fuel injection
NACA	National Advisory Committee for Aeronautics (USA)
NAMMA	NATO MRCA Management Agency
NAMMO	NATO MRCA Management Organization
NASMO	NATO Starfighter Management Office
NATO	North Atlantic Treaty Organization
NKF	*Neues Kampfflugzeug* (New Combat Aircraft)
Ob. d. L.	*Oberbefehlshaber der Luftwaffe* (Commander-in-Chief of the Air Force)
OCA	offensive counterair (operations)
ODIN	operational data interface
OEO	operational equipment objectives
OER	operational equipment requirements
OKL	*Oberkommando der Luftwaffe* (Air Force High Command)
OKW	*Oberkommando der Wehrmacht* (Armed Forces High Command)
PAH	*Panzerabwehrhubschrauber* (anti-armor helicopter)
PeilG	*Peilgerät* (RDF system)
R. d. L.	*Reichsminister der Luftfahrt* (Minister of Aviation)
R-Gerät	*Rauchgerät* (rocket assisted takeoff pack)
RATO	rocket-assisted takeoff
Revi	*Reflexvisier* (reflective gunsight)
RMK	*Reihenmeßkamera* (mapping camera)
RTOL	rolling vertical take-off and landing
SAHR	secondary attitude and heading reference
SEAD	suppression of enemy air defenses
SK	*Störkörper* (jamming device)
SKAV	*Störkörperausbringungsvorrichtung* (jamming device extrusion system)
Stuvi	*Sturzflugvisier* (dive bombsight)
TACAN	tactical air navigation
TL	*Turboluftstrahltriebwerk* (turbojet engine)
TNT	trinitrotuluol
TOT	time on/over target
TSA	*Tief- und Sturzfluganlage* (low level/dive system)
TSR	tactical strike reconnaissance
TTTE	Trinational Tornado Training Establishment
USAAF	United States Army Air Forces
USAF	United States Air Force
USSR	Union of Soviet Socialist Republics
VG	variable geometry
VOR	VHF omnidirectional radio range
V/STOL	vertical/short take-off and landing
WB	*Waffenbehälter* (weapons canister/pod)
WEU	West-European Union
WP	Warsaw Pact
WSO	weapons systems officer

Unit Conversion Table

Preferred units are in bold print.
Weight
SI: **kg** (kilogram), also mg, g, t (metric ton)

Density
SI: kg/m^3, also g/cm^3, **kg/dm^3**

Time
SI: s (second), also min (minute), h (hour)

Speed
SI: m/s, also **km/h**

Revolutions per minute
SI: s^{-1}, also **min^{-1}** or 1/min

Power
SI: N (Newton), also **kN**
1 kp = 9.8066 N   1 N = 0.1019 kp
1 kp = 0.0098 kN 1 kN = 101.9 kp
1 kp = 0.89 daN   1 daN = 1.019 kp

Pressure
SI: bar, mbar, Pa (Pascal)
1 bar = 1000 mbar
      = 10,197 kg/cm^2 = 750.1 mmQS (Torr)
1 Pa = 10^{-5} bar = 10^{-2} mbar

Technical Atmospheric Pressure (at):
1 ata = 1 kg/cm^2 = 0.9806 bar absolute value
      = 10,000 mm WS = 735.5 mmQS (Torr)

Physical Atmospheric Pressure (Atm):
1 Atm = 1.0332 kg/cm^2 = 1.0133 bar
      = 10.332 mm WS = 760.0 mmQS (Torr)

Torque
SI: Nm, also kNm
1 mkg = 9.8066 Nm   1 Nm = 0.1020 mkg

Performance
SI: W (Watt), also **kW**, MW
1 kW = 1.3596 hp   1 hp = 07.355 kW

Temperature
SI: K (Kelvin, also °C (absolute value zero point of T_0 = 273.15 K)

Energy, Work, Heat
SI: J)Joule), also **kJ**, MJ, kWh (1 J = 1 WS = 1 Nm)
1 m kp = 9.8066 J   1 J = 0.1020 mkp
1 hph = 0.7355 kWh 1 kWh = 1.3596 hph
1 kWh = 3600 kJ   1 kJ = 0.000278 kWh
1 kcal = 4.1868 kJ   1 kJ = 0.2388 kcal

Rate of Airflow
SI: **kg/s**, also g/s, g/h, kg/h

Until the advent of the SI system the manifold pressure of aircraft engines was given in *ata* as absolute pressure. Since that time the technical pressure has been listed in *bar*, or *mbar*. 1 ata = 1 bar was established for converting older manifold pressure values, since the conversion error is due to the measurement tolerance. A manifold pressure figure is considered an absolute value. Manifold pressure gauges for non-running engines indicate the local outside barometric pressure.

The following units and conversion factors are often used in the aviation community:

Airspeed
kilometers per hour (km/h)	in knots (k or kts)	x 0.539
kilometers per hour (km/h)	in statute miles per hour (m.p.h.)	x 0.621
knots (nautical miles per hour)	in kilometers per hour (km/h)	x 1.853

Mach(number), more accurately: flight Mach number = the ratio of an aircraft's true airspeed to the speed of sound under prevailing atmospheric conditions

miles per hour (m.p.h.)	in kilometers per hour (km/h)	x 1.609
miles per hour (m.p.h.)	in knots (k or kts)	x 0.868

Rate of Climb
feet per minute (ft/min)	in meters per second (m/s)	x 0.005
meters per second (m/s)	in feet per minute (ft/min)	x 0.019

Fuel Capacity
gallon (imperial) (imp. gall.)	in liters (l)	x 4.546
gallon (imperial) (imp. gall.)	in U.S. gallons	x 1.205
gallon (US) (US gall.)	in liters (l)	x 3.785
gallon (US) (US gall.)	in imperial gallons (imp. gall.)	x 0.833
liter (l)	in imperial gallons (imp. gall.)	x 0.220
liter (l)	in U.S. gallons (US gall.)	x 0.264

Flight Altitude/Level
foot (ft or ')	in meters (m)	x 0.305
	1,000 feet (ft)	= 305 meters (approx. 300 m)
meter (m)	in feet (ft)	x 3.28
	1,000 m	= 3,280 ft (approx. 3,300 ft)

Picture Credits

A total of 450 drawings/photos from the following sources:

Herr Erik Theodor Lässig: 108 drawings
Aviagraphica Company: cutaways on pp. 282, 283
Deutsches Museum, Munich. *Sondersammlung Luftfahrt*: 143 pictures and graphical drawings
Bundesarchiv, Coblenz: 28 photographs
Dornier and Panavia Companies: 14 photographs
Luftwaffe, particularly *Aufklärungsgeschwader 52*: 5 photographs
Author: remaining pictures/drawings as well as all tables

The author wishes to express his sincere gratitude to the *Deutsches Museum*, the companies of Dornier and Panavia, and the *Luftwaffe* for graciously providing their photographic materials free of charge.   R. C.

History of German Aviation: Bombers and Reconnaissance Aircraft

The Author

Roderich Cescotti was born in 1919 into a South Tirolean engineer's family.

After graduating in 1937 he began training as a pilot, and during the war flew bombers, combat transports, and finally fighters. Cescotti served as a technical officer at the squadron, group, wing, and command levels, focusing on the He 111, Ju 88, Fw 200 and other maritime aircraft, the Do 217, He 177, Fw 190, and Ta 152 programs. He was a squadron commander in both an He 111 bomber unit and a combat transport unit (He 111s with DFS 230s). He also served in the General Staff's operations planning section. The end of the war found Cescotti as commander of a fighter group (Fw 190s).

In 1945 he became a British prisoner of war and completed his foreign language studies. 1947 saw him working as an interpreter for the government and an export manager in the metals industry.

From June 1952 onward Cescotti was assigned to the Luf*twaffe* planning section of *Amt Blank*. From 1955 to 1980 he served with the *Bundesluftwaffe* as a pilot and staff officer: to 1957 as a training planning officer in the Ministry of Defense, to 1959 as the director of pilot training in Canada, and to 1965 as the commander of a heavy reconnaissance wing (RF-84Fs, RF-104Gs). In 1965 Cescotti attended the NATO Defense Academy in Paris, following which he served until 1969 with the General Staff's Defense Planning division in Washington and Brussels. Promoted to brigadier general in 1969, he served until 1973 as the defense and air force attaché in London, then in Lisbon as the director of operations. From 1974 to 1975 he was the NATO air defense commander. Cescotti's promotion to major general occurred in 1975, whereupon he served until 1977 as the vice commander of the Allied Tactical Air Fleet. From 1977 until 1980 he was the commander of the Allied Air Forces covering the Baltic approaches.

Since 1980 he has been on the advisory board of the D*eutsches Museum*. Cescotti is also an independent colleague of the European Security Studies (ESECS) science and politics foundation and the chairman of the *Luftwaffe* Memorial Foundation.

His previous works include: "Luftfahrt-Wörterbuch/Aeronautical Dictionary" German-English, English-German, (1954 and 1957), "Luftfahrtdefinitionen/Aeronautical Definitions" English-German, German-English (1956, 1969, and 1987), and the translations: "Aviation Cadet - T-33 -WILCO" (1957) and "Target Berlin - Angriffsziel Berlin" (1982).